T0401031

MATERIALS SCIENCE AND TECHNOLOGIES

SPRAYS: TYPES, TECHNOLOGY AND MODELING

MATERIALS SCIENCE AND TECHNOLOGIES

Additional books in this series can be found on Nova's website
under the Series tab.

Additional E-books in this series can be found on Nova's website
under the E-book tab.

ENGINEERING TOOLS, TECHNIQUES AND TABLES

Additional books in this series can be found on Nova's website
under the Series tab.

Additional E-books in this series can be found on Nova's website
under the E-book tab.

MATERIALS SCIENCE AND TECHNOLOGIES

SPRAYS: TYPES, TECHNOLOGY AND MODELING

MARIA C. VELLA
EDITOR

Nova Science Publishers, Inc.
New York

Additional color graphics may be available in the e-book version of this book.

Library of Congress Cataloging-in-Publication Data

Sprays : types, technology, and modeling / editor, Maria C. Vella.
p. cm.
Includes bibliographical references and index.
ISBN 978-1-61324-345-9 (hardcover : alk. paper) 1. Spraying. I. Vella, Maria C.
TP156.S6S65 2011
660'.294515--dc23

2011012537

Published by Nova Science Publishers, Inc. † New York

CONTENTS

PREFACE

In this book, the authors gather and present topical research in the study of the types, technology and modeling of sprays. Topics discussed include charged-spray technologies and their application in technology; spray drying to produce dried foods and vegetables; spray drying in the ceramic industry; atmospheric plasma spray; liquid flow structure in pressure swirl sprays and modeling a water-urea spray including mass and heat transfer.

Chapter 1 - Charged sprays are those two-phase systems of liquid as the dispersed phase (droplets) flowing within a gas as the continuous phase, usually remaining at rest whose droplets are electrically charged. Sprays are applied in many technologies, for example, for surface coating or painting, for scrubbing flue gases, for gas or surface cooling, in the production of chemicals, in manufacturing powder materials of different kinds including pigments, catalysts or reinforcing fillers. Charged sprays have many advantages over the uncharged ones, such as self-dispersing due to Coulomb repulsion, resulting in the absence of droplet coagulation, motion along the electric field lines, or higher deposition efficiency on objects. The processes and methods used for charged spray generation are: charging of droplets by ionic or electron beam, charging by ionic current, induction charging during liquid dispersion by mechanical atomizers, conduction charging by connection an atomizer to a high voltage source, or electrospraying. Charged sprays are, for example, applied for surface coating, thin-film deposition, spray forming, electroscrubbing, agriculture chemical deposition, fuel dispersion, or as propellant for space vehicle propulsion. The methods of charged sprays generation and their applications are presented in this chapter.

Chapter 2 - Consumption of fruits and vegetables has increased due to their high content of bioactive compounds. Incorporation of bioactive compounds into functional foods can decrease the occurrence of several human diseases such as cancer, cardiovascular disease, inflammatory disease, and oxidative stress. However, it is very difficult to store fruits and vegetables for long time periods even at low temperatures due to their perishable nature. Therefore, processing is necessary to prolong their shelf life. Drying is one of the oldest food preservation techniques and has been translated into a current method by modern technologies. Among all drying methods, spray drying is one of the best methods for the preservation of bioactive compounds. Spray dryers are widely used to produce dried fruits and vegetables. Spray drying effectively preserves heat-sensitive bioactive components, such as phenolic, carotenoid, and anthocyanin compounds, compared to other drying methods. During spray drying, different types of encapsulating agents such as starches, maltodextrins, corn syrups, inulin, and arabic gum are used in processing. Encapsulating agents protect bioactive compounds from oxygen, water, and light. Moreover, encapsulating agents can be used to increase the stability of bioactive compounds. In this report, the authors reviewed the current status of bioactive compounds in different fruits, vegetables, herbs, spices, and legumes from around the world and investigated the total contents and activities of vitamin C, phenolics, flavonoids, anthocyanins, β-carotene, lycopene, and betalain compounds using different types of encapsulating agents with spray dryers. This information gained in this review will be helpful to consumers and product developers as well as the food industry.

Chapter 3 - In this paper, the effect of variable density, viscosity and surface tension of the fluid on drop characteristics (i.e. drop size, drop velocity, number density and drop shape) and spray formation characteristics in terms of spray angle, drop size distribution, spray shape and penetration (mean resident time) due to variations of fluid temperature, differential pressure of fluid across nozzle orifice and fluid flow rate of highly temperature dependent high viscous salt solution was studied. This work was mainly conducted to study and validate the drop absorption concept developed to be utilized in absorption refrigeration cycles. This study investigated the characteristic of drops and sprays for two sets of pressure atomizers (i.e., swirl jet atomizers and full jet atomizers). In swirl jet atomizer testing, four different models (sizes) and in full jet atomizer testing, three different models (sizes) were tested. For each selected nozzle, by changing the fluid temperature and the nozzle pressure, an average of 30 different spray flow trials were conducted, which totaled to 200 tests.

In the process of designing and setting up the experimental test rig, several factors needed to be considered. This include building of a single nozzle spray absorber chamber that is able to attain a low pressure of 1.23 kPa, development of solution tempering system that is able to vary the solution temperature in the range of 60°C to 90°C, setting up of leak proof high vacuum system with leakage rate less than 4.67×10^{-6} scc/sec, assembling a Phase Doppler Particle Analyzer (PDPA) and high speed digital imaging system to measure drop and spray characteristics, and development of complete data acquisition and control system to control the process and collect all of the data. In addition, for comparison purpose, a spreadsheet program based on existing drop formation models was developed.

In this study, the ranges of input parameter variation were controlled by the development and expansion of the spray due to nozzle geometry of each individual nozzle, variation of properties of sprayed fluid and changing of the flow condition inside the vacuum chamber that limited by the diameter and length of the chamber. During the testing, input parameters included fluid density, viscosity ratio i.e. k-factor of dispersed to continuous phase flow and surface tension of fluid were varied in the ranges of 2502 kg/m^3 – 2524 kg/m^3, 1015 – 1708 and 0.2073 kg/s^2 – 0.2575 kg/s^2, respectively. The input parameters of mass flow rate and differential pressure across the nozzle related to different flow conditions were varied from 0.011 kg/sec to 0.045 kg/sec and 21.31 kPa to 519.33 kPa, respectively. The input parameters related to nozzle geometry that is nozzle constants and flow numbers were varied from 0.69 to 6.9 and 3.6×10^{-7} to 2.3×10^{-5}, respectively.

All the tested sprays showed the breaking of drops is mainly caused by viscous effect. The range of the droplet Reynolds number (Re) observed was between 2.38×10^2 and 2.14×10^3 and the range of Weber number (We) was between 1.22×10^{-5} and 3.05. The ranges of both Bond number (Bo) and Morton number (M) that define the shape of the droplets were between 0.0037 – 0.043 and 6.14×10^{-12} – 3.16×10^{-23}. The drops generated in the vacuum chamber were largely spherical drop in creeping and high-speed flows and drops with larger Re number and Bo number showed some “wobbling” effects. The mean volumetric diameter

(MVD) of the drops that generated from nozzles were varied from 200 to 500 micrometers and the drop velocities were varied from 4.5 m/sec to 16.75 m/sec. The distribution factor Q for drops varied from 4.6 to 6.8. Since the pressure atomizers used were limited to swirl jet and full jet atomizers due to the overall requirement of generating smallest possible droplets, the spray shapes were limited only to "full cone" and "hollow cone" spray patterns. The range of spray angles that were generated by selected nozzles varied from 32 degrees to 98 degrees. The drop mean resident times that depended on drop velocities and length of the absorber were range from 0.12 sec to 0.33 sec.

The experimental results agreed well with the spreadsheet program developed based on published drop formation models. The values were within the percentages of 79% to 94% and the experimental scatter was in the range of the error margins of the equipments.

Chapter 4 - Spray-drying is a well-known processing technique commonly used for granulating powder materials. In fact, spray drying has been developed for many industrial applications due to its ability to produce high volumes of uniform particles with identical characteristics. An example of this is its extensive use in the ceramic industry coupled with other processing stages like die-pressing.

In this report, the authors discuss the advantages of using the Spray Drying method (SD) in the synthesis and processing of advanced ceramics, through some cases of study which include the preparation of yttrium aluminum garnet -$Y_3Al_5O_{12}$- (YAG), α-alumina, $K_{0.5}Na_{0.5}NbO_3$ (KNN), $(K_{0.48}Na_{0.52})_{0.96}Li_{0.04}Nb_{0.85}Ta_{0.15}O_3$ (KNLNT), and Cu/Mg/Al mixed oxides. The applications of these ceramics include optical, piezoelectric, ferroelectric, structural, ion exchange/adsorption, pharmaceutical, photochemical, and electrochemical.

The synthesis of different advanced materials was performed by SD of aqueous solutions and organic compounds. For YAG, KNN, and KNLNT preparation, citric acid was used as a chelating agent, while the synthesis of α-alumina was achieved by using aluminum formate ($Al(O_2CH)_3$) as a metal-organic precursor. Besides, macroporous Cu/Mg/Al mixed oxides were fabricated by using layered double hydroxides as building blocks and submicrometric polystyrene spheres as pore former agents.

The results demonstrate that spherical agglomerates can be obtained in most cases after being calcined, depending upon the equipment operation conditions, composition, concentration, and organic compound content. Furthermore, the agglomerates with narrow size distribution formed by nanometric particles can be produced. This route shows that the calcination temperatures can be reduced considerably compared with the conventional ceramic method. These features demonstrate the feasibility of this method for the synthesis of advanced ceramics.

Chapter 5 - The microstructure and in-service properties of plasma-sprayed coatings are derived from an amalgamation of intrinsic and extrinsic spray parameters. These parameters are interrelated, which follow mostly non-linear relationships. The interactions among the spray parameters make the optimization and control of this process quite complex. Understanding relationships between coating properties and process parameters is mandatory to optimize the spray process and ultimate product quality. Process control consists of defining unique combination of parameter sets and maintaining them as a constant during the entire spray process. This unique combination must take the in-service coating properties into consideration. Artificial intelligence is a suitable approach to predict operating parameters of atmospheric plasma spray to attain required coating characteristics.

Chapter 6 - In this work, a detailed characterization of the liquid flow structure in sprays generated by pressure swirl nozzles was performed through experimental techniques. The overall droplet size distributions across the spray were obtained and the effects of the droplet collision phenomena on the spray development were studied. A Phase Doppler particle analyzer was used to obtain simultaneous measurements of size and droplet velocity and a data post-processing, applying the generalized integral method, was used to evaluate number concentrations and liquid volume fluxes for different droplet size classes. The parameters of radial mean spread and spatial dispersion of the sprays were also calculated.

A number of liquid injection conditions, which included two atomization regimes, were investigated. High collision rates were estimated in the spray densest zones and some collision outcomes, as the droplet coalescence and the separation with satellite droplet formation were detected. A liquid flow rate transfer between size classes, due to the coalescing collisions, was obtained as the sprays developed. This process, which favored the large size classes, caused the progressive growth of the droplet mean diameter along the spray. The separation was found a more important droplet collision outcome as droplet velocities increased.

Chapter 7 - Eulerian-Lagrangian simulations of an evaporating water-urea spray have been performed. Velocity, temperature, composition and size distribution are simulated. The chapter contains an extensive discussion of what forces and mass/heat transfer models that should be included.

Evaporation of water is included and evaporation and reaction of urea is modeled using a new model for the urea vapor pressure, previously developed by the authors [17].

The main objective of this chapter is to present a guide how to model a water-urea spray including mass and heat transfer. It is shown how the modeling of turbulent dispersion and the choices of injector position affects the model predictions in terms of spray uniformity and residence time.

Chapter 8 - Oxide and metal nanopowders were synthesized by spray pyrolysis using various types of atomizers and heating sources. In the spray pyrolysis process, the aerosols generate from the aqueous solution that contains dissolved precursors by using an atomizer such as an ultrasonic transducer and two-fluid nozzle. The advantages of spray pyrolysis are that the controls of particle morphology, size, size distribution, composition of resulting powders are possible. The diameter of aerosol is dependent on the frequency of the piezoelectric transducer or the size of two-fluid nozzle. Therefore, the particle size is controlled by adjusting them. In the pyrolysis, the aerosols were directly dried and heated to form spherical oxide nanoparticles at high temperatures. Homogenous, spherical oxide and alloy powders with narrow particle size distribution were successfully prepared by the spray pyrolysis. The particle microstructure of as-prepared powders was influenced by solution concentration, residence time in the furnace, pyrolysis temperature of the starting precursor. However, the stoichiometric composition of multi-component oxide or alloy was maintained regardless of particle microstructure. The particle size depended on the concentration of starting solution and flow rate of carrier air, but the particle size distribution was independent. The authors have tried various types of oxide and metal powders by the spray pyrolysis. In this paper, synthesis and characterization of oxide powders derived from spray pyrolysis were described. Furthermore, the mass production system was developed by flame spray pyrolysis using gas burners. The various types of cathode materials for an lithium ion battery were continuously produced. The excellent rechargeable performance of them was demonstrated by electrochemical measurement.

Chapter 9 - Fluorine doped tin dioxide (SnO2:F) films were grown using spray pyrolysis technique to obtain a high quality transparent conductor (TCO). During optimization of each process parameter it was observed that the film thickness and growth rate are crucial in assigning the structural properties, which thereby govern the electro-optical properties of the films. Presented efforts are to understand the structural evolution through the competitive roles of film thickness and growth rate when governed through the time of deposition and the precursor concentration, respectively. For this, the films were grown in three ways. Set A contains films of different thickness in a range ~50-1000 nm, grown using nearly constant growth rate ~100 nm/min. Set B consists of films of thickness ~500 nm, grown using different growth rates in a range ~20-200 nm/min. Set C consists of films of varied thickness up to ~1000 nm, grown using different growth rates in a range ~20-200 nm/min and with a fixed time of deposition. Competitive effects of both the film thickness and growth rate in set C films were isolated by comparing their results with those for set A and set B films. Film properties were investigated using x-ray diffraction technique, scanning electron microscopy and Hall effect measurements. The structural evolution helped to achieve a technologically important textured growth in these films viz. along [200] orientation. The growth rate induced effects accelerated this growth.

Chapter 10 - Various sets of new flamelet equations for spray combustion are derived in mixture fraction space. The spray droplets are assumed as points and described in an Lagrangian framework. The influence of a droplet on gas phase is represented by the mass, momentum, and energy source terms. The derived equations are different from the standard flamelet equations for gaseous combustion because of the droplet evaporation source term that was neglected in most previous studies. In addition, a general flamelet transformation for spray combustion is formulated, and the potential advantages of combined premixed and diffusion flamelet models for spay combustion is discussed.

In: Sprays: Types, Technology and Modeling
Editor: Maria C. Vella, pp. 1-100
ISBN 978-1-61324-345-9

Chapter 1

CHARGED SPRAYS GENERATION AND APPLICATION

A. Jaworek* and A. Krupa
Institute of Fluid Flow Machinery,
Polish Academy of Sciences, Poland

Abstract

Charged sprays are those two-phase systems of liquid as the dispersed phase (droplets) flowing within a gas as the continuous phase, usually remaining at rest whose droplets are electrically charged. Sprays are applied in many technologies, for example, for surface coating or painting, for scrubbing flue gases, for gas or surface cooling, in the production of chemicals, in manufacturing powder materials of different kinds including pigments, catalysts or reinforcing fillers. Charged sprays have many advantages over the uncharged ones, such as self-dispersing due to Coulomb repulsion, resulting in the absence of droplet coagulation, motion along the electric field lines, or higher deposition efficiency on objects. The processes and methods used for charged spray generation are: charging of droplets by ionic or electron beam, charging by ionic current, induction charging during liquid dispersion by mechanical atomizers, conduction charging by connection an atomizer to a high voltage source, or electrospraying. Charged sprays are, for example, applied for surface coating, thin-film deposition, spray forming, electroscrubbing, agriculture chemical deposition, fuel dispersion, or as propellant for space vehicle propulsion. The methods of charged sprays generation and their applications are presented in this chapter.

1. INTRODUCTION

Spray is a two-phase system of droplets flowing within a gas remaining at rest. Sprays differ from aerosols in two properties: (i) spray droplets can be larger than those required for the system to be quasi-stabile, and (ii) spray droplets are usually in motion whereas the continuous phase (gas) can remain principally in rest. In aerosols, the droplets have to be sufficiently small, that the time scale, over which the aerosol can be considered stable, depending on an industrial process or experimental conditions, is sufficiently long. The size

* Fiszera 14, 80-952 Gdańsk, Poland. E-mail: jaworek@imp.gda.pl.

of water aerosol droplets has to be smaller than 50 μm to obtain the quasi-stable system. Terminal velocity of water droplet of 10 μm is of the order of magnitude of 10 mm/s. In spray, the droplets can be as large as a couple of one mm. The aerosol droplets usually have the velocity close to the velocity of the continuous phase. In the case of spray, the droplets are injected into a quiescent or flowing gas and the difference in the velocity between both phases can be extremely high. The droplets can be accelerated/decelerated, deformed and undergo breakup following the interaction with the gas (Azzopardi and Hewitt [1997], Krzeczkowski [1980], Feng [2010]). Nevertheless, there are some parameters that characterize both of these two-phase systems in the same way. In practice, sprays are frequently used for aerosol formation.

Sprays have found many applications in technology, for example, for surface coating or painting, for scrubbing flue gases, for gas or surface cooling, in the production of chemicals, and in manufacturing powder materials of many different kinds including pigments, catalysts or reinforcing fillers. Charged sprays differ from electrically neutral ones in that there is an electric charge on the droplets' surface. The charge causes stronger mutual interactions between the droplets and between the droplets and nearby objects. Charged-spray technologies are those processes that utilize electrical effects for the process improvement via controlling the spray droplets motion. Charged sprays have been involved in such technological processes as, for example, surface coating, thin-film deposition, spray forming, or electroscrubbing. Charged droplets have also been utilized for agriculture chemical deposition, fuel dispersion before burning, or as propellant for space vehicle propulsion. The fact that droplets are electrically charged facilitates the control of their motion (including their deflection and focusing) by means of electric field. Charged droplets are self-dispersing in space due to mutual Coulomb repulsion, resulting in the absence of droplet coagulation, characteristic to uncharged sprays.

The following processes are used for charged spray generation:

1. Ion- or electron beam charging,
2. Charging by ionic current,
3. Induction charging during liquid atomization,
4. Conduction charging by connection atomizer to a high voltage source,
5. Electrospraying.

The paper discusses various methods of generating charged droplets and applications of charged spray.

2. Charged Spray Characterization

Un-charged spray is characterized by the volume, mass, or number concentration of the dispersed phase, volume fraction of the dispersed phase, and droplet size distribution (Bayvel and Orzechowski [1993], Hinds [1999], Lefebvre [1989]). Charged sprays are characterized by additional parameters such as charge distribution in the dispersed phase, space charge density, and current carried by the dispersed-phase (spray current).

1. *Total charge* Q_t of spray is the sum of charges q_i on all individual droplets:

$$Q_t = \sum_i Q_i \qquad (1)$$

2. *Surface charge density* on a droplet is the ratio of the charge Q on the droplet of aerodynamic radius R_d to the droplet's surface area:

$$\sigma_q = \frac{Q}{4\pi R_d^2} \qquad (2)$$

3. *Specific charge* of the spray is defined as the ratio of total charge Q of charged droplets to the mass m_d of all droplets. This quantity is referred to as the "*q over m*" ratio. This quantity is also used for characterizing electrohydrodynamic properties of a single "mean" droplet. Because the inertial force is proportional to the droplet's mass $F_i = am_d$ and the electric force to the droplet's charge $F_e = QE$, the ratio of these forces is proportional to the specific charge:

$$\frac{F_e}{F_i} = \frac{E}{a}\frac{Q}{m_d} \qquad (3)$$

where a is the acceleration of the droplets, and m_d is the droplet mass.

In charged spray, the total current of droplets is the integral of the product of charge distribution $Q(r)$, as a function of droplets' radius r, the droplet size distribution within the spray $f(r)$, and the number n of droplets generated per unit time:

$$I = \int_0^\infty Q(r) f(r) n \mathrm{d}r \qquad (4)$$

The total mass flow rate of the spray is:

$$\dot{m} = \frac{4}{3}\pi\rho_l \int_0^\infty r^3 f(r) n dr \qquad (5)$$

Assuming the charge Q proportional to the droplet's surface area, and the surface charge density σ_q:

$$Q(r) = 4\pi\sigma_q r^2 \qquad (6)$$

we can conclude that the current to mass flow rate ratio is:

$$\frac{I}{\dot{m}} = \frac{3\sigma_q \int_0^\infty r^2 f(r) n \mathrm{d}r}{\rho_l \int_0^\infty r^3 f(r) n \mathrm{d}r} = \frac{3\sigma_q}{\rho_l r_s} \tag{7}$$

where r_s is the Sauter mean radius of the droplets in the spray.

The ratio of spray current to the mass flow rate of the spray is the measure of specific charge of droplet of Sauter mean diameter r_s:

$$\frac{I}{\dot{m}} = \frac{4\pi\sigma_q r_s^2}{\frac{4}{3}\pi\rho_l r_s^3} = \frac{Q(r_s)}{m_d(r_s)} \tag{8}$$

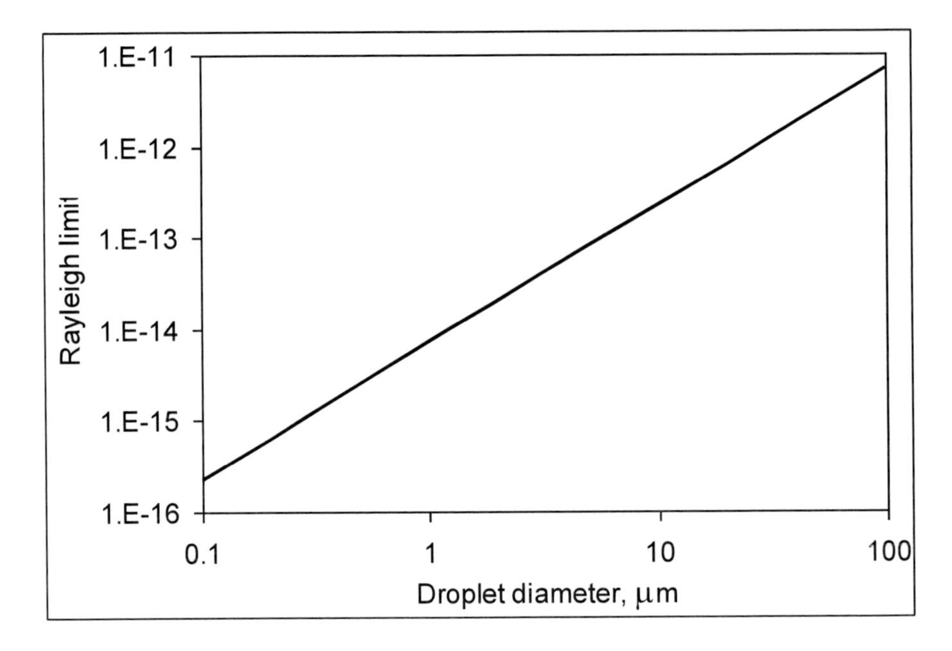

Figure 1. The magnitude of charge at Rayleigh limit for water droplets.

This parameter is frequently used for the estimation of mean droplet charge in a spray for which the Sauter mean droplet size was determined.

4. *The Rayleigh limit.* The Rayleigh limit is the magnitude of charge on the droplet, which produces the repulsive force on the surface larger than the surface tension force. In terms of surface charge density σ_{qmax} and the surface tension σ_l of the liquid, this condition can be represented by the following equation:

$$\frac{\sigma_{q\max}^2}{2\varepsilon_0} = \frac{2\sigma_l}{r_d} \tag{9}$$

in which ε_0 is the electric permittivity of the free space. From this equation results the maximum equilibrium charge Q_R, known as the Rayleigh limit (Rayleigh [1882]):

$$Q_R = 8\pi\sqrt{\varepsilon_0 \sigma_l r_d^3} \tag{10}$$

When the charge on a droplet exceeds the Rayleigh limit specific to the liquid, the droplet disrupts to smaller ones. The Rayleigh limit is used for the characterization of stability of the droplet. The charge larger than the Rayleigh limit causes changes in the size distribution of the spray. The magnitude of charge of Rayleigh limit for water droplets determined from Equation (10) is presented in Figure 1.

5. *The fraction of the Rayleigh limit* is the charge Q on a droplet of the radius r_d, related to the Rayleigh limit charge Q_R, and it equals:

$$\frac{Q}{Q_R} = \frac{\sigma_q}{2}\sqrt{\frac{r_d}{\varepsilon_0 \sigma_l}} \tag{11}$$

6. *Electric field* at the surface of charged droplet is also sometimes used as a parameter characterizing a charged spray. The electric filed is used for the determination of electrostatic interaction of the droplets or deposition the droplets on an object (Lapple [1970]):

$$E_d = \frac{Q}{4\pi\varepsilon_0 r_d^2} \tag{12}$$

7. *Charged-spray current I_S*. On-line measurement of charge on a single droplet and statistical distribution of charges in spray is a serious problem, which has not been solved satisfactorily (Brown [1997], Flagan [1998], Jaworek [1993]) due to high level of thermal and electromagnetic noise distorting the detection of low-level signals induced by small charges, smaller than femtocoulombs. Therefore, usually the total spray current of charged droplets is measured, and from this current the mean droplet charge is estimated:

$$Q_{1d} = \frac{4\pi r_S^3}{3Q_V} I_s \tag{13}$$

where r_S is the Sauter mean radius of the droplet, Q_V is the volume flow rate of the spray.

8. *Time constant of charged spray expansion*. Charged droplets are self-dispersing in the space due to mutual Coulomb repulsion that results in absence of droplets agglomeration. The expansion of a spherical cloud of charged droplets is given by the equation (Lapple [1970]):

$$-\frac{1}{c_d}\frac{dc_d}{dt} = \frac{2C_c \varepsilon_0 c_d E_{ds}^2}{\eta_g \rho_l} = \frac{c_d}{\tau_{exp}} \tag{14}$$

in which c_d is the initial mass concentration of the aerosol droplets, η_g is the gas dynamic viscosity, ρ_l is the liquid density, and C_c is the Cunningham slip correction factor.

E_{ds} is the electric field on the surface a single droplet charged to the magnitude Q.

In equation (14), τ_{exp} is the time constant of expansion of the cloud:

$$\tau_{exp} = \frac{\eta_g \rho_l}{2C_c \varepsilon_0 E_{ds}^2} \tag{15}$$

Due to the cloud expansion, the mean mass concentration of the spray decreases reciprocally with time:

$$c_d = c_{d0}\left(1 - \frac{t}{\tau_{\exp}}\right) \tag{16}$$

The radius of a charged spherical cloud of aerosol particles equals to:

$$R(t) = R_0 \sqrt{1 + \frac{\mu_p}{\varepsilon_0}\rho_0 t} \tag{17}$$

9. *Droplet size distribution*. Four droplet size distributions in sprays generated by various atomizers are used (Bayvel and Orzechowski [1993], Lefebvre [1989]):

- *Nukiyama-Tanasawa* (Lefebvre [1989], Azzopardi and Hewitt [1997]), referring to pneumatic atomizers with parallel flow:

$$\frac{dN}{dD} = aD^2 \exp\left(-bD^q\right) \tag{18}$$

where a, b, q are independent constants determined from measurements.

- *Rosin-Rammler* (Lefebvre [1989], Azzopardi and Hewitt [1997]):

$$1 - F = \exp\left(-\left(\frac{D}{X}\right)^q\right) \tag{19}$$

where X and q are independent constants, F is the fraction of total volume contained in drops of diameter less than D, q is a measure of the spread of size distribution of the droplets. The higher the value of q, the more uniform is the spray. For pneumatic atomizers q=1.5 ....4, and for rotary atomizers can be as high as 7 (Lefebvre [1989]). The value of q can be determined from experimental data as a slope of the line of ln(D) against $\ln(1\text{-F})^{-1}$.

For pressure-swirl atomizers the modified Rosin-Rammler distribution is used (Lefebvre [1989]):

$$\frac{dF}{dD} = q\frac{(\ln D)^{q-1}}{D(\ln X)^q}\exp\left(-\left(\frac{\ln D}{\ln X}\right)^q\right) \tag{20}$$

- *Kim-Marshall* ([1971]). The following equation gives the volume size distribution for pneumatic atomizer:

$$f_3(r) = \frac{16.7\exp(-2.18r_0)}{(1+6.67\exp(-2.18r_0))^2} \tag{21}$$

where $r_0 = r/r_{30}$ and r_{30} is given by the equation:

$$r_{30} = 2.68\times10^{-3}\frac{\sigma_l^{0.41}\eta_l^{0.32}}{(v^2\rho_a)^{0.57}A^{0.36}\rho_l^{0.16}} + 1.72\times10^{-3}\left(\frac{\eta_l^2}{\sigma_l\rho_l}\right)^{0.17}\frac{1}{v^{0.54}}\left(\frac{\dot{m}_a}{\dot{m}_l}\right)^k \tag{22}$$

$$k = \begin{cases} -1 & \text{for} \quad \dfrac{\dot{m}_a}{\dot{m}_l} < 3 \\ 1 & \text{for} \quad \dfrac{\dot{m}_a}{\dot{m}_l} > 3 \end{cases} \tag{24}$$

$\dot{m}_a$ and $\dot{m}_l$ are the mass flow rate of air and liquid, respectively, A is the cross section of annular gas nozzle, and v is the liquid-air relative velocity.

- *Log-normal distribution*

$$f(D) = \frac{dN}{dD} = \frac{1}{\sqrt{2\pi}Ds_g}\exp\left(-\frac{(\ln D - \ln D_{ng})^2}{2s_g^2}\right) \tag{25}$$

where D_{ng} is the number geometric mean drop size, and s_g is geometric standard deviation.

3. Charged Spray Generation

3.1. Charging by Electron or Ionic Beam

Spray charging by ionic or electron beam takes place when the spray droplets cross a beam of charged molecules (ions) or electrons, which collide with the droplets due to their kinetic energy. The droplet charge is proportional to the time the droplet remains within the beam (Moore [1973]):

$$Q = 2\pi \frac{J R_b r_d^2}{v_d} \quad (26)$$

where J is the ionic (electron) current density, R_b is the radius of the beam, r_d is the droplet radius, and v_d is the droplet velocity.

It was assumed in this equation that the kinetic energy of ions is sufficiently high to overcome the Coulomb repulsion of the charged droplet. Because ionic beam charging requires reduced pressure in order to avoid the beam dispersion, this method can only be effectively used for charging of solid nanoparticles (Seto et al. [2005], Vasil'ev et al. [2010], Walch et al. [1995]). Because of evaporation of the droplets, the ionic beam charging can be potentially applied only to the liquids of low-vapor pressure.

3.2. Charging by Ionic Current

Charging of droplets or solid particles by ionic current is accomplished via electric (corona) discharge generation between a pair of electrodes when the droplets/particles flow throughout the discharge region. Due to high electric field in the vicinity of the discharge electrode, free electrons are accelerated to energy sufficiently high to cause gaseous molecules ionization. For positive polarity of the discharge electrode, electrons drift to this electrode while the positive ions drift towards the counter electrode. For negative polarity of the discharge electrode, positive ions move to this electrode while electrons drift to the counter electrode. When electronegative molecules (O_2, CO_2) are also components of the gas, the electrons can attach to these molecules, forming negative ions. In either case, the ions or electrons collide with solid particles or liquid droplets thus charging them.

Due to an excess of gaseous ions of one polarity, the net charge deposited onto the droplets significantly differs from zero. The process of particle charging is described by the following equation in which the ionic current to the particle is a function of time and the electric field near the particle (due to the charge previously deposited on this particle), (Lawless [1996]):

$$\mathbf{j}(\mathbf{r}) = -eD\nabla n_i(\mathbf{r}) + e\mu_i n_i(\mathbf{r})\mathbf{E}(\mathbf{r}) \quad (27)$$

where j is the ionic current density, $\mathbf{r}$ is the position vector, D is the diffusion coefficient of ions, e is the elementary charge, n is the ion number density, μ_i is the ion mobility, and $\mathbf{E}$ is the electric field close to the droplet's surface. In the following it is assumed that only single-ionized gaseous ions are in the charging zone. Although corona chargers were principally developed for solid particles charging, most of these devices can virtually be used also for liquid-droplet sprays charging.

The deposition of gaseous ions onto a particle or droplet is governed by two principal mechanisms: diffusion charging (first term in equation (27)) and field charging (second term in equation (27)) (Arendt and Kallman [1926], Böhm [1982], Lapple [1970], Pauthenier and Moreau-Hannot [1932], Penny and Lynch [1957], White [1963]). For droplets larger than 1 μ m, the dominating mechanism is the field charging. In the field charging mechanism, the gaseous ions are deposited onto the droplet due to their kinetic energy taken from external electric field. This energy is required to overcome the repulsive force of the charge on the droplet, and increases with the charging time elapsing. The second mechanism, diffusion charging, is a process of ions deposition due to their kinetic energy of thermal motion. This mechanism becomes dominating for droplets smaller than 0.1 μm. In this case, the charging rate is independent of the electric field but only on the kinetic energy of ions.

In the transition region of size of the particles/droplets, roughly from 0.1 to 1 μm, neither of the models is adequate, and models that combine these mechanisms have been developed. To solve the problem of charging in this regime, a general model has been developed by Liu and Kapadia [1978]. This model is, however, complex and require sophisticated numerical procedures. Therefore, the sum of charges from diffusion and field charging models, assuming that both mechanisms operate independently, are most frequently used (Biskos et al. [2005a], Fjeld et al. [1983], Frank et al. [2004], Kirsch and Zagnit'ko [1990], Klett [1971], Lackowski et al. [2003a,b], Lawless [1996], Sato [1987]).

In order to determine the net charge transferred from ionized gas to a droplet or particle, the ionic current must be integrated over its surface. The gaseous ions do not flow to the droplet over its entire surface, but only at the area at which the electric field lines are towards the droplet surface.

The electric field distribution over the droplet surface is a superposition of the external electric field (produced by the electrodes) and the field due to the charge accumulated on the droplet. The electric field component perpendicular to the droplet's surface is:

$$E_n = 3\frac{\varepsilon_r}{\varepsilon_r + 2}E\cos\theta - \frac{Q}{4\pi\varepsilon_0 r_d^2} \tag{28}$$

where θ is the angle over the droplet surface measured from the external field lines.

The electric flux to the droplet is

$$\Psi_E = \int_0^{\theta_0} 2\pi r_d^2 \left(\frac{3\varepsilon_r}{\varepsilon_r + 2}E\cos\theta - \frac{Q}{4\pi\varepsilon_0 r_d^2} \right)\sin\theta d\theta = 3\pi E r_d^2 \frac{\varepsilon_r}{\varepsilon_r + 2}\left(1 - \frac{Q}{12\pi\varepsilon_0 \frac{3\varepsilon_r}{\varepsilon_r + 2} E r_d^2} \right)^2 \tag{29}$$

where θ_0 is the angle at which the magnitude of electric field (28) equals zero.

The ionic current to the droplet is:

$$I_i = \frac{dQ}{dt} = 3\pi ne\mu_i E r_d^2 \frac{\varepsilon_r}{\varepsilon_r + 2}\left(1 - \frac{Q}{12\pi\varepsilon_0 \frac{3\varepsilon_r}{\varepsilon_r + 2} E r_d^2}\right)^2 \tag{30}$$

where: t is the time of charging, r_d is the droplet radius, ε_r is the relative permittivity of the droplet, μ_i is the gaseous ion mobility, E_m is the electric field in the charging zone, and n is the mean value of the ion number density of gaseous ions in the charging zone.

In parallel-plate electrode DC or AC-field chargers (cf. the following sections), when one of the electrodes is the current emitter, the ion number density can be determined from the simplified equation:

$$n = \frac{Id}{\mu_i UA} \tag{31}$$

where I is the ionic current flowing through the charging zone, d is the distance between the electrodes (grids), μ_i is the gaseous ion mobility, A is the cross section of the charging zone (surface of the contour of emitter electrode), and U is the voltage between the electrodes.

Equation (30) describes the rate of charge increase on droplet due to the field charging mechanism. After the infinite time of charging, the charge on the droplet is saturated. The saturation charge is given by the Pauthenier equation:

$$Q_s = 12\pi\varepsilon_0 r_d^2 E \frac{\varepsilon_r}{\varepsilon_r + 2} \tag{32}$$

The saturation charge is frequently called "the Pauthenier limit". It is the maximal magnitude of charge, which can be accumulated on droplet or particle at a given electric field after infinitely long time of charging. The saturation charge is independent of ion concentration, provided $n>0$, but only on the electric field magnitude.

The process of diffusion charging is described by the Arendt-Kallmann equation (Arendt and Kallmann [1926], Lapple [1970]):

$$\frac{dQ}{dt}\left[1 + \frac{\pi\varepsilon_0 \hat{v}_i r_d^2}{Q\mu_i}\right] = \pi en\hat{v}_i r_d^2 \exp\left(-\frac{eQ}{4\pi\varepsilon_0 r_d kT}\right) \tag{33}$$

in which μ_i is the gaseous ion mobility, n is the number concentration of the gaseous ions, T is the absolute temperature of the gas, k is the Boltzmann's constant, and $\hat{v}_i$.is the rms velocity of the gaseous ions.

Equation (33) is nonlinear one and can be solved only numerically. In order to determine the charge on solid particles, White [1963] has linearized this equation by neglecting the second term in parentheses on the left side of the equation, and for this case obtained the following analytical solution:

$$Q(t) = \frac{4\pi\varepsilon_0 kTr_d}{e} \ln\left(1 + \frac{r_d \bar{v}_i e n_i t}{4\varepsilon_0 kT}\right) \tag{34}$$

Equation (34) is commonly used for approximate determination of charge on a droplet or particle due to the diffusion charging mechanism. However, for infinitely long time of charging, the acquired charge is also infinite. This problem has, however, not found a satisfactory solution yet.

In sprays, droplets smaller than 1 μm are not frequently produced, and therefore, the diffusion charging mechanism can be neglected, and will not be further considered.

The process of charging of droplets in ionic current differs from the charging of solid particles in that it depends on the polarity of the liquid (Inculet [1982], Javaid et al. [1980]). Non-polar dielectrics are those materials whose polarization is due only to the displacement of electrons in the molecule when placed in an electric field. The polar materials have permanent dipole moments. The time constant of particle/droplet charging is the product of dielectric constant of the material (resulting from the polarization of the material) and its conductivity. The time constant is a measure of the deposited charge to relax over the surface of the droplet. When a liquid droplet is exposed to ionic current for a very short time, the total charge acquired is substantially lower than the saturation charge.

$$Q(t) = Q_S \frac{t}{t + \tau_1} \tag{35}$$

where t is the time of exposure of the droplet on the ionic current, τ_1 is the charging time constant of the liquid material:

$$\tau_1 = \frac{4\varepsilon_0}{N_0 \mu_i} \tag{36}$$

where N_0 is the ionic current density, μ_i is the liquid ion mobility.

Experiments have shown that non-polar materials acquired from 10 to 36% less charge than equivalent water droplets and that polar materials acquired from 1 to 15% more charge than the water droplets (Inculet [1982], Javaid et al. [1980]). Inculet [1982] and Javaid et al. [1980] assumed that these differences, which are not predicted by the Pauthenier theory, result from the charge relaxation time depending on the droplet's material, which play fundamental role in the rate of charge distribution over the droplet's surface.

3.2.1. DC-Corona Chargers

The simplest type of charger is that which utilizes direct-current (DC) corona discharge between a sharp electrode and counter electrode, usually a plate or cylinder of larger radius of curvature. The sharp electrode can be in the form of wire or a set of needles.

A cylindrical charger with a thin wire placed in the axis of a grounded cylinder is shown in Figure 2a. Because of $1/r$ distribution of the electric field within this type of charger and decreasing ion concentration from the wire to the cylinder, the charge on a droplet depends on the radial coordinate of the droplet's trajectory. Another type of charger is a system of two parallel electrodes: a needle electrode facing a flat plate (Figure 2b). In this configuration, the field magnitude and ion concentration are nearly constant in the whole charging zone, and the charge acquired by a particle or droplet does not depend on its trajectory.

Charging in DC corona discharge is not an effective method for the charging of droplets when high penetration through the charger is required. In DC corona chargers, the trajectories of charged droplets are deflected towards the ground electrode due to electrostatic forces, and the droplets are deposited onto this electrode. This phenomenon is a serious disadvantage of this type of charger. Droplets, which are not deposited within the charger and leave it, have the electric charge considerable lower than the Pauthenier limit (Pauthenier and Moreau-Hannot [1937]).

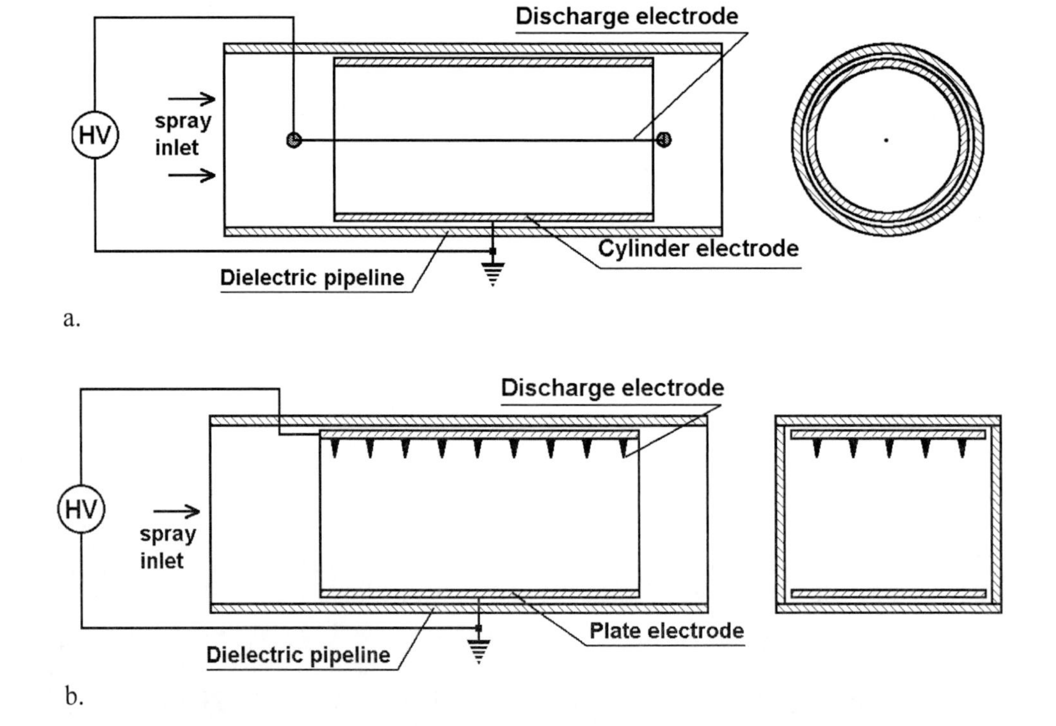

Figure 2. DC-electric field corona chargers: cylindrical corona charger (a), parallel plate corona charger (b).

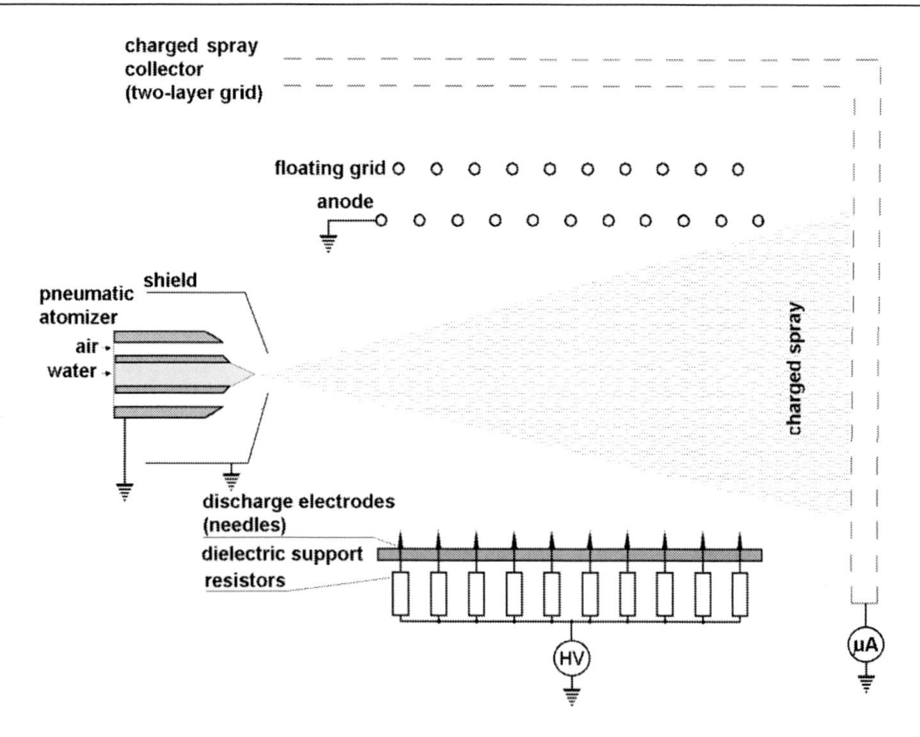

Figure 3. Schematic of a set-up for the measurement of efficiency of spray charging in corona discharge.

An experimental set-up for the measurement of efficiency of spray charging in a corona discharge is schematically shown in Figure 3.

The corona discharge is generated by a set of sharp needle electrodes supplied from a DC high voltage source via individual ballast resistors (14 MΩ). A higher electric field in the charging zone can be produced when individual ballast resistors are used. When a spark or streamer discharge occurs from a single needle, the voltage drops at the resistor, quenching the discharge only from this single needle and not from entire electrode. When a common ballast resistor would be used, the discharge would be stopped from the entire electrode, and the spray will not be charged. The ions generated from the needle electrodes flow towards the anode made of a set of rods. Behind the anode, a floating grid made of a set of rods is placed. The floating grid prevents the gaseous ions from flowing towards the charged-spray collector and minimizes the distortion of measurement of the charge of droplets. The ions of mobility higher than the droplets can be "reflected" towards the anode, while the low-mobility droplets can further flow to the collector. The potential of the floating grid is determined by the charge of ions and droplets deposited on the grid. A dense, two-layer grid collects the spray charge, and the current is measured with a moving-coil ammeter. Pneumatic atomizer generating the spray is placed behind a grounded shield in order to prevent inducing the charge on the surface of sprayed liquid by the electric field from the discharge electrode. In order to correct the spray current measurements by taking into account the droplets deposited on the anode, the volume of the liquid deposited on the anode was measured. The volume of liquid collected on the collector was corrected by subtracting the volume deposited on the anode from the total volume sprayed.

The results of measurements of spray current vs. air pressure for various voltages at the discharge electrode (average electric field between the charger electrodes) can be determined from theoretical considerations (Jaworek [1989]):

$$I_s = \frac{2.28\pi\varepsilon_0 E_0 Q_V}{r_S} \tag{37}$$

based on the Nukiyama-Tanasawa droplet size distribution for pneumatic atomizers (18).

The spray current increases linearly with the pressure due to decreasing size of the droplets in the spray. Smaller droplets form larger surface area of the spray that allows increasing the total spray current. The mean droplet charge in the spray can be estimated from the total spray current (37). The charge on a single droplet of Sauter radius vs. electric field obtained in DC corona charger, estimated from the equation:

$$Q_{1k} = \frac{4\pi r_S^3 I_s}{3Q_V} \tag{38}$$

is presented in Figure 4. The charge increases linearly with the electric field and is practically independent of the air nozzle pressure, for pressures higher than 0.6×10^5 Pa. In Figure 5, the charge on a single droplet is compared with the Rayleigh limit determined for the droplet of mean Sauter diameter. The mean charge is smaller than 10% of the Rayleigh limit.

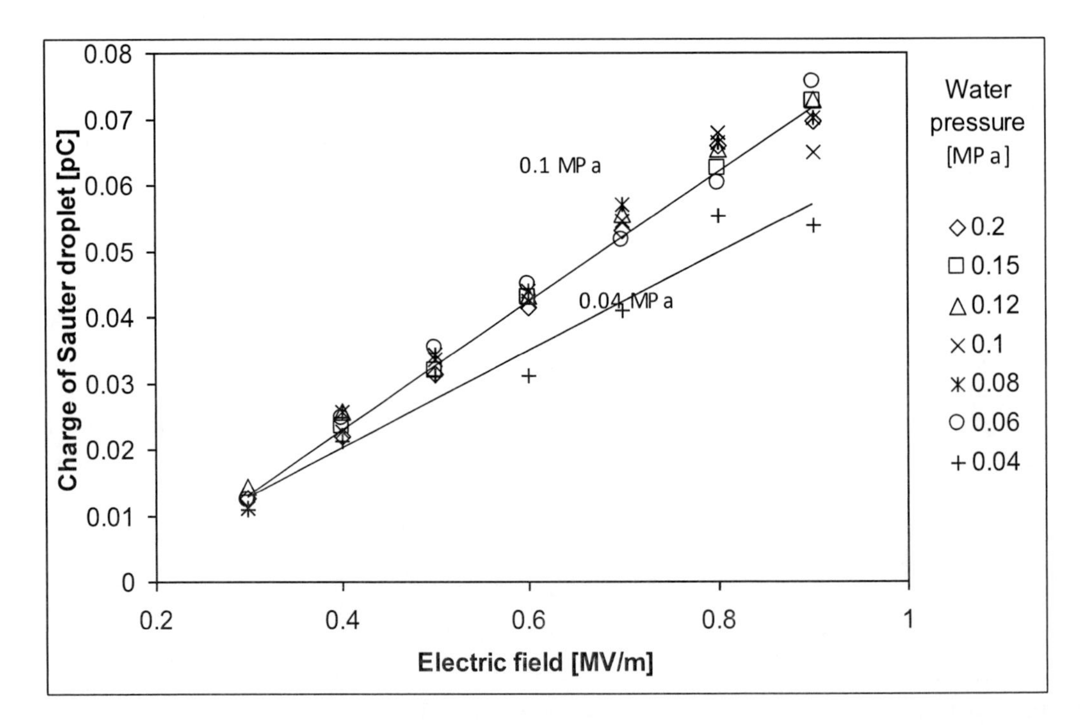

Figure 4. Charge on a single droplet generated by pneumatic atomizer and charged by ionic current in DC corona discharge.

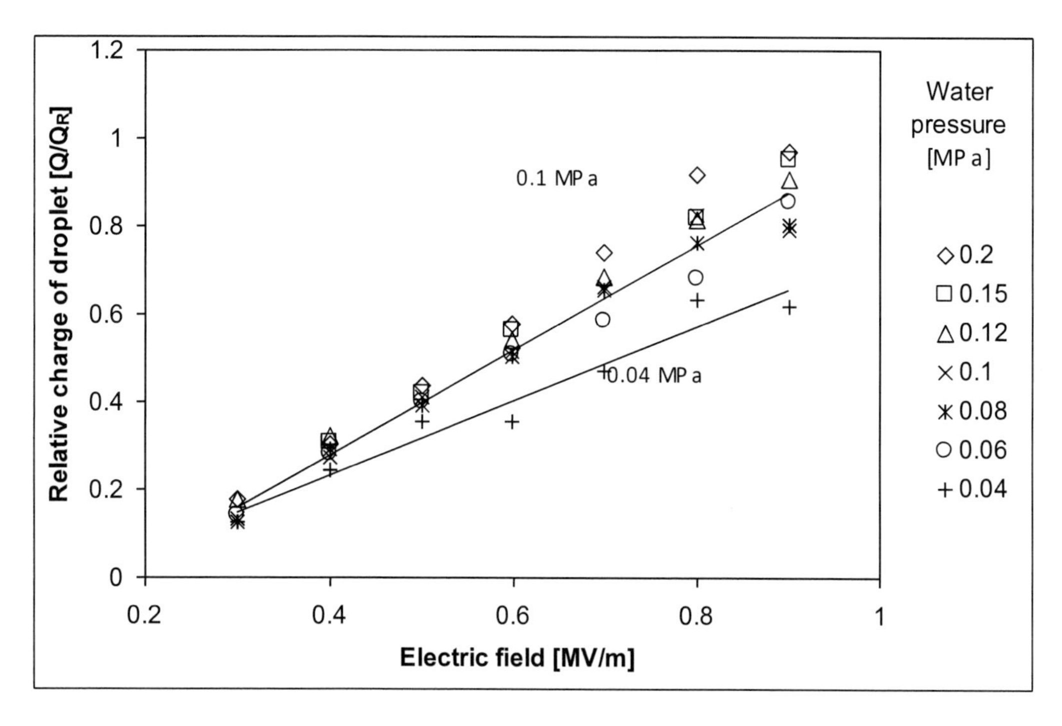

Figure 5. The fraction of the Rayleigh limit of charge on a single droplet generated by pneumatic atomizer and charged by ionic current in DC corona discharge.

A modification of DC corona charger is a device known as corona triode. In corona triode, the discharge electrode and grounded plate electrode are separated with a grid (Withers et al. [1978]). The charge imparted to the particles or droplets can be controlled by changing the grid potential and the current flowing through the charging zone between the grid and the plate electrode. However, in such a device the deflection of charged particles/droplets towards the plate electrode has not been eliminated.

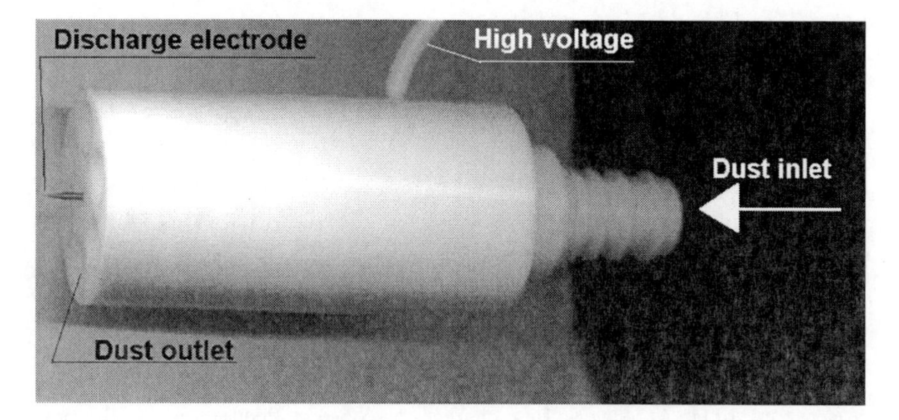

Figure 6. Photograph of co-flow corona charger.

In order to overcome the flaws inherent to DC-corona chargers, DC-corona co-flow chargers and chargers with alternating electric field in the charging zone have been invented. In the co-flow chargers, the gaseous ions are injected co-linearly with the flowing droplets. The ground electrode is in the form of a co-axial ring, and the droplets when of sufficient high kinetic energy are not bent to this electrode. A photograph of a co-flow corona charger is shown in Figure 6.

3.2.2. AC-Electric Field Chargers

In charging devices in which alternating electric field is applied, the particles move along oscillating trajectory between the electrodes, and their deposition on the charger elements is reduced. Edge or needle electrodes are used as the discharge electrodes in this type of chargers. A few types of chargers utilizing alternating electric field have been constructed.

Hewitt [1957] was probably the first who applied alternating electric field for charging aerosol particles. The ionic current was generated by corona discharge from a wire electrode, while additional electrode, excited with square-waveform voltage, produced the alternating electric field causing the oscillatory motion of the particles (Figure 7). In this type of charger, the particles were charged only in one half-cycle of the excitation voltage, because the ions were generated only in this half-cycle of the voltage.

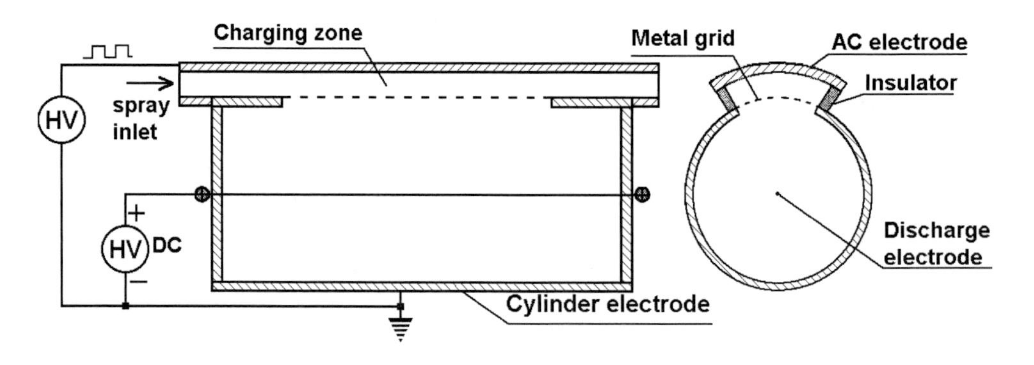

Figure 7. Schematic of the AC-electric field charger developed by Hewitt [1957].

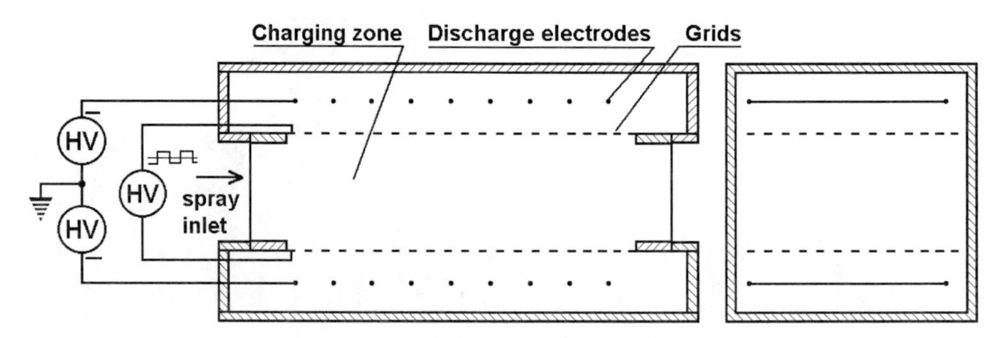

Figure 8. Schematic of the AC-electric field charger developed by Penney and Lynch [1957].

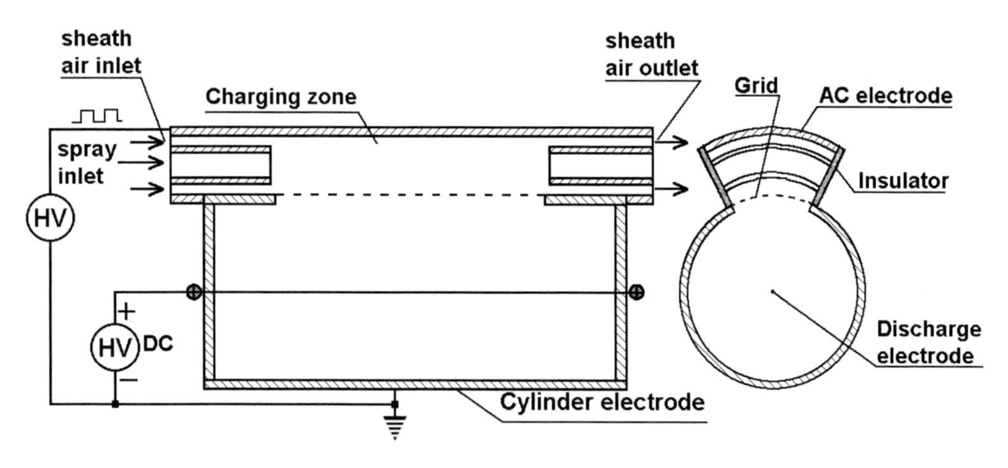

Figure 9. Schematic of the AC-electric field charger developed by Kirsch and Zagnit'ko [1981, 1990].

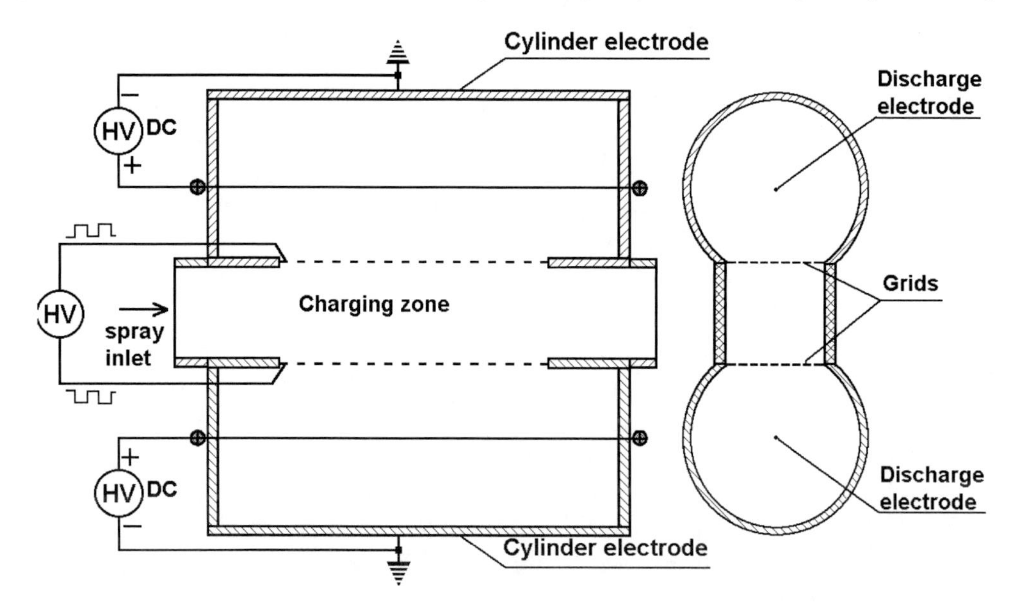

Figure 10. Schematic of the AC-electric field charger developed by Kruis et al. [1998, 2001].

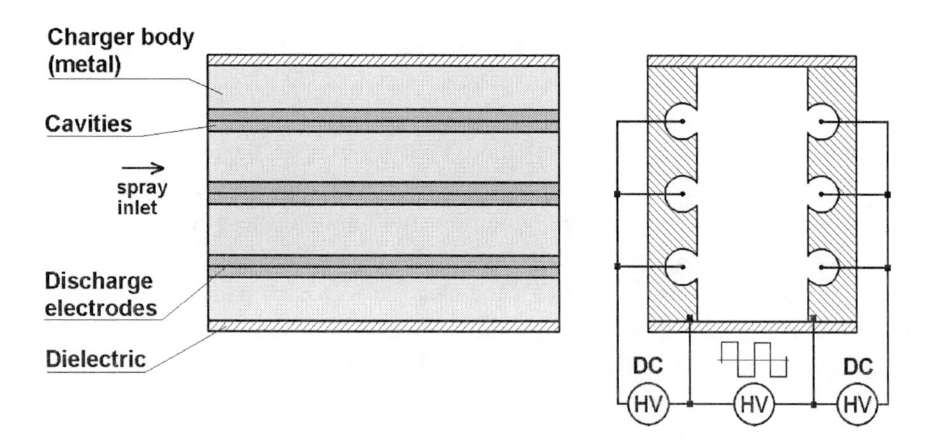

Figure 11. Schematic of the AC-electric field charger developed by Hutchins and Holm [1989].

Penney and Lynch [1957] have developed a plane-parallel electrode configuration (Figure 8) with charging zone formed by two parallel metal grids. The electric field between the grids was changed periodically because the electrodes were supplied with a square-waveform voltage. The ions were provided from the wire electrodes placed in caves behind the grids.

The Hewitt charger was modified by Kirsch and Zagnit'ko [1981, 1990] by adding a gas sheath parallel to the aerosol inlet (Figure 9), and by Kruis et al. [1998, 2001] by combining two symmetrical chargers of this type operating in anti-phase (Figure 10). The authors called this device "twin Hewitt charger" (Kruis and Fissan [2001]).

Hutchins and Holm [1989] have multiplied and scaled down the discharge zones of wire-cylinder configuration (Figure 11). The cylinder electrodes were bored in a piece of metal, which formed a kind of grid supplied with AC voltage. The ions generated by the wire electrodes supplied with DC voltage leave the cylinders via a slot and flow to the opposite electrode.

Büscher and Schmidt-Ott [1990, 1992, 1994] have developed a concentric alternating electric field charger in which the particles oscillate between the cylinders while flowing through the charger, but they are charged only in one half-cycle of the excitation voltage.

The disadvantage of wire-cylinder devices is that only small fraction of the discharge current flows through the charging zone that limits the energy efficiency of such devices.

Charger in which the ions are generated by high frequency (HF, 16 kHz) surface discharge, and the alternating electric field of low frequency (500 Hz) is produced by additional pair of electrodes, has been invented by Masuda [1978a,b, 1984], and is known as "Masuda boxer charger" (Figure 12). The HF voltage is imposed to the ceramic-embodied electrode when the stripe electrode is at negative potential and the counter stripe electrode is at positive. The role of electrodes is reversed in every half-cycle of the square-waveform voltage. The particles or droplets are charged by negative ions, which flow from the active electrode through the charging zone towards the counter electrode.

In the charger developed by Jaworek and Krupa [1989], the gaseous ions were generated from two discharge electrodes mounted at opposite sides of charging zone, and the alternating electric field was generated by two additional grids located between these discharge electrodes (Figure 13).

In this type of charger, the particles or droplets are charged by ionic current flowing from one of the discharge electrodes operating alternatingly in every half cycle of the AC voltage. The grids are made of rods of circular cross-section, and the corona-discharge electrodes, placed outside the grids, are made as a set of sharp needles. The electrodes are excited by two high voltage AC sources connected in anti-phase, through the circuit made of diodes and resistors. In one half-cycle of the excitation voltage, one of the discharge electrodes at negative potential becomes a source of negative ions. These ions flow throughout the plane of grounded nearby grid and the charging zone towards the opposite grid and opposite needle electrode, which are maintained at positive polarity at this time period. In the second half-cycle of the excitation voltage, the role of the electrodes in each pair is reversed. Therefore, for most of the time, the charging zone between the grids is filled with negative ions. During the charging process the particles undergo oscillations due to the alternating electric field between the grids. These oscillations prevent the particles from being precipitated at the charger electrodes. A scheme of electric potential distribution between the charger electrodes is shown in Figure 14.

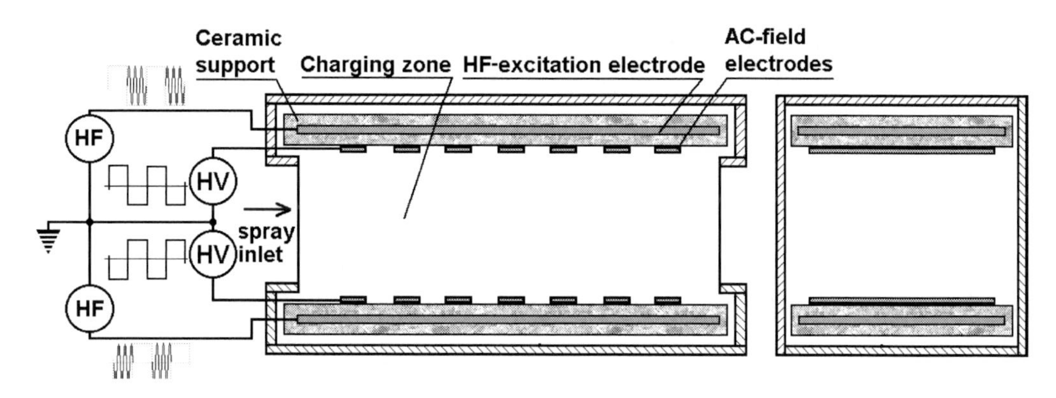

Figure 12. Masuda boxer charger (Masuda [1978a,b]).

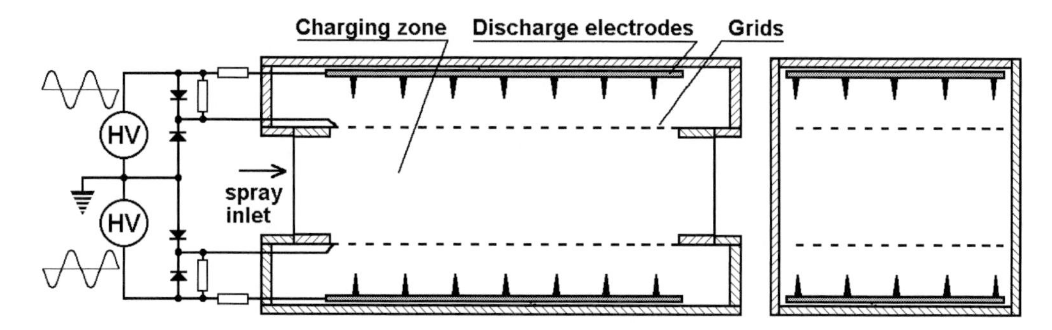

Figure 13. Schematic of the AC-electric field charger developed by Jaworek and Krupa [1989].

The electric field in the charging zone can be determined from Poisson equation:

$$\nabla E = -\frac{n_i e}{\varepsilon_r \varepsilon_0} \tag{39}$$

The ionic current density between the grids:

$$j = n_i e \mu_i E \tag{40}$$

The time intervals of the particle charging are given by the following equations:

$$\left(\arcsin\left(\frac{U_0}{U_m}\right) + k\pi\right) \langle \omega t \langle \left((k+1)\pi - \arcsin\left(\frac{U_0}{U_m}\right)\right) \tag{41}$$

$$\left(\arcsin\left(\frac{Q(t)}{Q_S}\right) + k\pi\right) \langle \omega t \langle \left((k+1)\pi - \arcsin\left(\frac{Q(t)}{Q_S}\right)\right) \tag{42}$$

where $k = 1,2,...\omega t_e/\pi$ number of half-cycles of ac field in the time t_e of the particle charging

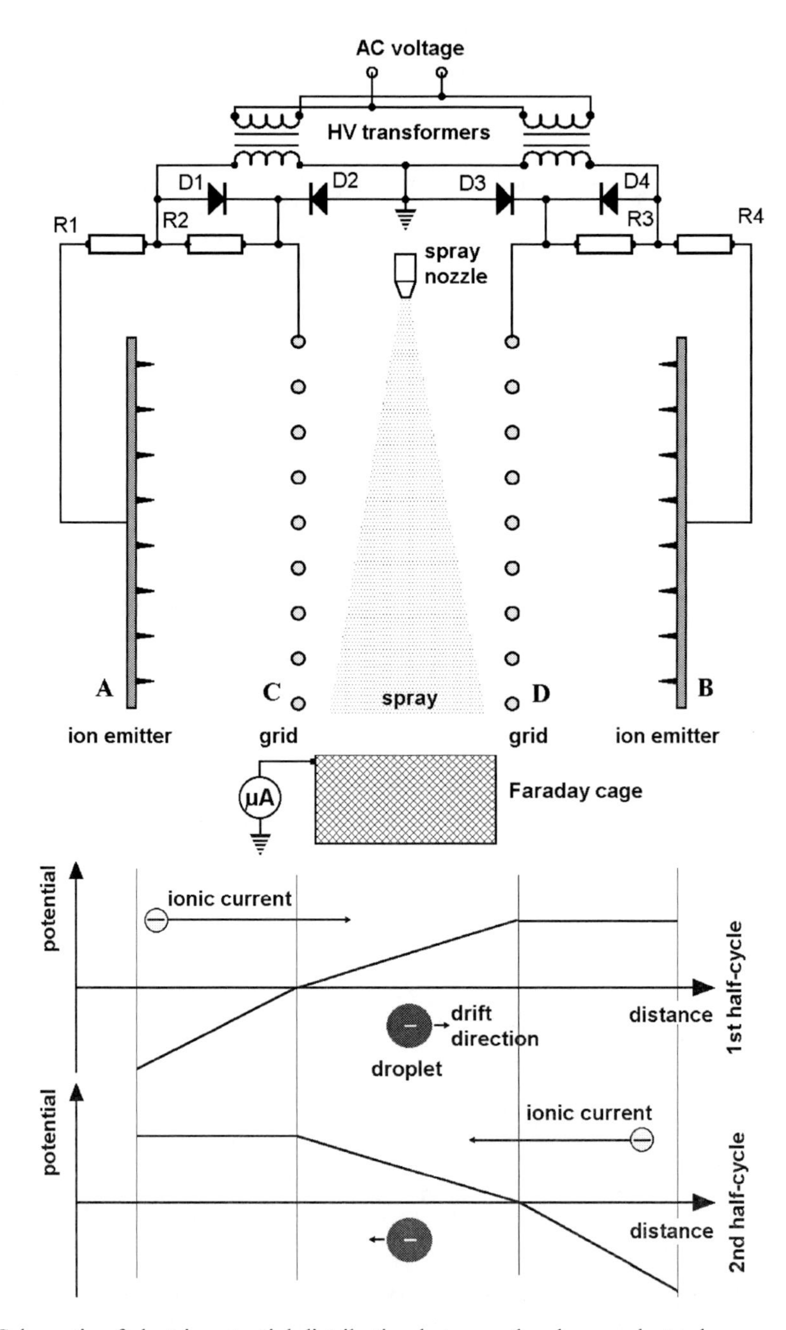

Figure 14. Schematic of electric potential distribution between the charger electrodes.

The first condition (41) refers to the intervals of ionic current generation, when the AC voltage is above the level of corona onset voltage. The second equation (42) describes the time intervals at which the gaseous ions are attracted towards the particle. The time intervals of droplet charging are shown in Figure 15. Initially the charging time is limited by the intervals of corona discharge. When the charge on droplet increases, the time interval is shorter because higher voltage (higher electric field) is required to overcome the ions repulsion (cf. Equation (41)). The charge accumulated on the droplet was determined from equation (32). It can be noticed that the charge attains about 80% of the Pauthenier limit in a few periods of AC voltage.

Figure 16 shows the two examples of measurement of the charge acquired by a single droplets (Figure 16a) and by glass beads simulating fine droplets (Figure 16b) in the AC charger vs. the excitation voltage.

Frequency of the AC electric field is limited to the value determined by the time of flight of gaseous ions between the electrodes within of half-period of AC voltage:

$$f_{\max} = \frac{\mu_i U_0}{2d^2} \tag{43}$$

where U_0 is the corona onset voltage, and d is the distance between the grids.

This type of charger is of higher energy efficiency than the former ones because at least 50% of ionic current emitted by discharge electrodes flows through the charging zone. Studies of the charging process and modeling of particle trajectories during the charging in this type of the charger were performed by Adamiak et al. [1995] and Lackowski [2001]. Dalley et al. [2005] have applied square-waveform voltage to this type of charger, and the particles followed the zig-zag paths. By square-wave excitation the dead period decreases practically to zero because the maximal voltage and maximal electric field between the grids is attained instantly after changing the voltage polarity.

Alternating electric field chargers are also particularly suitable for charging high-resistivity particles or droplets because the ion bombardment from both sides of the particle facilitates the distribution of the charge over the particle surface.

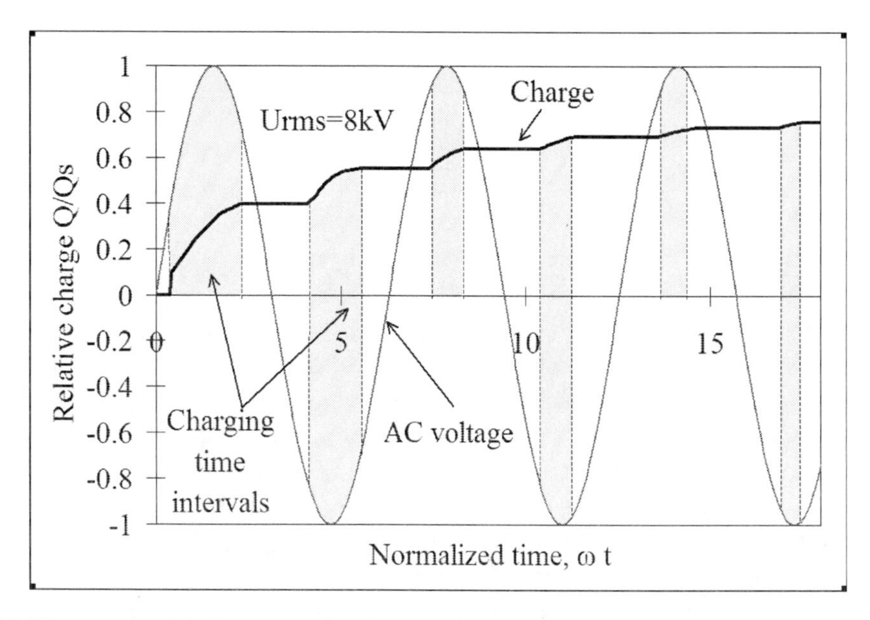

Figure 15. Time resolved charge accumulated on a droplet normalized to the saturation charge.

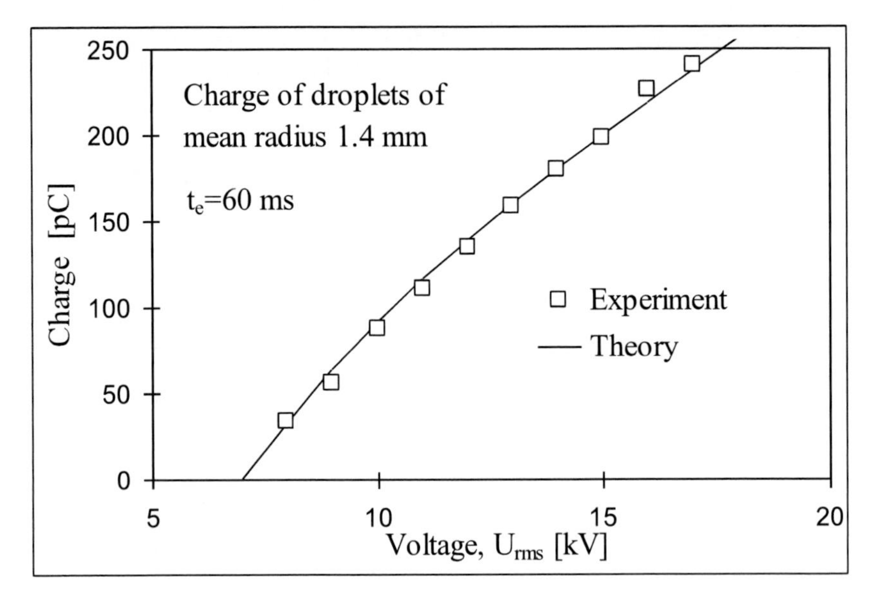

a.

b.

Figure 16. Charge acquired in the AC-electric field charger. Charging time: 60 ms (droplets), 175 ms (glass beads).

The distribution of ionic current between the electrodes was investigated and geometry optimization of the charger was carried out by Lackowski et al. [2001, 2003a,b, 2005, 2010]. For the type of charger considered by the authors, the optimal distance between the discharge electrode and the grid was 55 mm, and between the grids 40 mm. For these distances, the ratio of the charging current to the discharge current was maximized. It was also shown that with the voltage increasing, the charge imparted to the particles also increases, but the particle penetration becomes lower. The optimal supply voltage was that at which the product of penetration and the particle charge is maximized. For the charger considered, the voltage was about 8.5 kV. It was also shown that the particle penetration does not depend on the frequency of supply voltage, within the tested frequency range from 50 to 400 Hz. Above the frequency of 400 Hz, the charging current decreased due to limited ion mobility.

3.3. Induction Charging

Induction charging during liquid atomisation is based on the attraction of electric charges from ground to the surface of liquid jet, formed at the outlet of a nozzle, due to external electric field produced by high-voltage electrode placed in the vicinity of the nozzle. The droplets detaching the bulk of the liquid meniscus lift the electric charge on their surface. Rotary (Balachandran and Bailey [1984], Balachandran et al. [2001a, 2003], Sato [1991], Yamada et al. [2009]), pressure (Hensley et al. [2008], Higashiyama et al. [1999], Latheef et al. [2009], Moon et al. [2003]) pressure-swirl (Asano [1986]), Hall and Hemming [1992]), or pneumatic atomisers (Bailey and Cetronio [1978], Brown et al. [1994], Law [2001], Law and Cooper [1988], Law et al. [1999], Maski and Durairaj [2010a,b], Maynagh et al. [2009], Savage and Hieftje [1978], Scherm et al. [2007], Zhao et al. [2005a,b]) are usually used for the spraying of liquids with simultaneous induction charging of the droplets. The method of induction charging is effective for liquids of conductivity higher than 10^{-10} S/m. For lower conductivities, the concentration of charge carriers within the liquid is too low to build the surface charge on the liquid cone at the nozzle outlet. Droplets generated mechanically and charged by induction posses usually charge an order of magnitude lower than the Rayleigh limit, but the method is useful for the spraying of large amount of liquid.

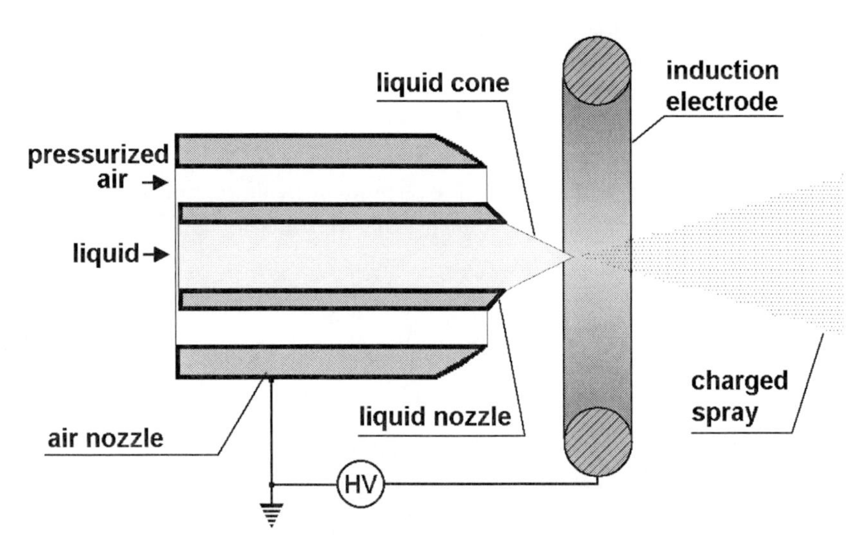

Figure 17. Pneumatic atomizer with induction electrode.

3.3.1. Pneumatic Atomizers

A scheme of pneumatic atomizer with induction-charging electrode is shown in Figure 17. The liquid is atomized due to shear forces on the meniscus, formed at the nozzle outlet, caused by the pressurized air flowing with high velocity from the annular nozzle. Due to these forces, the waves on the liquid jet surface are of increasing amplitude leading to detaching of the droplets. The electric field on the liquid surface is established by the ring induction electrode placed co-axially with the nozzle. Droplets produced by pneumatic atomisers can be of 30-50 μm in diameter, and can be charged to specific charge up to 8-10 mC/kg (Law [2001], Law and Cooper [1988], Law et al. [1999]). The induction electrode should be as close to the nozzle outlet as possible in order to increase the electric field at the liquid jet. However, electrode of too small diameter and placed too close to the nozzle can cause an electric discharge and increases the leakage current, which neutralize the induced charge.

Spray current is a product of mean charge of the droplets and the number of droplets produced per unit time:

$$I_s = n_d Q_{mean} \tag{44}$$

The mean charge of the droplets is determined from the following integral:

$$Q_{mean} = \int_0^\infty \int_0^h Q(r_d, z) f(r_d, z) \mathrm{d}z \mathrm{d}r_d \tag{45}$$

where $Q(r_d,z)$ and $f(r_d,z)$ are the charge and size distribution, respectively, of the droplet of radius r_d formed at the surface of liquid cone at the point of coordinate z, measured from the cone base.

The charge on droplet of radius r_d is equals to the charge on the fragment of cone of the same surface area as the droplet's cross section at the equator (Moore [1973]):

$$Q(r_d, z) = \pi r_d^2 \sigma(z) \tag{46}$$

Assuming that the droplet radius r_d is independent of the coordinate z, the size distribution function $f(r_d,z)$ can be presented as a product of number size distribution of the droplets $f(r_d)$ and the density of probability $f(z)$ that the droplet is formed at a point of z coordinate. This probability can be determined from the cone geometry:

$$f(z) = \frac{2(h-z)}{h^2} \tag{47}$$

where h is the height of liquid cone meniscus at the nozzle outlet (Bayvel and Orzechowski [1993]):

$$h = \frac{3.25a}{\lg\left(\frac{\rho_l}{\rho_a}\right)} \tag{48}$$

a is the inner radius of the liquid nozzle, and ρ_l and ρ_a are the mass density of liquid and air, respectively.

The number of droplets produced per unit time is:

$$n_d = \frac{Q_V}{\frac{4}{3}\pi \int_0^\infty r^3 f(r)\mathrm{d}r} \tag{49}$$

As a result, the spray current from pneumatic atomizer with induction charging is:

$$I_s = \frac{Q_V \int_0^\infty r^2 f(r)\mathrm{d}r \int_0^h \sigma(z) f(z)\mathrm{d}z}{\frac{4}{3}\pi \int_0^\infty r^3 f(r)\mathrm{d}r} \tag{50}$$

or

$$I_s = \frac{3Q_V\sigma}{4\pi r_S} \tag{51}$$

where r_S is the Sauter mean radius and

$$\sigma = \int_0^h \sigma(z)\frac{2(h-z)}{h^2}\mathrm{d}z \tag{52}$$

is the mean surface charge density at the cone surface.

An experimental set-up for the measurement of efficiency of charging by induction is schematically shown in Figure 18. The results of measurements of spray current vs. air pressure for various voltages at the induction electrode are presented in Figure 19. The induction electrode of the length of 40 mm and inner diameter of 40 mm was made of brass. The charge of the spray was measured with a moving-coil ammeter connected to the two-layer grid made of dense cooper mesh. Initially, the current increases nearly proportional to the pressure, but it departures from linearity for pressure higher than 0.15 MPa. This effect is more evident for lower voltages, indicating that the charge is not built on the meniscus before the liquid surface is broken-off.

The effect of voltage on induction electrode and the air nozzle pressure on specific charge of the spray, estimated from Equation (52), is presented in Figure 20. The charge initally increases up to its maximum at 10-12 kV, and next drops because of the electric discharge from the nozzle tip and the liquid surface. The specific charge for inductively charged sprays is between 5 and 15 mC/kg In Figure 21, the charge of a single droplet is compared with the Rayleigh limit determined for a droplet of mean Sauter diameter. The charge also increases with the voltage and starts to decrease because of electric discharge. Sprays charged by induction are, therefore, charged to levels similar to those charged in DC corona discharge. Unlike corona charging, the trajectories of the droplets are not deflected due to ionic wind and electric field and the deposition of droplets on the electrodes is lower. The induction charging is ineffective for liquids of conductivity lower than about 10^{-10} $(\Omega m)^{-1}$ due to too long electric time constant of building the charge on the liquid surface.

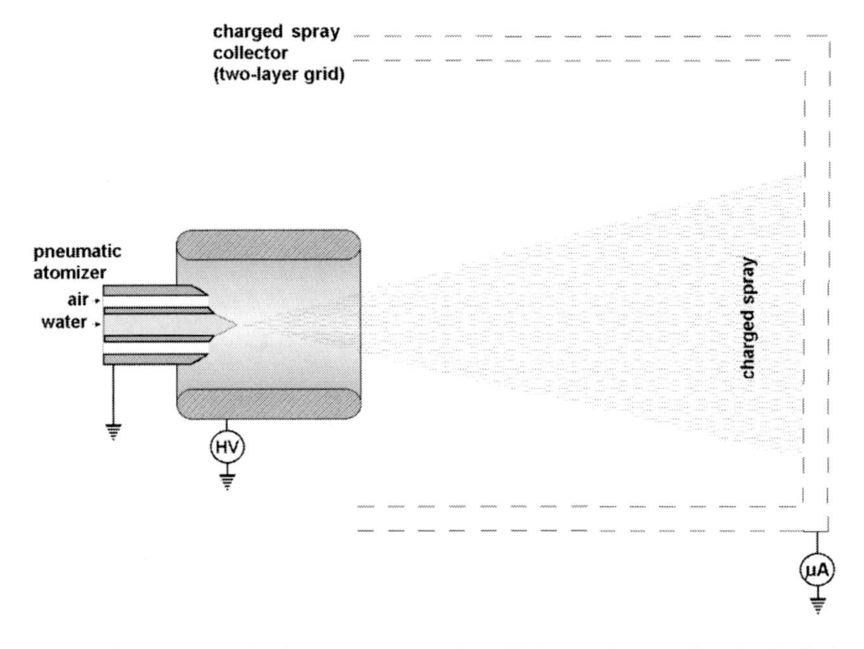

Figure 18. Schematic of a set-up for the measurement the efficiency of spray charging by induction.

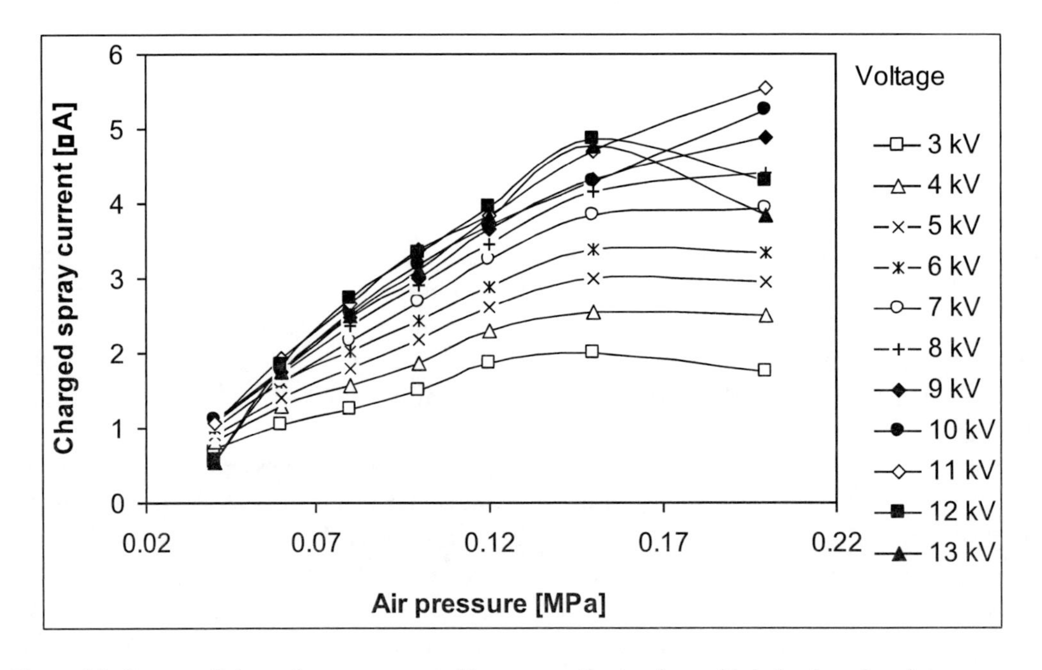

Figure 19. Current of charged spray generated by pneumatic atomizer with induction charging.

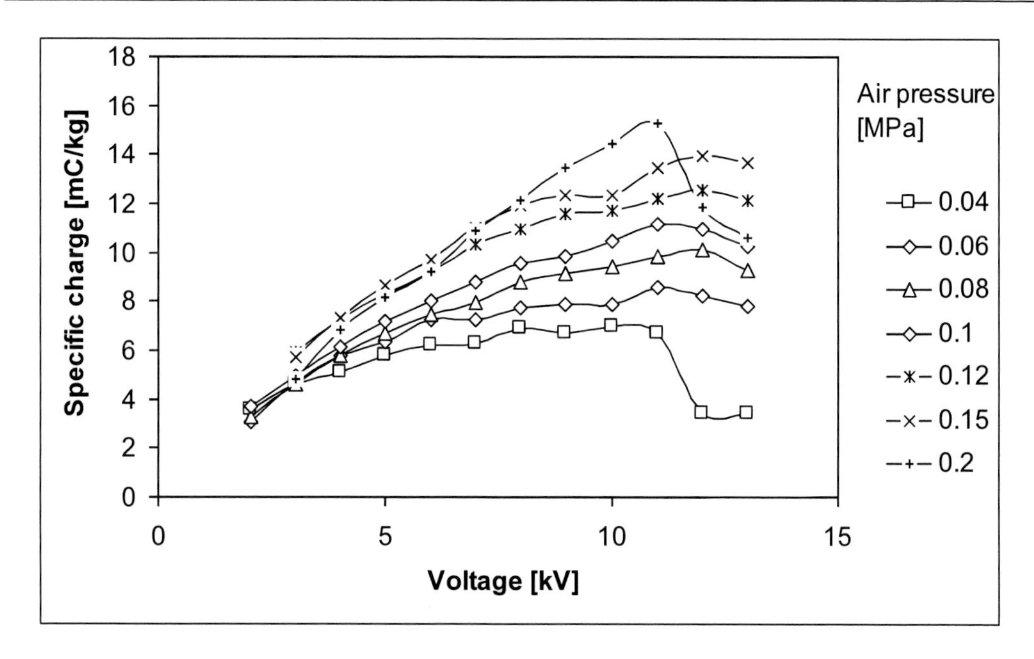

Figure 20. Specific charge of spray generated by pneumatic atomizer with induction charging.

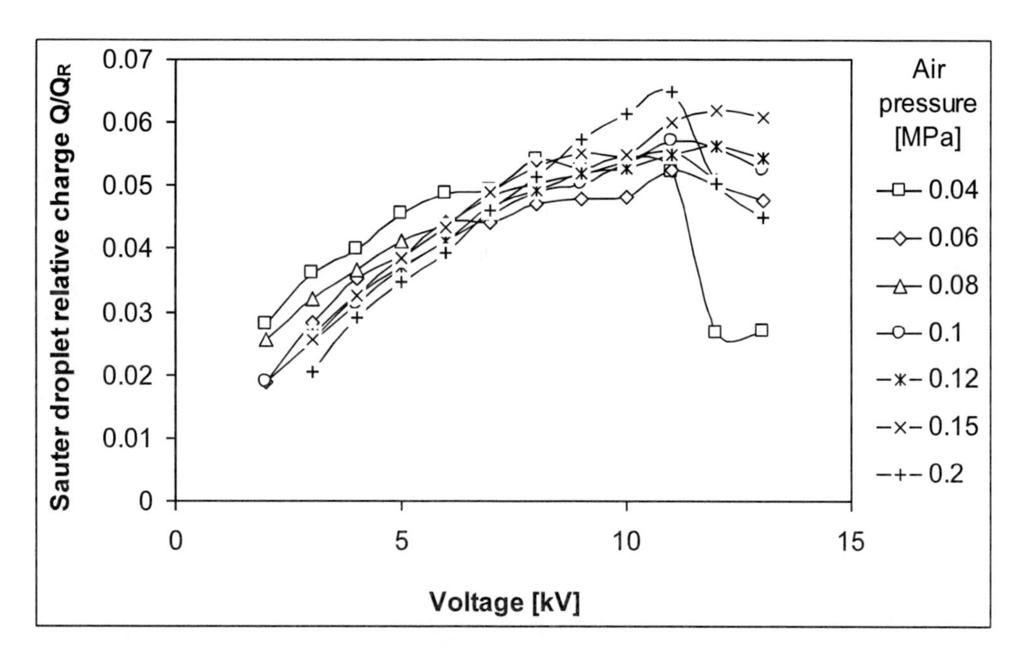

Figure 21. Ratio of the charge on droplet of Sauter mean diameter to the Rayleigh limit for the charged spray generated by pneumatic atomizer with induction charging.

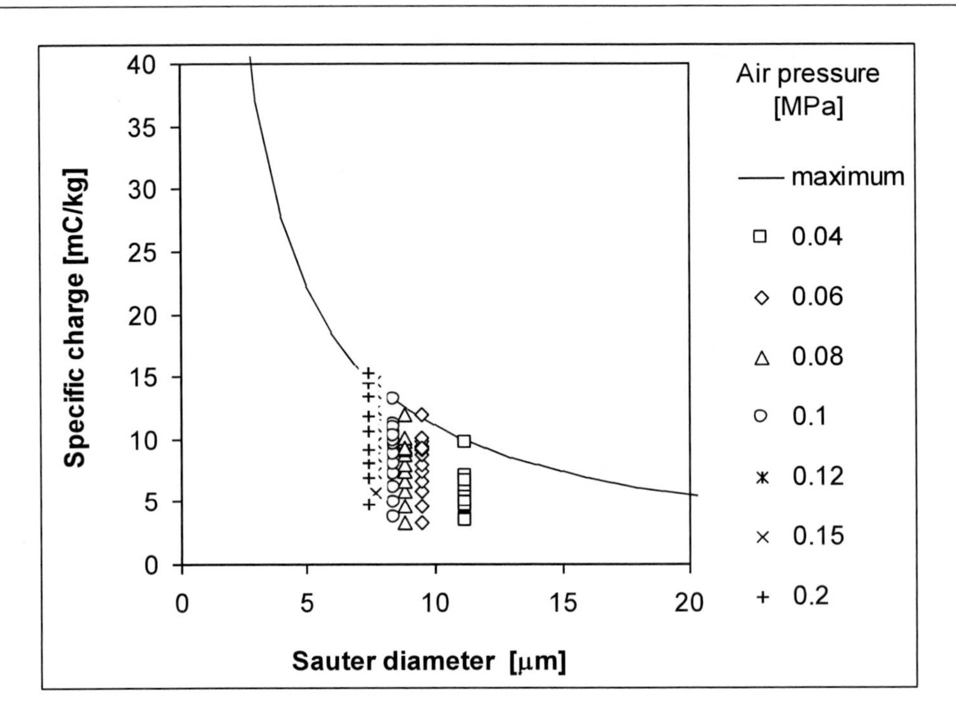

Figure22. Specific charge of spray generated by pneumatic atomizer with induction charging versus Sauter mean diameter (the continuous line is the specific charge determined from the Rayleigh limit).

3.4.2. Pressure Atomizers

In pressurized atomizer (Figure 23), the liquid jet flows from the nozzle with high velocity under a high pressure (Bayvel and Orzechowski [1993], Lefebvre [1989]). In contact with the surrounding air, the jet is destabilized and disperses into droplets. If the nozzle is placed within a cylindrical electrode, the linear charge density build on the jet surface is (Moore [2003]):

$$\lambda = \frac{2\pi\varepsilon_0 U}{\ln\dfrac{b}{a}} \tag{53}$$

where U is the voltage of the induction electrode, a is the jet diameter, b is the inner diameter of the cylinder electrode.

The charge of a single droplet of radius r_0 is:

$$Q = \lambda\Delta x = \frac{8\pi\varepsilon_0 U r_0^3}{3a^2 \ln\dfrac{b}{a}} \tag{54}$$

where Δx is the length of the jet from which the droplet is formed.

The spray current for a flat pressure atomizer with two parallel bar electrodes was determined by Hensley et al. [2008].

3.3.3. Pressure-Swirl Atomizers

In swirl atomizers the liquid flows tangentially into the swirl chamber in which it swirls towards the orifice (Bayvel and Orzechowski [1993], Lefebvre [1989]). The rotational speed increases in the confusor part before the liquid leaves the atomizer. The kinetic energy of the rotating liquid is converted to spreading a thin liquid film and disintegration of this film into droplets because of centrifugal forces. A scheme of pressure-swirl atomizer with induction charging electrode is shown in Figure 24.

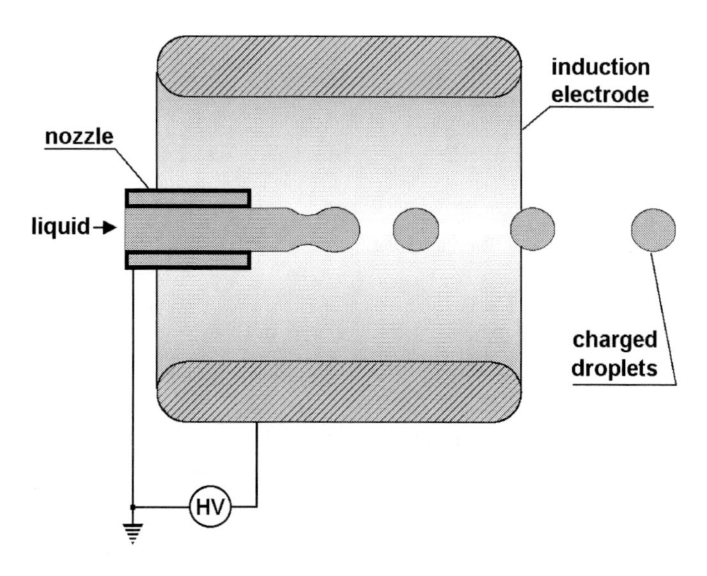

Figure 23. Pressure atomizer with induction-charging electrode.

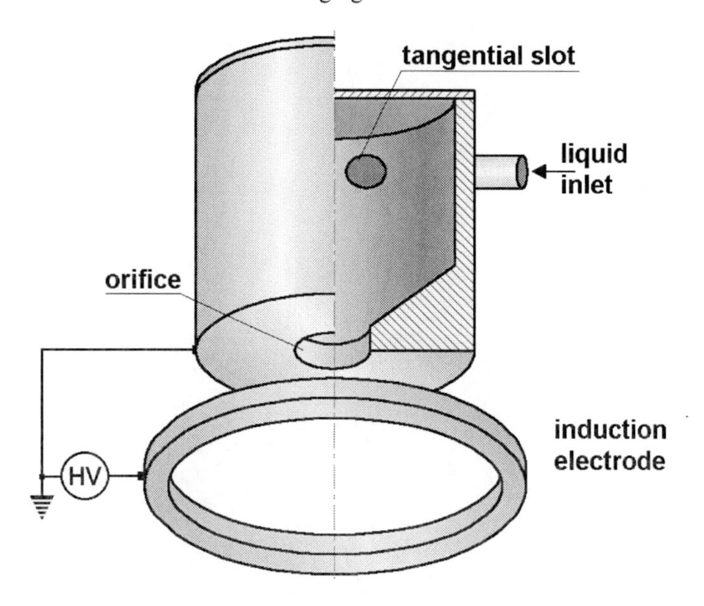

Figure 24. Pressure swirl atomizer with induction-charging electrode.

The surface charge induced on a liquid film at the nozzle outlet is

$$\sigma(z) = \varepsilon_0 E(z) \tag{55}$$

where z is the axial coordinate at the liquid film, and $E(z)$ is the electric field at this point. The electric field depends on the induction electrode potential, the electrode geometry, and the distance of the electrode from the nozzle outlet.

The charge on a droplet of radius r_d detaching from the liquid film depends on the surface charge density on the fragment of film it was formed from:

$$Q(r_d, z) = \pi r_d^2 \sigma(z) \tag{56}$$

The total spray current for all droplets is:

$$I = \int_0^\infty \int_S Q(r_d, z) f(r_d, z) n \mathrm{d}z \mathrm{d}r_d \tag{57}$$

where $Q(r_d,z)$ and $f(r_d,z)$ are the charge and size distribution, respectively, of droplet of radius r_d formed from the liquid film at the point of coordinate z, and n is the number of droplets produced per unit time.

In the following it is assumed that the size of droplet r_d is independent of the coordinate z, and the size distribution function $f(r_d,z)$ is a product of number size distribution of spray droplets $f(r_d)$ and the density of probability $f(z)$ of the formation of droplet at a point of z coordinate. With (56), equation (57) becomes:

$$I = \pi \int_0^\infty r^2 f(r_d) n \mathrm{d}r_d \int_S \sigma(z) f(z) \mathrm{d}z \tag{58}$$

The first integral represents the total volume of all droplets in the spray, and the second - the mean surface charge density:

$$\sigma = \int_S \sigma(z) f(z) \mathrm{d}z \tag{59}$$

The specific charge of a droplet of the Sauter radius r_s is:

$$\frac{Q}{m} = \frac{3\sigma}{4\rho_l r_S} \tag{60}$$

and the fraction of the Rayleigh limit:

$$\frac{Q}{Q_R} = \frac{\sigma}{4}\sqrt{\frac{r_S}{2\varepsilon_0\sigma_l}} \tag{61}$$

The Sauter mean radius r_S of the droplets for the pressure-swirl atomizer can be find in Bayvel and Orzechowski [1993]. for the measurement of efficiency of charging by induction. The results of measurements of specific charge for the pressure-swirl atomizer, and the fraction of the Rayleigh limit, obtained at an experimental set-up similar to that shown in Figure 18, are presented in Figs. 25 and 26, respectively. The charge increases with the voltage up to the voltage magnitude at which the corona discharge starts.

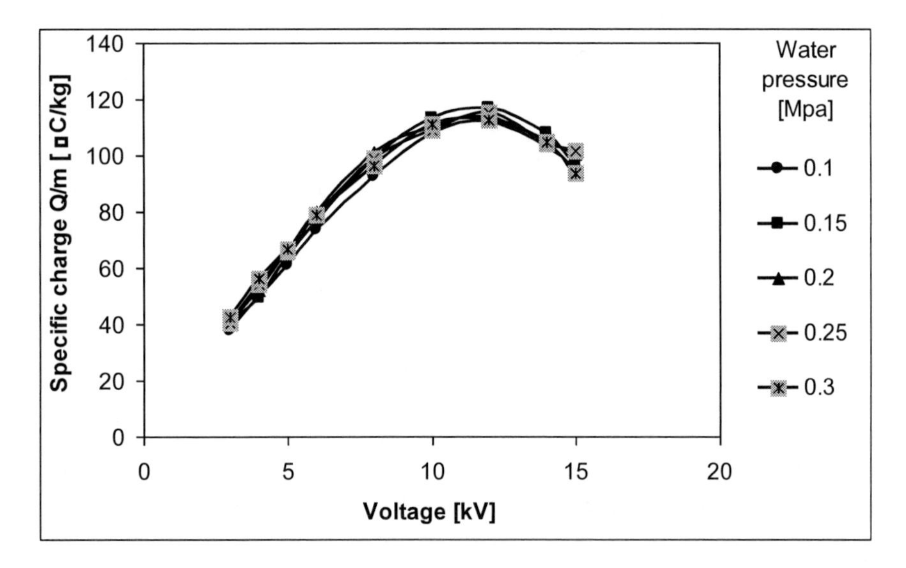

Figure 25. Specific charge of the spray generated by a pressure-swirl atomizer with induction charging.

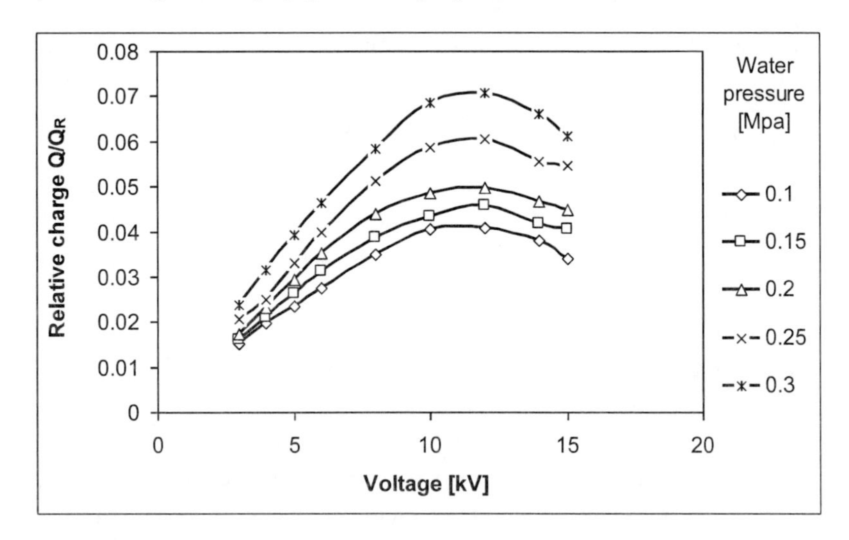

Figure 26. Ratio of the charge of droplet of Sauter mean diameter to the Rayleigh limit for the spray generated by a pressure-swirl atomizer with induction charging.

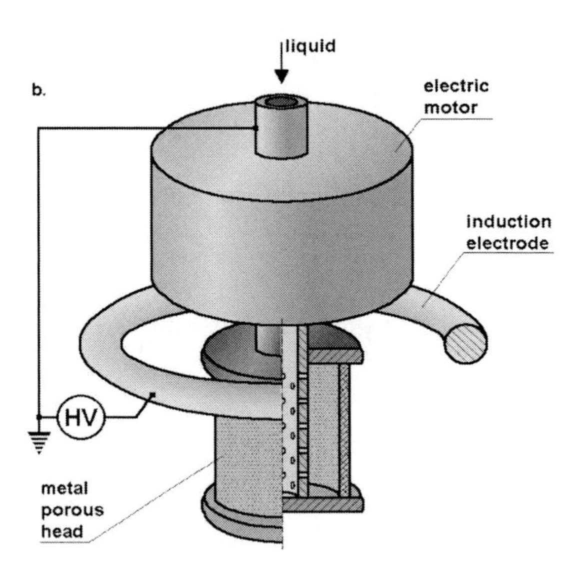

Figure 27. Rotary atomizer with induction-charging electrode.

3.3.4. Rotary Atomizers

In rotary atomizers, liquid is fed onto a rotating head where it spreads out under the action of centrifugal force (Bayvel and Orzechowski [1993], Lefebvre [1989]). The rotating head may take the form of flat disk, vaned disk, cup, bell, slotted wheel, or porous cylinder (drum). The induction electrode is placed coaxially with rotating element. A rotary atomizer with rotating porous drum is shown in Figure 27. The liquid issues through the pores in the drum and an electric charge is induced on the small jets formed at each pore in the drum. Droplets generated by rotary atomisers are in the range of 80-100 μm in diameter, and their specific charge is usually of about 1 mC/kg (Balachandran et al. [2001a, 2003]).

The mean size of the droplets produced by spinning-disk atomizer is (Saini et al. [2002]):

$$d = \frac{0.47}{\omega}\left(\frac{\sigma_l}{\rho_l R}\right)^{1/2} \tag{62}$$

where R is the disk radius, and ω is the rotational speed of the disk.

3.4. Conduction Charging

Conduction charging is a method of depositing electric charge on droplet by direct connection of atomizer nozzle to a high voltage source. The process is similar to the induction charging with an exception that the atomizer is not grounded but remains at high potential, and an induction (extractor) electrode may not be used. All the atomizers discussed in Section 3.3. can be adopted for the conduction charging. The disadvantage of this method is that the atomizer and the liquid container are at high potential that can be risky for personnel. The differences in charged spray parameters between induction and conduction charging have been investigated by Zhao et al. [2005a,b]. The authors concluded that for a similar geometry, conduction charging produces a larger target current, lower leakage current to the extractor electrode, and larger drift current.

3.5. Electrospraying

Electrospraying is a process of liquid atomisation by electrical forces. An electrospray device usually consists of a capillary nozzle maintained at high electric potential and a grounded counter electrode (Figure 28). Instead of using high potential to the nozzle, an extractor electrode, placed co-axially close to the nozzle and connected to high voltage, is used in order to steer the spraying process. In this case, the nozzle can take the ground potential. The extractor electrode induces an electric field at the capillary outlet, and an electric charge is built at the liquid meniscus. A photograph of an electrospray atomizer with extractor electrode is shown in Figure 29.

The liquid meniscus at the outlet of capillary nozzle maintained at high electric potential is subjected to shear stress caused by the electric field. The electric field at the tip of the capillary can be obtained from the equation determined for a paraboloid of revolution:

$$E_0 = \frac{\sqrt{2}U}{a \ln \frac{4h}{a}} \tag{63}$$

where U is the potential of the capillary, h is the distance between the tip of the capillary and the counter electrode, and a is the curvature radius of the meniscus, which frequently is assumed to be equal half of the capillary outer diameter.

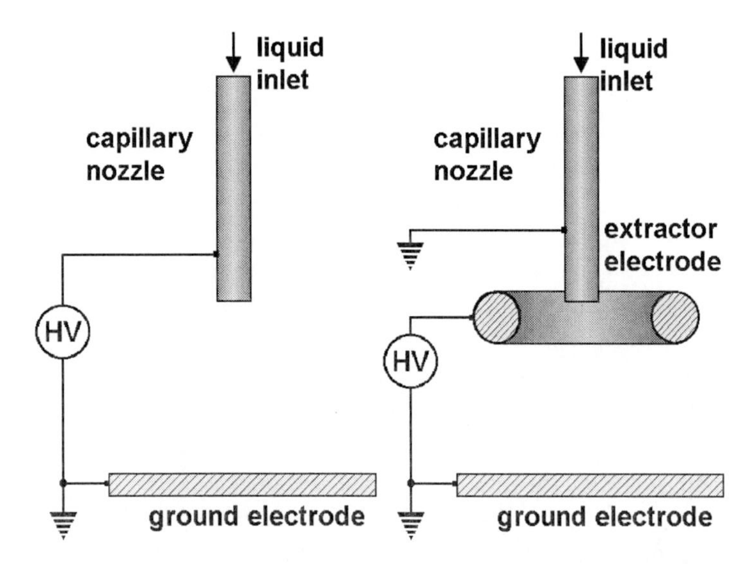

Figure 28. Electrospray systems: capillary nozzle and ground electrode (substrate, work-piece) (left), capillary-nozzle with extractor electrode (right).

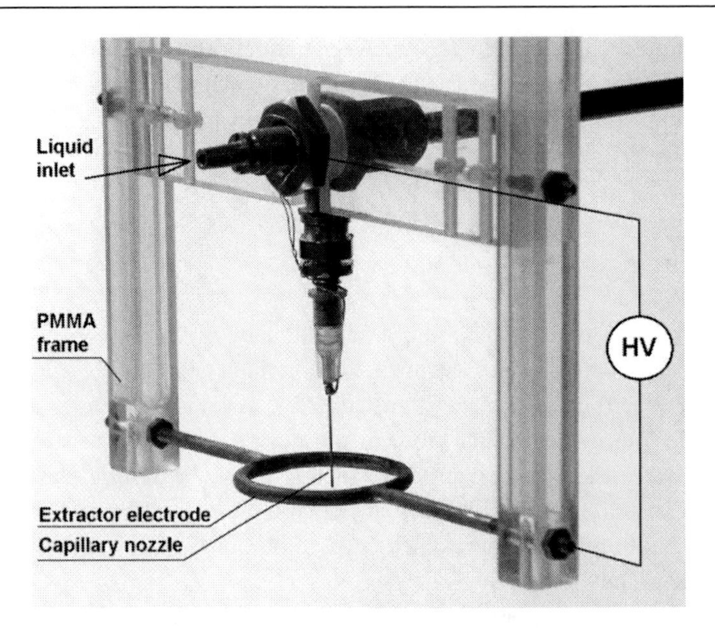

Figure 29. A photograph of an electrospray atomizer with extractor electrode.

Assuming the hyperboloid of revolution, Xiong et al. [2006] have determined the maximum electric field at the tip of the capillary emitter (applicable to colloid microthrusters). The equation has been presented in a simplified form for the capillary to extractor distance much larger than the capillary diameter:

$$E_0 = \frac{2U}{r_c \ln\left(2 + 2\frac{h}{r_c} - \sqrt{\left(2\frac{h}{r_c} + 1\right)\left(3 + 2\frac{h}{r_c}\right)}\right)} \tag{64}$$

where r_c is the radius of the meniscus base.

Because of the shear stress, a fine jet is stretched from tip of the meniscus, which disperses into fine droplets due to mechanical instabilities. The size of droplets of the electrostatically generated sprays is controlled by the potential at the capillary nozzle and the liquid flow rate. The process of meniscus and jet formation depends on physical parameters of the liquid (density, surface tension, electric conductivity, viscosity, permittivity), and on the process parameters such as capillary diameter, voltage and flow rate. The type of jet and drop formation is called the mode of spraying. Depending on the mode of spraying, the properties of the generated spray, droplet size distribution, spray plume pattern, and charge of the droplets can be significantly different.

In general, the size distribution of electrospray droplets is not monodisperse, and frequently it is bi-modal or multi-modal that is an effect of the process of droplet formation from liquid meniscus. During this process, droplet is connected with the meniscus with a thin liquid thread, which brakes at its both ends and disperse on few smaller droplets because of the electric charge accumulated on it. At higher electric fields, a similar thread can be expelled also from the front side of the droplet, producing an additional plume of fine droplets. In many cases, these processes are difficult to avoid and the main droplet of required size is accompanied with a fine mist of smaller, submicron droplets. In other cases, only long liquid jet exits the nozzle and this jet disperse for many droplets of different sizes. The two modes that generate nearly monodisperse spray are the cone-jet and microdripping modes.

The most commonly used classification of the modes of spraying is based on the geometrical forms of meniscus and jet observed during the electrospraying. Following the works of Hayati et al. [1987a,b], Cloupeau and Prunet-Foch [1990, 1994], Grace and Marijnissen [1994], and Jaworek and Krupa [1999a,b], the spraying modes can be categorized into two groups:

1. *Dripping modes*. These modes refer to the electrospraying process in which only fragments of liquid are ejected directly from the meniscus at the capillary outlet. These fragments can be in the form of regular large drops (dripping mode), fine droplets (microdripping mode), or elongated spindles (spindle or multispindle modes). Sometimes irregular fragments of liquid can be also ejected (ramified jet mode). At a distance from the nozzle outlet these fragments contract into regular spherical droplets.
2. *Jetting modes*. In a jetting mode, the liquid is elongated into a fine jet or a few jets, which disintegrate into droplets due to electrohydrodynamic instability. It was observed that the jet could be smooth and stable (cone-jet mode) or could move in a regular way: rotate around the capillary axis (precession mode) or oscillate in its plane (oscillating mode). When few are formed at the circumference of the capillary this mode is called the multijet mode.

Figure 30 presents example of these electrospraying modes. It should be noted that the droplets generated in the jetting modes are usually smaller than those produced by any dripping mode.

Other classifications of the spraying modes have been based on the current waveform recorded during the electrospraying (Juraschek and Röllgen [1998]) or Fourier analysis of current oscillations correlated with fast jet imaging (Marginean et al. [2004, 2006, 2007a, 2007b], Parvin et al. [2005]), which pursues quantitative analysis of electrospraying.

The *dripping mode* of electrospraying does not differ significantly from the dripping of liquid from an uncharged capillary. The drops are formed as regular spheres detaching from the capillary as the weight of the drop and the electric force overcome the surface tension force. With the voltage increasing, the meniscus elongates and the drop becomes smaller. For liquids of low viscosity, siblings resulting from the fine liquid thread connecting the droplet with the meniscus can also be generated. Because of electrostatic repelling, the sibling is ejected perpendicularly to the capillary axis. For higher voltages, the liquid thread becomes thinner and breaks off to a few smaller droplets after its detachment from the meniscus and main droplet, and forms fine mist following the main droplet. After the droplet emission, the meniscus contracts back to its hemispherical form.

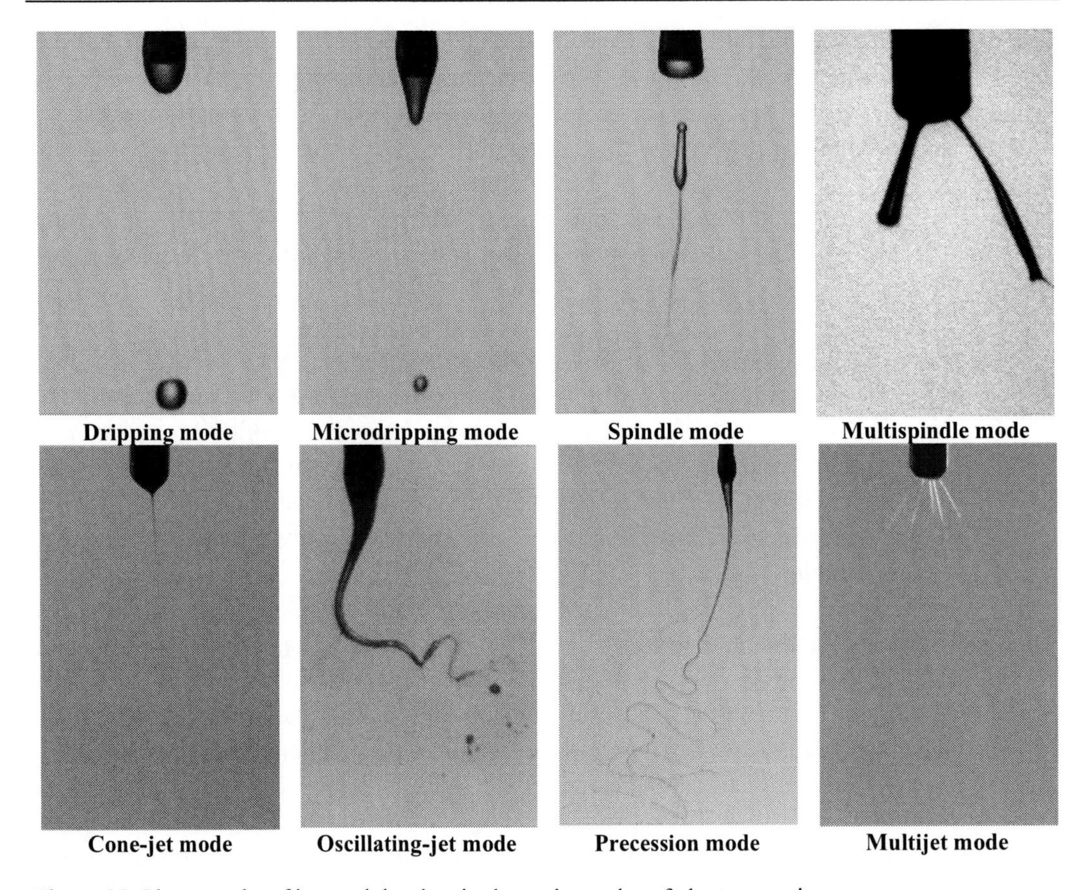

Figure 30. Photographs of jets and droplets in the main modes of electrospraying.

In *microdripping mode*, the liquid at the capillary outlet forms a stable cone-like meniscus from which a small droplet, much smaller than the capillary diameter, is ejected. This mode occurs only at low flow rates of liquid of low viscosity. The microdripping mode differs from the dripping mode in that the meniscus does not contract after the droplet detachment but only pulsates with small amplitude. Also in this mode, a short filament, a few micrometers in diameter, coupling the main droplet with the cone can be observed. For liquids of low viscosity (ethanol), this filament breaks off under the electrostatic and inertial forces into fine mist. For liquids of higher viscosity (ethylene glycol), the broken filament contracts into meniscus and its second fragment into the droplet, and no sibling droplet can be observed. The size of the droplets can range from a couple of microns up to tens of micrometers, and the droplet size distribution is usually monodisperse. The frequency of the emission of the droplets ranges from a few tens up to a few thousands of droplets per second.

The regular drop emission is not easy to accomplish with DC excitation voltage, and the dripping and microdripping modes are generated only in limited ranges of voltage and flow rate. Out of these ranges, usually the *spindle mode* can be generated. In the spindle mode, the meniscus of liquid elongates in the direction of electric field as a thick jet and detaches as a vast spindle-like fragment of liquid. After the liquid detachment, the spindle can disrupt into several smaller droplets of different sizes, which disperse off the capillary axis due to Coulomb repulsion. The meniscus contracts to its initial shape and a new jet start to be formed. The spindle can be sometimes connected with the capillary by a thick thread several mm long. The thread whips irregularly due to electrohydrodynamic instability. In the case of low viscosity liquids (ethanol) also a single thin thread can be ejected from the leading side of the spindle. This thread detaches from the main drop and next disrupts into several small fragments, forming siblings or fine polydysperse aerosol. The spindle mode differs from the dripping and microdripping modes in that irregular, elongated fragment of liquid is emitted from the capillary. The size distribution of droplets is polydisperse. The diameter of main droplets varies from 300 to 1000 μm, while siblings are smaller than 100 μm. In the spindle mode, the process of droplet formation is random, due to variations in the electric field at the liquid meniscus, and random waves generated on jet surface.

For higher voltages, the spindle mode can change into *multispindle mode* with two or more jets emitted from symmertically located points at the circumference of the capillary. The formation and shape of the droplets is similar to the spindle mode, but the spindles are smaller and not emitted along the capillary axis. After detachment, the liquid fragment can disintegrate into a few smaller droplets. The number of points of the spindle emission increases with the voltage increasing. A fine thin jet at its back and/or leading side can accompany each spindle.

In *cone-jet mode*, the liquid meniscus is in the form a stable, regular cone. From the apex of this cone, a stable thin jet of diameter smaller than 100 μm, in particular cases smaller than 10 μm, is formed. The jet flows along the capillary axis, and its free end undergoes instabilities of one of two types: varicose (regular contractions, which become narrower, leading to the jet fission), and kink (the end of the jet moves irregularly and breaks up into series of droplets). In the case of varicose instabilities, the axis of the jet is not deflected and the jet disintegrates into equal droplets, which flow along the capillary axis. In the case of kink instabilities, the droplets are dispersed randomly in each direction down the capillary, forming a cone-like plume. The droplets are nearly uniform in size, but not exactly monodisperse. The flow rate of liquid required for stable cone-jet mode generation is usually smaller than 10 ml/h. The size of droplets is below 50 μm, and in particular cases, droplets smaller than 1 μm can also be obtained.

In *precession mode*, the meniscus assumes the form of a skewed cone with a jet at its apex. The meniscus and jet rotate regularly around the capillary axis with 200 to 300 revolutions per second (for distilled water). The end of the jet assumes a form of spiral, which breaks up into droplets smaller than 100 μm. The plume of the spray assumes the form of cone and the droplets are nearly uniformly dispersed into this cone. The base of the spray plume is much larger than in the cone-jet mode. The precession mode is generated by relatively high flow rates, 100-200 ml/h. Because of these properties, the precession mode can be used for surface coating.

In *oscillating-jet mode*, the meniscus is also skewed, but the jet and the cone oscillate in one plane instead of rotating. The droplets are larger than in the precession mode and not monodisperse. The plume is in a form of nearly flat triangle with the base of elongated ellipse.

Table. Scaling laws for electrospray mean droplet diameter obtained in cone-jet mode

Scaling law	Liquid (properties)	Definitions	References
$d = 2\left(\frac{3Q_V \varepsilon_0^{1/2} \varepsilon_r \sigma_l^{1/2}}{4\pi\kappa_l E_0}\right)^{2/7}$			Pfeifer and Hendricks [1968]
$d = \alpha\kappa_l^{-0.3}\eta_l^{0.4}$	polyvinyl alcohol and water solution of nitrium acrylate	α - constant	Hagiwara (cf. Ogata et al. [1978a])
$\frac{d_{vs}}{D_c} = \alpha C_a^{2/3} T_i^{2/9} P_o^{-10/9} S_e^{2/5}$	0.5<*We*<30,	$\alpha = 11.4$ for $\sigma_l/\eta_l \le 10$ $Ca = \frac{v_j \eta_l}{\sigma_l}$, $T_i = \frac{\varepsilon_l}{\kappa_l}\frac{\sigma_l}{D_c \eta_l}$, $P_o = \frac{U}{\left(D_c \frac{\sigma_l}{\varepsilon_0}\right)^{1/2}}$, $S_e = \frac{h}{D_c}$, l_j/D_c=15...50	Ogata et al. [1978a,b]
$\frac{d}{D_c} = \alpha\delta^{0.05}(D_0/D_c)^{0.025} N_Q^{-0.88} We^{0.21} \mathrm{Re}^{0.25}$		$\alpha = 0.25$, $\delta = \kappa_l/\kappa_0$, $\gamma_0 = 10^{-2} S/m$ $N_Q = \frac{\varepsilon_0 E^2 D_0}{\sigma_l}$	Tomita et al. [1986]
$d = \alpha\left(\frac{Q_V \varepsilon_0 \varepsilon_r}{\kappa_l}\right)^{1/3}$		α - constant	Fernandez de la Mora and Loscertales [1994]
$d^* = Q_V^{*2/3} \times$ $\times\left(0.37 + 0.49 Q_V^{*-0.11} \exp\left(-0.56 Q_V^{*1/2} U^*\right)\right)$	heptane	$d^* = d\left(\frac{\rho_l \kappa_l^2}{\sigma_l (\varepsilon_0 \varepsilon_r)^2}\right)^{1/3}$ $Q_V^* = Q_V \frac{\rho_l \kappa_l}{\sigma_l \varepsilon_0 \varepsilon_r}$ $U^* = \frac{U - U_0}{U_0}$ U_0- cone-jet mode onset voltage	Tang and Gomez [1996]
$d = \alpha(\varepsilon_r)\left(\frac{Q_V \varepsilon_0 \varepsilon_r}{\kappa_l}\right)^{1/3}$	ε_r=12.5-182; $\left(\frac{\sigma_l^2 \varepsilon_0 \varepsilon_r \rho_l}{\mu_l^3 \kappa_l}\right)^{1/3}$ $\in (0.12 \ldots 10)$	$\alpha(\varepsilon_r) =$ $10.87\varepsilon_r^{-6/5} + 4.08\varepsilon_r^{-1/3}$	Chen and Pui [1997]
$d = \alpha\left(\frac{Q_V^3 \varepsilon_0 \rho_l}{\pi^4 \sigma_l \kappa_l}\right)^{1/6}$		α - constant	Gañan-Calvo [1997, 1999]
$d = \alpha\left(\frac{Q_V^3 \varepsilon_0 \rho_l}{\sigma_l \kappa_l}\right)^{1/6}$		α - constant	Hartman et al. [2000]

Scaling law	Liquid (properties)	Definitions	References
$d = \alpha Q_V^{0.5} I_{glow}^{-0.04} \kappa_l^{-0.2}$	water (glow-stabilized cone-jet)	α - constant, Q_l=35-200 ml/h, U=20-30 kV	Borra et al. [2004]
$d = \alpha\left(\frac{Q_V\varepsilon_0\varepsilon_r}{\rho_p}\frac{yM}{\lambda_0 - bC^{1/2}}\right)^{1/3}$ $d = \alpha\left(\frac{Q_V\varepsilon_0\varepsilon_r}{\rho_p}\frac{yM}{\lambda_0}C^{1/2}\left(\frac{xy}{K_a}\right)^{1/2}\right)^{1/3}$	strong electrolyte weak electrolyte	$K_a \approx \delta^2 xyC$	Basak et al. [2007]

U is the voltage applied to the nozzle, Q_V is the volume flow rate of the liquid, D_c is the outer diameter of the capillary, η_l κ_l σ_l ε_l are the liquid viscosity, conductivity, surface tension, and permittivity, respectively, v_j is the jet velocity, h is the distance between the capillary outlet and the counter electrode, I_{glow} is the glow current, ρ_p is the density of solid precursor, x,y are the charges on anion and cation, respectively, λ_0 is the equivalent conductivity of the solution at infinite dilution, M is the molecular weight of the precursor, C is the solution concentration in M, b is the empirical constant for Kohlrausch expression, K_a is the equilibrium constant, δ is the dissociation constant of electrolyte.

The term in denominator in P_o predicts a minimum voltage necessary to liquid spraying (Ogata et al. [1978a,b]).

The dimensionless number N_Q, is the ratio of the Coulomb force the surface tension force (Tomita et al. [1986]).

In *multijet mode*, several, uniformly distributed, jets are ejected simultaneously from the circumference of the capillary. The number of jets increases with the voltage increasing. Maximum eight such jets have been reported in the literature. Due to kink instabilities similar to those in the cone-jet mode, the jets are dispersed into droplets smaller than 10 μm, forming fine aerosol (mist) around the capillary. At certain distance from the capillary, the aerosol fills the whole volume of the spray plume.

The diameter d of droplets produced by electrospraying cannot be determined in a simple manner. Droplet diameter depends on the spray mode, which in turn depends on the electrosprayed liquid properties, flow rate, and voltage. Most of the equations for mean droplet diameter are based on experimental results and are valid only for a specific mode of spraying, mainly the cone-jet mode. The scaling laws proposed in the literature are summarized in Table .

The scaling laws proposed by Fernandez de la Mora and Loscertales [1994], Gañan-Calvo [1999], and Hartman et al. [2000], Fernandez de la Mora [2007] were confirmed by many experiments for a single, coaxial jet (Bocanegra et al. [2005a,b], Mei and Chen [2007, 2008]), and recently, for a jet generated within an insulating liquid: heptane (Barrero et al. [2004]), or sunflower oil (Lopez-Herrera et al. [2003]). The scaling laws were modified for strong ionic solutions and dry particles by Basak et al. [2007]. Regardless of an effect of other parameters, all the equations indicate that the droplet diameter increases with the liquid flow rate, and is inversely proportional to the liquid conductivity to a certain power. The droplet size can be decreased via decreasing the liquid flow rate and increasing liquid conductivity or surface tension.

The minimum flow rate at which the cone-jet mode can operate stably is determined by liquid parameters (Barrero and Loscertales [2007], Chen and Pui [1997], Fernandez de la Mora and Loscertales [1994], Gañan-Calvo et al. [1997]):

$$Q_{V\min} \approx \frac{\sigma_l \varepsilon_0 \varepsilon_r}{\rho_l \kappa_l} \tag{65}$$

The onset voltage for the cone-jet mode resulting from the balance of the electric-field force and the surface tension force is (Smith [1986]):

$$U_{onset} = A\left(\frac{d_c \sigma_l}{\varepsilon_0}\cos\theta\right)^{1/2} \ln\frac{4h}{r_c} \tag{66}$$

where θ is the half angle of the liquid cone at the tip of the capillary, and *A* is a dimensionless constant. The term $\ln(4d/r_c)$ determines the electric field at the tip of capillary (*h* is the distance between capillary tip and plate electrode). This model predicts the onset voltage for long slender capillaries.

For colloid microthrusters, the hyperboloid of revolution was assumed by Xiong et al. [2006] and the emission onset voltage was determined as follows:

$$U_{onset} = 0.312\times10^6\sqrt{\sigma_l r_c}\,\ln\left(2+2\frac{h}{r_c}-\sqrt{\left(2\frac{h}{r_c}+1\right)\left(3+2\frac{h}{r_c}\right)}\right) \tag{67}$$

where r_c is the radius of the meniscus base.

The dependence of droplet diameter on voltage in the dripping mode is (Poncelet et al. [1999]):

$$d(U) = d_0\left(1-\left(\frac{U}{U_{crit}}\right)^2\right)^{1/3} \tag{68}$$

where d_0 is the droplet diameter for U=0, and U_{crit} is the voltage at which the dripping mode ceases.

For polymer solutions, the critical voltage U_{crit} was determined by Poncelet et al. [1999]:

$$U_{crit} = \left(\frac{d_c \sigma_l}{k\varepsilon_0}\right)^{1/2} \tag{69}$$

where d_c is the inner diameter of the capillary, *k* is the parameter whose value depends on the ratio of the characteristic time of adsorption of the surfactant and the time of formation of the liquid drop. At this voltage, unstable liquid jets occurs, and microscopic charged droplets are generated.

The critical jet radius for which the jet becomes unstable and breaks up into droplets is (Kim and Turnbull [1976]):

$$r_{j(crit)} = \left(\frac{4\varepsilon_0 \sigma_l Q_V^2}{I^2} \right)^{1/3} \tag{70}$$

where Q_V is the volume flow rate, and I is the spray current. The jet diameter is of the order of magnitude of 10-40 μm for the flow rate of 0.1 mm^3/s and for the spray current of 10^{-8} A. The equation refers to straight linear jet and cannot be applied to the precession and oscillating modes, because the jet assumes more complex shapes.

For the cone-jet mode, and for liquids with viscosity in the range from 0.73 to 10.2 mPa×s, the frequency of droplets generation is in good agreement with the Weber [1931] equation, determined for electrically neutral sprays (Mutoh et al. [1979]):

$$f = \frac{\pi}{4} \rho_l d_j^2 v_j \left[\frac{\pi^2}{4} \rho_l d_j^3 \left(2\left(1 + 3\left(\frac{\eta_l^2}{\rho_l d_j \sigma_l} \right)^{1/2} \right) \right)^{1/2} \right]^{-1} \tag{71}$$

where d_j is the jet diameter, and v_j is its velocity.

Chen et al. [2006] have determined the scaling law for the pulsation frequency of the droplets emission:

$$f = \alpha \frac{\kappa_l E^2}{\varepsilon_0 \varepsilon_r \eta_l L} \left(\frac{\rho_l d_i^5}{\sigma_l} \right)^{1/2} \tag{72}$$

where d_i and L are the capillary inner diameter and length, respectively, and validated this equation experimentally.

In the microdripping mode, the frequency of droplets generation is (Tomita et al. [1986]):

$$f \frac{D_c}{v_j} = 29\delta^{-0.32} \left(D_0 / D_c \right)^{-0.43} N_Q^{0.65} We^{-0.39} \operatorname{Re}^{-0.26} \tag{73}$$

$$N_Q = \frac{\varepsilon_0 E^2 D_0}{\sigma_l} \tag{74}$$

This equation is in a good agreement with experimental results for droplet of diameter d<200μm.

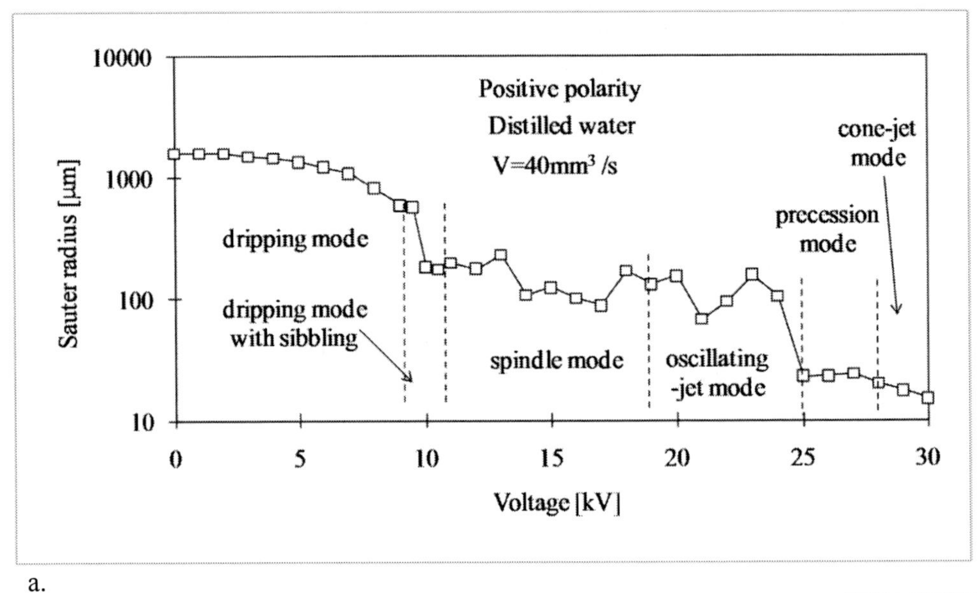

a.

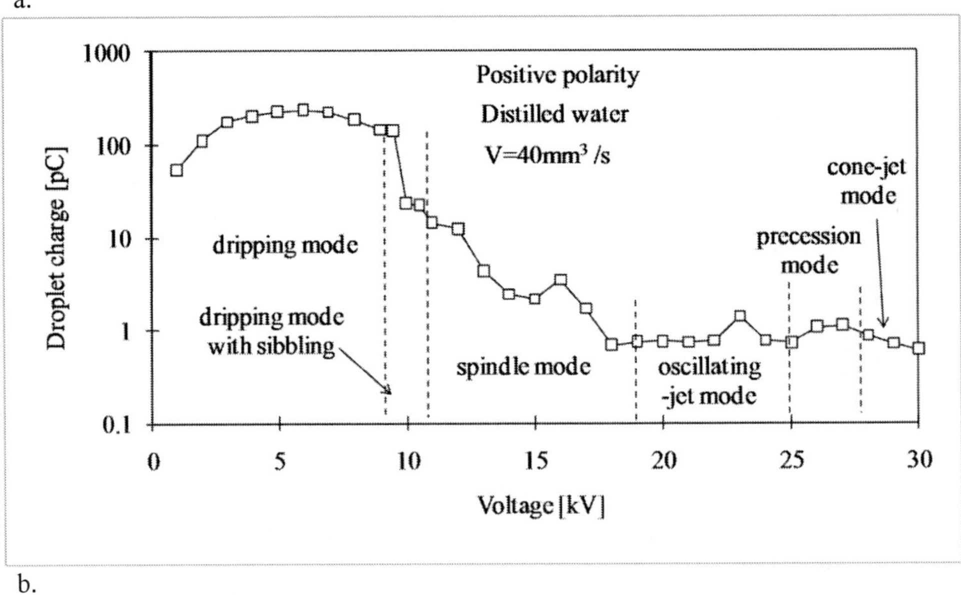

b.

Figure 31. Sauter mean radius of electrospray droplets (a) and the mean droplet charge (b) for distilled water (nozzle - substrate distance 50 mm).

The size of the capillary has also an effect on the droplet diameter. In general case, changing liquid physical properties independently each other is not an easy task and the droplet size control by this way is only limited.

The charge on droplet detaching from a capillary can be determined from the following equation (Hines [1966]):

$$Q = \frac{9\sqrt{2}\pi\varepsilon_0 d^2 E_0}{4} \tag{75}$$

with electric field E_0 determined from equation (63).

An example of evolution of Sauter mean diameter of electrosprayed water droplets with increasing voltage is shown in Figure 31a. The measurements of charge of water droplets related to the Rayleigh limit determined for the Sauter mean diameter of the droplet vs. voltage are shown in Figure 31b. In the dripping mode, the size of droplets decreases but the charge on the droplets increases nearly proportional with the voltage, up to the value of $0.5Q_R$. When the dripping mode changes to the spindle or multispindle modes, the droplets' charge drops suddenly below $0.1Q_R$. The charge increases again to the value $0.5Q_R$ for the precession and cone-jet modes.

There are some limitations in electrospraying modes, regarding the voltage, flow rate and liquid properties. There lack general recipes for which values of parameters, the specific mode can be generated. Nowadays, it is commonly assumed that semi-conducting liquids, which can be sprayed in the cone-jet mode by electrohydrodynamic method, are those of conductivity ranging from 10^{-4} to 10^{-8} S/m. Other parameters of the liquid have to be chosen experimentally. For example, the addition of a surfactant to an electrosprayed liquid can alter the operating range of spraying modes in terms of voltage and flow rate. For example, the addition of 0.1 wt.% of dodecyl sulfate sodium salt decreases the onset potential of the cone-jet (axial emission) and multijet (rim emission) modes from 3 to about 1.5 kV, and from about 5 kV to 3.5 kV, respectively (Lüttgens et al. [1992]).

4. Charged Sprays Application

4.1. Surface Coating and Spray Forming

Charged sprays have been used for surface coating, paint spraying, or thick film deposition (Anestos et al. [1977, 1986], Balachandran and Bailey [1981], Elmoursi and Garner [1993], Elmoursi and Speck [1991], Hines [1966], Im et al. [2001], Inculet [1998], Inculet and Klein [1993], Kleber [1963a,b], Meesters et al. [1990b], Plunkett et al. [1986], Sickles and Anestos [1979], Snyder et al. [1989], Yamada et al. [2009]). Because high flow rate of charged paint or other coating agent is required for surface coating, mechanical atomizers, such as pneumatic, rotary, centrifugal or pressure with induction or conduction charging are used. Charged droplets are attracted to an object due to image forces, covering its front and back sides. The cumbersome spraying from all directions can be therefore avoided.

Water-based paints cannot be atomized by electrospraying because conductivity of such paints is too high (1-10^3 S/m) for stable cone-jet mode generation. Mechanical atomizers with induction charging have therefore been used for such paints, because the conductivity is sufficiently high for building a charge at the jet surface. The reason for using water-based paints, for example, for car body finishing, is that they provide a more glamorous finish and contain down to 35% less solvents as compared to paints based on organic components (Elmoursi and Speck [1991]).

In pneumatic atomizer (Figure 32a), the paint is fed through a central nozzle and disrupts to fine droplets by aerodynamic forces of the air flowing with high velocity through the annular-slit nozzle. An induction electrode placed in front of the atomizer induces an electric charge on the surface of the paint. Pneumatic atomizers with spray electrification produce droplets of relatively high velocity and electrostatic deposition process is not as effective as could be expected, particularly for small objects and short distances. Better results are obtained for centrifugal or rotary atomizers with induction electrode placed near the rotating element (Figure 32b). In such devices, the paint is fed on a rotating element, and due to the centrifugal forces it is accelerated and spread on the atomizer head, made in a form of a flat disk, bell or porous cylinder. The jets flowing out from the head are atomized as only the surface tension and viscose forces are overcame by centrifugal and electrostatic forces. The size of the paint droplets generated by mechanical atomizers is usually in the range from 10 μm to 100 μm. The atomization with electrostatic assistance requires about 12 times less energy than by pure centrifugal spraying (Balachandran and Bailey [1981]).

Charged paint can be sprayed by maintaining an atomizer (Figure 33a) or induction electrode (Figure 33b) at high potential, whereas the work piece is grounded (Inculet [1998]). Alternatively, high voltage is applied to a work piece while the atomizer is grounded (Figure 33c). In the latest case, the charge is induced on the droplets by the electric field produced by the work piece. The attraction of the spray cloud to the work piece is higher, even outside the spray zone, because the oversprayed droplets can also be captured. The spray cloud is expanded, and the wraparound effect is enhanced. This method of spraying was tested in automobile industry, where the reduction of a few percents in paint loss can substantially reduce the material costs.

Electrostatic painting of plastic surfaces presents some problems because the charge accumulated on the surface repels new incident droplets, preventing their deposition on the object. The dissipation of the accumulated charge can persists up to minutes or even hours. Electrostatic painting of insulating surfaces requires, therefore, pre-coating of the surface with a conducting film, which allows dissipation of the accumulated charge. A couple of solutions to this problem have been proposed and tested. Elmoursi and Garner [1993] placed the substrate to be covered onto a grounded metal plate in order to facilitate the surface charge removal. Bouguila et al. [1991] used ionised air flowing on the backside of the substrate in order to neutralize the accumulated charge and decrease the electric field generated by this charge. With this method the thickness of the paint layer was increased up to twice. Inculet and Klein [1993] used an additional ultrasonic atomizer for spraying oppositely charged water droplets, which were deposited on the backside of the substrate. The deposited droplets produced an electric field, which inductively charged the paint sprayed from the front side from a grounded nozzle. This method was used for the painting of plastic surfaces in automobile industry.

The advantage of painting with charged sprays is that droplets are dispersed more uniformly in space and more than 80-90% of paint is deposited on the work piece, while for conventional spray systems as much as 30-50% of the paint evades it (Biallas [2002], Hines [1966], Im et al. [2001], Inculet [1998], Snyder et al. [1989], Yamada et al. [2009]). The electrostatic forces cause also an increased uniformity of the coating on flat surfaces, and can produce thicker layer at corners and sharp edges because the electric field is higher at those points. Charged paint can also penetrate the cavities of the work piece. The electrified sprays allow reduction of volatile organic compounds emission and unwanted paint sludge production. However, the surface quality can sometimes be degraded by the current flowing through the layer, especially for metallic paints (Im et al. [2001]). The efficiency of spray deposition can be decreased by the back-flow, observed in the middle of the spray close to the object (Biallas [2002]).

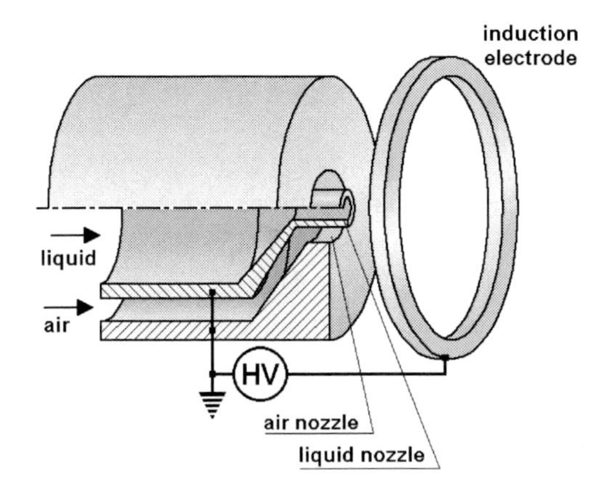

a.

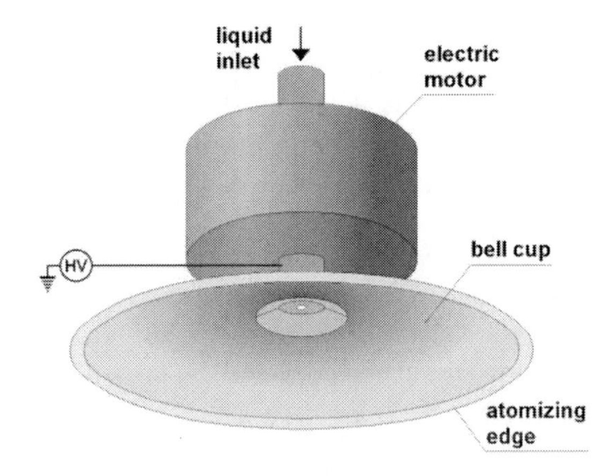

b.

Figure 32. Pneumatic atomizer with induction charging (a), and bell-type rotary atomizer with conduction charging (b) for charged paint spraying.

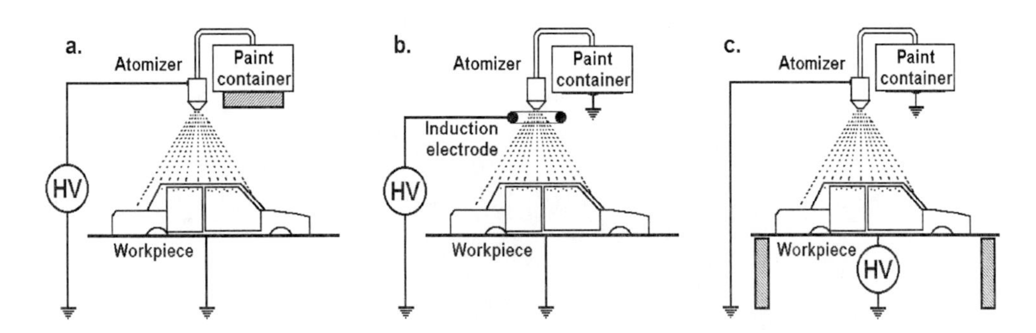

Figure 33. Electrostatic paint spraying, a. atomizer nozzle at high potential, b. induction electrode at high potential, c. work piece at high potential.

Spray forming is a similar process to the surface coating, but in this case, semi-solid droplets of a material are sprayed layer-by-layer onto a substrate in order to form a thick coating or bulk product (Chaudhury et al. [2004], Gopalakrishnan et al. [2003], Jaworek [2007a], Markus et al. [2002, 2004], Mesquita and Barbosa [2004], Schneider et al. [2004], Srivastava et al. [2001], Wang et al. [2005, 2006], Yamada et al. [2008], Yu et al. [2001]). Metal sheets, pipes, rolls or huge billets are formed from steel alloys, cooper alloys or aluminium alloys directly from the molten metals (Lawley and Leatham [1999]). The product of spray forming can be made as a composite one when consecutive layers are repeatedly deposited from various materials. This technology has been developed mainly for the manufacturing of composites based on metal-metal, ceramic-ceramic, metal-ceramics, or metal-IVth-group elements (for example, alumina-titania, aluminium-rutile, aluminium-silicon).

The composition of layers can be varied on demand during the production process by changing the feeding liquid (melt, solution, suspension, precursor). During the process of spray forming, fine droplets impacting a surface mix with previously deposited layer, penetrating its pores to form a tight product. The final result is similar to sintering process but the spray-formed materials can be less porous, and their microstructure can be more uniform than those produced by other methods, for example, casting (Chaudhury et al. [2004], Mesquita and Barbosa [2004], Srivastava et al. [2001], and Yu et al. [2001]). The particles in the matrix are distributed uniformly, and the hardness of the spray-formed composites or alloys can be relatively great (Chaudhury et al. [2004], Mesquita and Barbosa [2004], Srivastava et al. [2001]).

Spray forming from nanometer metal droplets allows the production of alloys of controlled composition and microstructure down to molecular layers. An advantage of spray formed alloys is that the bulk product is cooled faster without shrinkage, the strength within the material is relatively low, and the number of cracks is decreased.

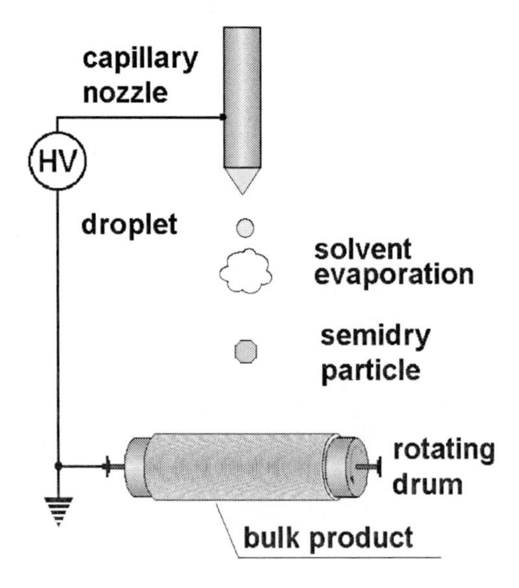

Figure 34. Schematic of the electrospray forming of bulk product via layer-by-layer deposition on a rotating drum.

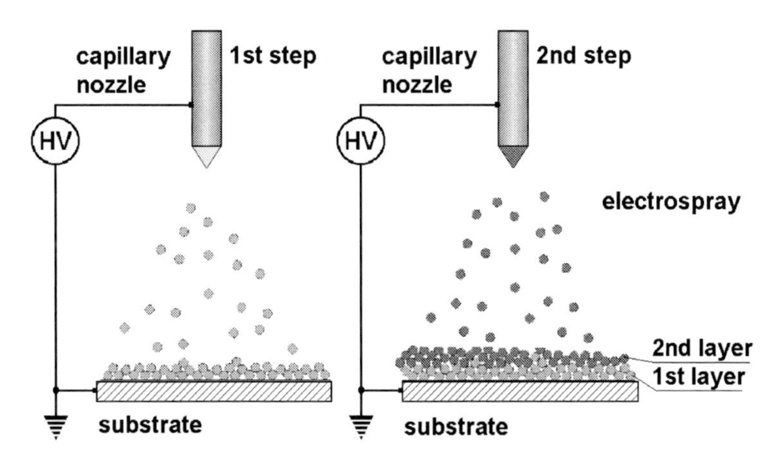

Figure 35. Scheme of electrospray deposition of micro- and nanocomposite film: a) 1st layer deposition, b) 2nd layer deposition.

Voids, pores or other imperfections, which can be found in the spray-formed product, are caused by various droplets sizes and their thermodynamic states. These faults can be removed via post-forming processes, for example, annealing at elevated temperatures. Spray forming with pneumatic atomizer produces gaseous entrapments of the size as large as 40-80 μm (Chaudhury et al. [2004]), and the overall solid fraction can be between 40 and 90% (Cantor et al. [1997]). Pressure-swirl atomisation in vacuum gives better results (Smallman et al. [1997]) but its application is limited to large surfaces and requires expensive vacuum systems.

Small products, of a size of a couple of millimeters can be formed by the ink-jet, drop-by-drop, printing of a solution or suspension of the material under controlled destination of each droplet (Teng et al. [1997], Wang et al. [2005]).

In un-charged spray forming systems, 40-50% of the material is oversprayed (Schneider et al. [2004]), but the efficiency can be increased to 80% when charged spray is applied. Particularly useful for the spray forming is the method of electrospraying (Chen et al. [1996a], Jaworek [2007a], Jayasinghe and Edirisinghe [2002, 2004, 2005b], Wang et al. [2005, 2006]). A schematic of electrospray forming process is illustrated in Figure 34. A solution or colloidal suspension of a material can be layer-by-layer deposited onto a substrate, forming, after drying or solidification, a thick coating or small bulk product. An advantage of electrospray forming over conventional spray-forming techniques is that the particles are of the same size, and have similar thermodynamic states. Because of this, the number and size of voids and cracks in the bulk product is reduced. Simultaneous, drop-by-drop and layer-by-layer deposition of two or more materials of different properties allows formatting a composite bulk product having novel and unique properties. The process of composite layer deposition is schematically illustrated in Figure 35.

4.2. Thin Solid Film Deposition

For the deposition of thin solid films, i.e., thinner than 10 μm, based on charged-spray technology, electrospraying is almost exclusively used. Thin solid films are required in microelectronic devices as insulating, conducting or semiconducting layers, in manufacturing micro- and nano- electromechanical systems (MEMS or NEMS), or as active layer in solar cells, electrodes or solid fuel in fuel cells, electrodes in supercapacitors or lithium batteries, gas-sensing layer in gas sensors, or as catalyst (Jaworek [2007a], Jaworek and Sobczyk [2008], Schoonman [2000]).

For industrial applications, the following conventional physical and chemical methods are commonly used for thin solid film deposition on a substrate (Bhushan [2004], Choy [2003b], Madou [2002]):

1. *Casting* which is based on spreading of a solution or colloidal suspension of an active layer on a substrate, for example by centrifugal forces. The material to be spread can be feed by dripping or from the aerosol phase, generated by pneumatic or rotary atomizers, but large amount of material is oversprayed. Frequently the film is not sufficiently homogeneous and the thickness can vary with the radial distance. The spreading process is followed by solvent evaporation and annealing, if required.
2. *Cathode spraying*. In the cathode spraying method, a target of the material to be deposited is placed at the cathode surface and bombarded with positive ions of inert gases, usually argon or xenon, created in an electrical discharge. Due to the momentum transfer from the ions, the atoms, molecules or small clusters are sputtered from the surface and partially deposited onto the substrate. In this process usually metal targets are used for layer preparation.
3. *Magnetron sputtering*. This is a process similar to the cathode sputtering, but the electrons created in the electrical discharge are subjected to a magnetic field in order to increase their paths near the cathode. Thus the number of collisions with gaseous atoms is larger and the production rate of ions sputtering the cathode increases.
4. *Laser ablation* deposition utilizes intense laser radiation to evaporate or sputter a target of the material to be deposited due to absorption of laser beam energy. The ejected atoms, molecule or clusters are deposited onto the substrate.
5. *Chemical vapour deposition* (CVD) is a process in which the film is deposited from one of the products of two reacting gases, which condense or grow epitaxially on the substrate. When the CVD process proceeds in plasma, it is called the plasma assisted/enhanced chemical vapour deposition.
6. *Physical vapour deposition* (PVD) is a process based on vaporization of the deposited material by thermal heating or electron sputtering. The vapours are transported to the substrate where they condense and solidify after the cooling, forming a thin film.
7. *Electroplating*. The process of material deposition takes place in an electrolyte with the material dissolved in a solution containing reducible ions of the metal. The method is applicable only to the metal films formation. The substrate is maintained at negative potential (cathode) and the anode is made of the material to be deposited. The ions are reduced at the substrate and remain on the cathode forming a film.

The deposition rates achieved by using cathode spraying, magnetron sputtering, CVD, PVD, laser ablation, or vapour condensation are relatively low. Only small amount of the material is deposited onto the substrate and the rest is lost to the chamber walls. The growth rate of film deposited by, for example, PVD is in the range from 0.006 to 0.06 μm/min, and by CVD from 0.02 to 0.05 μm/min, while for electrospray it is about 0.1 μm/min (Choy

[2003a]). The electrospray process is readily controlled via liquid flow rate and supply voltage. Electrospray deposition can proceed at or near atmospheric pressure, and is less expensive than chemical or physical vapour deposition, or plasma sputtering, which require high vacuum installations for the production of thin solid films of high quality. The electrospray systems used for thin film deposition usually operate in the cone-jet or multi-jet modes. The advantage of multi-jet mode is that the droplets are smaller than those obtained in the cone-jet mode, because the droplets are simultaneously dispersed from a few emission cones. The electrospraying technique allows generating fine droplets in micro- and submicron size range, with narrow size distribution and relatively uniformly distributed on the substrate that facilitates the deposition of high quality, smooth and even films with small number and size of voids, flaws and cracks.

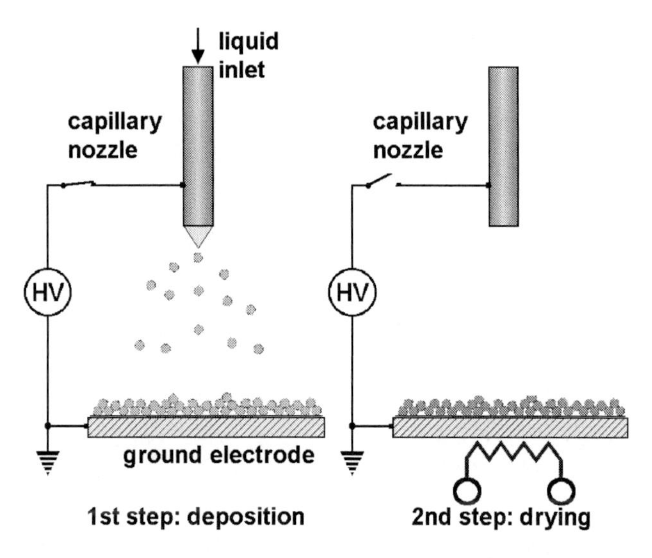

Figure 36. Scheme of electrospray deposition of thin solid film: a) 1st step: layer deposition onto a substrate, b) 2nd step: film drying and sintering.

Thin solid film deposited by electrospraying can be produced from a solution or colloidal suspension of a material forming the layer, or from a precursor. When a solution or suspension is used, the solid layer is obtained after solvent evaporation. Evaporation can be speed-up by heating the substrate. When a precursor is sprayed, it is decomposed at high temperature on the substrate or in the gaseous phase, and the layer is formed from a product of the decomposed precursor. Metal chlorides, nitrates, or acetates dissolved in water, methanol, ethanol, or their mixtures, or metal-organic salts dissolved in organic solvents are electrosprayed as precursors for metal-ceramic layer production. In order to improve mechanical properties of the produced film, the electrosprayed layer may be sintered at higher temperatures. A scheme of electrospray deposition of thin solid film is shown in Figure 36.

The formation of thin solid film deposited from charged spray is a complex and multi-step process. Morphology of the film depends, on the gas and substrate temperature, distance between spray nozzle and the substrate, kind of solvent used as precursor or for colloidal suspension (its boiling point), precursor or suspension concentration, rate of solvent evaporation from the film, liquid flow rate, and deposition time (Chen et al. [1997c]). Setting proper values of these parameters, the film morphology and other properties of electrospray-deposited film, for example, density, porosity, crystallinity, and structure can be tailored to some extent. The film thickness can be simple controlled by varying the concentration and flow rate of the colloidal suspension or precursor solution.

The substrate temperature and the critical temperature of precursor decomposition are two most important parameters, which influence the film morphology, i.e., film roughness, cracking, crystallinity, etc. (Chen et al. [1995, 1996b, 1998a, 1999a], Choy and Su [2001], Huang et al. [2003, 2004], Kim et al. [2000a,b], Lintanf [2007b], Neagu et al. [2006b,d], Nguyen and Djurado [2001], Perednis and Gauckler [2005], Perednis et al. [2005], Princivalle et al. [2004, 2005], Rhee et al. [2001], Wilhelm et al. [2005], Zaouk et al. [2000a,b]). Three temperature ranges, specific to deposited material, at which the film of high quality but different morphology is formed, can be distinguished:

1. *First (lower) temperature range*; the solvent evaporates to high extent during the flight of the droplet towards the substrate, and only dry or semi-dry particles of the precursor are deposited onto the substrate. The precursor decomposition takes place at the substrate. The film deposited in this temperature range is porous and amorphous, and firmly adheres to the substrate.
2. *Second (medium) temperature range*; the solvent and precursor evaporate before the droplet reaches the substrate, and only the precursor vapours diffuse to the substrate where it undergoes the process of decomposition. The adherence of the film to the substrate is also good because the crystallites are formed from the vapour phase.
3. *Third (higher) temperature range*; the droplet vaporizes short after its emission from the nozzle, and the precursor undergoes chemical reactions in the vapour phase. The product of these reactions in the form of solid particles is deposited onto the substrate.

When the substrate and gas temperatures are too low, the solvent evaporates slowly, and the droplet splashes on the substrate. The precursor is decomposed after solvent evaporation, however, the film is of low quality because too much solvent remains in the incident droplet, and the cracks are formed due to mechanical stresses. The solvent evaporates first from the surface of the layer and this layer shrinks faster that the part at the base. As a result, differential shrinkage between the layers occur, and the tensile stress breaks the layer adhered to the substrate, and delaminates it (Neagu et al. [2006b,d]). When the gas temperature is too high, the chemical reactions proceed within the droplet during solvent evaporation, and only dry particles arrive at the substrate, forming a porous, fractal-like structure.

The ranges of these temperatures vary with the flow rate of the solution and the distance between nozzle and substrate (Ghimbeu et al. [2007a, 2008a], Hwang et al. [2007a], Neagu et al. [2006b,d], Perednis et al. [2005], Princivalle et al. [2004, 2005]). When the flow rate of the precursor is increased, the droplets are larger (cf. (12)) and are not able to evaporate before landing, forming thus a cracked film. A longer time is, therefore, needed for the droplets to evaporate that can be achieved by increasing the distance between the nozzle and substrate. However, for longer interelectrode distances a higher voltage is needed in order to onset the cone-jet mode. For larger droplets, the temperature of the substrate can also be increased to facilitate solvent evaporation. When the flow rate of the precursor is decreased and the distance to the substrate is longer, the electrosprayed droplets are smaller and evaporate completely before landing. Only dry particles are deposited onto the substrate, forming a porous film. At the optimal flow rate and distance between the nozzle and substrate, the film is smooth and dense.

The optimal substrate temperature is that at which the solvent evaporates close to the surface, and the decomposition and chemical reactions take only place at the substrate (Choy [2001], Choy and Su [2001], Su et al. [2001]). These conditions are fulfilled when the substrate temperature is slightly higher than the boiling point of the solvent, and the solvent evaporation rate is equal to the liquid deposition rate (Cao and Prakash [2002], Chen et al. [1999b], Huang et al. [2003, 2004]). Multiple depositions, followed by subsequent pyrolysis or sintering at elevated temperatures can also facilitate the production of dense and crack-free film (Choy et al. [1998], Huang et al. [2004]).

Similar processes occur when the film is deposited from a particle suspension. The film becomes cracked due to solvent accumulation on the substrate, when the substrate temperature is too low, the flow rate is too high, and the distance is too short. The optimal conditions are those at which solvent evaporation rate is equal to the liquid deposition rate. Because usually ceramic materials are electrosprayed in the suspension form, the problem of their decomposition is unimportant.

Electrospray has been applied for thin solid film deposition from various materials and for various purposes. The most important are listed in the following:

1. *Thin radioactive films* including α- or β particle sources or neutron emitters (e.g. U^{233}, Pu^{238}, Am^{241}, Cm^{242}, or Ru deposited from their nitrates), targets activated in particle accelerators or nuclear reactors, and neutron detectors (Bruninx and Rudstam [1961], Carswell and Milsted [1957], Gorodynskiy [1959], Lauer and Verdingh [1963], Michelson and Richardson [1963], Michelson [1968], Reifarth et al. [2000], Shorey and Michelson [1970]). Film thickness of such devices was between 1 $\mu g/cm^2$ and 10 mg/cm^2.

2. *Micro- and nanoelectronic devices*; dielectric or semiconductor layers:
   - Al_2O_3 deposited from colloidal suspension (Chen et al. [1999a], Jaworek et al. [2009], Jayasinghe et al. [2002, 2004a,b], Jayasinghe and Edirisinghe [2003, 2004, 2005a]),
   - Al_2O_3-ZrO_2 double layers deposited from colloidal suspension (Balasubramanian et al. [2006]),
   - CeO_2 deposited from $Ce(NO_3)_2$ precursor (Oh and Kim [2007], Wei and Choy [2005], Wei et al. [2006]),
   - CdS deposited from $CdCl_2$ + $(NH_2)_2CS$ precursors (Choy [2001], Choy and Su [2001], Smyntyna et al. [1995], Su and Choy [2000a,c], Su et al. [2001], Turetsky [1999], Turetsky et al. [2003,2004]),
   - CoO deposited from $Co(NO_3)_2$ precursor (Chen et al. [1999a]),
   - In_2O_3 deposited from $InCl_3$ precursor (Ghimbeu et al. [2008b, 2009]),
   - In_2O_3:Sn (ITO) deposited from $SnCl_4$ + $InCl_3$ precursors (Chandrasekhar and Choy [2001a], Raj and Choy [2003]),
   - MgO deposited from $Mg(C_{11}H_{19}O_2)_2$ or $Mg(CH_3COO)_2$ precursors or from colloidal suspension (Jaworek et al. [2009], Kim and Kim [1999], Kim et al. [2000a,b], Rhee et al. [2001]),

- SiC deposited from $(CH_3)(C_6H_5)SiCl_2$ + $(C_6H_5)SiCl_3$ + $(CH_3)(CH_2 = CH)SiCl_2$ precursors or colloidal suspensions (Balachandran et al. [2001b], Grigoriev et al. [2001, 2002], Miao et al. [1999, 2001a,b,c, 2002]),
- SnO_2 deposited from $SnCl_2$, $SnCl_4$, or $Sn(NO_3)_2$ precursors (Chandrasekhar and Choy [2001b], Cusano et al. [2006, 2007], Ghimbou et al. [2007a], Gourari et al. [1998, 1999], Matsushima et al. [2003, 2004, 2006], Mohamedi et al. [2001a], Pisco et al. [2006], Yu et al. [2007], Zaouk et al. [2000a,b, 2007]),
- SnO_2+LaOCl composite deposited from $SnCl_4$ + $LaCl_3$ precursor (Diagne and Lumbreras [2001]),
- Ta_2O_5 deposited from $Ta(OC_2H_5)_5$ precursor (Lintanf et al. [2007a]),
- ZnO deposited from $Zn(CH_3COO)_2$, $Zn(C_2F_3COO)_2$, or $Zn(NO_3)_2$ precursors or colloidal suspension (Chen et al. [1999a], Fujimoto et al. [2006b], Ghimbeu et al. [2007c], Hogan and Biswas [2008], Hwang et al. [2007b], Jaworek et al. [2009], Ryu and Kim [1995], Zaouk et al. [2006]),
- ZnS deposited from $ZnCl_2$ + $(NH_2)_2CS$ precursor (Choy [2001], Choy and Su [2001], Hou and Choy [2006], Su and Choy [2000a,c,d], Turetsky [1999], Wei and Choy [2002]),
- ZrO_2 deposited from $Zr(OCH_2\ CH_2CH_3)_4$ or $ZrO(NO_3)_2$ precursors or colloidal suspensions (Balachandran et al. [2001b], Chen et al. [1998a, 1999a], Ksapabutr et al. [2010], Miao et al. [1999, 2001a,b,c, 2002], Neagu et al. [2005, 2006c], Nguyen and Djurado [2001], Slamovich and Lange [1988], Teng et al. [1997], Wang et al. [2006]).

3. *Photovoltaic cells*:
   - CdS deposited from $CdCl_2$ + $(NH_2)_2CS$ precursors (Choy [2001], Choy and Su [2001], Su and Choy [2000a,c]),
   - CdSe deposited from $CdCl_2$ + $((NH_2)_2CSe$ precursors (Su and Choy [2000b,c]),
   - $CuInS_2$ deposited from $Cu(NO_3)_2$ + $In(NO_3)_2$ precursors (Hou and Choy [2005b], Kaelin et al. [2004]),
   - $CuInS_2$-CdS double layer deposited from $Cu(NO_3)_2$ + $In(NO_3)_2$ (for $CuInS_2$ layer) and $CdCl_2$ + CH_4N_2S (for CdS layer) precursors (Hou and Choy [2005b]),
   - SnO_2:F deposited from $Sn(CH_3COO)$ + HF or $SnCl_4$ + HF precursors (Chandrasekhar and Choy [2001b], Zaouk et al. [2000a,b]).
   - TiO_2 deposited from $Ti(i\text{-}C_3H_7O)_4$, $Ti(OCH(CH_3)_2)_4$, titanium diisopropoxide bis(2,4- pentamedionate) (TIPD) precursors or colloidal suspension (Aduda et al. [2004], Chen et al. [1997c, 1998a,b, 1999b], Choy [2001], Choy and Su [1999], Fujimoto et al. [2006c], Hou and Choy [2004, 2005a]).

4. *Fuel cells*:
   - CGO (gadolina-doped ceria) solid electrolyte deposited from $Gd(NO_3)_3$ + $Ce(NO_3)_3$ precursors (Taniguchi et al. [2003b]),
   - $La_xCa_yCr_zMg_{1-z}O_{3-\delta}$ cathode deposited from $La(NO_3)_3$ + $Ca(NO_3)_2$ + $Cr(NO_3)_3$ precursors (Jiang et al. [2007]),

- $La_xSr_yCo_zFe_{1-z}O_3$ cathode deposited from $La(NO_3)_3$ + $SrCl_2$ + $Co(NO_3)_2$ + $Fe(NO_3)_2$ or $La(NO_3)_3$ + $Sr(NO_3)_2$ + $Co(NO_3)_2$ + $Fe(NO_3)_2$ precursors (Fu et al. [2005], Taniguchi and Schoonman [2002], Taniguchi et al. [2003a]),
- $La_xSr_yGa_zMg_{1-z}O_3$ cathode deposited from alkoxides and nitrates or $La(NO_3)_3$ + $SrCl_2$ + $Mg(NO_3)_2$ + $Gd(NO_3)_3$ precursors (Choy et al. [1998], Taniguchi et al. [2003b]),
- $La_xSr_yMnO_3$ cathode deposited from alkoxides and nitrates or $La(NO_3)_3$ + $SrCl_2$ + $Mn(NO_3)_2$ precursors (Choy et al. [1998]), Princivalle et al. [2004, 2005], Princivalle and Djurado [2008], Taniguchi and Schoonman [2002], Yan et al. [2008]),
- $NiO:CeO_2$-Gd anode for SOFC deposited from $Gd(NO_3)_3$ + $Ce(NO_3)_3$ + $Ni(NO_3)_2$ precursors (Hwang et al. [2007a]),
- NiO-SDC composite anode for SOFC deposited from $Sm(NO_3)_3$ + $Ce(NO_3)_3$ + $Ni(NO_3)_2$ precursors (Taniguchi and Hosokawa [2008], Xie et al. [2008]),
- NiO:YSZ deposited from $Ni(NO_3)_2$ + $Y(CH_3COO)_2$ + $ZrO(NO_3)_2$ precursors (Jang et al. [2009]).
- SDC (samaria-doped ceria) solid electrolyte deposited from $Sm(NO_3)_3$ + $(NH_4)_2Ce(NO_3)_6$ + $Ce(NO_3)_3$ precursors (Ksapabutr et al. [2008], Taniguchi and Hosokawa [2008], Xie et al. [2008]),
- YSZ (yttria-stabilized zirconia) solid electrolyte deposited from $Zr(C_6H_7O_2)_4$ + YCl_3, $Zr(C_5H_7O_2)_4$ + YCl_3, $Zr(C_3H_7O)_4$ + YCl_3, $ZrO(NO_3)_2$ + YCl_3, $ZrCl_4$ + YCl_3, $ZrO(NO_3)_2$ + $Y(NO_3)_3$ or $Zr(C_5H_7O_2)_4$ + 0.05 M $Y(C_5H_7O_2)_3$ precursors or from colloidal suspension (Chen et al. [1996e], Choy et al. [1998], Kelder et al. [1994], Neagu et al. [2006a,b,d], Nguyen and Djurado [2001], Nomura et al. [2005a,b], Perednis and Gauckler [2004], Perednis et al. [2005], Wang and Kim [2008], Wilhelm et al. [2005], Will et al. [2000]),
- Tb-YSZ (terbia-doped yttria-stabilized zirconia) solid electrolyte deposited from $Zr(C_5H_7O_2)_4$ + $Y(C_5H_7O_2)_3)$ + $Tb(CH_3COO)_3$ precursors (Stelzer and Schoonman [1996]).

In the NiO:YSZ anode, the NiO-component is the electronic conductor and fuel catalyst, while the YSZ phase serves as a supporting matrix, which adheres the nickel to the electrolyte (Jang et al. [2009]). In YSZ solid electrolyte the ZrO_2=92 wt.%, Y_2O_3=8 wt.%.

5. Lithium batteries:
   - $BPO_4:Li_2O$ solid electrolyte deposited from $Li(CH_3COO)$ + $Mn(CH_3COO)_2$ + H_3BO_4 + P_2O_5 precursors (Chen et al. [1997b]),
   - $CoO-Li_2O$ composite anode deposited from $Co(NO_3)_2$ + $LiNO_3$ or $Co(CH_3COO)_2$ + $Li(CH_3COO)$ precursors (Yu et al. [2005a, 2006]),
   - $LiAl_{0.25}Ni_{0.75}O_2$ cathode deposited from $Li(CH_3COO)$ + $Ni(CH_3COO)_2$ + $Al(NO_3)_3$ precursors (Yamada et al. [1999]),
   - $LiCoO_2$ cathode deposited from $Co(CH_3COO)$ + LiOH, $Co(NO_3)_2$ + $Li(CH_3COO)$, $Co(NO_3)_2$ + $LiNO_3$, or precursors (Chen et al. [1995, 1996b,c,d 1997a], Fujimoto et al. [2006a], Koike and Tatsumi [2007], Shu et al. [2003b], Yoon et al. [2001]),

- $LiCo_xMn_{2-x}O_4$ cathode deposited from $LiNO_3$ + $Co(NO_3)_2$ + $Mn(NO_3)_2$, $Li(CH_3COO)$ + $Mn(NO_3)_2$ + $Co(NO_3)_2$ or precursors (Chung et al. [2004b, 2005], Dokko et al. [2003, 2004],),
- $LiCo_{0.5}Ni_{0.5}O_2$ cathode deposited from $Li(CH_3COO)$ + $Ni(CH_3COO)_2$ + $Co(NO_3)_2$ precursors (Yamada et al. [1999]),
- $LiMn_{0.4}Fe_{0.6}PO_4$ cathode deposited from $LiNO_3$ + $Fe(NO_3)_2$ + $Mn(NO_3)_2$ + P_2O_5 precursors (Ma and Qin [2005]),
- $Li_xMn_2O_4$ cathode deposited from $Li(CH_3COO)$ + $Mn(NO_3)_2$, $Li(CH_3COO)$ + $Mn(CH_3COO)_2$, $Mn(CH_3COO)_2$ + LiCl, $LiNO_3$ + $Mn(NO_3)_2$, $Mn(CH_3COO)_2$ + LiOH, $Mn(CH_3COO)_2$ + $LiNO_3$, $Mn(CH_3COO)_2$ + Li_2CO_3 precursors (Anzue et al. [2003], Cao and Prakash [2002], Chen et al. [1997b, 1998b], Chung and Kim [2002], Chung et al. [2004a, 2006], Kobayashi et al. [2002], Koike and Tatsumi [2005], Mohamedi et al. [2001b, 2002a,b], Nishizawa et al. [1998], Shu et al. [2003a], Shui et al. [2004, 2006], Uchimoto et al. [2005], Uchiyama et al. [2000], van Zomeren et al. [1994], Yoon et al. [2003, 2006]),
- $LiNi_xMn_{2-x}O_4$ cathode deposited from $Li(CH_3COO)$ + $Mn(NO_3)_2$ + $Ni(CH_3COO)$ precursors (Chung et al. [2004b, 2005]),
- $LiNiO_2$ cathode deposited from $Li(CH_3COO)$ + $Ni(CH_3COO)_2$ precursors (Yamada et al. [1999]),
- $Li_4Ti_5O_{12}$ anode deposited from $Li(CH_3COO)$ + $Ti(OC_4H_9)_4$ precursors (Yu et al. [2005b]),
- SnO_2, SnO_2 + CoO, SnO_2 + Li_2O, SnO_2 + CuO, SnO_2 + Li_2O + CuO composite anodes deposited from $Sn(CH_3COO)_4$, $Sn(NO_3)_2$, $Sn(NO_3)$, $SnCl_4$, $Co(CH_3COO)_2$, $LiNO_3$, $Cu(NO_3)_2$ precursors (Mohamedi et al. [2001a], Yu et al. [2007, 2009], Zhu et al. [2009]),
- V_2O_5 cathode deposited from $[(CH_3)_2CHO]_3VO$ precursor (Kim et al. [2003]).

$Li_xMn_2O_4$ can be used simultaneously as the cathode for $0.3<x<1$, and as the anode, for $1<x<2$, due to the electrochemical potential difference between these two compounds (Chen et al. [1997b]).

6. *Electrochemical capacitors (Supercapacitors).* The electrodes of high specific surface area in pseudocapacitors are made of conducting polymers or transition-metal oxides:
   - $RuO_2 \bullet xH_2O$ deposited from $RuCl_3 \bullet xH_2O$ precursor (Kim and Kim [2001, 2004, 2006]),
   - Mn_3O_4 deposited from $KMnO_4$, $Mn(CH_3COO)_2$, or $MnCl_2$ precursors (Dai et al. [2006, 2007], Nam and Kim [2006], Shui et al. [2006]).

In double-layer capacitors, the carbon electrodes (for example CNT) of high specific surface area (up to 2000 m^2/g) filled with electrolyte are used (Zhang and Zhao [2009], Zhang et al. [2009]). The distance between the electrodes is of the order of 1 nm (Chen et al. [2009]).

7. Gas sensors:
   - $Cr_{2-x}Ti_xO_3$ (CTO) film deposited from chromium acetate + titanium diisopropoxide bis(acetylacetonate) precursors (Du et al. [2006]) used as a gas sensor for the detection of combustible and toxic gases,
   - In_2O_3 film deposited from $InCl_3$ (Ghimbeu et al. [2008b, 2009]), used as gas sensor for O_3, NO_x, or CO/H_2 detection,
   - SiO_2 thin-film gas sensor was produced for volatile organic compounds detection (Lubert et al. [2007]),
   - SnO_2, SnO_2:CuO, SnO_2 + LaOCl, SnO_2 + Mn_2O_3, and SnO_2 + ZnO films deposited from $SnCl_2$, $SnCl_4$, or $Sn(NO_3)_2$ precursors with the additives $Cu(NO_3)_2$, $LaCl_3$, $Mn(CH_3COO)_2$, and $Zn(CH_3COO)_2$, respectively for sensing the flammable and toxic gases (Chen et al. [1999a], Gourari et al. [1998, 1999]), CO_2 (Diagne and Lumbreras [2001]), H_2 (Ghimbeu et al. [2007a, 2008a], Gourari et al. [1998], Matsushima et al. [2003, 2004, 2006]), NH_3 (Pisco et al. [2006]), H_2S (Ghimbeu et al. [2007a, 2008a, 2009]), SO_2 (Ghimbeu et al. [2008a]), or NO_2 (Ghimbeu et al. [2007a, 2008a]),
   - $SrTi_{1-x}Fe_xO_{3-\delta}$ (perovskite (x=0.1-0.5)) deposited from $Sr(NO_3)_2$ + $Fe(NO_3)_3$ + $Ti(CH_3)_2CHO$ precursors (Sahner et al. [2007]) is a **p**-type material for the detection of hydrocarbons, ammonia, NO, CO, and H_2.
   - WO_3 deposited from $W(C_2H_5O)_6$ precursor (Ghimbeu et al. [2007b, 2009]) is **n**-type semiconductor used as gas sensor for NO_x, H_2S, or NH_3.
   - ZnO film deposited from $Zn(CH_3COO)_2$ precursor (Ghimbeu et al. [2007c]) for the detection of NO_2 and H_2S.

8. *Electrocatalysts* made from $NiCo_2O_4$ (Lapham et al. [2001]), catalyst made of platinum (Lintanf et al. [2007b]).

9. *MEMS actuators*; piezoelectric microactuators made of zirconium n-propoxide, $Zr(C_3H_7O)_4$, titanium tetraisopropoxide, $Ti((CH_3)_2CHO)_4$, or lead acetate, $Pb(CH_3COO)_2$ (Lu et al. [2003]), piezoelectric transducers of $PbTiO_3$ (Huang et al. [2003, 2004], Lu et al. [2002], Wang et al. [2007]), or $BaZrO_3$ (Su and Choy [1999b], or polymer ferroelectric films on an Si wafer (Rietveld et al. [2006]).

4.3. Fine Particles Production

Fine particles, of size smaller than 10 μm are, for example, applied for ceramic coating or thin solid film deposition, paint production, emulsion fabrication, or used in cosmetic or pharmaceutical industries. Micron and submicron solid particles can be readily produced from a solution of the particle material after atomisation of the liquid and solvent evaporation (Jaworek [2007a,b]). Precursor-based particles have to be decomposed at high temperatures in the aerosol phase before their collection on a substrate or filter.

Conventional methods for fine solid particles production are based on crushing grains of a solid material, condensation of vapours of the material in a gaseous phase, or pyrolysis of a spray (Gomez [1993], Kruis et al. [1998], Liu and Pui [1974]). Charged sprays can also be applied as a means for fine particle production (Borra et al. [1997, 1999], Deotare and Kameoka [2006], Erven et al. [2005], Hendricks and Babil [1972], Jaworek [2007b], Lenggoro et al. [2000, 2002], Lewis et al. [1994], Meesters et al. [1990a,b, 1991, 1992], Murtomaa et al. [2005], Nakaso et al. [2003], Park and Burlitch [1996], Sato [1991], Sato et al. [1988, 1999], Wang et al. [2005]).

All types of atomizers generating charged spray can be used for the production of fine solid particles, but electrospraying allows production smallest particles with narrow size distribution. A diagram showing the process of micro- and nanoparticle production from charged sprays is shown in Figure 37. The particles are produced from a solution or suspension of a solid material. The solvent from the electrosprayed droplets evaporates, and the remaining solid material forms a fine powder. The solvent evaporation can be fastened when stream of a hot gas is used. For the solution-based droplets, the remaining substance solidifies or crystallizes forming solid particles. When colloidal suspension is used, the nanosized particles suspended in the solvent agglomerate forming a tight cluster after the droplet is dried. Size of the particles can be controlled via changing the concentration of the dissolved or suspended substance.

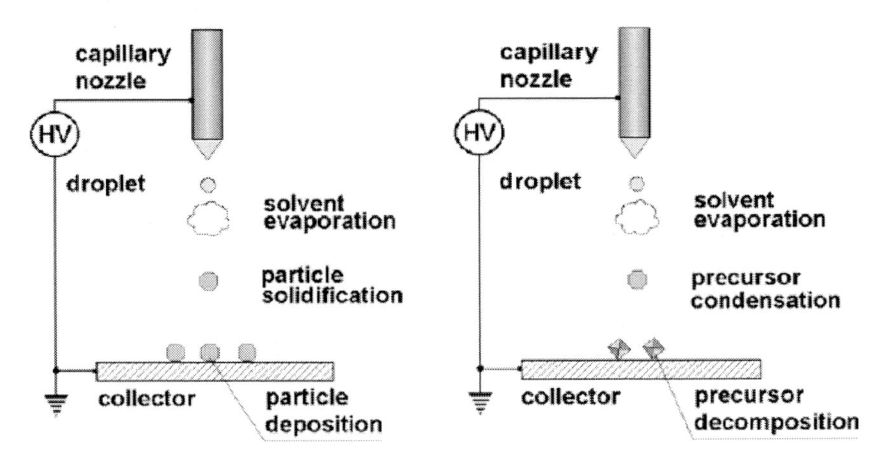

Figure 37. Particle production from electrospray droplets, a. from a solution after solvent evaporation, b. from a precursor after the precursor decomposition.

The size of solid particle after solvent evaporation can be determined from the following equation (Basak et al. [2007]):

$$d_p = \alpha \left(\frac{Q_V \varepsilon_0 \varepsilon_r}{\kappa_l \rho_p} \frac{CM}{1000} \right)^{1/3} \tag{76}$$

where ρ_p is the density of solid precursor, κ_l is the solution conductivity, M is the molecular weight of the precursor, C is the solution concentration in M.

The production of particles of exactly uniform size can be accomplished by generation of the cone-jet mode with varicose instabilities or the microdripping mode (Cloupeau and Prunet-Foch [1994]; Jaworek and Krupa, [1999a,b]).Such particles can be as small as 10 nm (Basak et al. [2007], Chen et al. [1995], Choy [2001], Dudout et al. [1999], Lenggoro and

Okuyama [1997]). The production rate of the particles is usually in the range of 10^5 to 10^4 particles/s, for particles of about 10 μm, and it increases to 10^{10} to 10^{11} particles/s for particles smaller than 10 nm. Stable operation of the cone-jet mode is, however, sensitive to any change in liquid properties, and the droplet size can vary unexpectedly with these parameters changing, for example, with temperature.

Better results are obtained when pulsed or AC superimposed on DC bias voltage (AC/DC) is used for liquid jet excitation (Balachandran et al. [1994b], Bollini et al. [1975], Jaworek et al. [2000], Sample and Bollini [1972], Sato [1984], Sato et al. [1983, 1996a,b, 1999], Vonnegut and Neubauer [1952]). For properly tuned AC voltage frequency, when the frequency of excitation voltage is close to the frequency of natural mechanical vibration of the jet (Sakai et al. [1991]), a droplet is expelled from the jet when the voltage approaches the peak value (Balachandran et al. [1994b]). The size of the droplets can be controlled by the DC and AC voltage magnitudes, AC voltage frequency, and volume flow rate of the liquid. The operating range of synchronous electrospraying in terms of volume flow rate and AC voltage frequency depends on liquid viscosity (Jaworek et al. [2000, 2003]). For liquids of low viscosity (water), the synchronous droplet generation is easier, and it occurs in wider range of flow rate and frequency than for more viscose liquids (glycol). Liquids of higher viscosity can be electrosprayed in the synchronous mode for lower frequencies than liquids of lower viscosity (Sato et al. [1988]; Jaworek et al. [2000]). When the flow rate of liquid increases, the AC excitation frequency has also to be increased.

For the production of large amount of uniformly sized droplets, a multinozzle rotary atomizer, with 100 capillaries mounted on a rotating disk, has been developed (Sato et al. [1988], Sato [1991]). The jets flowing out from these capillaries were synchronously excited with a high voltage applied to a ring induction electrode placed around the disk. The formation of jets at high rotational speed of the disk is independent of the electric field but only on centrifugal forces (Sato et al. [1988], Sato [1991]). Ultra-uniform silica particles from soluble glass were produced by this method, via dehydration of water from the droplets in 60% alcohol solution, and converting them to solid particles by removing sodium from the glass in a sulfuric acid solution (Sato et al. [1999]).

It should be emphasized that the electrospraying allows the production of fine particles without significant change in chemical composition or physical properties of the electrosprayed. This process seems to be useful for industries requiring fine powders of material that must remain unaffected during the production process, including biological samples sensitive to high temperatures.

Charged sprays were also utilized for the production of micro- and nanocapsules. Capsules are double-component particles made of a solid material, forming envelope (shell) enclosing another solid, liquid or gaseous material (core) (Ciach [2006, 2007], Jaworek [2008]). Encapsulation, as a process of capsules production, can be considered as the conversion of liquid phase to powder particles having the physical and chemical properties of the shell for easier transport and processing of the core material, which, under certain conditions (heating, diffusion, or shell dissolution), can be released from the shell. Encapsulation is, for example, used for the protection of environment-sensitive core materials, such as cultures, vitamins, flavours, dyes, enzymes, salts, sweeteners, acidulates, nutrients, or preservatives from adverse conditions. The shells used in the process are starch, gums, fats, waxes, oils, dextran, polysaccharides, proteins, or glucoses. Electrospraying is particularly useful as an encapsulation method.

4.4. Fuel Combustion

Conventional techniques used for fuel atomization rely on mechanical break-up of liquid jet into fine droplets in order to achieve high dispersion level before ignition. Mechanical dispersion does not allow controlling the properties of the droplets, and as a consequence the spray has usually wide polydisperse size distribution and, in many cases, droplets are too large for optimal combustion. Optimization of the combustion process has to be based on the following parameters of the fuel/gas phase mixture: droplet size distribution, droplet concentration, droplet velocity, droplets temperature, fuel concentration, uniformity of mixing of aerosol droplets in air, velocity of flame propagation.

For the combustion process, the aerosol droplet size distribution is characterized by the Sauter Mean Diameter (SMD), which is the ratio of the sum of the diameters cubed, times the number of all droplets of the same volume, divided by the sum of the diameters squared, times the number of all droplets of the same surface:

$$d_{32} = \frac{\sum_{i=1}^{m} d_i^3 \Delta N_i}{\sum_{i=1}^{m} d_i^2 \Delta N_i} \tag{77}$$

The Sauter mean diameter represents the diameter of a droplet whose volume and surface times the number of droplets equals the total volume and surface area of all droplets. It characterizes the contact area between the fuel and oxygen prior to the combustion. The decrease of the droplet size causes an increase in the Sauter mean diameter and improves the combustion efficiency.

Complex chemical reactions and nucleation processes take place in the unburned fuel that lead to the formation of soot particles (Bellan [1983]). Soot particles have a complex structure, usually in the form of large aggregates of the size up to a few hundreds of nanometers, consisting of large number of nano-sized primary particles (Mikhailov et al. [1999]). Diesel engines produce particularly large amount of soot, which is formed in the regions of engine where the fuel accumulates and undergoes only pyrolytic reactions without burning due to deficit of oxygen (Kwack et al. [1989]). Many methods have been developed in order to reducing soot emission from combustion. One of the methods is swirling the flow for better mixing fuel droplets with air. By this method, the regions rich of fuel are reduced, but not eliminated completely. Another method is charging the fuel droplets in order to better their dispersion within the combustion chamber. Shrimpton [2003] proved via numerical modeling that experimentally realizable amounts of electric charge are sufficient to cause spray expansion throughout the engine cylinder within the timescale comparable with engine stroke.

Charged sprays can offer an advantage of faster fuel evaporation due to lower gas phase partial pressures at the drop surface, and more uniformly dispersion and mixing of droplets within the combustion chamber, due to Coulomb forces. Particularly, electrospraying, which utilizes electric forces in the stage of droplets production, can generate smaller droplets, quite monodisperse in size, and highly charged. Charging the droplets can also cause secondary fission of the droplets when their charge reaches the Rayleigh limit. This process leads to increased surface to volume ratio that further increases the droplet evaporation rate and aids stabilization of the flame (Efimov et al. [1979]). Electrification prevents also agglomeration of smaller droplets due to their mutual repulsion (Gomez [1993], Kelly [1999]). Phase Doppler anemometer measurements have shown (Shrimpton and Yule [1999]) that electrosprayed hydrocarbon aerosol contained a large number of small droplets, which were repelled off the spray axis, whereas larger drops remained near the centerline of the spray. This radial distribution of the droplets' size can also be advantageous for fuel combustion. The electrostatic dispersion of fuel enhances its evaporation rate and, by this way, reduces soot nucleation (Bellan and Harstad [1997a,b, 1998]).

Two groups of applications of charged sprays can be distinguished:

1. *Charged-fuel microburners* (Behrens et al. [2010], Chen and Gomez [1992], Deng et al. [2007], Gomez and Chen [1994], Gomez et al. [2007], Jido [1989], Jones and Thong [1971], Kelly [1999], Kyritsis et al. [2002, 2004a,b], Laryea and No [2004], Shrimpton [2003], Shrimpton and Yule [1999, 2001], Thong and Weinberg [1971]),
2. *Diesel engine charged-spray injectors* (Anderson et al. [2007a,b], Bailey and Cetronio [1978], Balachandran et al. [1994a], Bankstone et al. [1989], Kwack et al. [1989], Lehr and Hiller [1993], Romat and Badri [2001], Selenou-Ngoms et al. [1999], Yule et al. [1995]).

The following techniques of charged spray generation for fuel microburners or internal combustion engines have been employed:

1. *Electrospraying* for diesel fuel (Balachandran et al. [1994a], Bankstone et al. [1989]), ethanol and 1-butanol (Behrens et al. [2010]), heptane (Chen and Gomez [1992], Gomez and Chen [1994]), JP-8 fuel (Deng et al. [2007], Gomez et al. [2007], Kyritsis et al. [2002, 2004a,b]), kerosene (Balachandran et al. [1994a], Jido [1989], Jones and Thong [1971]), paraffin oil (Thong and Weinberg [1971]), or PNF fuel (Lian et al. [2010]).
2. *Kelly's spray triode* for diesel fuel (Kwack et al. [1989], Romat and Badri [2001]), jet-A fuel (Kelly [1999]), or kerosene (Shrimpton [2003], Shrimpton and Yule [1999, 2001]).
3. *Pressure or pneumatic atomizers with induction charging* for diesel fuel (Selenou-Ngoms et al. [1999]), or petrol (Bailey and Cetronio [1978]).
4. *Pressure atomizer or fuel injector with charge injector* for gasoline (Anderson et al. [2007a,b], kerosene, diesel oil, or ethanol (Laryea and No [2004], Yule et al. [1995]).

The main problem with electrostatic charging of fuel is the transfer of the electric charge to the droplets. Almost all fuels are insulating liquids of typical electrical conductivity of the order of magnitude of 10^{-12}-10^{-15} S/m, whereas the best spraying results are obtained for the liquids of conductivity of 10^{-5} to 10^{-7} S/m. Two main solutions to this problem are used: increasing the fuel conductivity by doping with an anti-static agent (Chen and Gomez [1992], Gomez and Chen [1994], Deng et al. [2007], Gomez et al. [2007], Kyritsis et al. [2002, 2004a,b], Selenou Ngoms et al. [1999], Thong and Weinberg [1971]), and injection of electric charge into the fuel with a charge injector, usually made of tungsten or stainless steel wire, maintained at high potential (Anderson et al. [2007a,b], Jido et al. [1989], Kelly [1984], Yule et al. [1995]).

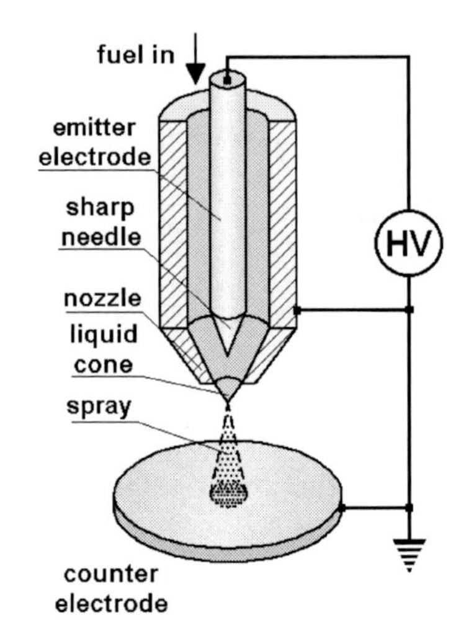

a.

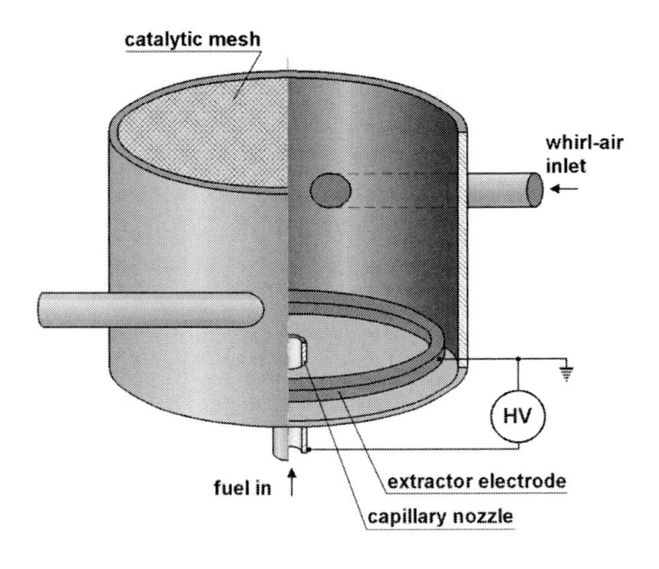

b.

Figure 38. Electrospray fuel burners: a. the Kelly's spray triode (Kelly [1984]), b. multijet electrospray burner (Kyritsis et al. [2004a]).

Charge injector for direct charging of liquids of low conductivity (lower than 10^{-12} S/m) without addition of an antistatic agent was developed by Kelly [1984]. This device, known as "Kelly's spray triode", is shown schematically in Figure 38. The spray triode consists of a central, sharply pointed emitter electrode (charge injector) positioned co-axially within the nozzle close to its outlet. The charge injector is at high potential, usually negative, and the nozzle body serves as an anode. The third electrode is a grounded collector placed outside the nozzle. High electric field, of about of $5*10^9$ V/m between the charge injector and the anode enables charge emission from the emitter tip. The electrons attach to hydrocarbon molecules, which act as charge carriers. The mobility of charged molecules is very low, and for sufficiently high fuel velocity, they can flow out through the orifice before they attain the nozzle walls (anode) where they could be discharged. Due to the repelling Coulomb forces, the expansion velocity of the charged spray can be as high as 10 m/s, whereas the fuel injection velocity is usually 20-40 m/s (Kwack et al. [1989]). Devices of this type were tested many times for fuel atomization (Balachandran et al. [1994a], Bankston et al. [1989], Kwack et al. [1989], Lehr and Hiller [1993], Shrimpton and Yule [1999, 2001], Romat and Badri [2001], Yule et al. [1995]). Negative potential of the charge injector, which causes electron emission, is advantageous over the positive potential, causing field ionization, because the same ionization current is generated by the electric field of an order of magnitude lower (Yule et al. [1995]).

Tungsten fibers placed at the tip of emitter electrode in the Kelly's spray triode, allows increasing the discharge current that makes the wider spatial dispersion of the droplets, and improves the combustion of Diesel fuel (Bankston et al. [1989]). Due to secondary fission of the droplets, the number of smaller droplets increases in core of the spray plume, and the spray evaporation is faster. The number density of fibers used by Bankston et al. [1989] was about $10^6/cm^2$.

The spray triode was used in a fuel burner, called "pocked stove" (Kelly [1999]). Although the combustion process was similar to the gas flame, by charging the spray it was possible to achieve complete combustion of different fuels even under extremely stringent conditions, such as low temperature. The method allowed the fuel injection at a cold soak, at temperatures down to -40°C.

A fuel burned in an electrostatic fuel burner provides flame similar to a gaseous one because the fuel droplets are not in direct contact with the flame before their evaporation (Chen and Gomez [1992], Gomez [1993], Gomez and Chen [1994]).

Catalytic micro-combustors supplied with liquid fuels, operating as energy conversion devices, are the attractive alternative to batteries (Behrens et al. [2010], Kyritsis et al. [2002, 2004a,b]). With the electrospray-atomized fuel, a catalytic reactor can be miniaturized, because the distance required for complete droplet vaporization can be reduced from typical 35 mm to 1.5 mm (Behrens et al. [2010]). With Pt catalyst (mesh) or rhodium-coated alumina foam (of large surface area), the conversion of ethanol can be as high as 75% or close to 100%, respectively. The micro-combustor can generate electricity in conjunction with thermoelectric device (Behrens et al. [2010]) or with a free-piston Stirling engine (Gomez et al. [2007]).

In the catalytic micro-combustor, schematically shown in Figure 38b, the fuel is electrosprayed upwards, toward the catalytic mesh. The electric field is produced by the extractor electrode placed at the chamber base. This location causes the droplets trajectories to be bent toward this electrode instead of to the catalytic mesh, and the residence time of the fuel in the mixing chamber becomes longer that is favourable to its evaporation. As a result, the necessary preheat chamber height can be decreased by a factor of 3, with the same vaporization rate by a given flow rate of fuel (Kyritsis et al. [2004a]). Additionally, air whirling caused by its tangential injection through two pipes, improves temperature uniformity on the catalytic surface (Gomez et al. [2007]). A micro-combustor operating on JP-8 jet fuel (Figure 38b), which was electrosprayed at a flow rate of the order of 10 g/h, achieved a temperature of 900-1300 K, the power production of the order of 100 W, and the combustion efficiency of about 97%. The flue gases were free of soot and NO_x, and the CO/CO_2 emission ratio was far under 1% (Deng et al. [2007], Kyritsis et al. [2002, 2004a,b]).

Internal combustion engine charged-spray injector has been thoroughly investigated by Anderson et al. [2007a,b]. Based on the comprehensive studies, authors have noticed the similarities and differences between charged and un-charged sprays, which can be summarized in the following table:

charged	uncharged
a fuel ligament at the leading edge of the spray are formed	poorly atomised droplets ahead of the main spray cone are injected
velocity field is less coherent due to the Coulombic repulsion, and the spray has vortical structure	stream of droplets following the head-core is well alined
pockets can be observed in the chamber (volumes where the liquid is absent)	vortical motions of the droplets develops, decreasing the pockets
droplets are dispersed faster	droplets are dispersed slower
droplets size can only slowly increase	droplet size increases due to coalescence (after 8 ms past injection)
low standard deviation of the median droplet diameter	high standard deviation of the median droplet diameter
similar median diameter	
similar spray width and penetration	

The differences between the electrostatically charged and un-charged sprays diminish with increasing injection pressure. The vortical motion of charged spray and pockets of air can enhance vaporization of the spray at elevated temperatures. The authors concluded that the morphology of the spray is not affected drastically by the application of the electric field because the inertial forces mainly determine the process of liquid dispersion. The electrostatic field can only affect the atomisation mechanism, repeatability of droplet size, and velocity of the atomised droplets (Anderson et al. [2007a,b]).

The advantages of electrostatically assisted fuel injection have been considered by Anderson et al. [2007b], and can be summarized as follows:

1. The momentum of the fuel droplets is decoupled from the mass of the injected fuel;
2. The droplet size and fuel dispersion can be controlled independently;
3. The injection pressure can be reduced, potentially reducing the parasitic losses due to high pressure fuel pumps;
4. Good atomization during cold-start operation, preventing port wall wetting.

The main shortcoming in large-scale practical applications of electrospray nozzle as fuel injector is a relatively small flow rate, which, for example, in actual diesel engines should be higher than 30 l/h. An annular-slit nozzle developed by Chen and Gomez [1992] allows generating simultaneously over one hundred jets that improves operational properties of such nozzle. The nozzle can also be multiplexed via using a set of electrospray capillary nozzles. The multiplexing allows generating droplets of smaller size than from a single nozzle, by the same total flow rate, because the droplet size decreases with the flow rate decreasing. Additionally, evaporation time for smaller droplets becomes shorter, and, the droplets can be dispersed more uniformly in the combustor chamber than for a single nozzle used (Kyritsis et al. [2002, 2004a,b]). Another alternative is a single capillary with many grooves at the circumference, operating in the multijet mode (Kyritsis et al. [2004a,b]). Such capillary nozzle can be of larger diameter that makes it much less susceptible to clogging by solid or highly viscous deposits in the fuel or to the obstruction after burner switch-off.

Flammability of spray mainly depends on the mean droplet size and droplet volume concentration. Aerosols of smaller droplet size tend to have higher flammability, while lower droplet volume concentration may reduce the aerosol flammability (Lian et al. [2010]). Multiplexing of electrospray nozzles or generating the multijet mode allow production droplets of smaller size than from a single nozzle and improves spray flammability. The gaseous pollutant emission can be reduced when the combustion process is optimized with respect to the spatial uniformity of the spray dispersion and a mean droplet size appropriate for the burned fuel. Electrostatic dispersion of fuel enhances its evaporation and, in this way, reduces soot nucleation that improves operational properties of tactic missiles because lower soot emission can reduce the visibility of the weapon (Bellan and Harstad [1997a,b, 1998]).

4.5. Colloid Thrusters for Space-Vehicle Propulsion

Conventional rocket engines convert chemical energy of a fuel into heat in the process of burning. Next, the heated gas expands through a nozzle, due to increased pressure, converting heat into kinetic energy. The thrust of such engines is limited by the gas temperature, which, in turn, depends on the thermal capacity of the fuel. Chemical thrusters provide high thrust, of the order of magnitude of MN, at high available power, which can be higher than 1 GW, and also relatively high power per unit mass of fuel. The jet velocity created by chemical engines is in the range of 3000 - 4500 m/s (Grigoryan [1996]). The most energetic thrusters in which liquid hydrogen is combusted in liquid oxygen give an exhaust velocity of 4700 m/s (Fearn and Smith [1998]). Higher jet velocities, up to 50000 m/s, can be obtained by electric propulsion.

Electric propulsion is based on the thrust produced by a beam of ions or charged microparticles accelerated by an electric or magnetic field (Baker [2005]). The most important electric propulsion systems are resistojets, plasma propulsion systems, ion engines, and colloid thrusters.

In resistojets, the propellant (usually Xe) is heated to high temperature by a resistive heater and next expands through a nozzle producing a thrust. Xe thruster can give specific impulse of 45 s after heating the gas to 600°C (Baker et al. [2005]).

In plasma propulsion systems, the propellant is ionized in a high voltage electric discharge. The positive ions are extracted from the plasma, accelerated to a high velocity (10^4-10^5 m/s) in an electric field, and ejected to the space. The thrust level in such engines can be of the order of magnitude of 300 mN (Fearn [1997]). In ion engines, liquid metal ions or charged clusters are pulled out from the Taylor cone formed at the tip of a nozzle in an electric field higher than 10^9 V/m. The ions are next accelerated to velocities of about 15 km/s, producing thrust transferred to space vehicle. The most appropriate propellants for ion engines are alkali metals because the ions can be easier generated due to high conductivity and high surface tension of these metals.

In colloid thrusters (Figure 39), submicron droplets or multimolecular clusters charged close to the Rayleigh limit are pulled out from a capillary nozzle or a tip of needle due to a high electric field produced by the potential between the nozzle and an extractor electrode.

The droplets accelerated by the electric field produced by additional electrodes transfer the thrust to the vehicle. Electrical neutralization of the spacecraft, in all types of ion-emitting electric propulsion systems is however required. It can, for example, be accomplished by ejection of electrons from electrically heated hot-wire tungsten cathode or from an e-beam accelerator.

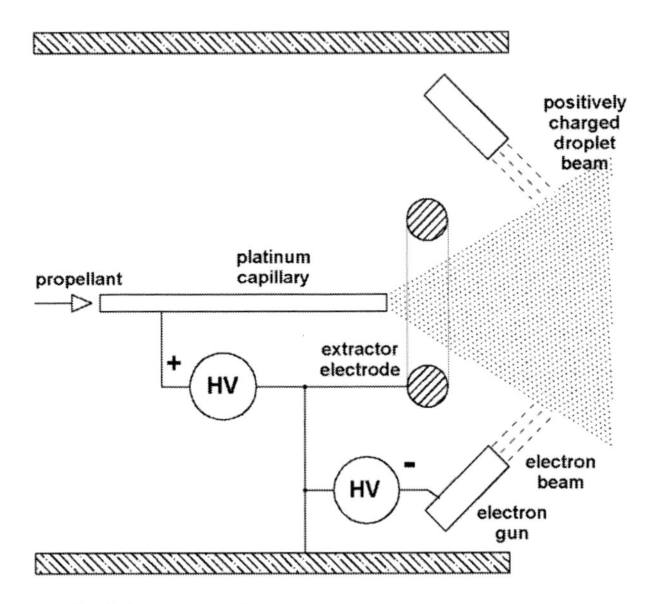

Figure 39. Principle of colloid-thruster engine.

The chemical rocket engines are used for space vehicle launching but colloid thrusters were developed for near earth missions, station keeping and satellites repositioning. They are of high efficiency in the specific impulse range of 500 - 1500 s, and they are simple in operation. The thrust level delivered by a single nozzle is in the range from 0.1 to 100 μN. In ion-emission regime the energy smaller than 10 eV is required to produce a single ion, and the emitted ionic beam have only a small dispersion of velocity.

The performances of ion engines and colloid thrusters are characterized by the following three parameters:

1. *Thrust*, which is the force required for accelerating the beam of particles to a velocity v (or rate of propellant use dm/dt multiplied by rocket velocity v):

$$Tr = \dot{m}v = mv\frac{i}{e} \tag{78}$$

where m and $\dot{m}$ are the mass and mass flow rate of the charged droplets, particles or ions, respectively, i is the electric current of the ion or charged particle beam and e is the elementary charge.

2. *Specific impulse*, measured in seconds, which is the ratio of the thrust to the weight flow rate of the propellant for sea-level gravity:

$$I_{sp} = \frac{\dot{m}v}{\dot{M}g} \tag{79}$$

where $\dot{M}$ is the total mass flow rate of the propellant, including uncharged particles, g is the Earth gravity constant.

3. *Specific charge* defined as the total charge of a beam of particles to the mass of this beam, known as the q/m ratio:

$$\frac{Q}{m} = \frac{i}{\dot{m}} \tag{80}$$

The specific charge can be determined as the beam current to mass flow rate ratio.

Ion engines and colloid thrusters have a few advantages over other electric propulsion systems, for example, low-pressure discharge or surface ionization plasma sources. They are simple in operation, can produce higher acceleration force per unit power, have high power efficiency, and higher specific impulse (Perel [1968]). The thrust and specific impulse are easier to control, and the beam direction can be changed over a wide range without lost of thruster performances and efficiency (Huberman et al. [1968], Kidd [1968]). High power efficiency can be attained because colloid thrusters operate at lower temperatures than plasma systems and no thermal ionization, electron bombardment or magnetic fields (used for plasma confinement), which are energy consuming, are required. The propellant particle formation, its charging and acceleration take place in the same electrohydrodynamic process. Because of high velocity of ions only a small amount of propellant is needed, and a reservoir with total amount of propellant can be integrated with the ion engine that reduces the engine mass and costs (Saccoccia et al. [2000]). The difficulty with colloid thrusters is that they require controlled temperature in order to keep the propellant viscosity constant (Kaufman [1990]).

Ejection of particles with specific charge of 10^2 - 10^5 C/kg is necessary for effective operation of a colloid propulsion system (Pfeifer and Hendricks [1968]). This can be attained only for particles smaller than 1 μm. Greater thrust can be obtained for heavier particles or ions, such as single ionized Hg (4.77×10^5 C/kg) or Cs (7.21×10^5 C/kg). The higher the velocity of the ejected mass the smaller the quantity of the propellant, which has to be taken on-board.

Electric thrusters for space vehicle propulsion were developed in 1960s (Bailey [1973, 1988], Hogan and Hendricks [1965], Huberman [1970], Huberman et al. [1968], Krohn [1961], Perel et al. [1969, 1971], Pfeifer and Hendricks [1968]), however, those devices were not implemented practically. The concept of colloid propulsion was first proposed by Preston-Thomas [1952] (cf. Hogan and Hendricks [1965]), and tested by Cohen in 1961 (cf. Huberman [1970]). Colloid thrusters were developed for near earth missions, station keeping or satellites repositioning. Various values of the specific charge were obtained for different liquids working as propellant: for octoil it was of 1 C/kg (Hendricks [1962]), for oil 4.5 C/kg (Hendricks [1962]), for glycerol doped with $ZnCl_2$ Wineland and Hunter obtained 3000 C/kg (cf. Makin and Bright [1969]), for glycerol doped with $SbCl_3$ - about 500 C/kg (Krohn [1974]), and for glycerol doped with NaI about 10^4 C/kg (Cohen, cf. Krohn [1974]). For liquid metals Krohn [1974] obtained specific charge of 4.6×10^4 C/kg, and Swatick and

Hendricks [1968] 7×10^5 C/kg. From mid 1970s until the end of 20th century, the interest in colloid thrusters was only marginal.

At the beginning of 21st century the attention was turned to the microthrusters operating on the same principle as colloid thrusters but designed for microdroplets/ microparticle emission and acceleration. Telecommunication satellites or satellites used for scientific missions, like, for example those for the detection of gravitational waves, require positioning the satellite with micrometer and 10 μarcsec accuracy, in order to compensate any satellite drift from its desired orbit, as a result of external forces, such as gravity, atmospheric drag or solar pressure (Franks [1998]). Only small engines of a power not higher than 5 kW, and thrust in the range from 1 to 1000 μN, emitting ions or particles, can accomplish this goal. The colloid thrusters generate a specific impulse in the range of 500 - 1500 s, and the thrust level delivered by a single nozzle is in the range from 0.1 to 100 μN. Due to the compact construction and simple operation the colloid thrusters are particularly suitable for small satellites, of the mass between 10 and 100 kg. The electric or magnetic energy required for acceleration of charged particles or droplets can be derived from an on-board electric supply or from solar batteries.

The following types of colloid thrusters were developed and tested:

1. *Capillary nozzle* (Figure 40a) (Kidd [1968], Lear et al. [1972]),
2. *Annular-slit nozzle* (Figure 40b) (Bailey et al. [1972], Perel et al. [1971]),
3. *Cylindrical-rim nozzle* (Figure 40c) (Lear et al. [1972]),
4. *Linear-slit nozzle* (Figure 40d) (Bartoli et al. [1984], Franks et al. [1998], Perel et al. [1971]),
5. *Liquid metal ion source* (LMIS) (Figure 41a) (Bollini et al. [1975], Fehringer et al. [1998], Krohn [1961, 1974], Swatik and Hendricks [1968]).
6. *Ionic liquid ion source* (ILIS) (Lozano et al. [2004], Lozano and Martinez-Sanchez [2005a]).
7. *Co-extrusion nozzle* (Castro and Bocanegra [2006]),
8. *Micronozzle arrays* for droplets emission (Figure 41b) (Krpoun and Shea [2007, 2008, 2009], Krpoun et al. [2008, 2009], Lozano et al. [2004], Paine et al. [2004], Si et al. [2007], Velasquez-Garcia et al. [2006a], Xiong et al. [2002]),
9. *Externally-wetted microneedle arrays* for ionic liquid emission (Figure 41c) (Castro and Fernandez de la Mora [2009], Gassend et al. [2007, 2008], Velasquez-Garcia et al. [2006b]).

The advantages of a capillary nozzle are: high thrust density leading to minimized thruster size and weight, two-dimensional electrostatic thrust vectoring capability and reduced number of components (Lear et al. [1972]). However, high electric fields at the nozzles of small diameter can cause electric breakdown between the electrodes and subsequent nozzle material erosion. The annular-slit nozzle has reduced susceptibility to breakdown due to larger diameter of the nozzle, and focusing properties better than linear nozzle (Perel et al. [1971]). The linear-slit nozzle requires higher voltage for proper operation than a single-capillary nozzle to produce the required extraction field, but it allows obtaining relatively high thrust density and higher specific impulse for a given charge to mass ratio (Lear et al.

[1972]). In the cylindrical-rim nozzle the angle of the beam is too large, causing unwanted particle dispersion that reduces the effective thrust (Lear et al. [1972]).

In order to increase the total thrust produced by electrostatic engine, multiple-nozzle systems have been constructed (Huberman et al. [1968], Huberman [1970], Kidd [1968], Zafran et al. [1973]). Multiple systems allow also controlling the desired thrust via activation only a fraction of the nozzles. Perel et al. [1969], for example, used an emitter array of 4x4 stainless steel capillaries, and also a larger one of 37 positive and 36 negative capillaries. The efficiency of this thruster ranged from 50% to 80% depending on supply voltage and mass flow rate of the propellant. Huberman et al. [1968], and Huberman [1970] constructed a 60-needle colloid thruster operating with total efficiency of 70%. A large multinozzle system consisting of 432 stainless steel capillaries was tested by Zafran et al. [1973].

In order to avoid using a large number of nozzles, annular-slit, cylindrical rim, and linear slit nozzles were constructed (Bailey et al. [1972], Bartoli et al. [1984], Lear et al. [1972], Perel et al. [1971]). The annular thruster has linear dependence of the thrust on the mass flow rate in wider flow rate range than a capillary nozzle and also indicates more stable operation and good focusing properties. A cylindrical-rim nozzle forms "meniscus well" in which the propellant was hold and also operates with high stability (Lear et al. [1972]). Linear slit nozzle (Bartoli et al. [1984]) is another alternative for multiple nozzle constructions. The nozzle is characteristic of a high ratio of the ion mass emission rate to the total mass flow rate (about 80% when operating with Cs as propellant). A fine slit emitter of slit width of 1 μm requires only a few grams of Cs for several years of operation (Mitterauer [1991], Franks et al. [1998]).

Bipolar or alternating-current thrusters are used in order to eliminate the need of charge neutralization at the spacecraft by electron emission. A bipolar thruster generates simultaneously positively and negatively charged droplets, from two sets of nozzles (Huberman et al. [1968], Perel et al. [1969]).

Alternating-current colloid thruster is made in a form of a single or multiple nozzle emitter, connected to an alternating voltage source, which generates positive and negative pulses of sine- or square-waveforms of frequency of hundred of hertz to kilohertz (Burson and Herren [1971]). Supplying the nozzle by square-wave was more effective because there lack zero-emission intervals. The efficiency of such emitter was however only of about 40%. The magnitude of emission current for negative polarity is higher than for positive one, and different positive/negative pulse amplitudes are required to obtain zero net charge on the spacecraft. Burson and Herren [1971] have also constructed a thruster composed of two needles supplied with two AC voltages in antyphase, but of different flow rates to compensate the current differences. Bollini et al. [1975] proposed harmonic electrical excitation of the liquid jet in order to synchronous spraying of a beam of monodisperse, uniformly charged molten metal particles in vacuum, which could be applied in a space vehicle thruster. Particles generated by this method were uniform in size and charge, and could be easily accelerated and deflected by electric fields.

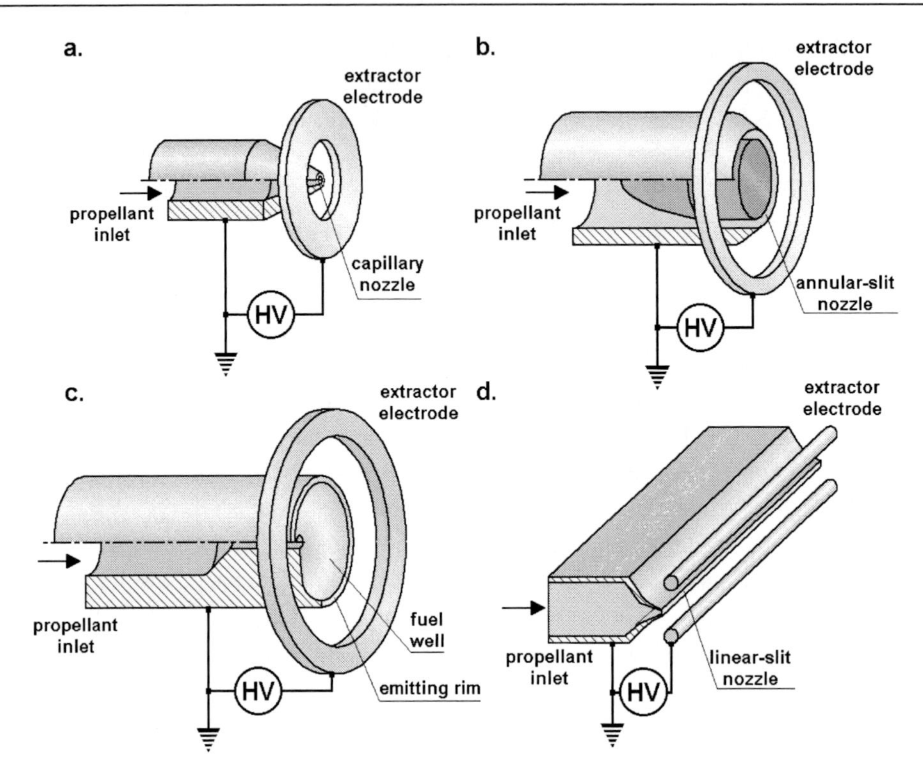

Figure 40. Types of colloid thruster nozzles, a. capillary nozzle, b. annular-slit nozzle, c. cylindrical-rim nozzle, d. linear-slit nozzle.

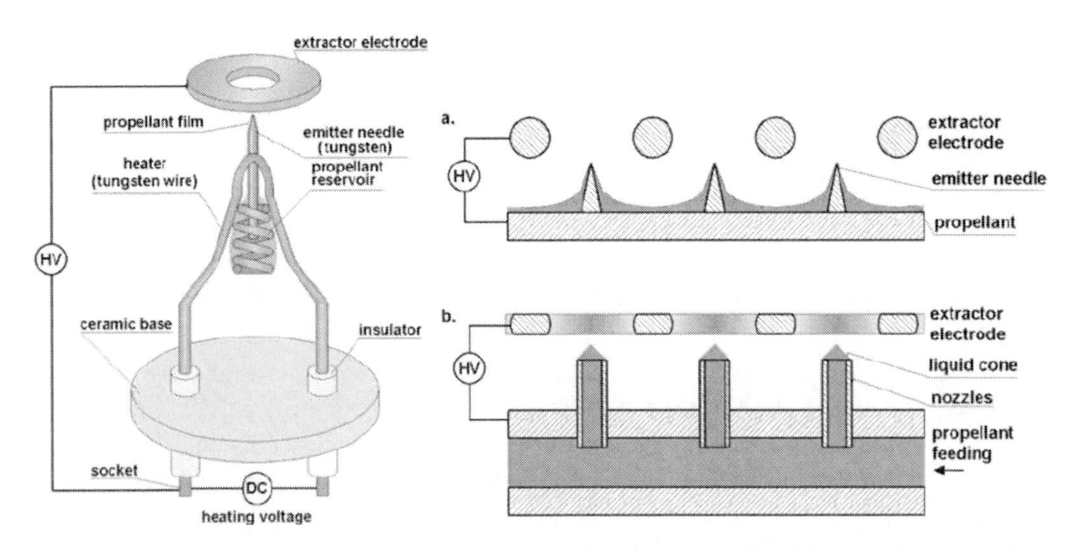

Figure 41. Types of micronozzle colloid thrusters: liquid-metal-ion-source thruster (left), needle-type array of ion emitters (right-up), capillary-nozzle array of charged droplet emitter (right-bottom).

Due to periodic reversing of emitter polarity, polarization of the material surface and the subsequent loss of the electric field, and electrochemical decomposition in some ionic liquid propellants can be avoided (Krpoun et al. [2009], Lozano et al. [2004], Lozano and Martínez-Sánchez [2004, 2005b]). The voltage polarity has to be reversed in the time shorter than the time constant of double layer formation on the electrode (capillary nozzle). The time can be estimated from a plate-parallel capacitor model (Lozano and Martinez-Sanchez [2004]):

$$t_w = \varepsilon_r \varepsilon_0 \frac{V_w^{\pm}}{I} \frac{A}{\delta} \tag{81}$$

where $V_w^{\pm}$ electrochemical window (potential difference between electrodes), I is the current flowing to the double layer capacitor, A and δ are the effective double layer interface area and thickness, respectively. For EMI-BF4 propellant, the time constant was estimated to 11 s for both polarities. The measurements with a sharp needle ion emitter showed that a frequency of the order of 1 Hz is sufficient to obtain clean electrodes and reliable ion emission (Lozano and Martinez-Sanchez [2004]).

From the beginning of the 21st century, the research is aimed at the development of new types of propellant and novel propulsion microthrust systems operating in the micronewton thrust range for precise positioning (drag force, solar pressure, and magnetic torque compensation, orbit phasing and adjustment) and manoeuver a satellite, and in long-term (ten years or longer) orbital missions of small satellites, of mass smaller than 20 kg (Baker et al. [2005], Gamero-Castaño [2004]). Printed-circuit-board (PCB), microelectronic and MEMS technologies are employed to the construction of micronozzle arrays. These emitters are usually fabricated in silicon (Alexander et al. [2006], Gamero-Castaño [2004], Gassend et al. [2007, 2008], Krpoun and Shea [2007, 2009], Krpoun et al. [2008, 2009], Lozano et al. [2004], Lozano and Martinez-Sanchez [2004], Velasquez-Garcia et al. [2006a,b, 2007], Xiong et al. [2002, 2003, 2006]).

Micronozzle arrays for fine droplets emission can be in the form of linear or planar emitter (Krpoun and Shea [2007, 2008, 2009], Krpoun et al. [2008, 2009], Lozano et al. [2004], Paine et al. [2004], Si et al. [2007], Velasquez-Garcia et al. [2006a], Xiong et al. [2002]). Due to small dimension of the capillaries (<100 μm) such a system can operate also as ion emitter.

Bailey et al. [1972] showed that the thrust and exhaust velocity are the linear function of the square root of the emission current. The thrust produced by single nozzle electrospraying droplets is (Alexander et al. [2006]):

$$Tr = \rho_l Q_V \left(2U \frac{Q}{m} \right)^{1/2} \tag{82}$$

When the capillary nozzle operates in purely ionic emission mode, the thrust is (Krpoun and Shea [2009]):

$$Tr = I \left(2\Phi_B \frac{m_0}{Q} \right)^{1/2} \tag{83}$$

where I is the beam current, m_0 is the particle mass, Q is the droplet charge, Φ_B is the energy of the droplet gained between the nozzle and the extractor electrode divided by its charge.

The ion current density emitted in the process of ion evaporation from a liquid surface subjected to a sufficiently strong electric field has been determined from the Iribarne and Thompson [1976] theory (Gamero-Castaño and Hruby [2001], Gassend et al. [2007]):

$$j = \frac{\sigma_q kT}{h} \exp\left(-\frac{q_i \phi - \sqrt{\dfrac{q_i^3 E}{4\pi\varepsilon_0}}}{kT} \right) \tag{84}$$

where k is the Boltzmann constant, T is the absolute temperature, h is the Planck constant, ϕ is the free energy of ion evaporation (in eV), E is the electric field, q_i is the charge of the emitted ions, σ_q is the surface charge density.

The specific impulse for charged droplets emitter is (Gamero-Castaño and Hruby [2001], Velasquez-Garcia et al. [2006a]):

$$I_{sp} = \frac{1}{g}\sqrt{2U\frac{Q}{m}} \tag{85}$$

Because the specific charge Q/m decreases with the flow rate increasing (Velasquez-Garcia et al. [2006a]):

$$\frac{Q}{m} = \frac{\alpha}{\rho_l}\sqrt{\frac{\sigma_l \kappa_l}{\varepsilon_0 \varepsilon_r Q_V}} \tag{86}$$

the flow rate should be decreased down to the value at which steady-state operation can be sustained, as determined by Fernandez de la Mora and Loscertales [1994]:

$$Q_{V\min} = \alpha \frac{\varepsilon_0 \varepsilon_r \sigma_l}{\rho_l \kappa_l} \tag{87}$$

where α is the dimensionless parameter for which a stable cone-jet operation is sustained. It is the ratio of the inertial force to the capillary force evaluated at the point where the electric field is the maximum (at the cone apex) and was determined experimentally to be nearly equal to 1.

Externally-wetted microneedle arrays (Figure 41c) (Castro and Fernandez de la Mora [2009], Gassend et al. [2007, 2008], Velasquez-Garcia et al. [2006b]) have been developed for ionic liquid emission and to avoid capillary clogging. An advantage of externally wetted ion emitter over capillary emitters is the lower flow rate achievable and lack of unwanted droplets emission at these flow rates. The control of ion emission can be accomplished via adjusting the emitter-extractor voltage, and also via the emitter temperature, which influences the liquid viscosity (Lozano and Martinez-Sanchez [2005a,b], Romero-Sanz et al. [2003],

Velasquez-Garcia et al. [2006b]). For example, the viscosity of EMI-BF_4 propellant can be reduced to about half and the ionic current of emitter can increased by a factor of about 2 for a temperature increasing from 25 to 50°C. This effect suggests that the ion emission is limited by the viscous transport of charge carriers, which is inversely proportional to the liquid viscosity (Lozano and Martinez-Sanchez [2005a,b], Velasquez-Garcia et al. [2006b]). The angular dispersion of the ions emitted by ILIS from the axis of an emitter is lower than 18-20° (Lozano and Martinez-Sanchez [2004], Velasquez-Garcia et al. [2006b]).

Another emitters were also proposed and tested. For example, Lozano et al. [2004] have proposed an emitter in a form of small holes bored in a dielectric, nonwetting material (polyethylene or PTFE). Electrostatic modeling has shown that for an emitter material of low dielectric constant, the local electric field near the emission site is enhanced in a very similar way as with sharp metallic needle, and helps anchoring the Taylor cone to the edge of the hole. By the microfabrication technique, an array of large numbers of individual emitters of this configuration can be produced. Castro and Bocanegra [2006] have used co-extrusion electrospray nozzle for testing the feasibility of water as propellant for colloid thruster. The advantage of using water as a propellant is its longer ion evaporation delay and relatively high conductivity enabling higher specific impulses and propulsion efficiency. Additionally, high surface tension of water enables obtaining higher charges and lower m/q ratio. In order to decrease the evaporation rate of water droplets their surface was covered with thin layer of oil flowing from the outer nozzle. Krpoun and Shea [2007] and Krpoun et al. [2008] have built a porous structure by filling a silicon capillary with 5 μm silica beads. Via this technique, the pressure drop across the capillary can be increased and the flow rate can be reduced to a value below 1 nL/s, and a pure ionic emission from the capillary can be obtained at these flow rates.

Traditionally, cone-ended needles are used as ion emitters in colloid thruster. Velasquez-Garcia et al. [2006b] have compared two shapes of ion emitter: with sharp tip ("pencil") and flat-ended ("volcano"), which were arranged in a planar array of 32×32 on silicon wafer. The volcano emitter had eight sharp corners at the top surface, and the propellant was emitted from each one of these corners. Authors demonstrated the feasibility of obtaining substantially larger emission currents at the same extraction voltage by controlling the propellant temperature.

Nozzleless electrostatic colloid thruster has been developed by Song and Shumlak [2008]. In this device, charged droplets were generated from a crest of capillary standing waves excited with piezoelectric transducer in a thin liquid film. The charge was induced on the detaching droplets by an electric field produced by an extractor electrode. The vibration amplitude is kept at the level required for the critical stable condition, i.e., the threshold of atomization inception. The electrostatic pressure caused by an electric field leads to an instability of the capillary waves and thereby produces charged droplets. The capillary waves in unstable conditions allows reducing the electric field required to extract charged droplets from the wave's crest and a more uniform q/m distribution in the electrospray was expected by the authors. The ion evaporation from the liquid meniscus is unlikely to take place. The current density carried by the electrospray can be higher due to high surface number density of the emitting points because a typical wavelength is about 10 μm. The capillary clogging will be avoided. The authors noticed that the size of the droplets decreases when the electric field is applied.

The propellants used for space vehicle propulsion should be characterized by low volatility, low viscosity, and high electrical conductivity in order to obtain optimal *m*/*q* ratio (Castro and Bocanegra [2006]). Low values of *m*/*q* imply high specific impulse but at the same time the power consumption per unit thrust increases. Broad statistical distribution of *m*/*q* values results in low propulsion efficiency (Castro and Bocanegra [2006]). The mean values of *m*/*q* and its narrow statistical distribution can be attained for propellants of high conductivity and low viscosity. Too high liquid conductivity, however, causes ion evaporation that decreases the propulsion efficiency (Castro and Bocanegra [2006], Gamero-Castaño and Hruby [2002]). Low volatility of a propellant is required because the liquid evaporated from the Taylor cone does not produce any thrust (Castro and Bocanegra [2006]).

Glycerol or other organic liquids of low vapour pressure and high surface tension (0.065 N/m for glycerol) that allow a stable interface in vacuum were usually tested as propellants. Glycerol is doped with NaCl, NaI, H_2SO_4, (for positive ion emission) and $FeCl_3$, $MgCl_2$, $SnCl_4$, or $SbCl_3$ (for negative ions) in order to adjusting liquid conductivity. When NaI is used as depend, such complexes as, for example, Na^+, $Na^+{+}C_3H_8O_3$ and $Na^+{+}2C_3H_8O_3$ are detected in the beam while the remaining I^- ions are neutralized at the capillary surface (Huberman et al. [1968], Huberman [1970]). Glycerol is also a good polar solvent allowing dissolving salts or other electrolytes. Strong dependence of glycerol viscosity on the temperature makes it possible to control the spraying process by changing the liquid temperature. Glycerol has also low aggressiveness to the materials used for thruster construction. The colloid thrusters operate in the cone-jet, microdripping or multijet modes (Huberman [1970], Kidd [1968], Krohn [1963]). At negative polarity of the emitter, erosion of capillary was seen after 200 hours of operation (Kidd [1968]).

Liquid metals like Bi+Pb+Sn+Cd (Woods metal), Bi+Pb+Sn (Krohn [1961]), Bi+Pb (Krohn [1974]), Ga+In eutectic alloys (Swatik and Hendricks [1968]), In (Fehringer et al. [1998]), or Cs (Franks et al. [1998]) were also used as propellants, allowing higher thrust than glycerol due to its high atomic weight. Cs has also low melting point, low ionization potential and good wettability to other metals from which a nozzle is made. The disadvantage of caesium is its high chemical reactivity. Single and doubly charged atoms and single charged clusters of two to ten atoms can be detected in a liquid metal beam (Krohn [1964]).

From the beginning of the 21st century, alternative sources of ions used as propellants for space missions, replacing the liquid metal ion sources, have also been developed and tested. Ionic liquids are salts formed with large organic cations and anions that remain in the liquid phase at or near room temperature. Ionic liquids ion sources (ILIS) are used as propellants, due to their high conductivity, lower surface tension than LMIS, capability of wetting a solid metal emitter at room temperature, and relatively low vapor pressure. Lower surface tension requires lower voltages for ion emission, while lower melting point and high wettability allow operation of ILIS at room temperature. The advantage in using ionic liquid instead of LMIS is that ILIS can operate in bipolar mode, emitting positive cation or negative anion beams, while LMISs are restricted to operating as positive metal-ion beams (Lozano and Martinez-Sanchez [2005a,b]). This property opens a possibility of periodic changes of emitter polarity for efficient charge neutralization of vehicle. Besides, the number of ionic liquids available is larger than metals and their alloys of low melting point required for efficient ion emission.

The most frequently tested ILIS propellants, for example, are: formamide doped with LiCl, NaI, (Gamero-Castaño and Hruby [2001], Lozano and Martinez-Sanchez [2000], Velasquez-Garcia et al. [2006a]), thributyl phosphate + NaI (Gamero-Castaño and Hruby

[2001]), TEG+NaI (Alexander et al. [2006], Paine [2009], Paine et al. [2007], Stark et al. [2005]), Bmim-BF_4 (Lozano and Martinez-Sanchez [2004, 2005b], or BF_4 dissolved in 1-ethyl-3-methylimidazolium tetrafluoroborate (EMI-BF_4) (Gamero-Castaño [2004], Gassend et al. [2007, 2008], Krpoun and Shea [2007], Krpoun et al. [2008], Lozano and Martínez-Sánchez [2004, 2005a], Velasquez-Garcia et al. [2006b, 2007]). For positive and negative polarity of an emitter, the EMI^+ and (EMI-BF_4)EMI^+, and BF_4^- and (EMI-BF_4)BF_4^- ions are emitted, respectively (Romero-Sanz and Fernandez de la Mora [2004]). The EMI-BF_4 ions, similarly to LMIS (Mair et al. 1983]), are emitted with energies in eV close to the potential of the emitter, and the energy deficit was estimated only to about 2.8% (Lozano and Martinez-Sanchez [2005a,b]). A disadvantage of ILIS is its contamination of the emitter electrode or electrochemical reaction with the electrode material by ions opposite to those extracted from the emitter. The liquid propellant and/or electrode can be degraded decreasing the performances and lifetime of the emitter. This problem has been solved by alternating electric field excitation of the emitter with relatively low frequency (Lozano and Martinez-Sanchez [2004, 2005a,b]).

4.6. Charged Sprays in Agriculture

Chemicals used in agriculture such as pesticides, herbicides, fungicides or fertilizers, which are used in the form of aerosols for crop protection can also be sprayed as charged droplets. The charged droplets flow to the plant leaves due to the repulsive force of the electric field generated by the space charge of the droplets and the image force of the charge induced on the leaves' surface. The electrostatic repulsive force of the cloud of charged droplets improves also the spatial dispersion of the aerosol over the crop.

The history of crop spraying with electrified aerosols began in 1950s at the Michigan State University where dusty pesticides were deposited onto plants (cf. Law [2001]). One of the first experiments on crop spraying with charged-liquid aerosols was that conducted by Göhlich [1958], who charged the spray in a corona discharge. The spraying of charged droplets allowed increasing the deposition efficiency, measured as the number of droplets per unit surface, up to 10 times, as compared to uncharged sprays. In 1960s Law and Bowen carried out research on liquid pesticides spraying with charging by induction (Law [2001], Law and Bowen [1975]).

Pesticides dispersed by conventional techniques are in a great fraction lost because of drift of the droplets away from the crop field by wind or settling mainly on the soil due to the gravity. Uncharged droplets falling due to gravity can cover only the adaxial (upper) side of the leaves, and therefore the deposition efficiency of pesticide sprayed by conventional methods is usually lower than 25% (Law [1983]). By windy weather about 50 to 70% of the spray drifts off the plant (Laryea and No [2003]).

Spraying of charged aerosol over a crop allows increasing the efficiency of insects control because charged droplets are deposited not only at the upper side of the leaves but also can penetrate their abaxial (underside) surface, where the insects usually reside. In the case of electrostatic crop spraying, it is essential that the droplets be smaller than 100 μm. For larger droplets the gravitational force is larger than the electrostatic force, which drives the droplets towards the leaves, and the electrostatic effects are negligible. Smaller droplets fall with low velocity and deflection of their trajectories is easier. From numerical modeling of charged droplets' trajectories near an object results that charged sprays containing small droplets can be deposited more effectively on the object due to the space charge and image forces and thus reduce the amount of pesticide use by about 50% (Zhao et al. [2008]). The results have also shown that the deposition efficiency increases with the charge-to-mass ratio increasing, however, the radial drift of droplets may also increase.

Due to mutual Coulomb repulsion of charged droplets in a cloud, the droplets located at the upper part of the cloud could expand upwards and hence some of the smaller droplets would drift. In order to counteract the drift of insecticide-loaded droplets, Inculet and Castle [1985] have generated two-layer charged aerosol cloud in which the upper half of a charged cloud has contained clean-water droplets and the lower half the insecticide aerosol. By this way only the clean water, which is environmentally neutral, would drift off the cloud, whereas the active insecticide or pesticide droplets are repelled towards the plant.

Transport and deposition of electrically charged sprays on a leaf is dependent on the charge on the droplets (the charging method and voltage), spray system velocity over the plants, the crop structure, target height and orientation relative to the spray, wind velocity, electric properties of the plant (conductivity of stem and leaves), and transport mechanisms of the droplets to the plant (Law [2001], Maski and Durairaj [2010a,b], Zhao et al. [2008]). The nozzle-to-target distance and the distance to ground are critical in electrostatic spraying. Large nozzle-to-target distances may cause loss of charged droplets because of the space charge effects (Zhao et al. [2008]). The conductivity of the soil plays also a crucial role in spray transfer to a plant. The charge accumulated on the low-conducting surface prevents deposition of the droplets on the ground and repels the droplets back to the more conductive leaves, upwards against gravity. In laboratory tests it was proved that charged-spray deposition efficiency increases when the soil is of low conductivity (Giles and Law [1990]). In a plant with dense leaves, charged spray penetrates mainly upper part of the canopy, and for the improvement of spraying process, covering also the lower-located leaves, an assistance of airblast is usually necessary (Law and Cooper [1988]).

Maski and Durairaj [2010a] have investigated the effect of induction charging voltage, velocity of the spray system, target height and orientation on the efficiency of charged spray deposition upon leaf abaxial and adaxial surfaces. The authors concluded that increasing specific charge of the droplets (charging voltage, 0-4 kV) causes an increase of deposition efficiency on abaxial surface of leaves, for medium spray system velocities (0.417 m/s). For lower system velocities (0.278 m/s), the adaxial deposition decreases with increasing charging voltage. The deposition efficiency on adaxial and abaxial surfaces is nearly the same for the maximum specific charge of the droplets, >5 mC/kg, and is in the range from about 1 to 1.5 μg/cm^2. The medium velocity with higher charging voltage was recommended by the authors as optimal for abaxial and adaxial deposition.

The magnitude of charge-to-mass ratio (q/m) is of primary importance in electrostatic aerial spraying. For small values of q/m, <0.3 mC/kg, there is only marginal effect of charging on spray deposition (Inculet and Fischer [1989]). For the q/m ratio increasing from 2 mC/kg to 8.2 mC/kg the efficiency systematically increases from 1.4 to 4.4 fold, compared to conventionally atomized uncharged sprays (Law and Lane [1981], Kirk [2003]). Large deposition efficiency has been observed for the q/m values larger than 3.8 mC/kg (Kirk [2003]).

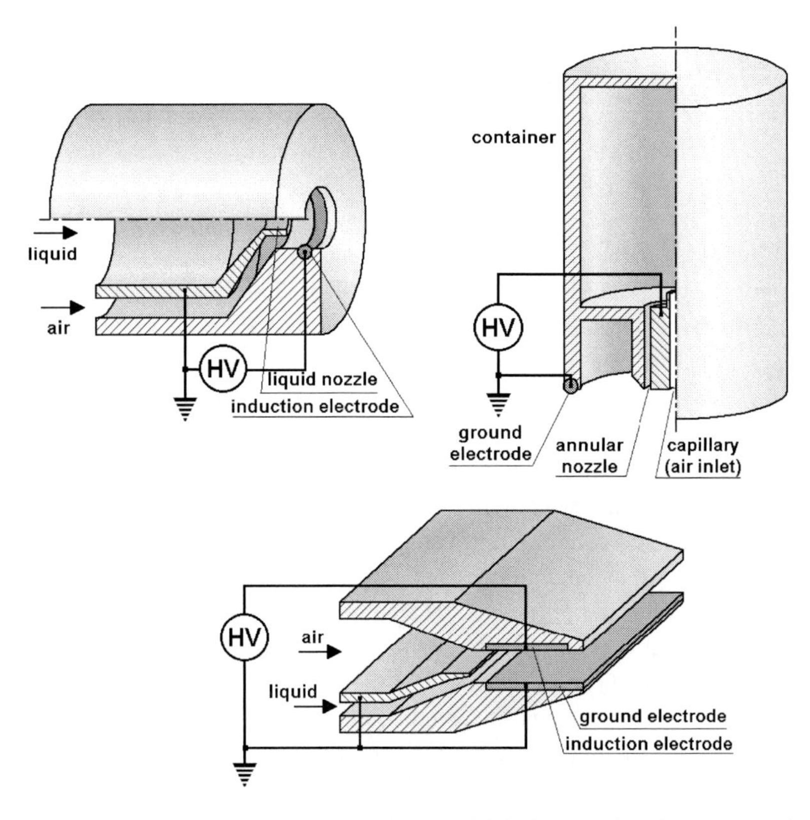

Figure 42. Agriculture atomizers, a. pneumatic atomizer with induction charging (Law et al. [1997]), b. hand-held electrostatic atomizer with annular-slit nozzle (Coffee [1981]), c. flat-nozzle pneumatic atomizer with induction charging (Bologa and Makalsky [1989], Marchant and Green [1982]).

The maximal value of q/m can be obtained with an electrospraying nozzle. However, crop spraying requires large quantity of pesticide, which cannot be sprayed by a single electrospray nozzle, annular slit nozzle (Coffee [1981]) or even a set of such nozzles. For this reason, pneumatic (Bologa and Makalsky [1989], Law [2001], Law et al. [1997], Machowski et al. [1998], Machowski and Balachandran [1997], Marchant and Green [1982], Maski and Durairaj [2010a,b]), rotary (Asano [1986], Bayat and Bozdogan [2005], Cooper et al. [1998]), pressurized (Kirk [2003], Latheef et al. [2009], Law and Scherm [2005], Metwally et al. [1995], Moon et al. [2003], Scherm et al. [2007]), or pressure-swirl-nozzle (Laryea and No [2003, 2004]) atomizers equipped with induction electrode are usually used.

Example of various types of crop-spraying atomizers are shown in Figure 42. A pneumatic atomizer equipped with an electrode for electrostatic induction charging is presented in Figure 42a. Liquid is fed to the central nozzle from a grounded reservoir.

The air flows with near-sonic velocity through the annular-slit nozzle. The shear stress on the liquid jet causes atomisation of the liquid to fine droplets. A ring induction electrode placed around the water nozzle induces an electric charge on the liquid surface, which is next taken off by the droplets. In a device of this type, a relatively low potential, a few hundred volts, is needed in order to induce the charge. The flow of the air keeps the induction electrode dry that prevents short-circuits between this electrode and the water jet (Law [1983]). In order to eliminate current leakage from the induction electrode to the grounded nozzle or prevent corona discharge, the electrode is isolated with a dielectric cylinder or coated with a varnish (Machowski et al. [1998], Moon et al. [2003]). With the induction electrode placed outside an insulating cylinder ("capacitive electrostatic spraying nozzle"), which was supplied with pulsed voltage (AC sinewave of 7.5 kHz, with DC bias at the level of AC voltage amplitude), the leakage current was eliminated and the crop spraying efficiency of was further increased (Moon et al. [2003]).

The investigations carried out on a plant-field, with pneumatic atomizers indicated two- to seven-fold increase in deposition efficiency for charged droplets as compared to similar uncharged spray (Law [1983, 1995]; Law et al. [1985], Law and Cooper [1988], Law and Scherm [2005], Scherm et al. [2007]). The size of the droplets was 30-50 μm VMD and their specific charge up to 10 mC/kg. Therefore, only 1/2 down to 1/7 amount of pesticides recommended for the control the insects by conventional spraying methods can be used with the same end effect. This type of atomizer can operate simultaneously with air blast assistance for increasing canopy penetration (Law and Cooper [1988]).

In a hand-held, integrated nozzle-container system utilizing electrospraying for pesticide atomization, developed and patented by Coffee (Coffee [1981], Bailey [1986]) (Figure 42b), the pesticide flows from the container through narrow annular-slit nozzle at the bottom of the container under the hydrostatic pressure. The charge of the electrosprayed droplets is 50-75% of the Rayleigh limit. This type of atomizer offers low energy consumption, and it can be supplied from solar batteries. Because smaller droplets can be generated by the electrospraying than by conventional mechanical atomizers, the insect control efficiency can be higher than for other methods. The atomizer was tested on cotton fields (Coffee [1981]), and about 90% deposition efficiency was obtained, that was 2.5 times higher than with uncharged droplets. The amount of chemicals sprayed by this method can be reduced from 2-4 L/ha down to 0.5 L/ha with similar effects in pest control (Coffee [1981]).

The linear-slit nozzle atomizer with induction charging is equipped with two parallel-plate electrodes (Figure 42c) (Bologa and Makalsky [1989], Marchant and Green [1982]). The liquid flows with high velocity through the linear-slit nozzle and is spread over the grounded electrode while the induction electrode is maintained at high potential. The charge induced on the liquid surface is taken-off during the liquid film disintegration by the air flowing out the upper nozzle. Experiments indicated that charging the spray enhances the droplet deposition efficiency by two to four times, and the droplets distribution on the leaves was more uniform than those sprayed without charging (Marchant and Green [1982]). The flow rate of water-based pesticides sprayed by pneumatic atomizer with a slit nozzle can be as high as 100 kg/h (Bologa and Makalsky [1989]).

Spinning disk atomizers with droplets charging were also applied for pesticides spraying. Cooper et al. [1998] used spinning disk atomizer, and the ratio of number of droplets deposited on leaves for charged and uncharged sprays varied between 1.3 and 2.9, and volume ratio was 0.84 to 3.62. Charged droplets were easily deposited underleaf.

Water or vegetable oils (cottonseed or soybean oil) are usually used as carriers for pesticides. Because, however, vegetable oils have low electrical conductivity (10^{-11} - 10^{-12} S/m), antistatic agents are used for enhancing their conductivity in order to charging the spray by induction. Optimal conductivity of liquids charged by induction is in the range between 10^{-4} and 10 S/m (Law [1983]). All aqueous-based pesticides fall into this range without special preparation. Satisfactory results can also be obtained for water-in-oil emulsions, with 3% of water in cotton oil, or 6% in soybean oil (Law and Cooper [1987]).

Unwanted corona discharge from sharp leaf tips is a serious problem in electrostatic crop spraying. The droplets are recharged by opposite ions generated near the leaf tip and repelled off the plant (Cooper and Law [1987]). In order to minimize the detrimental effects caused by the corona discharge, bipolarly charged sprays are used (Cooper and Law [1987], Law et al. [1999]). In a device developed by Cooper and Law [1987], the polarity of the aerosol was changed with the frequency of up to 4 Hz, in order to obtain spray cloud of zero net space charge at long distances, however, no significant increase in the spray deposition efficiency, as compared to conventional uncharged spray, was noticed.

The main applications of charged sprays in agriculture are the trailer spraying of orchards, aerial spraying of crop fields, and ramp or hand-held spraying inside greenhouses or barns (Inculet [1998]). Spraying of orchards is assisted with air blower, which generates airflow with velocity of the order of magnitude of 300 km/h to facilitate charged droplets penetration through the canopy.

Aerial dispersion of electrically charged sprays for the purpose of crop spraying requires neutralization of the charge accumulated on the aircraft. Two nozzles placed at both sides of the airplane, generating simultaneously charged sprays of opposite polarities are used for these purposes (Inculet [1998], Inculet and Fisher [1989]). An electrode generating corona discharge removes an excess charge. At the altitudes the aircraft normally flies in aerial spraying, the charged sprays move preferentially towards the crop rather than to attract each other. The aerial electrostatic atomizers provide droplets with the specific charge of up to 1.5 mC/kg, but it is still too low for effective crop spraying (Law [2001]).

Kirk [2003] and Latheef et al. [2009] have investigated the efficiency of aerial spraying of crops using two linear arrays of nozzles, one of which generated positively charged droplets, and one negatively charged. Charged sprays exhibited increased deposition on plants and reduced drift. The effect of charging of droplets has, however, been marginal when the speed of the airplane was too high. When the nozzle arrays generating oppositely charged sprays are located close to each other, the adjacent, oppositely charged fine droplets can either coalesce into larger droplets, reducing their charge, or merge with larger droplets. In both cases, the charge is neutralized and the fine droplet component of the spray spectrum is reduced (Kirk [2003]).

For greenhouse applications, smaller droplets, of the size of about 10 μm, are desired in order to increase the time of deposition up to 20 min (Machowski et al. [1998]). This time is required for the droplets to penetrate the canopy due to the Brownian motion that facilitates droplet deposition on the under-sided of the leaves due to the Coulomb force, and decreases the volume of pesticide required for the same final effect.

Generation of charged aerosol for crop spraying allows increasing the efficiency of pest control due to smaller size of droplets, and their deposition on the down-side of the leaves, were uncharged droplets do not penetrate. Electrical forces increase mass-transfer efficiency of pesticide onto plant surfaces. Long-term investigations shown that the residual deposits on the leaves, measured after 2, 5 and 8 days, were in average 35% larger than for mechanically deposited uncharged sprays (Inculet [1998]). Induction charging of aerosol is more practical for agriculture applications due to higher flow rates available. The droplets should be of the size of 30-50 μm volume median diameter, because for this size, the electrostatic force dominates over gravity. The specific charge of the spray should be at least 2 mC/kg for effective canopy penetration (Law [2001]). Any electric fields should be avoided to minimize the leaf tip corona and excessive deposition onto the outer canopy. The increased deposition efficiency has not only economic advantages but also is environment-friendly because smaller insecticide volume is required for the same pest-control effect.

4.8. Electroscrubbing for Gas Cleaning

The removal of particles in the size range from 0.01 to 2 μm, known as the 'Greenfield gap', from indoor or industrial gases presents still a serious problem, and an effective control of such particles remains still a great challenge for engineers. Particles of this size, to which smoke, soot, fine solid powders, oil vapours, spores, viruses or bacteria belong, are particularly hazardous for human lungs, because they penetrate the lower airways. Conventional devices used for the gas cleaning, such as cyclones, inertial wet scrubbers, fibrous filters or electrostatic precipitators remain still ineffective for such particles. In cyclone, the motion of fine particles is mainly affected by drag and molecular forces, which are stronger in magnitude than centrifugal force. Tighter fibrous filters can help control finer particles but they cause higher pressure drop. In inertial wet scrubbers, the inertial force causing the particle deposition on the droplet plays a diminishing role with decreasing particle size. Nozzle or Venturi scrubbers can offer higher collection efficiency in the submicron size range, but require highly dispersed liquid droplets of high velocity, that consumes more energy due to higher pressure drop. Water consumption in nozzle scrubber is about 0.05 l/m^3, and in Venturi scrubbers is in the range of 0.5-1.5 l/m^3 (Gemci and Ebert [1992]). Electrostatic precipitators require higher energy consumption for effective particle charging in submicron range, in order to remove them with sufficiently high efficiency, comparable with larger particles. Re-entrained of fine particles from the collection electrode, which was observed in electrostatic precipitators in submicron range, additionally decreases the collection efficiency.

Device, which could do this task, is the wet electrostatic scrubber (electroscrubber), which combines advantages of electrostatic precipitators and wet inertial scrubbers via employing electrostatic forces between charged fine particles and droplets. In electroscrubbers, an electrically charged spray scavenges the gas with the particles flowing through the chamber. Wet electrostatic scrubbers are recommended for exhaust or industrial gas cleaning, after the last stage of conventional electrostatic precipitator, as well as for indoors gas cleaning in small scale air-conditioning equipments.

Three main constructions of electrostatics scrubbers can be distinguished:

1. *Vertical electroscrubber* (Figure 43) in which the gas flows upwards or downwards through the scrubber chamber, and charged spray generated at the upper part of the chamber.
2. *Horizontal electroscrubber* (Figure 44) in which the gas flows horizontally, and the charged spray crossing this gas flow,
3. *Venturi-type electroscrubber* (Figure 45) positioned vertically or horizontally, with charged spray injected into the throat of the Venturi nozzle.

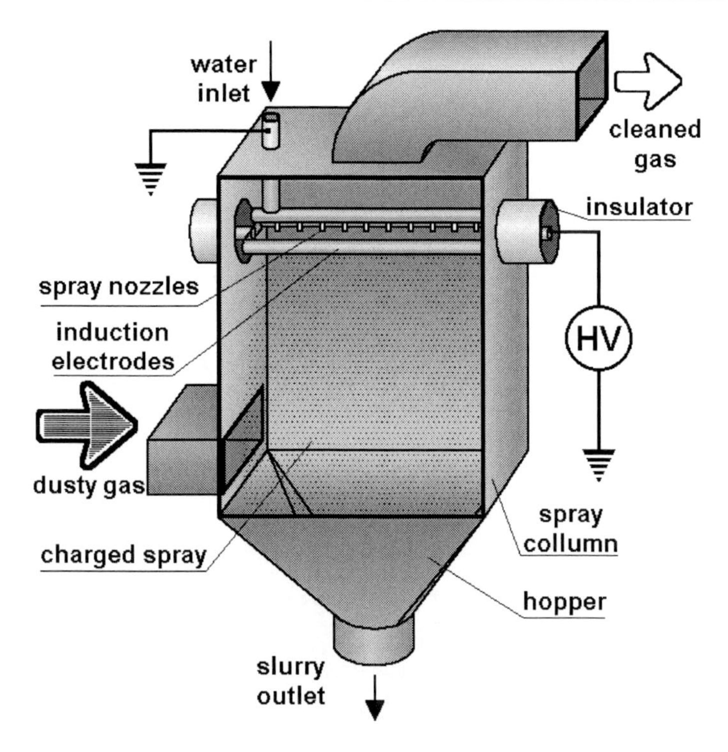

Figure 43. Vertical electroscrubber.

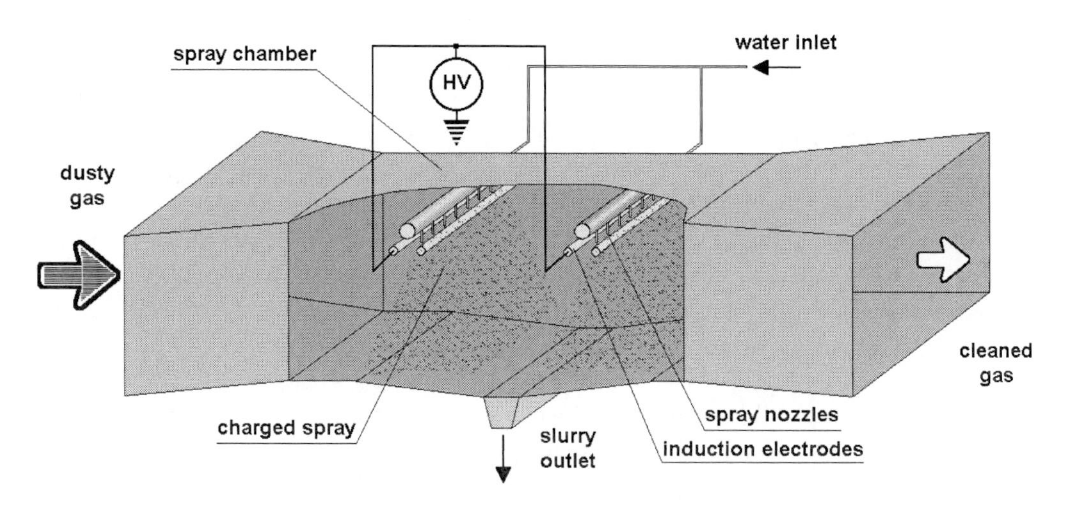

Figure 44. Horizontal electroscrubber.

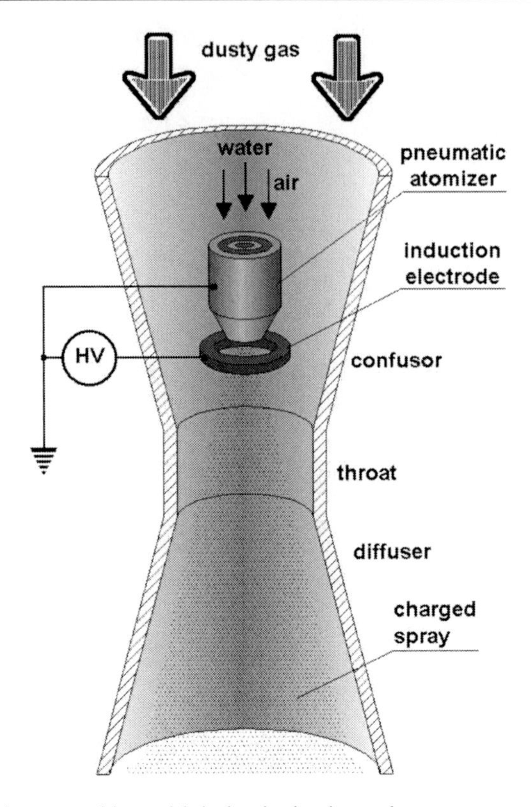

Figure 45. Venturi-type electroscrubber with inductively charged spray.

Pressure or rotary atomizers with induction charging, or electrospray nozzles are used in vertical and horizontal electroscrubbers because low velocity of droplets is required for high collection efficiency. In Venturi electroscrubbers, on the other hand, pneumatic atomizers with induction charging, generating droplets of high velocity, close to the dusty-gas velocity, are preferred, in order to obtain low velocity between the particles and droplets.

Wet electrostatic scrubber was first proposed and patented by Penney [1944], and later modified by many authors. In the electrostatic scrubbers, dust particles and scrubbing droplets are electrically charged to the same or opposite polarities. Five scrubbing systems based on various combinations of particle and droplet charge polarities are known:

1. *Opposite-polarity electroscrubber* (OPE); Spray of droplets and dust particles are charged to opposite polarities (Adamiak et al. [2001], Balachandran et al. [2003], Ha et al. [2010], Jaworek et al. [1996, 1998a,b, 2001, 2006], Kraemer and Johnstone [1955], Laitinen et al. [2006], Metzler et al. [1995, 1997], Penny [1944], Ricci [1977], Smith [1976], Zhao and Zheng [2008]). The charged droplets are dispersed in scrubber chamber due to space charge effect, and the oppositely charged dust particles are deposited onto the droplets due to the Coulomb attraction forces.
2. *Same-polarity electroscrubber* (SPE); Spray and dust particles are charged to the same polarity (Krupa et al. [2003], Metzler et al. [1995, 1997]). The repulsive force of the space charge of droplet cloud causes the particles to be precipitated on the chamber walls where they are washed out by the droplets (the particles are not deposited onto the droplets).

3. *Double-polarity electroscrubber* (DPE); Two equal sprays of droplets are charged, one with positive and one with negative polarity, and injected into electroscrubber chamber (Melcher et al. [1977]). The negatively charged particles are deposited onto positively charged droplets due to attractive Coulomb forces, and finally the positive droplets coagulate with the negatively charged ones.
4. *Charged-droplet electroscrubber* (CDE); Droplets are charged to either polarity while the particles remain uncharged (Balachandran et al. [2001a], Bologa [1996], Inculet et al. [1989], Kim et al. [2010], Melcher et al. [1977], Zhao and Zheng [2008]). The particles are deposited on the droplets due to the image charge induced on the particle and dielectrophoretic force on the particle in inhomogeneous electric field close to the droplet..
5. *Charged-Particle electroscrubber* (CPE); The particles are charged to either polarity and the droplets are uncharged (Ha et al. [2010], Laitinen et al. [2006], Melcher et al. [1977], Ricci [1977], Sheppard [1985], Zhao and Zheng [2008]). The particles are deposited on the droplets due to the image charge induced on the droplet.
6. *Charge-transfer electroscrubber* (CTE); The particles are uncharged and the droplets are charged to either polarity (Inculet et al. [1989], Tepper et al. [2007]). The evaporating droplets produce free ions or charged water clusters, which condense onto the smoke particles. The smoke becomes charged, expands and is deposited onto the chamber walls due to the space charge force.

The particles to be removed can be charged in a corona-discharge charger of DC or AC type, or sometimes by tribocharging during their pneumatic transportation through a channel. In order to maximize the particle charge and to reduce precipitation of the particles on the charger electrodes, two types of chargers were proposed and tested in the literature: the Masuda boxer charger (Masuda [1984], Masuda et al. [1978a,b]), and Jaworek and Krupa alternating electric field charger (Adamiak et al. [1995], Jaworek and Krupa [1989], Jaworek et al. [2001], Krupa et al. [2005], Lackowski [2001]). On the particle charging techniques and considerations on charging mechanisms, readers are referred to the comprehensive studies of Chang et al. [1995 (Chapter 3)], and Biskos et al. [2005a,b].

A trajectory of charged dust particle of mass m_p near a charged droplet (collector) is given by the following Newton vector differential equation:

$$m_p \frac{\mathrm{d}\vec{w}}{\mathrm{d}t} = \vec{F}_S + \vec{F}_e + \vec{F}_B + \vec{F}_T + \vec{F}_D + m_p \vec{g} \tag{88}$$

where m_p is the particle mass, $\vec{w}$ is the particle velocity, $\vec{F}_S$ is the aerodynamic drag, $m_p g$ is the gravitational for on the particles, and $\vec{F}_e, \vec{F}_B, \vec{F}_T, \vec{F}_D$ are the electrical, Basset, thermophoresis, and diffusiophoresis forces, respectively. The left-hand side of Eq. (88) is the inertial force on the particle.

Diffusiophoresis and thermophoresis are important only for small dust particles (<0.1 μm to 1 μm) and lower relative velocities (Deshler [1985], Viswanathan [1999]). Particles smaller than <0.01 μm in diameter can be deposited on the collector mainly due to the Brownian motion (Lear et al. [1975], Wang [1983], Shahub and Williams [1988]), whereas the relative velocities are of minor importance in this type of deposition. For particles larger than 2 μm, the Basset, thermophoresesis, and diffusiophoresis forces can be neglected because the particle motion is governed mainly by the aerodynamic drag F_d, electrical F_e and gravitational $m_p g$ forces.

The motion of a charged droplet within the spray can be described by similar equation:

$$m_c \frac{\mathrm{d}\vec{w}}{\mathrm{d}t} = \vec{F}_S + \vec{F}_e + m_c \vec{g} \tag{89}$$

In this equation only the drag, electrical, and gravitational forces on the particle are considered.

The electrostatic force on the particle has three components:

$$F_e = -\frac{Q_p Q_c}{4\pi\varepsilon_0 r^2} + \frac{R_p Q_c^2}{4\pi\varepsilon_0 r^3} \frac{\varepsilon_p - 1}{\varepsilon_p + 2} \left(\frac{r^4}{(r^2 - R_p^2)^2} - 1 \right) + \frac{R_c Q_p^2}{4\pi\varepsilon_0 r^3} \frac{\varepsilon_c - 1}{\varepsilon_c + 2} \left(\frac{r^4}{(r^2 - R_c^2)^2} - 1 \right) \tag{90}$$

where: Q_p and Q_c are the charges on the aerosol particle and the droplet, respectively, R_p and R_c is the radius of the particle and the droplet, respectively, r is the distance between the particle and the droplet centers, ε_0 is the permittivity of the free space, ε_p, ε_c are the relative permittivities of the particle and the collector, respectively, and ρ_p is the particle mass density.

The first term in Eq. (90) is the Coulomb force between two point charges placed in the particle and the collector centers. The second term is the force on the particle due to the image charge induced on the particle by the charged collector. The third term is the force caused by the charge induced on the collector by the charged particle. The image forces in (90) are reduced only to the first two terms of much complex infinite recurrent series. For two spherical objects, the image forces act along the line between the objects' centers. The effect of image forces strongly depends on the distance between the particle and collector, and is only important when both objects are close to each other. For particles smaller than 3 μm, the image forces only slightly participate in the total force between the droplet and the particle (Jaworek et al. [2002], Tripathi and Harrison [2002]), and the collection efficiency for such particles and uncharged droplet sharply decreases. The image forces are important for the deposition of larger dust particles even if the particles and the droplet are charged with the same polarity (Jaworek et al. [2002], Tinsley et al., [2000]).

The aerodynamic drag force on the particle is:

$$\vec{F}_d = \frac{\pi C_d}{2C_c} \rho_g R_p^2 |\vec{u} - \vec{w}| (\vec{u} - \vec{w}) \tag{91}$$

where u is the gas velocity far from the collector, and ρ_g is the gas density. C_c is the Cunningham slip correction factor (Cunningham [1910], Rader [1990]), C_d is the drag coefficient.

Effectiveness of the removal of dust particles from gas is defined by the collection efficiency, which is the ratio of particle number concentration or their total mass penetrating a cleaning device, regardless of its type, to the number concentration or mass of particles at the device inlet. We have, therefore, number

$$K = 1 - \frac{n_{out}}{n_{in}} \tag{92}$$

where: n_{out} and n_{in} are the particle concentrations at the outlet and the inlet of the cleaning device, respectively.
and mass collection efficiency:

$$K_m = 1 - \frac{m_{out}}{m_{in}} \tag{93}$$

where: m_{out} and m_{in} are the particle concentrations at the outlet and the inlet of the cleaning device, respectively.

The parameter, called "penetration", is defined as:

$$\lambda = 1 - K = \frac{n_{out}}{n_{in}} \tag{94}$$

In the case of scrubbers, the concept of 'collision efficiency' refers also to a single, isolated collector (drop), and frequently is called the 'single-drop collection efficiency', or 'collision efficiency'. The collision efficiency is the ratio of the number N_p of particles deposited on the collector (droplet) to the number of particles N_s flowing through the projected area of the collector falling in the scrubber chamber:

$$K = \frac{N_p}{N_s} \tag{95}$$

For the electroscrubbing purposes, the spray can be charged with one of the following methods:

1. *Corona charging* (Bologa et al. [2001], Xu et al. [1998, 2003]),
2. *Induction charging* (Balachandran et al. [2001a, 2003], Bologa [1996], Chang et al. [1987], Metzler et al. [1995, 1997], Penney [1944], Pilat [1975], Pilat et al. [1974], Schmidt and Löffler [1992, 1994], Yang [2003]),
3. *Electrospraying* (Adamiak et al. [2001], Beizaie and Tien [1980], Ha et al. [2010], Hara et al. [1984], Jaworek and Krupa [1990], Jaworek et al. [1996, 1998b, 2003, 2006], Krupa et al. [2003], Lear et al. [1975], Wang et al. [1986]).

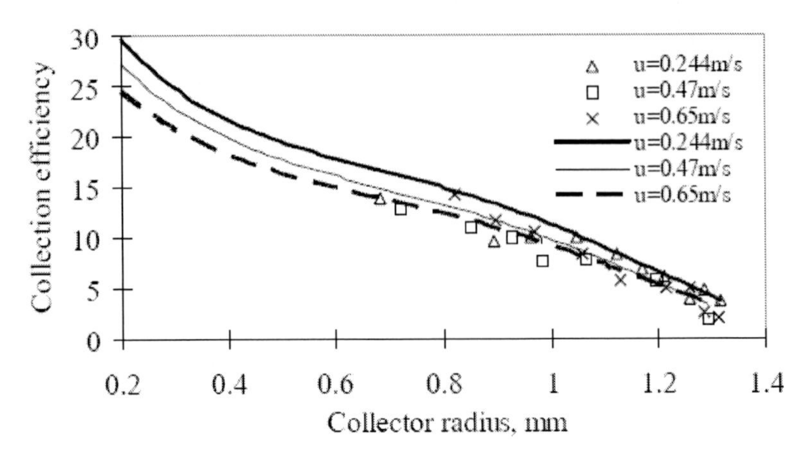

Figure 46. Collection efficiency of dust particles removal with charged spray (defined by Equation (28)) vs. droplet radius for various gas velocities (Jaworek et al. [1996]).

Charging of droplets in a corona discharge is not a very effective method for the electroscrubbing purposes because the charge acquired by a droplet is much lower than that charged by induction or obtained in the electrospraying. Another disadvantage of using the corona charging method is that the droplets are precipitated on the corona-electrode isolators. For these reasons the charging of sprays in corona discharge was not frequently used in electroscrubbing. The corona charger was used for droplets produced by pressure atomizer (Bologa et al. [2001]) or condensed steam (Xu et al. [1998, 2003]).

For induction charging of sprays in electroscrubbers, the pressure, pneumatic or rotary atomizers are used (Balachandran et al. [2001a, 2003], Bologa [1996], Metzler et al. [1995, 1997], Penny [1944], Pilat et al. [1974, 1975], Schmidt and Löffler [1992, 1994], Yang et al. [2003]).

From numerical simulation and experimental results it was concluded that the collection efficiency of electroscrubbers is higher for smaller droplets and lower relative velocities between the particles and droplets (Carotenuto et al. [2010], Deshler [1985], Dorman [1960], Jaworek et al. [1998b], Strauss [1971]). Vortices downstream of the collector, occurring for large Reynolds numbers can also participate in the particle deposition (Dau [1987], Dau and Ebert [1987]). The collection efficiency vs. the droplet size is shown in Figure 46. The collection efficiency for charged spray and particles can be many times higher than for uncharged species. The best cleaning results are obtained when the droplets and dust particles were charged to opposite polarities.

In Venturi scrubbers with charged droplets lower water consumption is required, and the collection efficiency is higher than in a scrubber with uncharged droplets (Bologa [1996], Yang et al. [2003]).

REFERENCES

Adamiak K., Jaworek A., Krupa A. *IEEE Trans. Ind. Appl.* 2001, 37, No.3, 734-750

Adamiak K., Krupa A., Jaworek A. *Inst. Phys. Conf. Ser.* No. 143, Bristol 1995, 275-278

Aduda B.O., Ravirajan P., Choy K.L., Nelson J. *Int. J. Photoenergy* 2004, 6, 141-147

Alexander M.S., Stark J., Smith K.L., Stevens B., Kent B. *J. Propuls. Power* 2006, 22, No.3, 620-627
Anderson E.K., Carlucci A.P., De Risi A., Kyritsis D.C. *Energy Conv. Manag.* 2007b, 48, 2762-2768
Anderson E.K., Carlucci A.P., De Risi A., Kyritsis D.C. *Int. J. Vehicle Design* 2007a, 45, Nos.1/2, 61-79
Anestos T.C. *IEEE Trans. Ind. Appl.* 1986, 22, No.1, 70-74
Anestos T.C., Sickles J.E., Tepper R.M. *IEEE Trans. Ind. Appl.* 1977, 13, No.2, 168-177
Anzue N., Itoh T., Mohamedi M., Umeda M., Uchida I. *Solid State Ionics* 2003, 156, 301-307
Arendt P., Kallmann H. *Z. Phys.* 1926, 35, 421-
Asano K. *J. Electrostatics* 1986, 18, 63-81
Azzopardi B.J., Hewitt G.F. *Multiphase Sci. Technol.* 9, No.2, 109-204
Bailey A.G. *J. Phys. D: Appl. Phys.* 1973, 6, No.2, 276-288
Bailey A.G. *Atomiz. Spray Techn.* 1986, 2, 95-134
Bailey A.G., Bracher J.G., Rohden H.J.von J. *Spacecraft* 1972, 9, No.7, 518-521
Bailey A.G., Cetronio A. *J. Phys. D: Appl. Phys.* 1978, 11, 123-127
Bailey A.G. *Electrostatic Spraying of Liquids*. J. Wiley and Sons 1988
Baker A.M., da Silva Curiel A., Schaffner J., Sweeting M., *Acta Astronautica* 2005, 57, 288-301
Balachandran W., Bailey A.G. *IEEE Trans. Ind. Appl.* 1984, IA-20, No. 3, 682-686
Balachandran W., Bailey A.G. *J. Electrostatics* 1981, 10, 189-196
Balachandran W., Hu D., Yule A.J., Schrimpton J., Watkins A.P. *IEEE Ind. Appl. Soc. Annual Conf., Denver*, 3-6 Oct. 1994, 1436-1441
Balachandran W., Jaworek A, Krupa A., Kulon J., Lackowski M. *J. Electrostatics* 2003, 58, No.3-4, 209-220
Balachandran W., Krupa A., Machowski W., Jaworek A. *J. Electrostatics* 2001a, 51-52, 193-199
Balachandran W., Machowski W., Ahmad C.N. *IEEE Trans. Ind. Appl.* 1994b, 30 No.4, 850-855
Balachandran W., Miao P., Xiao P. *J. Electrostatics* 2001b, 50, No.4, 249-263
Balasubramanian K., Jayasinghe S. N., Edirisinghe M. *J. Int. J. Appl. Ceram. Technol.* 2006, 3, No.1, 55-60
Bankstone C.P., Back L.H., Kwack E.Y., Kelly A.J. *Trans. ASME J. Eng. Gas Turbines Power* 1989, 110, No.7, 361-368
Barrero A., Lopez-Herrera J.M., Boucard A., Loscertales I.G.,Marquez M. *J. Colloid Interface Sci.* 2004, 272, 104-108
Barrero A., Loscertales I.G. *Annu. Rev. Fluid Mech.* 2007, 39, 89-106
Bartoli C., Rohden von H., Thompson S.P., Blommers J. *Vacuum* 1984, 34, No.1-2, 43-46
Basak S., Chen D.R., Biswas P. *Chem. Eng. Sci.* 2007, 62, 1263-1268
Bayat A., Bozdogan N.Y. *Crop Protection* 2005, 24, 951-960
Bayvel L., Orzechowski Z. *Liquid Atomization*. 1993, Taylor and Francis
Behrens D.A., Lee I.C., Waits C.M. *J. Power Sources* 2010, 195, 2008-2013
Beizaie M., Tien Ch. *Can. J. Chem. Eng.* 1980, 58, No 2, 12-24
Bellan J. *Combust. Flame* 1983, 51, No.1, 117-119
Bellan J., Harstad K. *Atomiz. Sprays* 1998, 8, No.6, 601-624
Bellan J., Harstad K. *JPL Techn. Rep.*, Monterey, California, Oct. 1997a

Bellan J., Harstad K. *JPL Techn. Rep.*, Seoul, South Korea, 18-22 Aug. 1997b
Bhushan B. (ed.) *Springer Handbook of Nanotechnology*. 2004, Spinger-Verlag Berlin, Heidelberg, New York
Biallas B., Stieber F. *Macromol. Symp.* 2002, 187, 731-737
Biskos G., Reavell K., Collings N. *J. Aerosol Sci.* 2005a, 36, 247-265
Biskos G., Reavell K., Collings N. *J. Electrostat.* 2005b, 63, 69-82
Bocanegra R., Galan D., Marquez M, Loscertales I.G., Barrero A. *J. Aerosol Sci.* 2005a, 36, 1387-1399
Bocanegra R., Gaonkar A.G., Barrero A., Loscertales I.G., Pechack D., Marquez M. *J. Food Sci.* 2005b, 70, No.8, E492-E497
Böhm J. *Electrostatic Precipitators,* 1982, Elsevier Sci. Publ.
Bollini R., Sample S.B., Seigal S.D., Boarman J.W. *J. Coll. Interface Sci.* 1975, 51, No 2, 272-277
Bologa A. *6th Int. Conf. Electrostatic Precipitation.* 18-21 June 1996, Budapest, 227-231
Bologa A. *Filtr. Separ.* 2001, 38, No.10, 26-30
Bologa A.M., Makalsky L.M. *J. Electrostatics* 1989, 23, 227-233
Borra J.P., Camelot D., Chou K.L., Kooyman P.J., Marijnissen J.C.M., Scarlett B. *Aerosol Sci.* 1999, 30, No.7, 945-958
Borra J.P., Camelot D., Marijnissen J.C.M., Scarlet B. *J. Electrostatics* 1997, 40&41, 633-638
Borra J.P., Ehouarn P., Boulaud D. *J. Aerosol Sci.* 2004, 35, 1313-1332
Bouguila N., Coelho R., Navarre D. *Rev. Gen. Electr.* 1991, No.8, 25-28
Brown J.D., Dobbins R.B., Inculet I.I. *IEEE Ind. Appl. Soc. Annual Conf.*, 3-6 Oct. 1994, Denver, USA, 1442-1444
Brown R.C. J. *Aerosol Sci.* 1997, 28, 1373-1391
Bruninx E., Rudstam G. *Nucl. Instrum. Methods* 1961, 13, 131-140
Burson W.C.Jr., Herren P.C., Jr. *J. Spacecraft* 1971, 8, No.6, 606-611
Büscher P., Schmidt-Ott A. *J. Aerosol Sci.* 1990, 21, Suppl.1, 567-570
Büscher P., Schmidt-Ott A. *J. Aerosol Sci.* 1992, 23, Suppl.1, 385-388
Büscher P., Schmidt-Ott A., Wiedensohler A. *J. Aerosol Sci.* 1994, 25, No.4, 651-663
Cantor B., Baik K.H., Grant P.S. *Progress Mater. Sci.* 1997, 42, 373-392
Cao F., Prakash J. *Electrochimica Acta* 2002, 47, 1607-1613
Carotenuto C., Di Natale F., Lancia A. *Chem. Eng. J.* 2010, 165, 35-45
Carswell D.J., Milsted J. *J. Nucl. Energy* 1957, 4, 51-54
Castro S., Bocanegra R. *Appl. Phys. Lett.* 2006, 88, paper No. 123105, 3 pp.
Castro S., Fernandez de la Mora J. *J. Appl. Phys.* 2009, 105, Paper No. 034903
Chandrasekhar R., Choy K.L. *J. Crystal Growth* 2001b, 231, 215-221
Chandrasekhar R., Choy K.L. *Thin Solid Films* 2001a, 398-399, 59-64
Chang J.S., Kelly A.J., Crowley J.M., eds. *Handbook of electrostatic processes.* 1995
Chang M.B., Leong K.H., Stukel J.J. *Aerosol Sci. Technol.* 1987, 6, 53-61
Chaudhury S.K., Sivaramakrishnan C.S., Panigrahi S.C. *J. Mater. Proc. Technol.* 2004, 145, 385-390
Chen C.A., Acquaviva P., Chun J.H., Ando T. *Scripta Materialia* 1996a, 34, No.5, 689-696
Chen C.H., Buysman A.A.J., Kelder E.M., Schoonman J. *Solid State Ionics* 1995, 80, 1-4
Chen C.H., Emond M.H.J., Kelder E.M., Meester B., Schoonman J. *J. Aerosol Sci.* 1999a, 30, No.7, 959-967

Chen C.H., Kelder E.M., Jak M.J.G., Schoonman J. *Solid State Ionics* 1996b, 86, 1301-1306
Chen C.H., Kelder E.M., Schoonman J. *J. Electrochem. Soc. Lett.* 1997c, 144, No. 11, L289-L291
Chen C.H., Kelder E.M., Schoonman J. *J. Europ. Ceramic Soc.* 1998a, 18, 1439-1443
Chen C.H., Kelder E.M., Schoonman J. *J. Mater. Sci.* 1996c, 31, No.20, 5437-5442
Chen C.H., Kelder E.M., Schoonman J. *J. Mater. Sci. Lett.* 1997a, 16, 1967-1969
Chen C.H., Kelder E.M., Schoonman J. *J. Power Sources* 1997b, 68, 377-380
Chen C.H., Kelder E.M., Schoonman J. *Thin Solid Films* 1999b, 342, 35-41
Chen C.H., Kelder E.M., vander Put P.J.J.M., Schoonman J. *J. Mater. Chem.* 1996d, No.5, 765-771
Chen C.H., Nord-Varhaug K., Schoonman J. *J. Mater. Synth. Process.* 1996e, 4, No.3, 189-194
Chen C.H., Saville D.A., Aksay I.A. *Appl. Phys. Lett.* 2006, 89, No.12, Paper No. 124103
Chen C.H., Yuan F.L., Schoonman J. *Eur. J. Solid State Inorg. Chem.* 1998b, 35, 189-196
Chen D-R. Pui D.Y.H. *Aerosol Sci. Techn.* 1997, 27, No.3, 367-380
Chen G., Gomez A. *24 Int. Symp. Combustion*, 1992, Sydney, Australia, 1531-1539
Chen H., Cong T.N., Yang W., Tan Ch., Li Y., Ding Y. *Progress in Natural Science* 2009, 19, 291-312
Choy K., Bai W., Charojrochkul S., Steele B.C.H. *J. Power Sources*. 1998, 71, 361-369
Choy K.L. *Mater. Sci. Eng.* C 2001, 16, 139-145
Choy K.L. *Mater. World* 2003b, June, 2-4
Choy K.L. *Progress Mater. Sci.* 2003a, 48, 57-170
Choy K.L., Su B. *J. Mater. Sci. Lett.* 1999, 18, 943-945
Choy K.L., Su B. *Thin Solid Films* 2001, 388, 9-14
Chung K.Y., Kim J.H., Kim K.B. *Electrochem. Solid-State Lett.* 2006, 9, No.4, A186-A189
Chung K.Y., Kim K.B. *J. Electrochem. Soc.* 2002, 149, No.1, A79-A85
Chung K.Y., Ryu Ch.W., Kim K.B. *J. Electrochem. Soc.* 2005, 152, No.4, A791-A795
Chung K.Y., Shu D., Kim K.B. *Electrochim. Acta* 2004a, 49, 887-898
Chung K.Y., Yoon W.S., Kim K.B., Yang X.Q., Oh S.M. *J. Electrochem. Soc.* 2004b, 151, No.3, A484-A492
Ciach T. *Int. J. Pharm.* 2006, 324, 51-55
Ciach T. *J. Drug Del. Sci. Tech.* 2007, 17, No.6, 367-375
Cloupeau M., Prunet-Foch B. *J. Aerosol Sci.* 1994, 25, No 6, 1121-1136
Cloupeau M., Prunet-Foch B. *J. Electrostatics* 1990, 25, 165-184
Coffee R.A. *Outlook Agricult.* 1981, 10, No.7, 350-356
Coffee R.A. *Static Electrification Conf.* London, May 1971, 200-209
Cooper J.F., Jones K.A., Moawad G. *Crop Protection* 1998, 17, No. 9, 711-715
Cooper S.C., Law S.E. *IEEE Trans. Ind. Appl.* 1987, 23, No.2, 217-223
Cunningham E. *Proc. Royal Soc.* 1910, 83A, 357-365
Cusano A., Consales M., Pisco M., Pilla P., Cutolo A., Buosciolo A., Viter R., Smyntyna V., Giordano M. *Appl. Phys. Lett.* 2006, 89, Paper No. 111103
Cusano A., Pilla P., Consales M., Pisco M., Pilla P., Cutolo A., Buosciolo A., Giordano M. *Optics Express* 2007, 15, No.8, 5136-5146
Dai Y., Wang K., Xie J. *Appl. Phys. Lett.* 2007, 90, paper No. 104102
Dai Y., Wang K., Zhao J., Xie J. *J. Power Sources* 2006, 161, 737-742

Dalley J.E.J., Greenaway R.S., Ulanowski Z., Hesse E., Kaye P.H. *J. Aerosol Sci.* 2005, 36, 1194-1209

Dau G. *Chem. Eng. Technol.* 1987, 10, 330-337

Dau G., Ebert F. *J. Aerosol Sci.* 1987, 18, No. 2, 147-157

Deitzel J.M., Kleinmeyer J.D., Hirvonen J.K., Beck Tan N.C. *Polymer* 2001, 42, 8163-8170

Deng W., Klemic J.F., Li X., Reed M.A., Gomez A. *Proc. Combustion Institute* 2007, 31, 2239-2246

Deotare P.B., Kameoka J. *Nanotech.* 2006, 17, 1380-1383

Deshler T. *J. Aerosol Sci.* 1985, 16, No 5, 399-406

Diagne E.H.A., Lumbreras M. *Sensors Actuators B* 2001, 78, 98-105

Dokko K., Anzue N., Makino Y., Mohamedi M., Itoh T., Umeda M., Uchida I. *Electrochem.* 2003, 71, No.12, 1061-1063

Dokko K., Anzue N., Mohamedi M., Itoh T., Uchida I. *Electrochem. Comm.* 2004, 6, 384-388

Dorman R.G. *Int. J. Air Pollution* 1960, 3, No.1-3, 112-125

Du J., Wu Y., Choy K.-L. *Thin Solid Films* 2006, 497, 42-47

Dudout B., Marijnissen J.C.M., Scarlett B. *J. Aerosol Sci.* 1999, 30, Suppl.1, 687-688

Efimov N.A., Zvonov V.A., Efimova L.Ya. *Elektron. Obrab. Mater.* 1979, 1, 1429-1433

Elmoursi A.A., Garner D.P. *IEEE Trans. Ind. Appl.* 1993, 29, No.6, 1053-1057

Elmoursi A.A., Speck C.E. *IEEE Trans. Ind. Appl.* 1991, 27, No.2, 311-315

Erven J. van, Moerman R., Marijnissen J.C.M. *Aerosol Sci. Technol.* 2005, 39, 941-946

Fearn D.G. *Proc. Inst. Mech. Eng. Pt.G, J. Aerospace Eng.* 1997, 211, No.G2, 103-112

Fearn D.G., Smith P. *IAF Paper*, 1998, IAF-98-S.4.01, (Sept/Oct), 1-11

Fehringer M., Ruedenauer F., Steiger W. *2nd Int. Symp. Detection and Observation of Gravitational Waves in Space*, Pasadena, July 1998, 207-213

Feng J.Q. *J. Fluid Mech.* 2010, 658, 438-462

Fernandez de la Mora J. *Annual Rev. Fluid Mech.* 2007, 39, 217-243

Fernandez de la Mora J., Loscertales I.G. *J. Fluid Mech.* 1994, 260, 155-184

Fjeld R.A., Gauntt R.O., McFarland A.R. *J. Aerosol Sci.* 1983, 14, 541-556

Flagan R.C. *Aerosol Sci. Techn.* 1998, 28, 301-380

Frank G.P., Cederfelt S.-I., Matrinsson B.G. *J. Aerosol Sci.* 2004, 35, 117-134

Franks A., Luty M., Robbie C.J., *Stedman M. Nanotechn.* 1998, 9, No.2, 61-6

Fu Ch.-Y., Chang Ch.-L., Hsu Ch.-Sh., Hwang B.-H. *Mater. Chem. Phys.* 2005, 91, 28-35

Fujimoto K., Kato T., Ito S., Inoue S., Watanabe M. *Solid State Ionics* 2006a, 177, 2639-2642

Fujimoto K., Takahashi H., Ito S., Inoue S., Watanabe M. *Applied Surface Sci.* 2006b, 252, 2446-2449

Fujimoto M., Kado T., Takashima W., Kaneto K., Hayase S. *J. Electrochem. Soc.* 2006c, 153, No.5, A826-A829

Gamero-Castano M. J. Propuls. Power 2004, 20, No. 4, 736-741

Gamero-Castano M., Hruby V. *J. Fluid Mech.* 2002, 459, 245-276

Gamero-Castano M., Hruby V. *J. Propuls. Power.* 2001, 17, No.5, 977-987

Gañan-Calvo A.M. *J. Aerosol Sci.* 1999, 30, No.7, 863-872

Gañan-Calvo A.M. *J. Fluid Mech.* 1997, 335, 165-188

Gassend B., Velásquez-García L.F., Akinwande A.I., Martinez-Sánchez M. *30th Int. Electric Propulsion Conf.*, Florence, 2007, Italy, Sept. 17-20

Gassend B., Velásquez-García L.F., Akinwande A.I., Martinez-Sánchez M. *IEEE MEMS Conf.*, 2008, Tucson, AZ, USA, Jan. 13-17, 976-979

Gemci T., Ebert F. *8th Annual European Conf. Liquid Atomisation and Spray Systems*. 30 Sept.-2 Oct. 1992, Amsterdam, 197-204
Ghimbeu C.M., Lumbreras M., Schoonman J., Siadat M., *Sensors* 2009, 9, 9122-9132
Ghimbeu C.M., Lumbreras M., Siadat M., van Landschoot R.C., Schoonman J. *Sensors Actuators B: Chemical* 2008a, 133, 694-698
Ghimbeu C.M., Schoonman J., Lumbreras M. *Ceramics Int.* 2008b, 34, 95-100
Ghimbeu C.M., Schoonman J., Lumbreras M., Siadat M. *Appl. Surf. Sci.* 2007c, 253, 7483-7489
Ghimbeu C.M., van Landschoot R.C., Schoonman J., Lumbreras M. *J. European Ceramic Soc.* 2007a, 27, 207-213
Ghimbeu C.M., van Landschoot R.C., Schoonman J., Lumbreras M. *Thin Solid Films* 2007b, 515, 5498-5504
Giles D.K., Law S.E. *Trans. ASAE* 33, 1990, No.1, 2-7
Göhlich H. *VDI-Forschungsheft* 1958, 467, Ausgabe B, Band 24
Gomez A. *3rd World Conf. Experimental Heat Transfer, Fluid Mechanics and Thermodynamics*, 1993Honolulu, Hawaii
Gomez A., Berry J.J., Roychoudhury S., Coriton B., Huth J. *Proc. Combustion Institute* 2007, 31, 3251-3259
Gomez A., Chen G. *Combust. Sci. Techn.* 1994, 96, 47-59
Gopalakrishnan M.V., Metzgar K., Rosetta D., Krishnamurthy R. *J. Mater. Proc. Technol.* 2003, 135, 228-234
Gorodinsky W.A., Romanov Ju.F., Sorokina A.W., Yakunin M.I. *Prib. Techn. Exper.* 1959, No.5, 128-130
Gourari H., Lumbreras M., Van Landschoot R., Schoonman J. *Sensors Actuators* B 1998, 47, 189-193
Gourari H., Lumbreras M., Van Landschoot R., Schoonman J. *Sensors Actuators* B 1999, 58, 365-369
Grace J.M., Marijnissen J.C.M. *J. Aerosol Sci.* 1994, 25, No.6, 1005-1019
Grigoriev D.A., Edirisinghe M., Bao X. *J. Mater. Res.*2002, 17, No.2, 487-491
Grigoriev D.A., Edirisinghe M., Bao X., Evans J.R.G., Luklinska Z.B. *Philos. Mag. Lett.* 2001, 81, No. 4, 285-291
Grigoryan V.G. *Rev. Sci. Instrum.* 1996, 67, No.3, 1126-1131
Ha T.H., Nishida O., Fujita H., Wataru H. *J. Mar. Sci. Technol.* 2010, 15, 271-279
Hall A., Hemming M. *Circuit World* 1992, 18, No.2, 32
Hara M., Sumiyoshitani S., Akazaki M. *2nd Int. Conf. Electrostatic Precipitation*, Kyoto, Nov. 1984
Hartman R.P.A., Brunner D.J., Camelot D.MA., Marijnissen J.C.M., Scarlett B. *J. Aerosol Sci.* 2000, 31, No.1, 65-95
Hayati I., Bailey A.I., Tadros Th.F. *J. Coll. Interface Sci.* 1987a, 117, No 1, 205-221
Hayati I., Bailey A.I., Tadros Th.F. *J. Coll. Interface Sci.* 1987b, 117, No 1, 222-230
Hendricks C.D. *J. Colloid Sci.* 1962, 17, 249-259
Hendricks C.D., Babil S. *J. Phys. E: Sci. Instrum.* 1972, 5, No.9, 905-910
Hensley J.L., Feng X., Bryan J.E. *J. Electrostatics* 2008, 66, 300-311
Hewitt G.W. *Trans. Am. Inst. Electr. Eng.* 1957, 76, No.7, 300-306
Higashiyama Y., Tanaka S., Sugimoto T., Asano K. *J. Electrostatics* 1999, 47, No. 3, 183-195

Hinds W.C. *Aerosol Technology: Properties, Behavior, and Measurement of Airborne Particles*. 2nd ed., 1999, Wiley, New York

Hines R.L. *J. Appl. Phys*. 1966, 37, No.7, 2730-2736

Hogan Ch.J. Jr., Biswas P. *Aerosol Sci. Technol*. 2008, 42, 75-85

Hogan J.J., Hendricks C.D. *AIAA J*. 1965, 3, No.2, 296-301

Hou X., Choy K.L. *Chem. Vap. Deposition* 2006, 12, 631-636

Hou X., Choy K.-L. *Materials Sci. Eng*. 2005a, C25, 669-674

Hou X., Choy K.-L. *Surface Coat. Technol*. 2004, 180-181, 15-19

Hou X., Choy K.-L. *Thin Solid Films* 2005b, 480–481, 13-18

Huang H., Yao X., Wu X., Wang M., Zhang L. *Microelectr. Eng*. 2003, 66, 688-694

Huang H., Yao X., Wu X.Q., Wang M.Q., Zhang L.Y. *Thin Solid Films* 2004, 458, 71-76

Huberman M.N. *J. Appl. Phys*. 1970, 41, No.2, 578-584

Huberman M.N., Beynon J.C., Cohen E., Goldin D.S., Kidd P.W., Zafran S. *J. Spacecraft* 1968, 5, No.11, 1319-1324

Hutchins D.K., Holm J. *Aerosol Sci. Technol*. 1989, 11, No.3, 244-253

Hwang B.-H., Chang Ch.-L., Hsu Ch.-Sh., Fu Ch.-Y. *J. Phys. D: Appl. Phys*. 2007a, 40, 3448-3455

Hwang K.-S., Jeong J.-H., Jeon Y.-S., Jeon K.-O., Kim B.H. *Ceramics Int*. 2007b, 33, 505-507

Im K.S., Lai M.Ch., Liu Y., Sankagiri N., Loch T., Nivi H. *Trans. ASME J. Fluid Eng*. 2001, 123, No.2, 237-245

Inculet I.I. *IEEE Trans. Electrical Insul*. 1982, EI-17 No.2, 168-171

Inculet I.I. *Part. Sci. Technol*. 1998, 16, No.1, 7-24

Inculet I.I., Castle G.S.P., Ting J. *IEEE Ind. Appl. Soc. Annual Meeting* 1989, 2144-2147

Inculet I.I., Castle G.SP. *IEEE Trans. Ind. Appl*. 1985, 21, No.2, 507-510

Inculet I.I., Fisher J.K. *IEEE Trans. Ind. Appl*. 1989, 25, No.3, 558-562

Inculet I.I., Klein R.G. *28th Ann. Meeting, IEEE Ind. Appl. Conf*. 2-8 Oct. 1993, Toronto, Canada, 1902-1904

Iribarne J.V., Thomson B.A. *J. Chem. Phys*. 1976, 64, No.6, 2287-2294

Jang S., Hwang I., Im J., Park I., Shin D. *J. Ceramic Processing Res*. 2009, 10, No. 6, 798-802

Javaid A., Castle P., Inculet I.I., Shelstad K.A., Crum G.W. *IEEE Trans. Ind. Appl*. 1980, 16, No.2, 292-296

Jaworek A. *J. Mater. Sci*. 2007a, 42, No.1, 266-297

Jaworek A., *J. Microencapsulation* 2008, 25, No. 7, 443-468

Jaworek A. *Powder Technol*. 2007b, 176, No.1, 18-35

Jaworek A. *Trans IFFM* 1993, 95, 131-146

Jaworek A. *Trans. IFFM* 1989, 89, 127-134 (in Polish)

Jaworek A., Adamiak K., Balachandran W., Krupa A., Castle P., Machowski W. *Aerosol Sci. Technol.*, 2002, 36, No.9, 913-924

Jaworek A., Adamiak K., Krupa A. *3rd Int. Conf. Multiphase Flow*, June 8-12, 1998a, Lyon, Paper No. 438

Jaworek A., Adamiak K., Krupa A. *IEEE Ind. Appl. Soc. Annual Meeting*, 5-10 Oct. 1996, San Diego, 2036-2043

Jaworek A., Adamiak K., Krupa A., Castle P. *J. Electrostatics* 2001, 51-52, 603-609

Jaworek A., Balachandran W., Krupa A., Kulon J., Machowski W. *11th Int. Conf. ELECTROSTATICS 2003*, 23-27 March 2003, Edinburgh, *Inst. Phys. Conf.* Ser. No. 178, 181-186
Jaworek A., Balachandran W., Lackowski M., Kulon J., Krupa A. *J. Electrostatics* 2006, 64, No.3-4, 194–202
Jaworek A., Krupa A. *Exp. Fluids* 1999a, 27, No.1, 43-52
Jaworek A., Krupa A. *J. Aerosol Sci.* 1999b, 30, No.7, 873-893
Jaworek A., Krupa A. *J. Electrostatics* 1989, 23, 361-370
Jaworek A., Krupa A. *Materials Sci.* 1990, 16, No 1-3, 33-38
Jaworek A., Krupa A., Adamiak K. *6th Int. Conf. Electrostatic Precipitation*. 18-21 June 1996, Budapest, 124-129
Jaworek A., Krupa A., Adamiak K. *IEEE Trans. Ind. Appl.* 1998b, 34, No.5, 985-991
Jaworek A., Machowski W., Krupa A., Balachandran W. *35th IEEE Ind. Appl. Conf., Rome*, 8-12 Oct. 2000, 770-776
Jaworek A., Sobczyk A.T. *J. Electrostatics* 2008, 66, No.3-4, 197-219
Jaworek A., Sobczyk A.T., Krupa A., Lackowski M., Czech T. *Bull. Polish Acad. Sci. Technical Sciences* 2009, 57, No.1, 63-70
Jayasinghe S.N., Edirisinghe M.J. *Appl. Phys.* A 2005a, 80, 399-404
Jayasinghe S.N., Edirisinghe M.J. *Appl. Phys.* A 2005b, 80, 701-702
Jayasinghe S.N., Edirisinghe M.J. *J. Europ. Ceramic Soc.* 2004, 24, 2203-2213
Jayasinghe S.N., Edirisinghe M.J. *J. Porous Mater.* 2002, 9, 265-273
Jayasinghe S.N., Edirisinghe M.J. *Mat. Res. Innovat.* 2003, 7, 62-64
Jayasinghe S.N., Edirisinghe M.J., DeWilde T. *Mat. Res. Innovat.* 2002, 6, No.3, 92-95
Jayasinghe S.N., Edirisinghe M.J., Kippax P.G. *Appl. Phys.* A 2004a, 78, 343-347
Jayasinghe S.N., Edirisinghe M.J., Wang D.Z. *Nanotechnology* 2004b, 15, 1519-1523
Jiang Y., Yu Y., Sun W., Chen Ch., Meng G., Gao J. *J. Electrochem. Soc.* 2007, 154, No.8, E107-E111
Jido M. *IEEE Ind. Appl. Soc. Annual Meeting*, San Diego, 1-5 Oct. 1989, 2058-2065
Jones A.R., Thong K.C. *J. Phys. D: Appl.* Phys. 1971, 4, No 8, 1159-1166
Juraschek R., Röllgen F.W. *Int. J. Mass Spectrom. Ion Proc.* 1998, 177, No.1, 1-15
Kaelin M., Zogg H., Tiwari A.N., Wilhelm O., Pratsinis S.E., Meyer T., Meyer A. *Thin Solid Films* 2004, 457, 391-396
Kaufman H.R. *Rev. Sci. Instrum.* 1990, 61, No.1, 230-235
Kelder E.M., Nijs O.C.J., Schoonman J. *Solid State Ionics* 1994, 68, No.1-2, 5-7
Kelly A.J. *Inst. Phys. Conf. Ser. No. 163*; 10th Int. Conf. ELECTROSTATICS '99, 28-31 March 1999, Cambridge, 99-107
Kelly A.J. *J. Inst. Energy*. 1984, 431, 312-320
Kidd P.W. *J. Spacecraft* 1968, 5, No.9, 1034-1039
Kim I.H., Kim K.B. *J. Electrochem. Soc.* 2004, 151, No.1, E7-E13
Kim I.H., Kim K.B. *J. Electrochem. Soc.*, 2006, 153, No.2, A383-A389
Kim I.H., Kim K.B. *Solid-State Lett.*, 2001, 4, No.5, A62-A64
Kim J.H., Lee H.S., Kim H.H., Ogata A. *J. Electrostatics* 2010, 68, 305-310
Kim K., Turnbull R.J. *J. Appl. Phys.* 1976, 47, No.5, 1964-1969
Kim K.Y., Marshall W.R. *AICHE J.* 1971, 17, No.3, 575-
Kim S.G., Choi K.H., Eun J.H., Kim H.J., Hwang Ch.S. *This Solid Films* 2000a, 377, 694-698

Kim S.G., Kim H.J. *J. Korean Phys. Soc.* 1999, 35, S180-S183

Kim S.G., Kim J.Y., Kim H.J. *This Solid Films* 2000b, 378, 110-114

Kim Y.T., Gopukumar S., Kim K.B., Cho B.W. *J. Power Sources* 2003, 117, 110-117

Kirk I.W. *Beltwide Cototn Conferences*, Nashville, TN, 2003, Jan. 6-10, 642-649

Kirsch A.A., Zagnit'ko A.V. *Aerosol Sci. Technol.* 1990, 12, 465-470

Kirsch A.A., Zagnit'ko A.V. *J. Coll. Interface Sci.* 1981, 80, 111-117

Kleber W. *Plaste Kautschuk* 1963a, 10, No.7, 441-447

Kleber W. *Plaste Kautschuk* 1963b, 10, No.8, 502-508

Klett J. D. *J. Atmosph. Sci.* 1971, 28, 78-85

Kobayashi Y., Miyashiro H., Takeuchi T, Shigemura H, Balakrishnan N., Tabuchi M., Kageyama H, Iwahori T. *Solid State Ionics* 2002, 152-153, 137-142

Koike S., Tatsumi K. J. *Power Sources* 2005, 146, 241-244

Koike S., Tatsumi K. J. *Power Sources* 2007, 174, 976-980

Kraemer H.F., Johnstone H.F. *Ind. Eng. Chem.* 1955, 47, No 12, 2426-2434

Krohn V.E. *J. Appl. Phys.* 1974, 45, No.3, 1144-1146

Krohn V.E. Jr. *Electric Propulsion Development.* Ed.: Stuhlinger E., Academic Press, New York, London, 1963, 435-440

Krohn V.E. Jr. *Liquid metal droplets for heavy particle propulsion.* Electrostatic Propulsion. Eds.: Langmuir D.B., Stuhlinger E., Sellen J.M. Jr., 1961, Academic Press, New York, London, 73-80

Krpoun R., Räber M., Shea H.R. *IEEE MEMS Conf.*, Tucson, AZ, USA, Jan. 13-17, 2008, 964-967

Krpoun R., Shea H.R. *14th Int. Workshop Physics of Semiconductor Devices.* Indian Inst Technol, Mumbai, INDIA, Dec. 17-20, 2007, 652-655

Krpoun R., Shea H.R. *J. Appl. Phys.* 2008, 104, paper No. 064511

Krpoun R., Shea H.R. *J. Micromech. Microeng.* 2009, 19, paper No. 045019

Krpoun R., Smith K.L., Stark J.P.W., Shea H.R. *Appl. Phys. Lett.* 2009, 94, paper No. 163502

Kruis F.E., Fissan H. *J. Nanopart. Res.* 2001, 3, 39-50

Kruis F.E., Fissan H., Peled A. *J. Aerosol Sci.* 1998, 29, No.5/6, 511-535

Krupa A., Jaworek A., Czech T., Lackowski M., Luckner J. *Inst. Phys. Conf. Ser. No.178*; 11th Int. Conf. Electrostatics 2003, 23-27 March 2003, Edinburgh, 349-354

Krupa A., Jaworek A., Lackowski M., Czech T., Luckner J. *J. Electrostatics* 2005, **63**, No.6-10, 673-678

Krzeczkowski S.A. *Int. J. Multiphase Flow* 1980, 6, 227-239

Ksapabutr B., Chalermkiti T., Wongkasemjit S., Panapoy M. *Thin Solid Films* 2010, 518, 6518-6521

Ksapabutr B., Panapoy M., Choncharoen K., Wongkasemjit S., Traversa E. *Thin Solid Films* 2008, 516, 5618-5624

Kwack E.Y., Back L.H., Bankston C.P. *Trans. ASME, Eng. Gas Turb. Power* 1989, 111, No.7, 578-586

Kyritsis D.C, Roychoudhury S., McEnally Ch.S, Pfefferle L.D., Gomez A. *Exp. Thermal Fluid Science* 2004b, 28, 763-770

Kyritsis D.C., Coriton B., Faure F., Roychoudhury S., Gomez A. *Combustion and Flame* 2004a, 139, 77-89

Kyritsis D.C., Guerrero-Arias I., Roychoudhury S., Gomez A. *Proc. Combustion Institute* 2002, 29, 965-972

Lackowski M. *J. Electrostatics* 2001, 51-52, 225-231
Lackowski M., Adamiak K., Jaworek A., Krupa A. *Powder Technol.* 2003a, 135-136, 243-249
Lackowski M., Jaworek A., Krupa A. *J. Aerosol Sci.* 2001, 32, Suppl. 1, 951-952
Lackowski M., Jaworek A., Krupa A. *J. Electrostatics* 2003b, 58, No.1, 77-89
Lackowski M., Jaworek A., Krupa A. *Turbulence* 2005, 11, 169-180
Lackowski M., Krupa A., Jaworek A. *European Phys. J. D* 2010, 56, 377-382
Laitinen A., Vaaraslahti K., Keskinen J. *10th Int. Conf. Electrostatic Precipitators*, Australia, 2006, Paper No. 6A4, 10 pp
Lapham D.P., Colbeck I., Schoonman J., Kamlag Y. *Thin Solid Films* 2001, 391, 17-20
Lapple C.E., *Electrostatic Phenomena with Particulates.* Advances in Chemical Engineering, Drew T.B., Cokelet G.R., Hoopes J.W.Jr., Vermeulen T. (eds.), 1970, vol.8, Academic Press, New York, London
Laryea G.N., No S.Y. *J. Electrostatics* 2003, 57, No.2, 129-142
Laryea G.N., No S.Y. *J. Electrostatics* 2004, 60, No.1, 37-47
Latheef M.A., Carlton J.B., Kirk I.W., Hoffmann W.C. *Pest Manag. Sci.* 2009, 65, 744-752
Lauer K.F., Verdingh V. *Nucl. Instrum. Methods* 1963, 21, 161-166
Law S.E. *IEEE Trans. Ind. Appl.* 1983, 19, No.2, 160-168
Law S.E. *Inst. Phys. Conf. Ser. No. 143*; 9th Int. Conf. ELECTROSTATICS '95, 2-5 April 1995, York, 1-12
Law S.E. *J. Electrostatics* 2001, 51-52, 25-42
Law S.E., Bowen H.D. *Trans. ASAE* 1975, 18, No.1, 35-39, 45
Law S.E., Cooke J.R., Cooper S.C. J. Electrostatics 1997, 40&41, 603-608
Law S.E., Cooper S.C. *Trans. ASAE* 1987, 30, No.1, 75-79
Law S.E., Cooper S.C. *Trans. ASAE* 1988, 31, No.4, 984-989
Law S.E., Cooper S.C., Law W.B. *Inst. Phys. Conf. Ser. No. 163*; 10th Int. Conf. ELECTROSTATICS '99, 28-31 March 1999, Cambridge, 243-248
Law S.E., Lane M.D. *IEEE Trans. Ind. Appl.* 1982, 18, No.6, 673-679
Law S.E., Marchant J.A., Bailey A.G. *IEEE Trans. Ind. Appl.* 1985, 21, No.4, 685-693
Law S.E., Scherm H. *J. Electrostatics* 2005, 63, 399-408
Lawless Ph.A. *J. Aerosol Sci.* 1996, 27, No.2, 191-215
Lawley A., Leatham A.G. *Adv. Powder Technol., Mater. Sci Forum* 1999, 299, No.3, 407-415
Lear C.W., Huberman M.N., Geist J.W. *J. Spacecraft* 1972, 9, No.5, 307-310
Lear C.W., Krieve W.F, Cohen E. *J. Air Poll. Control Assoc.* 1975, 25, No.2, 184-189
Lefebvre A.H. *Atomization and Sprays.* Taylor and Francis 1989
Lehr W., Hiller W. *J. Electrostatics* 1993, 30, 433-440
Lenggoro I.W., Okuyama K. *J. Aerosol Sci.* 1997, 28, Suppl.1, S351-S352
Lenggoro I.W., Okuyama K., Fernandez de la Mora J., Tohge N. *J. Aerosol Sci.* 2000, 31, No.1, 121-136
Lenggoro I.W., Xia B., Okuyama K., Fernandez de la Mora J. *Langmuir* 2002, 18, No. 12, 4584-4591
Lewis K.C., Dohmeler D.M., Jorgenson J.W., Kaufman S.L., Zarrin F., Dorman F.D. *Anal. Chem.* 1994, 66, No.14, 2285-2292
Lian P., Mejia A.F., Cheng Zh., Mannan M.S., (2010), J. Loss Prevention in the Process Industries 23, 337-345

Lintanf A., Mantoux A., Blanquet E., Djurado E. *J. Phys. Chem.* C 2007a, 111, 5708-5714
Lintanf A., Neagu R., Djurado E. *Solid State Ionics* 2007b, 177, 3491-3499
Liu B.Y.H., Kapadia A. *J. Aerosol Sci.* 1978, 9, 227-242
Liu B.Y.H., Pui D.Y.H. *J. Coll. Interface Sci.*, 1974, 47, No.1, 155-171
Lopez-Herrera J.M., Barrero A., Lopez A., Loscertales I.G., Marquez M. *J. Aerosol Sci.* 2003, 34, 535-552
Lozano P., Martínez-Sánchez M. *3rd Int. Conf. Spacecraft Propulsion*, Cannes, 10-13 Oct. 2000, 451-458
Lozano P., Martínez-Sánchez M. *J. Coll. Interface Sci.* 2004, 280, 149-154
Lozano P., Martínez-Sánchez M. *J. Coll. Interface Sci.* 2005b, 282, 415-421
Lozano P., Martínez-Sánchez M. *J. Phys. D: Appl. Phys.* 2005a, 38, 2371-2377
Lozano P., Martínez-Sánchez M., Lopez-Urdiales J.M. *J. Coll Interface Sci.* 2004, 276, 392-399
Lu J., Chu J., Huang W., Ping Z. *Jpn. J. Appl. Phys.* 2002, 41, Part 1, No. 6B, 4317-4320
Lu J., Chu J., Huang W., Ping Z. *Sensors Actuators* A 2003, 108, 2-6
Lubert K.H., Barie N., Rapp M. *Sensors Actuators* B 2007, 123, 218-226
Lüttgens U., Dülcks Th., Röllgen F.W. *Surface Sci.* 1992, 266, 197-203
Ma J., Qin Q.-Z. *J. Power Sources* 2005, 148, 66-71
Machowski W., Balachandran W. *IEEE Ind. Appl. Soc. Annual Meeting*, New Orleans, Louisiana, October 5-9, 1997, 1784-1789
Machowski W., Balachandran W., Huneiti Z. *14th ILASS Conf. Europe*, Manchester 6-8 July 1998, 505-510
Madou M.J., *Fundamentals of Microfabrication*. CRC Press 2002
Makin B., Bright A.W. *AIAA J.* 1969, 7, No.10, 2020-2022
Marchant J.A., Green R. *J. Agric. Eng. Res.* 1982, 27, 309-319
Marginean I., Kelly R.T., Page J.S., Tang K., Smith R.D. *Anal. Chem.* 2007a, 79, No.21, 8030-8036
Marginean I., Nemes P., Vertes A. *Phys. Rev.* E 2007b, 76, No.2, Paper No. 026320
Marginean I., Nemes P., Vertes A. *Phys. Rev. Lett.* 2006, 97, No.6, Paper No. 064502
Marginean I., Parvin L., Heffernan L., Vertes A. *Anal. Chem.* 2004, 76, 4202-4207
Markus S., Cui C., Fritsching U. *Materials Sci. Eng.* 2004, A383, 166-174
Markus S., Fritsching U., Bauckhage K. *Materials Sci. Eng.* 2002, A326, 122-133
Maski D., Durairaj D. *Crop Protection* 2010a, 29, 134-141
Maski D., Durairaj D. *J. Electrostatics* 2010b, 68, 152-158
Masuda S. *2nd Int. Conf. Electrostatic Precipitators*, Kyoto, Nov.1984, 177-185
Masuda S. *CSIRO Conf. Electrostatic Precipitation*, 21-24 Aug. 1978b, Leura, New South Wales, 54-90
Masuda S. *Int. Symp. Transfer and Utilization of Particulate Control Technology*. 24-28 July 1978a, Denver, Colorado, 19-31
Matsushima Y., Nemoto Y., Yamazaki T., Maeda K., Suzuki T. *J. Electroceramics* 2004, 13, 765-770
Matsushima Y., Nemoto Y., Yamazaki T., Maeda K., Suzuki T. *Sensors Actuators* B 2003, 96, 133-138
Matsushima Y., Yamazaki T., Maeda K., Noma T., Suzuki T. *J. Ceramic Soc. Japan* 2006, 114, No.12, 1121-1125

Maynagh B.M., Ghobadian B., Jahannama M.R., Hashjin T.T. *J. Agric. Sci. Techol.* 2009, 11, 249-257
Meesters G.M.H., Vercoulen P.H.W., Marijnissen J.C.M., Scarlett B. *J. Aerosol Sci.* 1990a, 21, Suppl.1, 669-672.
Meesters G.M.H., Vercoulen P.H.W., Marijnissen J.C.M., Scarlett B. *J. Aerosol Sci.* 1991, 22, Suppl.1, S11-S14
Meesters G.M.H., Vercoulen P.H.W., Marijnissen J.C.M., Scarlett B. *J. Aerosol Sci.* 1992, 23, No.1, 37-49
Meesters G.M.H., Zevenhoven C.A.P., Brons J.F.J., Verheijen P.J.T. *J. Electrostatics* 1990b, 25, 265-275
Mei F., Chen D.R. *Aerosol Air Quality Res.*, 2008, 8, No.2, 218-232
Mei F., Chen D.R. *Phys. Fluids* 2007, 19, paper No. 103303
Melcher J.R., Sachar K.S., Warren E.P. *Proc. IEEE* 1977, 65, No.12, 1659-1669
Mesquita R.A., Barbosa C.A. *Mater. Sci. Eng.* A 2004, 383, 87-95
Metwally I.A., Abed Y.A., El-Sahrigi A.F., Abdel-Mageed H.N., Noor H.M. *Conf. Electrical Insulation and Dielectric Phenomena*, 22-25 Oct. 1995, Virginia, USA
Metzler P, Weib P., Büttner H., Ebert F. *J. Electrostatics* 1997, 42, No.1-2, 123-141
Metzler P., Weib P., Büttner H., Ebert F., Krames J. *9th Int. Symp. High Voltage Engineering*. Graz, Austria, 28 Aug.-1 Sept. 1995, 7860-1-4
Miao P., Balachandran W., Wang J.L. *J. Electrostatics* 2001a, 51-52, 43-49
Miao P., Balachandran W., Xiao P. *IEEE Ind. Appl. Mag.* 2001c, 46-52
Miao P., Balachandran W., Xiao P. *IEEE Trans. Ind. Appl.* 2002, 38, No.1, 50-56
Miao P., Balachandran W., Xiao P. *J. Mater. Sci.* 2001b, 36, 2925-2930
Miao P., Huneiti Z.A., Machowski W., Balachandran W., Xiao P., Evans J.R.G. *Electrostatics 1999, Inst. Phys. Conf. Ser.* No. 163, 119-122
Michelson D. *J. Fluid Mech.* 1968, 33, Pt.3, 573-575
Michelson D., Richardson O.W. *Nucl. Instrum. Methods* 1963, 21, 355
Mikhailov E.F., Vlasenko S.S., Krämer L., Niessner R. *J. Aerosol Sci.* 1999, 30, Suppl.1, 443-444
Mitterauer J. *IEEE Trans. Plasma Sci.* 1991, 19, No.5, 790-799
Mohamedi M., Lee S.J., Takahashi D., Nishizawa M., Itoh T., Uchida I. *Electrochimica Acta* 2001a, 46, 1161-1168
Mohamedi M., Takahashi D., Itoh T., Uchida I. *Electrochimica Acta* 2002a, 47, 3483-3489
Mohamedi M., Takahashi D., Itoh T., Umeda M., Uchida I. *J. Electrochem. Soc.* 2002b, 149, No.1, A19-A25
Mohamedi M., Takahashi D., Uchiyama T., Itoh T., Nishizawa M., Uchida I. *J. Power Sources* 2001b, 93, 93-103
Moon J.D., Lee D.H., Kang T.G., Yon K.S. *J. Electrostatics* 2003, 57, No.3-4, 363-379
Moore A.D. *Electrostatics and its applications*. 1973, J. Wiley
Murtomaa M., Kivikero N., Mannermaa J.P., Lehto V.P. *J. Electrostatics* 2005, 63, 891-897
Mutoh M., Kaieda S., Kamimura K. *J. Appl. Phys.* 1979, 50, No 5, 3174-3179
Nakaso K., Han B., Ahn K.H., Choi M., Okuyama K. *J. Aerosol Sci.* 2003, 34, 869-881
Nam K.-W., Kim K.-B. *J. Electrochem. Soc.* 2006, 153, No.1, A81-A88
Neagu R., Djurado E., Ortega L., Pagnier T. *Solid State Ionics* 2006a, 177, 1443-1449
Neagu R., Perednis D., Princivalle A., Djurado E. *Chem. Mater.* 2005, 17, 902-910
Neagu R., Perednis D., Princivalle A., Djurado E. *Solid State Ionics* 2006c, 177, 1451-1460

Neagu R., Perednis D., Princivalle A., Djurado E. *Solid State Ionics* 2006d, 177, 1981-1984
Neagu R., Perednis D., Princivalle A., Djurado E. *Surface Coatings Techn*. 2006b, 200, 6815-6820
Nguyen T., Djurado E. *Solid State Ionics* 2001, 138, 191-197
Nishizawa M, Uchiyama T, Dokko K, Yamada K, Matsue T, Uchida I. *Bull. Chem. Soc. Japan* 1998, 71, No.8, 2011-2015
Nomura H., Parekh S., Selman J.R., Al-Hallaj S. *J. App. Electrochem*. 2005a, 35, 61-67
Nomura H., Parekh S., Selman J.R., Al-Hallaj S. *J. App. Electrochem*. 2005b, 35, 1121-1126
Ogata S., Hara Y., Shinohara H. *Int. Chem. Eng*. 1978a, 18, No. 3, 482-488
Ogata S., Hatae T., Shoguchi K., Shinohara H. *Int. Chem. Eng*. 1978b, 18, No.3, 488-493
Oh H., Kim S. *J. Aerosol Sci*. 2007, 38, 1185-1196
Paine M.D. *Microfluid. Nanofluid*. 2009, 6, 775-783
Paine M.D., Alexander M.S., Stark J.P.W. *J. Coll. Interface Sci*. 2007, 305, 111-123
Paine M.D., Gabriel S., Schabmueller C.G.J., Evans A.G.R. *Sensors Actuators* A 2004, 114, 112-117
Park D.G., Burlitch J.M. *J. Sol-Gel Sci. Technol*. 1996, 6, 235-249
Parvin L., Galicia M.C., Gauntt J.M., Carney L.M., Nguyen A.B, Park Y., Heffernan L., Vertes A. *Anal. Chem*. 2005, 77, 3908-3915
Pauthenier M.M., Moreau-Hanot M. *J.Physique* 1937, 5, No.5, 193-196
Penney G.W. *Electrified liquid spray dust-precipitators*. 1944, US Patent 2,357,354, Sept. 5
Penney G.W., Lynch R.D. *AIEE Trans*. 1957, 76, 294
Perednis D., Gauckler L.J. *J. Electroceramics* 2005, 14, 103-111
Perednis D., Gauckler L.J. *Solid State Ionics* 2004, 166, 229-239
Perednis D., Wilhelm O., Pratsinis S.E., Gauckler L.J. *Thin Solid Films,* 2005, 474, 84-95
Perel J. *J. Electrochem. Soc*. 1968, 115, No.12, 343-350
Perel J., Mahoney J.F., Moore R.D., Yahiku Y. *AIAA J*. 1969, 7, No.3, 507-511
Perel J., Yahiku A.Y., Mahoney J.F., Daley H.L., Sherman A. *J. Spacecraft* 1971, 8, No.7, 702-709
Pfeifer R.J., Hendricks C.D. *AIAA J*. 1968, 6, No. 3, 496-502
Pilat M.J. *J. Air Poll. Contr. Ass*. 1975, 25, No.2, 176-178
Pilat M.J., Jaasund S.A., Sparks L.E. *Envir. Sci. Techn*. 1974, 8, No.4, 360-362
Pisco M., Consales M., Campopiano S., Viter R., Smyntyna V., Giordano M., Cusano A. *J. Lightwave Technol*. 2006, 24, No.12, 5000-5007
Plunkett R.T., Inculet I.I., Klein R.G. *IEEE Trans. Ind. Appl*. 1986, 22, No.3, 523-526
Poncelet D., Babak V.G., Neufeld R.J., Goosen M.F.A., Bugarski B. *Adv. Coll. Interface Sci*. 1999, 79, 213-228
Preston-Thomas H. *J. Brit. Interplanet. Soc*. 1952, 11, 173
Princivalle A., Djurado E. *Solid State Ionics* 2008, 179, 1921-1928
Princivalle A., Perednis D., Neagu R.,Djurado E. *Chem. Mater*. 2004, 16, 3733-3739
Princivalle A., Perednis D., Neagu R.,Djurado E. *Chem. Mater*. 2005, 17, 1220-1227
Rader D.J. *J. Aerosol Sci*. 1990, 21, No.2, 161-168
Raj E.S., Choy K.L. *Mater. Chem. Phys*. 2003, 82, 489-492
Rayleigh F.R.S. *Phil. Mag*. 1882, 14, Ser.5, 184-186
Reifarth R., Schwarz K., Käppeler F. *Astrophys. J*. 2000, 528, No.1, 573-581
Rhee S.H., Yang Y., Choi H.S., Myoung J.M., Kim K. *Thin Solid Films* 2001, 396, No.1-2, 23-28

Ricci L.J. *Chem. Eng.* 1977, 52-58
Rietveld I.B., Kobayashi K., Yamada H., Matsushige K. *Coll. Interface Sci.* 2006, 298, 639-651
Romat H., Badri A. *J. Electrostatics* 2001, 51-52, 481-487
Romero-Sanz I., Bocanegra R., Fernandez de la Mora J., Gamero-Castaño M. *J. Appl. Phys.* 2003, 94, No.5, 3599-3605
Romero-Sanz I., Fernandez de la Mora J. *J. Appl. Phys.* 2003, 95, No.4, 2123-2129
Ryu Ch.K., Kim K. *Appl. Phys. Lett.* 1995, 67, No.22, 3337-3339
Saccoccia G., Gonzalez del Amo J., Estublier D. *ESA Bull.* 2000, 101, No.2, 62-71
Sahner K., Gouma P., Moos R. *Sensors* 2007, 7, 1871-1886
Saini D., Yurteri C.U., Grable N., Sims R.A., Mazumder M.K. *IEEE* 2002, 2451-2453
Sakai T., Sadakata M., Sato M., Kimura K. *Atomiz. Sprays* 1991, 1, No.2, 171-185
Sample S.B. Bollini R. *J. Coll. Interface Sci.* 1972, 41, No 2, 185-193
Sato M. *IEEE Trans. Ind. Appl.* 1991, 27, No.2, 316-322
Sato M. *J. Electrostatics* 1984, 15, 237-247
Sato M., Kato S., Saito M. *IEEE Trans. Ind. Appl.* 1996a, 32, No.1, 138-145
Sato M., Kudo N., Saito M. *IEEE Ind. Appl. Soc. Annual Meeting*, 5-10 Oct. 1996b, San Diego
Sato M., Miyazaki H., Sadakata M., Sakaki T. *4th Int. Conf. Liquid Atomization and Spray Systems*, 22-24 Aug. 1988, Sendai, Japan, 161-165
Sato M., Miyazaki S., Kuroda M., Sakai T. *Int. Chem. Eng.* 1983, 23, No.1, 72-77
Sato M., Takahashi H., Awazu M., Ohshima T. *J. Electrostatics* 1999, 46, No.2-3, 171-176
Sato T. *Trans. I.E.E. Japan* 1987, 107, 155-161
Savage R.N., Hieftje G.M. *Rev. Sci. Instrum.* 1978, 49, 1418-1424
Scherm H., Savelle T., Law E. *Biocontrol Sci. Techn.* 2007, 17, No.3, 285-293
Schmidt M., Löffler F. *Chem. Ing. Tech.* 1994, 66, No.4, 543-546
Schmidt M., Löffler F. *J. Aerosol Sci.* 1992, 23, Suppl.1, 773-777
Schneider A., Uhlenwinkel V., Harig H., Bauckhage K. *Mater. Sci. Eng.* A 2004, 383, 114-121
Schoonman J. *Solid State Ionics* 2000, 135, 5-19
Selenou Ngoms A., Romat H., Baudry K. *Inst. Phys. Conf. Ser.* No. 163; 10th Int. Conf. ELECTROSTATICS '99, 28-31 March 1999, Cambridge, 33-36
Seto T., Orii T., Sakurai H., Hirasawa M., Kwon S.-B. *Aerosol Sci. Technol.* 2005, 39, 750-759
Shahub A.M., Williams M.M.R. *J. Phys. D: Appl. Phys.* 1988, 21, No.2, 231-236
Sheppard S.V. *Mech. Eng.* 1985, 107, No.7, 72-75
Sheppard S.V. *Mech. Eng.* 1985, 107, No.7, 72-75
Shorey J.D., Michelson D. *Nuclear Instrum. Meth.* 1970, 82, 295-296
Shrimpton J.S. *Int. J. Numer. Meth. Engng* 2003, 58, 513-536
Shrimpton J.S., Yule A.J. *Atomiz. Sprays* 2001, 11, No.4, 365-396
Shrimpton J.S., Yule A.J. *Exp. Fluids.* 1999, 26, No.5, 460-469
Shu D., Chung K.Y., Cho W.I., Kim K.B. *J. Power Sources* 2003a, 114, 253-263
Shu D., Kumar G., Kim K.B., Ryu K.S., Chan S.H. *Solid State Ionics* 2003b, 160, 227-233
Shui J.L., Jiang G.S., Xie S, Chen C.H. *Electrochimica Acta* 2004, 49, 2209-2213
Shui J.L., Yu Y., Chen C.H. *Appl. Surf. Sci.* 2006a, 253, 2379-2385
Si B.Q.T., Byun D., Lee S. *J. Aerosol Sci.* 2007, 38, 924-934

Sickles J.E., Anestos T.C. *IEEE Trans. Ind. Appl.* 1979, 15, No.3, 273-276
Slamovich E.B., Lange F.F. *Better Ceramics Through Chemistry III. Symposium. Mater. Res. Soc.*, 1988, Pittsburgh, PA, USA, 5-8 April, 257-262
Smallman R.E., Harris I.R., Duggan M.A. *J. Mater. Proc. Technol.* 1997, 63, 18-29
Smith D.P.H. *IEEE Trans. Ind. Appl.* 1986, 22, No 3, 527-535
Smith M.H. *J. Appl. Meteorology* 1976, 15, 275-281
Smyntyna V., Golovanov V., Kaciulis S., Mattogno G., Righini G. *Sensors Actuators* B 1995, 24-25, 628-630
Snyder H.E., Senser D.W., Lefebvre A.H., Coutinho R.S. *IEEE Trans. Ind. Appl.* 1989, 25, No.4, 720-727
Song W., Shumlak U. *J. Propuls. Power* 2008, 24, No.1, 139-141
Srivastava V.C., Mandal R.K., Ojha S.N. *Mater. Sci. Eng.* 2001, A304-306, 555-558
Stark J., Stevens B., Alexander M., Kent B. *J. Spacecraft Rockets* 2005, 42, No. 4, 628-639
Stelzer N.H.J., Schoonman J. *J. Mater. Synth. Proc.*1996, 4, No.6, 429-438
Strauss W. (ed.) *Air pollution control.* Part I. J. Wiley & Sons, 1971
Su B., Choy K.L. *J. Mater. Chem.* 1999b, 9, No.7, 1629-1633
Su B., Choy K.L. *J. Mater. Chem.* 2000d, 10, No.4, 949-952
Su B., Choy K.L. *J. Mater. Sci. Lett.* 2000b, 19, 1859-1861
Su B., Choy K.L. *Thin Solid Films* 2000a, 359, 160-164
Su B., Choy K.L. *Thin Solid Films* 2000c, 361-362, 102-106
Su B., Wei M., Choy K.L. *Mater. Lett.* 2001, 43, 83-88
Swatik D.S., Hendricks C.D. *AIAA J.* 1968, 6, No.8, 1596-1597
Tang K., Gomez A. *J. Coll. Interface Sci.* 1996, 184, 500-511
Taniguchi I., Hosokawa T. *J. Alloys Compounds* 2008, 460, 464-471
Taniguchi I., Schoonman J. *J. Mater. Synth. Proc.* 2002, 10, No.5, 267-275
Taniguchi I., van Landschoot R.C., Schoonman J. *Solid State Ionics* 2003a, 156, 1-13
Taniguchi I., van Landschoot R.C., Schoonman J. *Solid State Ionics* 2003b, 160, 271-279
Teng W.D., Huneiti Z.A., Machowski W., Evans J.R.G., Edirisinghe M.J., Balachandran W. *J. Mater. Sci. Lett.* 1997, 16, 1017-1019
Tepper G., Kessick R., Pestov D. *J. Appl. Phys.* 2007, 102, paper No. 113305, 6 pp.
Thong K.C., Weinberg F.J. *Proc. Roy. Soc. London* 1971, A324, No.1557, 201-215
Tinsley B.A., Rohrbaugh R.P., Hei M., Beard K.V. J. Atmos. Sci. 2000, 57, No.13, 2118-2134
Tomita Y., Ishibashi Y., Yokoyama T. *Bull. JSME* 1986, 29, No. 257, 3737-3743
Tripathi S.N., Harrison R.G. *Atm. Res.* 2002, 62, 57-70
Turetsky A.Ye. *J. Aerosol Sci.* 1999, 30, Suppl.1, 689-690
Turetsky A.Ye., Chemeresyuk G.G., Chernova A.E. *J. Aerosol Sci.* 2003, 34, Suppl.1, S1283-S1284
Turetsky A.Ye., Chemeresyuk G.G., Chernova A.E. *J. Aerosol Sci.* 2004, 35, Suppl.1, S259-S260
Uchimoto Y., Amezawa K., Furushita T., Wakihara M., Taniguchi I. *Solid State Ionics* 2005, 176, 2377-2381
Uchiyama T., Nishizawa M., Itoh T., Uchida I. *J. Electrochem. Soc.* 2000, 147, No.6, 2057-2060
Vasil'ev M.N., Vorona N.A., Gavrikov A.V., Petrov O.F., Sidorov V.S., Fortov V.E. *Techn. Phys. Lett.* 2010, 36, No.12, 1143-1145

Velásquez-Garcia L.F., Akinwande A.I., Martinez-Sánchez M. *J. Microelectromech. Syst.* 2006a, 15, No.5, 1260-1271
Velásquez-Garcia L.F., Akinwande A.I., Martinez-Sánchez M. *J. Microelectromech. Syst.,* 2006b, 15, No.5, 1272-1280
Velásquez-Garcia L.F., Akinwande A.I., Martinez-Sánchez M. *J. Microelectromech. Syst.*, 2007, 16, No.3, 598-612
Viswanathan S. *Ind. Eng. Chem. Res.* 1999, 38, 4433-4442
Vonnegut B., Neubauer R.L. *J. Coll. Sci.* 1952, 7, 616-622
Walch B., Horanyi M., Robertson S. *Phys. Rev. Lett.* 1995, 75, No.5, 838-841
Wang D., Edirisinghe M.J., Dorey R.A. *J. European Ceramic Soc.* 2008, 28, 2739-2745
Wang D.Z., Edirisinghe M.J., Jayasinghe S.N. *J. Am. Ceram.* Soc. 2006, 89, No.5, 1727-1729
Wang D.Z., Jayasinghe S.N., Edirisinghe M.J. *J. Nanoparticle Res.* 2005, 7, 301-306
Wang D.Z., Jayasinghe S.N., Edirisinghe M.J., Luklinska Z.B. *J. Nanoparticle Res.* 2007, 9, 825-831
Wang H.C., Stukel J.J., Leong K.H. *Aerosol Sci. Techn.* 1986, 5, No.4, 409-421
Wang P.-K. *J. Coll. Interface Sci.* 1983, 94, No 2, 301-318
Wang Z.Ch., Kim K.-B. *Materials Letters* 2008, 62, 425-428
Weber C. *Z. Angew. Math. Mech.* 1931, 11, No.2, 136-154
Wei M., Choy K.L. *Chem. Vapor Depos.* 2002, 8, No.1, 15
Wei M., Choy K.L. *J. Crystal Growth* 2005, 284, 464-469
Wei M., Zhi D., Choy K.L. *Mater. Lett.* 2006, 60, 1519-1523
White H.J. *Industrial Electrostatic Precipitation.* Addison-Wesley, 1963
Wilhelm O., Pratsinis S.E., Perednis D., Gauckler L.J. *Thin Solid Films* 2005, 479, 121-129
Will J., Mitterdorfer A., Kleinlogel C., Perednis D., Gauckler L.J. *Solid State Ionics* 2000, 131, 79-96
Withers R.S., Melcher J.R., Richmann J.W. *J. Electrostatics* 1978, 5, 225-239
Xie Y., Neagu R., Hsu Chg.-Sh., Zhang X., Decès-Petit C. *J. Electrochem. Soc.* 2008, 155, No.4, B407-B410
Xiong J., Sun D., Zhou Zh., Zhang W. *IEEE/ASME Trans. Mechatronics*, 2006, 11, No. 1, 66-74
Xiong J., Zhou Z., Ye X., Wang X., Feng Y., Li Y. *Microelectronic Eng.* 2002, 61-62, 1031-1037
Xiong J., Zhou Z., Ye X., Wang X., Feng Y., Li Y. *Sensors Actuators* A 2003, 108, 134-137
Xu D., Li J., Wu Y., Wang L., Sun D., Lin Z., Zhang Y. *J. Electrostatics* 2003, 57, No.3-4, 217-224
Xu D., Liu Z., Wu G. *7th Int. Conf. Electrostatic Precipitation.* 20-25 Sept 1998, Kyongju, Korea, 224-229
Yamada A., Niikura F., Ikuta K. *J. Micromech. Microeng.* 2008, 18, paper No. 025035 (9pp)
Yamada K., Sato N., Fujino T., Lee Ch.G., Uchida I., Selman J.R. *J Solid State Electrochem.* 1999, 3, 148-153
Yamada Y., Imanishi T., Yasuda Sh., Yoshida O., Mizuno A. *IEEE Trans. Diel. Electr. Insul.* 2009, 16, No.3, 641-648
Yan J., Hou X., Choy K.-L. *J. Power Sources* 2008, 180, 373-379
Yang H.T., Viswanathan S., Balachandran W., Ray M.B. *Environ. Sci. Technol.* 2003, 37, No.11, 2547-2555

Yoon W.S., Ban S.H., Lee K.K., Kim K.B., Kim M.G., Lee J.M. *J. Power Sources* 2001, 97-98, 282-286
Yoon W.S., Chung K.Y, Nam K.W., Kim K.B. *J. Power Sources* 2006, 163, 207-210
Yoon W.S., Chung K.Y, Oh K.H., Kim K.B. *J. Power Sources* 2003, 119-121, 706-709
Yu F., Cui J., Ranganathan S., Dwarakadasa E.S. *Mater. Sci. Eng.* 2001, A304-306, 621-626
Yu Y., Chen Ch.-H., Shi Y. *Adv. Mater.* 2007, 19, 993-997
Yu Y., Chen Ch.-H., Shui J.-L., Xie S. *Angew. Chem. Int. Ed.* 2005a, 44, 7085-7089
Yu Y., Gu L., Dhanabalan A., Chen Ch.H., Wang Ch. *Electrochimica Acta* 2009, 54, 7227-7230
Yu Y., Shi Y., Chen Ch.-H. *Chem. Asian J.* 2006, 1, 826-831
Yu Y., Shui J.L., Chen C.H. *Solid State Comm.* 2005b, 135, 485-489
Yule A.J., Shrimpton J.S., Watkins A.P., Balachandran W., Hu D. *Fuel* 1995, 74, No.7, 1094-1103
Zafran S., Beynon J.C., Kidd P.W., Shelton H., Jackson F.A. *J. Spacecraft* 1973, 10, No.8, 531-533
Zaouk D., al Asmar R., Podlecki J., Zaatar Y., Khoury A., Khoury A., Foucaran A. *Microelectronics J.* 2007, 38, 884-887
Zaouk D., Zaatar Y., Asmar R., Jabbour J. *Microelectr. J.* 2006, 37, 1276-1279
Zaouk D., Zaatar Y., Khoury A., Llinares C., Charles J.P., Bechara J. *Microelectr. Eng.* 2000a, 51-52, 627-631
Zaouk D., Zaatar Y., Khoury A., Llinares C., Charles J.P., Bechara J. *J. Appl. Phys.* 2000b, 87, 7539-7543
Zhang L.L., Zhao X.S. *Chem. Soc. Rev.* 2009, 38, 2520-2531
Zhang Y., Feng H., Wu X., Wang L., Zhang A., Xia T., Dong H., Li X., Zhang L. *Int. J. Hydrogen Energy.* 2009, 34, 4889-4899
Zhao H., Zheng Ch. *Chem. Eng. Technol.* 2008, 31, No.12, 1824-1837
Zhao S., Castle G.S.P., Adamiak K. *J. Electrostatics* 2005a, 63, 261-272
Zhao S., Castle G.S.P., Adamiak K. *J. Electrostatics* 2005b, 63, 871-876
Zhao S., Castle G.S.P., Adamiak K. *J. Electrostatics* 2008, 66, 594-601
Zhu X.J., Guo Z.P., Zhang P., Du G.D., Zeng R., Chen Z.X., Li S., Liu H.K., *J. Mater. Chem.*, 2009, 19, 8360-8365
Zomeren van A.A., Kelder E.M., Marijnissen J.C.M., Schoonman J. *J. Aerosol Sci.* 1994, 25, No.6, 1229-1235

In: Sprays: Types, Technology and Modeling
Editor: Maria C. Vella, pp. 101-134
ISBN 978-1-61324-345-9

Chapter 2

APPLICATIONS OF SPRAYDRYER TO PRODUCTION OF BIOACTIVE COMPOUND-RICH POWDERS FROM PLANT FOOD MATERIALS: AN OVERVIEW

Maruf Ahmed[1, 2] and Jong-Bang Eun[1*]
[1]Department of Food Science and Technology and
Functional Food Research Institute, Chonnam National University,
Gwangju, South Korea
[2]Department of Food Processing and Preservation,
Hajee Mohammad Danesh Science and Technology University,
Dinajpur, Bangladesh

ABSTRACT

Consumption of fruits and vegetables has increased due to their high content of bioactive compounds. Incorporation of bioactive compounds into functional foods can decrease the occurrence of several human diseases such as cancer, cardiovascular disease, inflammatory disease, and oxidative stress. However, it is very difficult to store fruits and vegetables for long time periods even at low temperatures due to their perishable nature. Therefore, processing is necessary to prolong their shelf life. Drying is one of the oldest food preservation techniques and has been translated into a current method by modern technologies. Among all drying methods, spray drying is one of the best methods for the preservation of bioactive compounds. Spray dryers are widely used to produce dried fruits and vegetables. Spray drying effectively preserves heat-sensitive bioactive components, such as phenolic, carotenoid, and anthocyanin compounds, compared to other drying methods. During spray drying, different types of encapsulating agents such as starches, maltodextrins, corn syrups, inulin, and arabic gum are used in processing. Encapsulating agents protect bioactive compounds from oxygen, water, and light. Moreover, encapsulating agents can be used to increase the stability of bioactive compounds. In this report, we reviewed the current status of bioactive compounds in

* Corresponding Author: Jong-Bang Eun, Department of Food Science and Technology, College of Agriculture and Life Science, Chonnam National University, 77 Yongbong-ro, Buk-gu, Gwangju, S. Korea. Tel: +82-62-530-2145 (2145); Fax: +82-62-530-2149 email: jbeun@jnu.ac.kr.

different fruits, vegetables, herbs, spices, and legumes from around the world and investigated the total contents and activities of vitamin C, phenolics, flavonoids, anthocyanins, β-carotene, lycopene, and betalain compounds using different types of encapsulating agents with spray dryers. This information gained in this review will be helpful to consumers and product developers as well as the food industry.

INTRODUCTION

Fruits, vegetables, and herbs are essential to human health. Consumption of these has increased due to their high content of bioactive compounds. The most common bioactive compounds are vitamin C, phenolics, flavonoids, anthocyanins, β-carotene, lycopene, and betalain compounds. These components are most likely involved in the reduction of degenerative human diseases due to their antioxidative and free radical scavenging properties [Stintzing and Carle, R. 2004; Neill et al, 2002; Lee and Collins, 2001; Kim et al., 2006; Krishnaiah et al., 2010]. Usually, synthetic antioxidants such as butylated hydroxytoluene (BHT) and butylated hydroxyanisole (BHA) are used as antioxidants in the food industry. However, the use of synthetic antioxidants in foods is discouraged due to high levels of toxicity [Buxiang and Fukuhara, 1997] and carcinogenicity [Hirose *et al.*, 1998]. Therefore, natural antioxidants from plant extracts have attracted considerable attention due to their safety. Natural antioxidants can be added to different food products in different ways. They can be used to increase stability by preventing lipid peroxidation, thereby increasing the shelf life of food products.

In recent years, there has been a global trend toward the use of phytochemicals from natural resources, such as vegetables, fruits, oilseeds, and herbs, as antioxidants and functional ingredients [Elliott, 1999 J.G. Elliott, Application of antioxidant vitamins in foods and beverages, Food Technology 53 (1999), pp. 46–48. View Record in Scopus | Cited By in Scopus (38)Elliott, 1999; Kaur and Kapoor, 2001; Larson, 1988; Namiki, 1990]. However, it is very difficult to store fruits and vegetables for a long time due to their perishable nature, even at low temperature. Therefore, processing is necessary to prolong their shelf life. Drying is one of the oldest food preservation techniques and has been translated into a current method by modern technologies.

Among all drying methods, spray drying is one of the best methods for preserving bioactive compounds. Spray dryers are widely used to produce dried fruits and vegetables and can more effectively preserve heat-sensitive bioactive components such as phenolic compounds, carotenoids, and anthocyanins compared to other drying methods. Spray drying is most commonly used in the food industry for encapsulation. Different types of encapsulating agents such as starches, maltodextrins, corn syrups, inulin, and arabic gum are used in spray drying.

Encapsulation technology is used in the food industry in order to develop liquid and solid ingredients as an effective barrier against environmental parameters such as oxygen, light, and free radicals, etc [Desai and Park, 2005]. Bioactive compounds can be improved by using encapsulation techniques, which entrap sensitive ingredients inside a coating material [Saenz, Tapia, Chavez, and Robert, 2009]. Moreover, encapsulating agents can be used to increase the stability of bioactive compounds. Many researchers [Saenz, et al., 2009; Tonon et al., 2008; Ersus andYurdagel, 2007; Grabowski et al., 2006; Ahmed et al., 2010a; Ahmed et al., 2010b;

Kha et al., 2010; Cai and Corke, 2000; Moreira et al., 2010; Pitalua et al., 2010; Rascón et al., 2010; Kosaraju, 2006: Liu et al., 2010; Georgetti, et al., 2008; Quek et al., 2007; Robert et al., 2010; Osorio et al., 2010] have used various encapsulation materials such as starches, maltodextrins, corn syrups, inulin, arabic gum, and other materials to improve bioactive compounds during spray drying. Therefore, the objective of this review was to discuss existing research on bioactive compounds, mainly vitamin C, phenolics, flavonoids, anthocyanins, β-carotene, lycopene, and betalain compounds, from different fruits, vegetables, herbs, spices, and legumes from around the world in order to assess their total contents and activities using different types of encapsulating agents with spray dryers.

Most Common Encapsulating Agents Used during Spray Drying

Different types of encapsulating agents are used as wall materials during spray drying. In this review, we concentrated on those encapsulating agents that are commonly used during spray drying. These are given below:

Maltodextrins

Maltodextrins are polysaccharides and water-soluble materials that appear as a white powder. They consist of a α-(1-4)-linked D-glucose produced by the acid or enzymatic hydrolysis of corn starch [Regan and Mulvihill, 2009]. Maltodextrins are usually classified according to their dextrose equivalency (DE), which is a measure of the number of reducing sugar groups per sample weight [Desobry et al., 1999]. Usually, maltodextrins with DE values of 4, 10, 15, 20, 25, 30, and 42 are available in markets. The molecular formula of maltodextrin is $C_{6n}H_{(10n+2)}O_{(5n+1)}$ and its chemical structure is shown in Figure 1.

Maltodextrins are widely used as encapsulating agents for spray drying. Usually, maltodextrin is added to puree and acts as a drying aid, raising the glass transition temperature of the product and reducing stickiness [Quek *et al.*, 2007]. On the other hand, it has also been found to be more capable of retaining certain properties such as nutrient content, color, and flavor during spray drying [Rodríguez-Hernández *et al.*, 2005]. Maltodextrins are often used in various sugar-rich foods such as blackcurrant, raspberry, and apricot juice during spray drying.

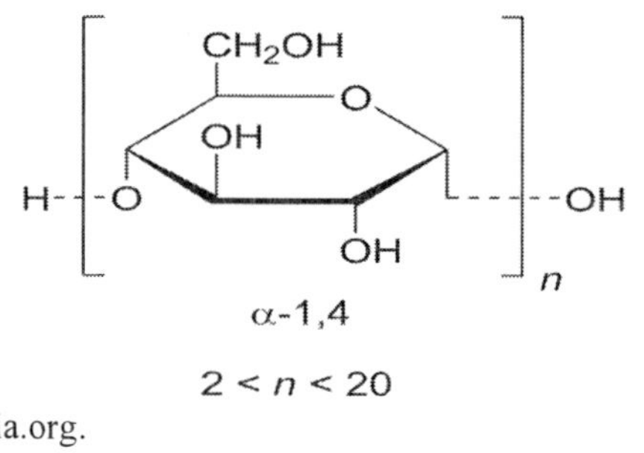

CH2OH, O, OH, H-O, OH, OH, n

α-1,4

$2 < n < 20$

From www://upload.wikimedia.org.

Figure1. Chemical structure of maltodextrin.

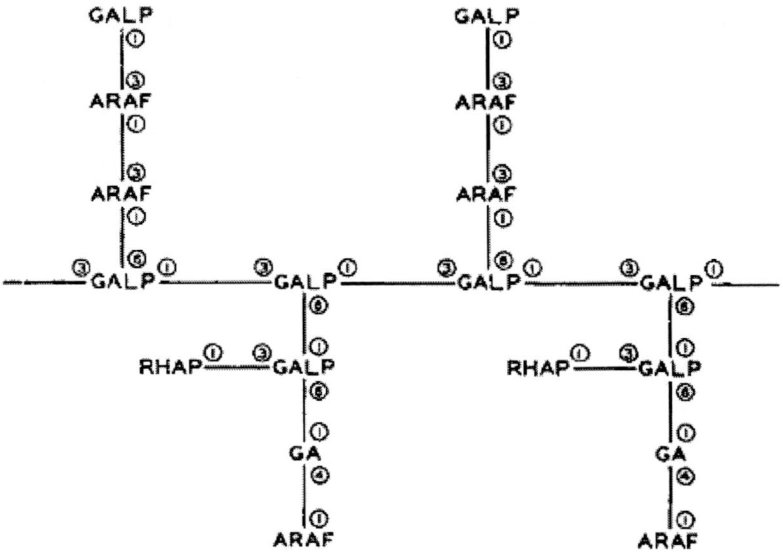

GALP= D-GALACTOPYRANOSE ARAF= L-ARABOFURANOSE
GA= D-GLUCURONIC ACID RHAP=L-RHAMNOPYRANOSE

From Glicksman and Schachat, 1959.

Figure 2. Structure of gum arabic.

Gum Arabic

Gum arabic is a mixture of polysaccharides and glycoproteins and is also known as acacia gum. Gum arabic forms a visible film at oil interfaces and shows stable emulsion with most oils over a wide pH range. It is often used as an effective encapsulating agent due to its protective colloid functionality [Krishnan et al., 2005]. D-galactose, L-rhamnose, L-arabinose, and D-glucuronic acid and approximately 2% protein are the main components [Dickinson, 2003]. Usually, gum arabic is used in the food industry as a stabilizer. Figure 2 shows the structure of gum arabic.

During spray drying, gum arabic is used as an encapsulating agent due to its emulsifying capacity and low viscosity in aqueous solution. Qi and Xu [1999] reported that gum arabic can protect orange oil against oxidation. Bixin encapsulated with gum arabic is 3 to 4 times more stable than that encapsulated with maltodextrin [Barbosa et al., 2005]. However, another study reveled that gum arabic is not an effective wall material compared to others such as citral, linalool, β-myrcene, limonene, and β-pinene [Bertolini et al., 2001].

Chitosan

Chitosan is a linear polysaccharide that is soluble in acidic aqueous media. It is obtained by the N-deacetylation of chitin, a β-(1-4)-linked N-acetyl-D-glycan [Tharanathan and Kittur, 2003] [Figure 3]. Chitosan has potential in many biotechnological applications due to its non-toxic, biodegradability, and biocompatible properties [Miles, 1992]. It is often used as a potential polysaccharide resource as well as in the food, cosmetics, biomedical, and pharmaceutical fields [Rinaudo, 2006].

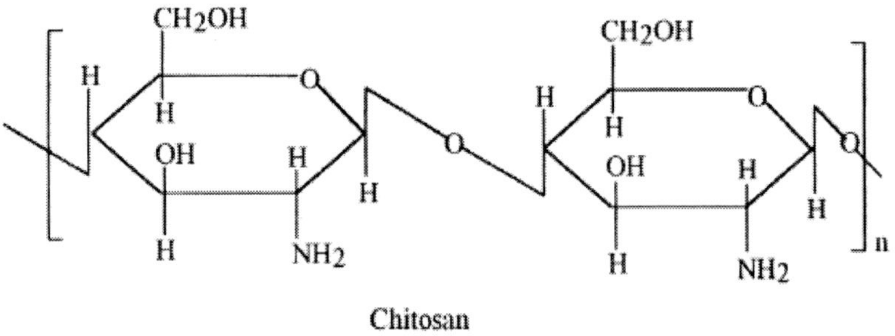

Chitosan

From Majeti and Kumar, 2000.

Figure 3. Structure of chitosan.

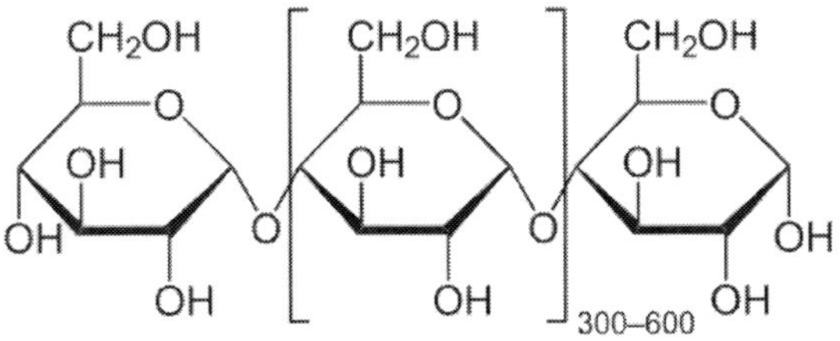

From www://upload.wikimedia.org.

Figure 4. Structure of amylose.

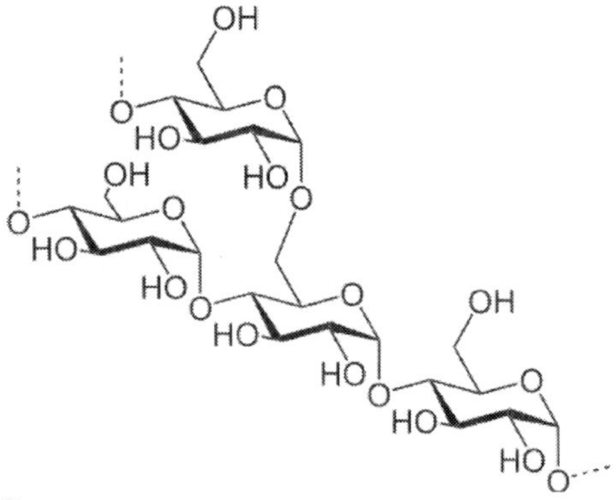

From www://upload.wikimedia.org.

Figure 5. Structure of Amylopectin.

Chitosan has gained interest as a coating material due to its biocompatibility, low toxicity, and biodegradability [Grenha et. al., 2007].

Starch

Starch is a semi-crystalline biopolymer composed of amylose and amylopectin macromolecules. Starch becomes soluble in water when heated and can be separated into two fractions, amylose and amylopectin. Amylose (Figure 4), an unbranched type of starch, consists of glucose residues connected by α-(1-4)-linkages. Amylopectin (Figure5), the branched form, is composed of on α-(1-6)-linkage per 30 α-(1-4)-linkages.

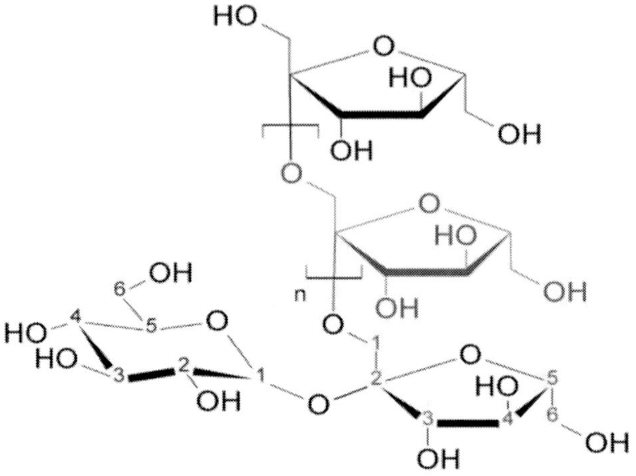

From www://upload.wikimedia.org.

Figure 6. Structural formula of inulin.

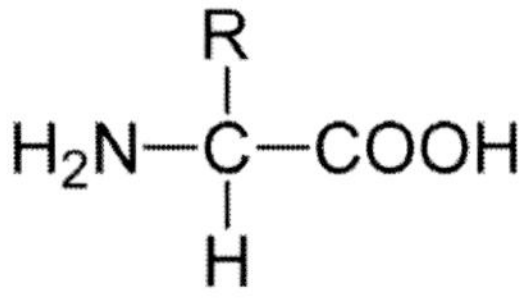

$$H_2N-\underset{H}{\overset{R}{C}}-COOH$$

Figure 7. Structure of amino acid.

Starch is widely used as a wall material due to its abundant availability, low cost, and emulsifying property during spray drying. Starch has also been used to enhance the storage stability of flavors [Partanen et al., 2002]. Different types of starch are used, such as modified tapioca starch, native tapioca starch, and waxy maize starch, as encapsulating agents during spray drying.

Inulin

Inulin is a polysaccharide produced by many types of plants, such as chicory (*Cichorium intybus)* root, Dahlia (*Dahlia pinuata* Cav.), and Jerusalem artichoke (*Helianthus tuberosus*). It is composed of β-(2-1)-linked d-fructose molecules (Figure 6) and can be used to replace sugar, fat, and flour.

Inulin is an interesting possible encapsulating agent due to its nutritive properties [Stevens, Meriggi and Booten, 2001], and it has also many health benefits. Inulin may receive even more attention as an encapsulating agent due to its low cost.

Proteins

Proteins are polymers of amino (Figure 7). Different types of proteins such as soy protein, whey protein, and gelatin are used in the food industry. Among them, soy protein is one of the most popular plant protein sources and is abundant and inexpensive. The major components of soy protein are globulin, glycinin, and β-conglycinin. Soy protein is produced from defatted soy meal by alkali extraction, followed by acid precipitation at pH 4.5 [Choa et al., 2007].

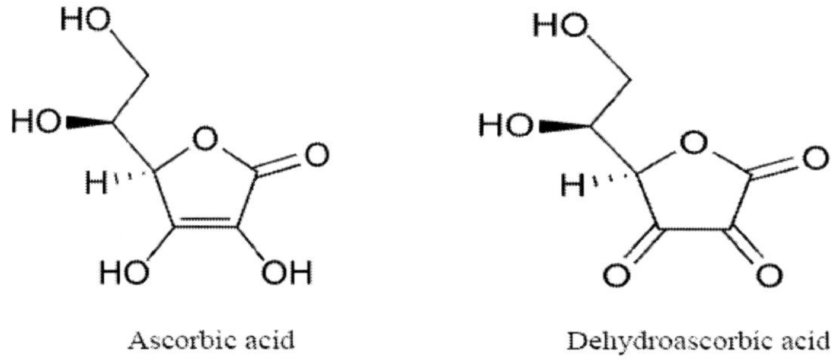

From www://upload.wikimedia.org.

Figure 8. Structure of ascorbic acid.

During spray drying, proteins are used as an encapsulating material. Protein has the capability to protect against oxidation [Bylaite et al., 2001] and high-binding properties [Landy et al., 1995].

Ascorbic Acid

Ascorbic acid is commonly known as vitamin C. Its appearance is a white or light-yellow powder. It is common in fruits and vegetables such as guava, orange, apple, strawberry, kiwi fruit, and cauliflower. Ascorbic acid is very unstable in the presence of air moisture, light, heat, and oxygen. Structures of ascorbic acid are shown in Figure 8.

Many researchers have been used ascorbic acid as a coating material during spray drying [Ahmed *et al.*, 2010b; Desai and Park, 2005]. Encapsulated acids reduce hygroscopicity, and dusting and provide a high degree of flow ability without clumping [Shahidi and Han, 1993]. It can be used as a vitamin supplement and to protect nutritive quality in food during processing [Kirby *et al.*, 1991].

The above-mentioned wall materials are widely used during spray drying. However, some researchers use different wall materials such as carrageenen, silicon dioxide, sodium caseinate, soy lecithin, and sodium alginate.

Most Common Bioactive Compounds from Fruits, Vegetables, and Herbs

Especially, we focused on the most common bioactive compounds [Table 1-3], which are those that have been shown to prevent or retard various types of human diseases such as cancer, aging, diabetes, and heart disease. Below are some of the most important bioactive compounds:

Anthocyanins

Anthocyanins are a large group of water-soluble pigments that are responsible for the attractive orange, red, purple, and blue colors of fruits, vegetables, and flowers [Prior and Wu, 2006]. These compounds constitute a sub-group within flavonoids and are characterized by a C-6-C-3-C-6-skeleton (Figure 9). More than 400 anthocyanins have been found in nature. Among them, the most common are pelargonidin, cyaniding, delphinidin, peonidin, petunidin, and malvidin [Mazza and Miniati, 1993]. Anthocyanins vary depending on the number of hydroxyl groups as well as their hydroxylation and methoxylation patterns.

Table 1. Bioactive contents of vegetables and fruits subgroups per 100 g

Bioactive components	Dark green leafy vegetables	Cabbage family vegetables	Lettuces	Legumes	Deep fruits tubers	Orange/yellow roots	Citrus family fruits	Red/purple/ blue berries	Other fruits and vegetables
Beta-carotene (mcg)	5325	846	2406	52	2644	364	120	43	93
Lycopene (mcg)	0	3	0	0	0	1539	177	0	0
Anthocyanidins (mg)	0.03	9.13	0.51	1.74	1.55	10.23	0.00	84.77	3.90
Vitamin C (mg)	59	51	13	130	18	50	40	22	18
Flavones-3-ols (mg)	0.24	0.00	0.00	4.82	2.37	3.88	2.67	0.00	25.72
Flavonones (mg)	0.00	0.02	0.00	0.00	0.00	0.00	0.09	36.13	0.04
Flavones (mg)	25.66	0.31	0.41	0.00	0.23	0.50	0.57	3.19	0.03
Flavonol (mg)	11.67	3.45	6.88	16.66	0.49	0.61	0.61	0.39	6.56
TAC	3082	1548	1108	5047	972	2157	1301	6319	1576

Source (adapted from Pennington and Fisher, 2010).

TAC= Total antioxidant capacity measured in Trolox equivalents.

Table 2. Total phenolic acid contents in some vegetables (mg/100 g of fresh weight) as aglycones

Vegetables name	Scientific name	A	B	C	D	E	F	G	H	Σ
Tomato	*Lycopersicon esculentum*	2.0	0.29	nd	0.14	nd	1.0	0.06	nd	3.5
White cabbage	*Brassica oleraceavar. capitata ''f.alba''*	0.29	0.27	2.8	Nd	Nd	0.18	0.19	0.11	3.8
Cauliflower	*Brassica oleraceavar. botrytis*	0.38	0.35	1.8	0.11	Nd	1.2	0.42	0.32	4.6
Red cabbage	*Brassica oleraceavar. capitata ''f.rubra''*	1.6	6.4	22	1.0	0.21	9.3	0.27	nd	44
Broccoli	*Brassica oleracea var. italica*	0.42	1.1	8.0	0.13	0.25	0.85	0.41	nd	15
Avocado	*Persea americana*	0.46	1.2	0.97	nd	0.21	0.81	0.13	0.15	3.8
Green bean/fresh	*Phaseolus vulgaris, viciafaba*	0.13	0.26	nd	nd	0.49	1.2	0.11	0.08	3.5
Carrot	*Daucus carota*	26	1.5	nd	nd	0.98	0.69	5.0	nd	34
Red beet	*Beta vulgaris*	nd	25	nd	0.35	0.34	0.65	0.05	0.51	27
Soya bean	*Glysine max*	0.33	12	12	nd	10	12	1.5	25	73

Source [adapted from Mattila and Hellstrom, 2007].

A: caffeic acid; B: ferulic acid; C: sinapic acid; D: protocatechuic acid; E: vanillic acid.

F: p-coumaric acid; G: p-hydroxybenxoic acid; H: syringic acid. nd= Not detected.

Table 3. Total Phenolic acid contents in some fruits (mg /100 g of fresh weight) as Aglycones

Fruits name	Scientific name	A	B	C	D	E	F	G	H	I	J	Σ
Straw berry	*Fragaria ananassa*	0.171	nd	nd	nd	nd	4.6	4.4	nd	1.07	3.3	14
Black currant	*Ribes nigrum*	3.5	1.33	1.17	4.5	0.70	4.7	1.4	nd	nd	5.5	23
Blueberry	*Vaccinium spp.*	59.1	1.29	0.70	2.0	2.0	1.65	nd	15.6	0.41	2.7	85
cloudberry	*Rubus chamaemorus*	1.8	2.0	0.34	nd	0.80	5.6	0.93	nd	2.6	26.0	40
peach	*Prunus persica*	4.9	0.11	nd	nd	0.25	0.52	0.22	0.26	-	nd	6
apple, Lobo, whole fruits	*Malus domestica*	4.3	0.27	0.080	0.45	0.092	0.66	nd	nd	-	7.2	13
banana	*Musa sapientum*	0.20	5.4	nd	nd	0.445	0.46	0.12	0.22	-	nd	7
grape, red	*Vitis vinifera L.*	3.4	0.43	nd	nd	1.07	3.8	nd	6.8	-	3.1	19
grape, green	*V. vinifera L.*	3.4	nd	nd	nd	nd	1.17	nd	nd	-	2.85	7
cherry	*Prunus avium,*	17.1	0.46	nd	3.0	1.17	5.1	0.88	nd	-	nd	28
pear	*Pyrus communis*	6.5	0.29	0.104	nd	0.27	0.70	nd	nd	-	nd	8
orange	*Citrus sinensis*	3.3	9.4	2.2	nd	0.44	1.78	0.54	nd	-	nd	18
Mandarin (clementine)	*Citrus retilculata*	6.6	9.24	1.51	nd	0.64	0.88	nd	nd	-	nd	19
grapefruit	*Citrus paradisi*	5.5	11.6	0.99	nd	1.66	1.35	nd	nd	-	nd	21
kiwi fruit	*Actinidia chinensis*	1.5	0.19	nd	0.66	0.19	0.25	nd	nd	-	nd	3
watermelon	*Citrullus lanatus*	0.12	0.35	nd	nd	0.23	0.37	nd	0.86	-	nd	2

Source [adapted from Mattila et al., 2006].

A: caffeic acid; B:, ferulic acid; C: sinapic acid; D: protocatechuic acid; E: vanillic acid; F: p-coumaric acid.

G: p-hydroxybenxoic acid; H: syringic acid; I: cinnamic acid; J: gallic acid.

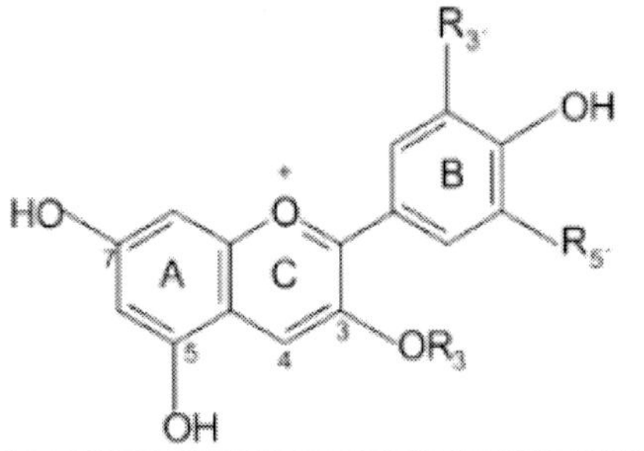

Anthocyanin	R_3	$R_{3'}$	$R_{5'}$
Pelargonidin	H	H	H
Cyanidin	H	OH	H
Delphinidin	H	OH	OH
Peonidin	H	OCH_3	H
Petunidin	H	OCH_3	OH
Malvidin	H	OCH_3	OCH_3
Pelargonidin 3-glucoside	Glc	H	H
Cyanidin 3-glucoside	Glc	OH	H
Delphinidin 3-glucoside	Glc	OH	OH
Peonidin 3-glucoside	Glc	OCH_3	H
Petunidin 3-glucoside	Glc	OCH_3	OH
Malvidin 3-glucoside	Glc	OCH_3	OCH_3

From Haslam, 1993; Giusti et al., 1999.

Figure 9. Basic structures of anthocyanins.

Anthocyanin stability is greatly affected by various conditions, including pH, temperature, light, oxygen, solvents, enzymes, flavonoids, proteins, and metallic ions. Figure 10 shows the structural transformations of anthocyanins upon pH increase, resulting in colorless hemiketal structures and bluish quinoidal bases [Mistry et al., 1991; Fossen et al., 1998; Cabrita et al., 2000].

Various studies have reported the relationship between consumption of anthocyanin-rich foods and improved health. Anthocyanins are often used due to their anti-mutagenic, anti-inflammatory, antioxidative, anti-cancer, and anti-diabetic activities [Chandra et al., 1992; Jayaprakasam et al., 2005; Wang et al., 1999]. Therefore, anthocyanin-rich foods have gained more attention from product developers and consumers.

Carotenoids

Carotenoids such as β-carotene, lycopene, lutein, and zeaxanthine are known to exhibit antioxidant activity. However, β-carotene is the most thoroughly studied. Certain carotenoids possess provitamin A activity. Citrus fruits and vegetables, including carrots, sweet potatoes, winter squash, pumpkin, papaya, mango, and cantaloupe, are rich sources of carotenoids [Khachik et al., 1991]. Lycopene sources are typically tomatoes, watermelon, pink grapefruit, apricot, and pink guava. On the other hand, carrot, and spinach are good sources of β-carotene and lutein. Common carotenoids structures are shown in Figure 11.

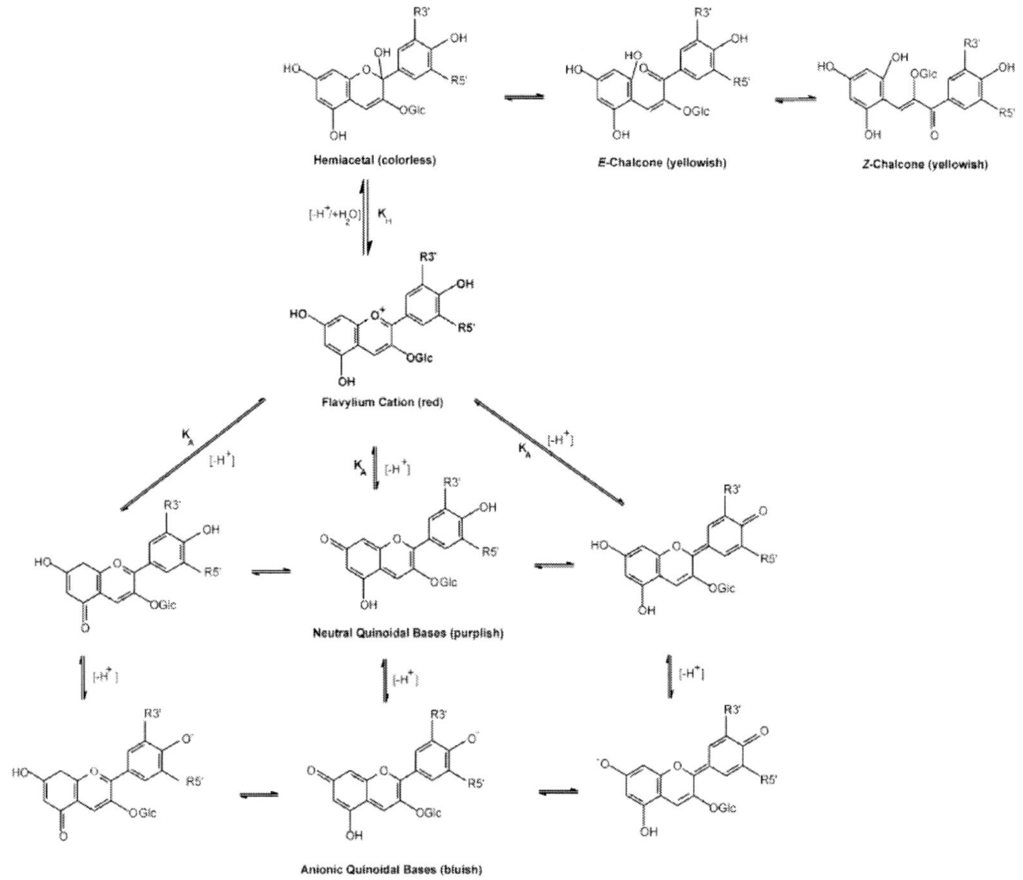

Hemiacetal (colorless)

E-Chalcone (yellowish)

Z-Chalcone (yellowish)

Flavylium Cation (red)

Neutral Quinoidal Bases (purplish)

Anionic Quinoidal Bases (bluish)

From Dangles et al., 1993.

Figure10. Structure transformations of anthocyanins upon pH changes.

There is much scientific evidence that carotenoids have important beneficial effects on human health. Johnson [2002] showed that β-carotene and lycopene are inversely correlated to the risk of cardiovascular disease and certain cancers. Another study reported that tomato products such as tomato sauce and tomato juice significantly reduce the levels of oxidized low-density lipoprotein [Fuhramn *et al.*, 1997]. In addition, carotenoids may play a role in enhancing cell-to-cell communication [Zhang *et al.*, 1991]. They also may act as anti-inflammatory and anti-tumor agents and induce detoxification of enzyme systems [Khachik *et al.*, 1999].

Flavonoids

Major sources of flavonoids are onion, apple, spinach, cauliflower, broccoli, carrot, plum, apricot, grape, different berries, tea beverages, and red wine [Aherne and O'Brien, 2002; Sultana and Anwar, 2008]. Flavonoids are classified into six classes, flavones, flavanones, flavonols, iso-flavones, anthocyanidins, and flavanols (or catechins), according to their molecular structure (Figure 12) [Peterson et al., 1998].

Recently, Salas et al. [2011] reported that flavonoids from citrus species, such as naringin, hesperidin, and neohesperidin compounds have anti-fungal activity against various fungi such as *Aspergillus parasiticus, Aspergillus flavus, Fusarium semitectum,* and

Penicillium expansum. Flavonoids have been shown to possess anti-carcinogenic, anti-inflammatory, anti-hepatotaxic, anti-bacterial, anti-viral, anti-allergic, anti-thrombic, and antioxidative effects [Meyer et al., 1998; Saint-Cricq de Gaulejac et al., 1999]. Many researchers suggest that cardiovascular disease could be reduced due to the consumption of flavonoids. Loke et al. [2008] showed that isolated flavonoids from tea can enhance nitric oxide status.

Vitamin C

L-ascorbic ($C_6H_8O_6$) and dehydroascorbic acids are the major forms of vitamin C. Its chemical name is 2-oxo-L-threo-hexono-1, 4-lactone- 2, 3-enediol [Moser and Bendich, 1990] (Figure 13). Vitamin C is a water-soluble antioxidant and is found in citrus fruits, green pepper, red pepper, strawberry, tomato, broccoli, brussel sprout, turnip, and other leafy vegetables. As the human body does not synthesize vitamin C, it must be supplied in the diet.

In a chemical sense, vitamin C serves as an electron donor that protects the body from radicals and pollutants [Iqbal, et al., 2004]. It is also capable of preventing the onsent of allergic rhinitis [Thornhill and Kelly, 2000], diabetes [Anderson et al., 2006], heart disease [Liu et al., 2002], and cancer [Enwonwu and Meeks, 1995]. High intake of ascorbic acid (>10 g/day) also may prevent cold infections and cancer [Douglas et al., 2000; Cameron and Pauling, 1979].

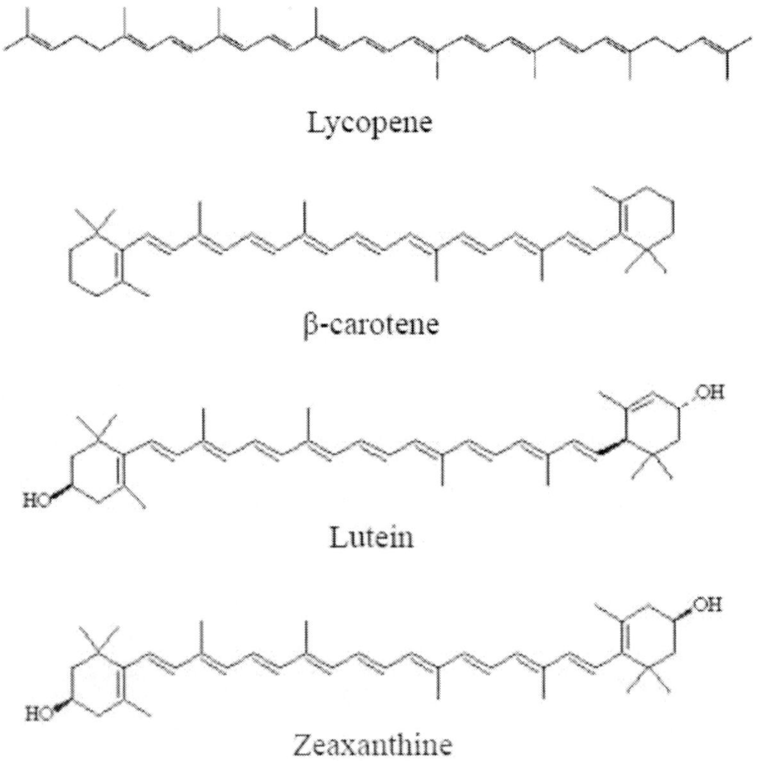

Lycopene

β-carotene

Lutein

Zeaxanthine

From Rao and Rao, 2007.

Figure 11. Carotenoid structures.

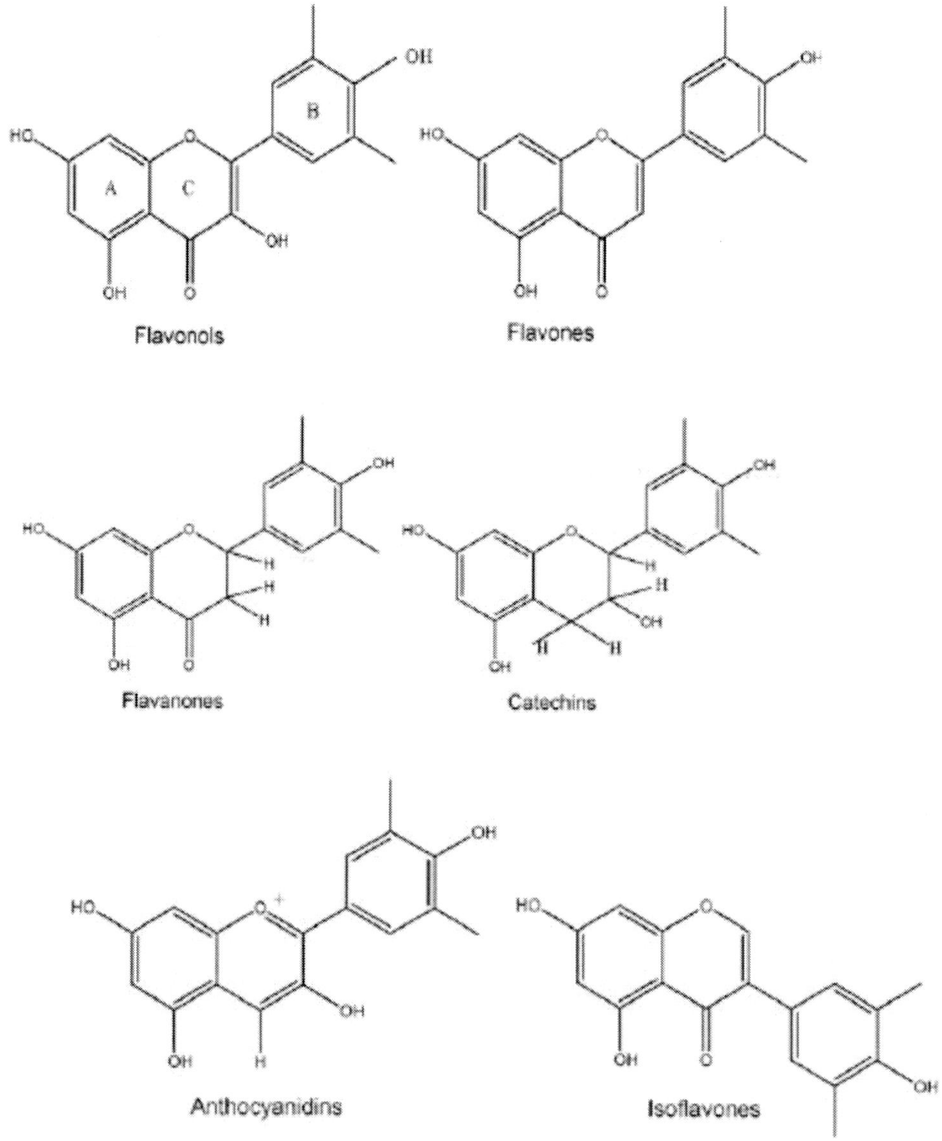

From Tripoli et al., 2007.

Figure 12. Molecular structures of flavonoids.

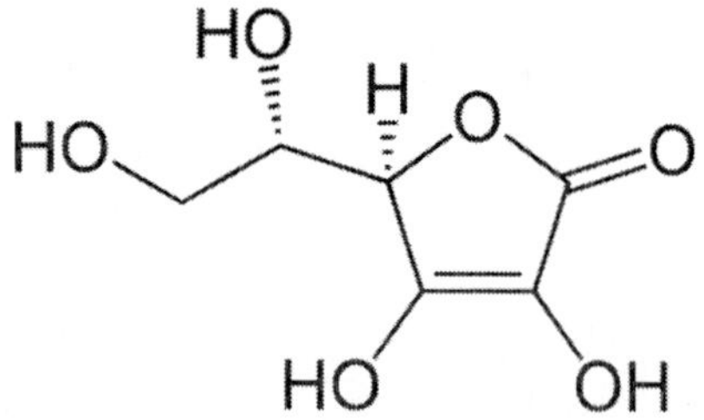

From www://upload.wikimedia.org.

Figure 13. Structure of L-ascorbic acid.

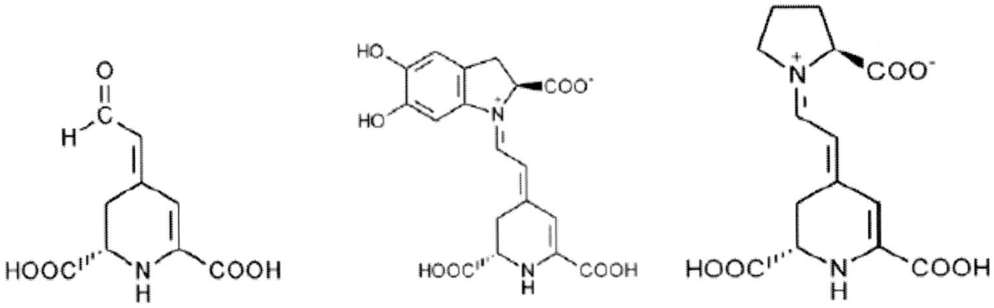

Betalamic acid Betanidin Indicaxanthin

From Wyler et al., 1963; Paiattelli et al., 1964.

Figure 14. Chemical structures of betalains.

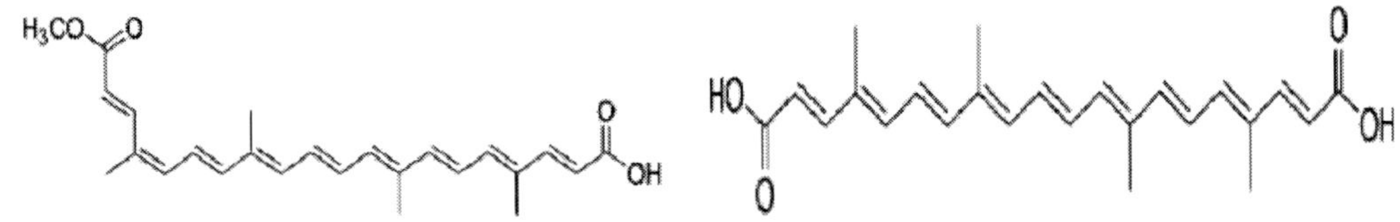

Cis-bixin norbixin

From www://upload.wikimedia.org.

Figure 15. Chemical structure of bixin.

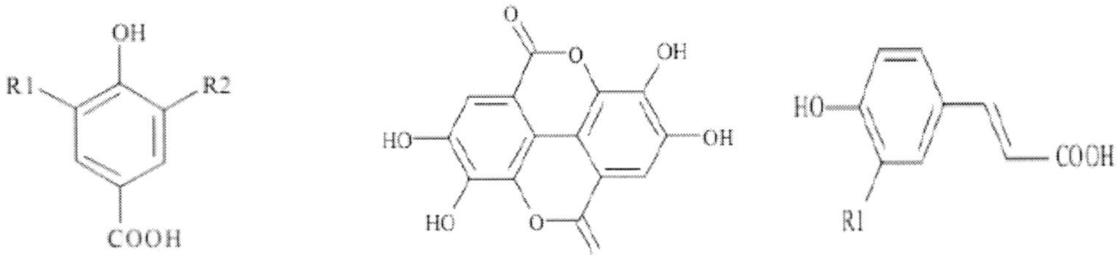

a) b) c)

From Sari, 2000.

Figure 16. Chemical structures of (a) hydroxybenzoic acids: p-hydroxybenzoic acid, R1=H, R2=H; gallic acid, R1=OH, R2=OH, and (b) ellagic acid. C) Hydroxycinnamic acids: p-coumaric acid, R1=H; caffeic acid, R1=OH; ferulic acid, R1=OCH_3.

Betalains

Betalains are water-soluble pigments that are stable between pH 3 and 7. Usually, yellow, red, and violet colors are exhibited by betalains. These compounds have gained interest due to their potential role as natural colorants. The most common representative is betanidin 5-O-β-glucoside (betanin) from red beet (*Beta vulgaris* subsp. v*ulgaris*) root. The structures of betalamic acid, betanidin, and indicaxanthin are shown in Figure 14.

Betacyanins have been shown to possess antioxidant and radical scavenging activities [Escribano et al., 1998; Kanner et al., 2001; Pedreno and Escribano, 2000]. Beet root applied to the skin of mice exhibits a significant inhibitory effect on lung cancer [Kapadia,et al., 1996]. Schwartz et al. [1983] reported that red beet did not exert hepatotoxic or mutagenic activity.

Bixin

Bixin is an apocarotenoid pigment obtained from seed coats. It is a fat-soluble pigment that contains a mixture of carotenoids, including norbixin, phytoene, and δ-carotene [Mercadante et al., 1996]. Some structures of bixin are shown in Figure 15.

Bixin colorant is considered to be a chemopreventive agent against oxidative DNA damage [Thresiamma et al., 1998; silva et al., 2001]. Recently, Agner et al. [2005] showed that annatto does not display any adverse effect on DMH-induced preneoplastic lesions and DNA damage in the colon of rat. Another study by Siva et al. [2008] reported that bixin dye could be used as an alternative tracking dye in gel electrophoresis. Bixin has also been found to act as an anti-inflammatory and anti-carcinogenic agent [Agner et al., 2005].

Phenolic Compounds

Phenolic compounds are secondary metabolites synthesized by plants. Vegetables, fruits, and herbs are good sources of phenolic acids. Different types of phenolic acids found in nature include caffeic, chlorogenic, ferulic, sinapic, *p*-coumaric, *p*-hydroxybenzoic, vanillic, and syringic acids [Dziedzic and Hudson, 1984; Larson, 1988]. Figure 16 shows several phenolic acid structures.

Phenolic compounds play an important role in adsorbing and neutralizing free radicals, quenching singlet and triplet oxygen, and decomposing peroxides [Osawa, 1994]. Phenolic acids have been reported to act as anti-bacterial, anti-viral, anti-carcinogenic, anti-inflammatory, and vasodilatory agents [Duthie et al., 2000; Breinholt, 1999]. They have also been used to prevent various diseases, such as cardiovascular and neurodegenerative diseases and cancer [Sari, 2000].

Influence of Encapsulating Agents on Bioactive Compounds in Fruits, Vegetables, and Herbs during Spray Drying

Several authors have used different kinds of encapsulating agents in order to stabilize the bioactive compounds in fruits, vegetables, and herbs (Table 4-6) during spray drying. In this section, we describe the different bioactive compounds in fruits, vegetables, and herbs that were encapsulated by spray drying using various types of wall materials.

Cactus Pear (Opuntia Ficus-Indica and Opuntia Streptacantha)

Cactus pear is a tropical fruit from America. Usually, it grows in arid and semi-arid regions [Saenz, 2000]. It is also a good source of bioactive compounds, including betacyanin, indicaxanthin, polyphenols, and vitamin C [Saenz et al., 2009; Rodrigue-Hernandez et al., 2005]. Cactus pulp and ethanolic extract of cactus were encapsulated with maltodextrin and inulin in a previous study [Saenz et al., 2009]. The contents of betacyanin and indicaxanthin were found to be almost similar using both encapsulated agents. However, the polyphenolic content was higher when encapsulated with inulin as compared to maltodextrin. These authors also showed that polyphenols are higher in both cactus pulp and the ethanolic extract of cactus encapsulated with maltodextrin and inulin during storage at 60^0C. Gandia-Herrero et al [2010] found that the stability of indicaxanthin can be highly increased by encapsulation with maltodextrin. Retention of vitamin C in cactus pear juice is higher in maltodextrin (10 DE) compared to maltodextrin (20 DE) [Rodrigue-Hernandez et al., 2005].

Acai (Euterpe Oleraceae Mart.)

Acai is a fruit native to Amazone, Brazil. It is considered to be an important source of anthocyanins and has high antioxidant activity [Tonon et al., 2008; Tonone et al., 2010]. Acai contains higher antioxidant capacity than other fruits, including high bush, blueberry, blackberry, and cranberry [Del Pozo-Insfran et al., 2004]. Tonon et al. [2008] prepared acai powder by spray drying using three types of maltodextrin (10, 20, and 30 DE). These authors reported that the various types of encapsulation powders have no significant effects on the retention of anthocyanins. More recently, the effects of maltodextrin (10 DE, 20 DE), gum arabic, and tapioca starch on the anthocyanin content and antioxidant activity of acai juice upon spray drying were studied by Tonon et al [2010]. These authors found that the anthocyanin content and antioxidant activity do not significantly differ between acai juice encapsulated with maltodextrin and gum arabic after spray drying. Meanwhile, tapioca starch has the lowest anthocyanin content and antioxidant activity. These authors also observed that maltodextrin (10 DE) has the highest anthocyanin content and antioxidant activity during storage, followed by maltodextrin (20 DE) and gum arabic.

Guava (Psidium Guajava L.)

Guava is native to tropical America and grows well in tropical and subtropical regions. It is a rich source of vitamin C (more than 100 mg/100 g of fruit, Perez Gutierrez et al., 2008). Chopda and Barrett [2001] reported that without maltodextrin, guava juice could not be spray dried due to its high sucrose content. According to Chopda and Barrett [2001], higher retention of ascorbic acid is made possible upon encapsulation with maltodextrin product of maltrin 100 compared to maltrin 500. Recently, Osorio et al. [2010a] prepared microencapsulated guava fruits using maltodextrin, gum arabic, and a mixture of the two. These authors revealed that encapsulation with maltodextrin results in higher vitamin C content compared to encapsulation with the mixture. Moreover, these authors showed that microencapsulation with arabic gum guava powder produces an undesirable taste and decreases thermal stability.

Watermelon (Citruluslanatus)

Watermelon is originally from southern Africa. It is often used as a popular snack during hot summer weather. Watermelon is an excellent source of lycopene, β-carotene, and vitamin C as well as minerals, such as potassium and magnesium [Quek et al., 2007]. In addition, Edwards et al [2003] showed that watermelon has higher lycopene content [4868 μg/100 g of watermelon fresh] than raw tomato (3025 μg/100 g of tomato). Watermelon powder was produced by encapsulation with 3% and 5% maltodextrin with spray drying [Quek et al., 2007]. These authors demonstrated that 5% maltodextrin results in better color than 3% maltodextrin. Moreover, the lycopene and β-carotene contents are much higher in encapsulated powder compared to raw watermelon juice.

Table 4. Major bioactive compounds of fruits and wall materials for encapsulation during spray drying

Sources	Bioactive compounds	Wall materials	References
Cactus pear (*Opuntia ficus-indica*)	Betacyanin,indicaxanthin, polyphenols	Maltodextrin , inulin	Saenz, et al., 2009
Cactus pear (*Opuntia Streptacantha)*	Vitamin C	Maltodextrin	Rodrigue-Hernandez et al., 2005
Acai (*Euterpe oleraceae Mart.*)	Total anthocyanin, antioxidant activity	Maltodextrin, arabic gum, tapioca starch	Tonon et al., 2008; Tonon et al., 2010
Guava (*Psidium guajava L.)*	Vitamin C	Maltodextrin, arabic gum	Osorio et al.,2010; Chopda and Barrett 2001
Water melon (*Citruluslanatus*)	Lycopene, β-carotene	Maltodextrin,	Quek et al., 2007
Pomegranate (*Punica granatum*)	Total anthocyanin, polyphenols	Maltodextrin, soybean protein isolates	Robert et al., 2010
Corozo fruit (*Bactris guineensis*)	Total anthocyanin	Maltodextrin	Osorio et al., 2010
Acerola pomace *(Malpighia punicifolia* L)	Ascorbic acid, total anthocyanin	Maltodextrin, cashew tree gum	Moreira et al., 2010
Gac fruit (*Momordica cochinchinensis*)	Total carotenoid, antioxidant activity	Maltodextrin	Kha et al., 2010
Camu-camu (*Myrciaria dubia*)	Vitamin C	Maltodextrin, arabic gum	Dib taxi, et al., 2003
Indian cherry (*Malpighia emarginata* DC)	Vitamin C	Maltodextrin, arabic gum	Righetto and Netto, 2006
Cashew Apple (*Anacardium occidentale*)	Vitamin C	Maltodextrin, cashew tree gum	De Oliveria et al., 2009
Grape seed (*Vitis vinifera*)	Procyanidin, polyphenols	Maltodextrin, arabic gum, Sodium caseinate-soy lecithin	Zhang et al., 2007;Kosaraju et al., 2008
Annatto (*Bixa orellana* L.)	Bixin	Maltodextrin, gum arabic	Barbosa et al., 2005
Olive leaf (*Olea europaea*)	polyphenols	Chitosan	Kosaraju, 2006

Table 5. Major bioactive compounds of vegetables and wall materials for encapsulation during spray drying

Sources	Bioactive compounds	Wall materials	References
Yellow-fleshed sweet potato*(Ipomoea batatas)*	Beta-carotene,vitaminC	Maltodextrin	Grabowski et al., 2008
purple-fleshed sweet potato*(Ipomoea batatas)*	Phenolic, ascorbic acid, flavonoid, anthocyanin content, antioxidant capacity	Maltodextrin, ascorbic acid	Ahmed et al., 2010a, 2010b
Black carrot (*Daucuscarota L.*)	Anthocyanin content, antioxidant activity	Maltodextrin	Ersus and Yurdagel, 2007
Carrot (*Daucuscarota L.*)	Beta-carotene,	Maltodextrin	Wagner and Warthesen, 1995
Amaranthus plant (*Amaranthus cruentus)*	Betacyanin	Maltodextrin , modified starch	Cai, and Corke, 2000
Beet root (*Beta vulgaris)*	Betacyanin, antioxidant activity	Maltodextrin, gum arabic	Azeredo et al., 2007; Pitalua et al., 2010
Tomato (*Solanum lycopersicum*)	Lycopene	Gelatin, sucrose	Shu et al., 2006
Soy bean (*Glycine max*)	Polyphenol, Antioxidant activity, Genistein	Maltodextrin, starch, silicon dioxide	Georgetti et al., 2008

Table 6. Major bioactive compounds of herbs, spices, and wall materials for encapsulation during spray drying

Sources	Bioactive compounds	Wall materials	References
Paprika oleoresin (*Capsicum annum*)	Carotenoid	Soy protein isolate, gum arabic	Rascón et al., 2010
Rosa mosqueta (*Rosa rubiginosa*)	Trans-β-carotene, trans-lycopene	Starch, gelatin	Robert et al., 2003
Yerba mate (*llex paraguariensis*)	Polypenols, antioxidant activity	Chitosan	Harris et al., 2010
Mengkudu (*Morinda citrifolia L*)	Antioxidant activity, phenolic and flavonoid content	Carrageenan	Krishnaiah et al., 2009
Pandan leaf (*Pandanus amaryllifolius*)	Antioxidant activity	Maltodextrin, gum Arabic, osa-modified starch	Porrarud and pranee, 2010
Quercus resinosa (*Pinus strobus*)	Phenolic content, antioxidant activity	Lactose-sodium caseinate	Rocha-Guzman et al., 2010
Ginger rhizomes (*Zingiber officinale Roscoe*)	6-gingerol	maltodextrin and liquid glucose	Phoungchandang and Sertwasana 2010
Turmeric (Curcuma longa)	Curcumin	Porous starch, gelatin, maltodextrin	Wang et al., 2009; Paramera et al., 2010

Pomegranate (Punica Granatum)

Pomegranate has gained popularity due to its high content of bioactive compounds. Usually, pomegranate grows in hot climate areas. Pomegranate juice is a good source of anthocyanins and flavonols. It also contains organic acids such as citric, malic, and oxalic acid [Robert et al., 2010]. Pomegranate polyphenols have been used to reduce the frequency of degenerative human diseases such as cardiovascular disease, cancer, and neurological damage [Lansky and newman, 2007; Mertens-Talcott et al., 2006]. Pomegranate juice and ethanolic extract of pomegranate was previously encapsulated with maltodextrin and soybean protein isolates by spray drying [Robert et al., 2010]. These authors showed that encapsulation with soybean protein isolates is more effective for protecting polyphenols, whereas anthocyanins are better protected by maltodextrin. It was also shown that maltodextrin microcapsules provide greater protection of polyphenols and anthocyanins than soybean protein isolate microcapsules during storage. These authors also observed that the stability of bioactive compounds in encapsulated and non-encapsulated yogurt remains the same, except in ethanolic extract encapsulated with maltodextrin.

Corozo (Bactris Guineensis)

Corozo fruit is a wild palm found in Central or South America. It is used in juices or alcoholic drinks [Osorio et al, 2010b]. Corozo fruit is a valuable source of anthocyanins and can be used as a natural colorant. Osorio et al. [2010b] prepared corozo fruit powder using maltodextrin by spray drying. It was found that microencapsulated fruit powder sample contains higher anthocyanin content than fresh juice. Microencapsulated powder is quite stable until 100°C.

Acerola (Malpighia Punicifolia L)

Acerola is commonly known as West Indian cherry or Barbados cherry and belongs to the Malpighiaceae family. It is also an important source of vitamin C and anthocyanins [Moreira et al., 2010; Righetto and Netto, 2006]. Acerola pomace has been encapsulated with maltodextrin and cashew tree gum by spray drying [Moreira et al., 2010]. Retention of anthocyanins and ascorbic acid does not significantly differ between encapsulation with maltodextrin and cashew tree gum. However, these authors demonstrated that maltodextrin may not be replaced by cashew tree gum due to changes in the contents of other compounds. Another study showed that vitamin C content is higher in West Indian cherry encapsulated with a combination of 5% maltodextrin and 15% gum arabic compared to that encapsulated with 20% maltodextrin and a combination of 15% maltodextrin and 5% gum arabic [Righetto and Netto, 2006].

Gac (Momordica Cochinchinensis)

Gac fruit contains high levels of β-carotene [more than 16 mg/100 g] and lycopene (more than 50 mg/100 g) [Aoki et al., 2002]. Spray-dried powder was previously produced from Gac fruit aril using different concentrations (10 to 30%) of maltodextrin [Kha et al., 2010]. These authors observed that the total carotenoid content is reduced as the maltodextrin content is increased. On the other hand, the total antioxidant activity remains the same between encapsulation with 10 to 20% maltodextrin. However, in the presence of 20 to 30% maltodextrin, the total antioxidant activity decreases. Therefore, these authors suggested that high quality Gac powder could be produced by spray drying at an inlet temperature of 120^0C and a maltodexdtrin concentration of 10%.

Camu-Camu (Myrciaria Dubia)

Camu-camu, a member of the *Myrtaceae* family, is native to the Amazonian region. Camu-camu fruits have an exotic flavor, high content of vitamin C [Dib taxi et al., 2003], and are traditionally used to deplete the immune system. It also helps promote brain function and circulatory system function (www.wholeworldbotanicals.com/info_royalcamu). Camu-camu juice was previously dried using different concentrations (5 to 35%) of gum arabic or maltodextrin as a wall material by spray drying [Dib taxi et al., 2003]. Both encapsulating materials resulted in similar retention of vitamin C. According to Dib taxi et al. [2003], 15% wall material and an inlet air temperature of 150^0C are the optimum conditions for encapsulation of camu-camu powder based on yield and retention of vitamin C.

Cashew Apple (Anacardium Occidentale)

Cashew apple is native to northeastern Brazil and is well known for its antioxidant properties. It is also considered as a rich source of vitamin C, carotenoids, phenols, and tannin [Assuncao and Mercadante, 2003]. It was previously reported that cashew apple has anti-tumor [Cavalcante, et al., 2005;], anti-microbial [Cavalcante et al., 2005], urease inhibitory [Kubo et al., 1999], and lipoxygenase activities [Ha and Kubo, 2005]. De Oliveria et al. [2009] prepared cashew apple powder using maltodextrin and cashew tree gum by spray drying. These authors also performed encapsulation by totally or partially replacing maltodextrin with cashew tree gum. De Oliveria et al. [2009] observed that ascorbic acid retention is not significantly affected by the replacement of maltodextrin with cashew tree gum. However, these authors reported that $\geq$ 50% of cashew tree gum could be replaced by maltodextrin. Kosaraju et al. [2008] reported significant retention of antioxidant activity after encapsulation using soy lecithin and sodium caseinate of apple plolyphenol extract.

Annatto (Bixa Orellana L.)

Annatto is commonly found in Central and South America. Bixin is the major coloring component in annatto. The seed coating of Annatto consists of more than 80% carotenoids [Balaswamy et al., 2006]. Annatto is used as a colorant in dairy products and soft drinks and as a thickener in soup, gravy, fabricated snacks, and bakery products. Bixin was previously encapsulated with gum arabic or maltodextrin by spray drying [Barbosa et al., 2005]. These authors found that bixin encapsulation increases stability 10 times compared to non-encapsulated sample in the absence of light. These authors also observed that bixin encapsulation with gum arabic increases stability 3 to 4 times more compared to that of encapsulation with matodextin.

Grape Seed (Vitis Vinifera L.)

Grapes and grapes seed are rich in antioxidant compounds, including polyphenols and procyanidin [Kosaraju et al., 2008; Zhang et al., 2007], which may reduce the risk of chronic diseases. Grape anthocyanins were encapsulated using different carrier agents, such as maltodextrin, γ–cyclodextrin, and arabic gum by spray drying [Burin et al., 2011]. In that study, a combination of maltodextrin and gum arabic were used to increase the anthocyanin half-life time and lower the degradation rate compared to other carrier agents. Zhang et al. [2007] encapsulated grape seed using gum arabic and maltodextrin by spray drying. They showed that the procyanidin content is not affected by encapsulation. These authors also reported that the stability of encapsulated product is sharply enhanced by storage. Another study showed that grape seed significantly increases the retention of antioxidant activity after encapsulation with soy lecithin and sodium caseinate by spray drying [Kosaraju et al., 2008].

Olive (Olea Europaea)

Olive fruits are usually grown in tropical and subtropical regions and belong to the Burseraceae family. Olives are used for treating faucitis, stomatitis, hepatitis, and toxicosis [He and Xia, 2007]. Several phenolic compounds have been found in olive fruits and olive leaf, including gallic acid, ellagic acid, hyperin, and andoleorupein [He and Xia, 2007; Kosaraju et al., 2008]. Olive leaf was extracted using chitosan by spray drying [Kosaraju et al., 2006]. These authors observed minor interactions between phenolic compounds in olive leaf extract and the polysaccharide matrix. The same authors also produced olive leaf powder using soy lecithin and sodium caseinate by spray drying. Significant retention of antioxidant activity was observed after encapsulation with soy lecithin and sodium caseinate by spray drying [Kosaraju et al., 2008].

Sweet Potato (Ipomoea Batatas)

Sweet potatoes are a highly nutritious vegetable that is rich in biologically active phytochemicals such as β-carotene, polyphenols, ascorbic acid, and anthocyanins [Van Hall, 2000; Yang and Gadi, 2008]. Yellow-fleshed sweet potato powder was prepared using maltodextrin by spray drying [Grabowski et al., 2008]. This author showed that the β-carotene content significantly differs between sweet potato treated with maltodextrin powder and untreated powder, whereas the vitamin C content does not significantly differ. Another author produced encapsulated flour from purple-fleshed sweet potato using combinations of various levels of maltodextrin and ascorbic acid by spray drying [Ahmed et al., 2010a]. According to Ahmed et al. [2010a], encapsulated flour contains higher total phenolic content and antioxidant capacity than non-encapsulated flour. However, the flavonoid content of encapsulated flour depends on the concentrations of ascorbic acid and maltodextrin. The same author demonstrated in another research paper that maltodextrin-treated, purple-fleshed sweet potato flour has higher phenolic content and antioxidant activity than untreated flour. On the other hand, the anthocyanin content is lower and the ascorbic acid content does not significantly differ between maltodextrin-treated and untreated flour [Ahmed et al., 2010b].

Carrot (Daucuscarota L.)

Carrots are an excellent source of antioxidants, including β-carotene, α-carotene, and anthocyanins. Usually yellow, red, black, and purple-colored carrots are available. Ersus and Yurdagel [2007] encapsulated black carrot by spray drying using different concentrations of maltodextrin such as maltodextrin (28-31 DE0), glucodry 210 (20-23 DE), and stardri 10 (10 DE). The results indicated that Glucodry 210 provides the highest anthocyanin content and stability compared to other wall materials. Other authors [Wagner and Warthesen, 1995] also used maltodextrin 4, 15, 25, and 36.5 (DE) to produce spray-dried carrot powder. They found that maltodextrin 36.5 (DE) is superior for retaining α-carotene and β-carotene compared to other maltodextrins. However, surface contents of α-carotene and β-carotene are higher in 25 (DE). The author also mentioned that encapsulation with maltodextrin powder improves shelf life 70-220 times compared to carrot juice spray-dried alone.

Amaranthus (Amaranthus Cruentus)

Amaranthus pigments are commonly known as betacyanins, which are widely used in the food industry [Hendry and Houghton, 1996]. Amaranthus pigments are often used as a source of betacyanin-type pigments. In China, Amaranthus pigments are legally used as food ingredients [Cai et al., 1998]. Very recently, another research group [Pasko et al, 2010] showed that amaranth seeds are more efficient for antioxidation of plasma, heart, and lungs of rats. Cai and Corke [2000] used maltodextrin 10, 20, and 25(DE) and native or modified starch to produce Amaranthus powder by spray drying. The authors reported that betacyanin is significantly reduced after adding carrier agents. In addition, treatment with native or modified starch reduces the content of betacyanins compared to treatment with maltodextrin. Moreover, betacyanin retention is higher upon treatment with carrier agents. This study revealed that betacyanin pigment produced with maltodextrin 15 (DE) by spray drying at 180^0C is superior to commercial red beet powder.

Beet Root (Beta Vulgaris)

Beet root is an important source of potent antioxidants and nutrients. The main pigment in beet root is betalain [Stintzing and Carle, 2004], which is commercially used very often. Pitalua et al. [2010] beet juice was encapsulated with gum arabic by spray drying. The results showed that there are no significant differences in betalain concentration, antioxidant activity, and redox potential between encapsulated beet juice samples during storage at different water activities.

In another study conducted by [Azeredo et al., 2007], it was reported that beet root extract encapsulated with gum arabic has lower betalain content than that microencapsulated with maltodextrin. These authors demonstrated that betalain degradation rates are decreased at a higher maltodextrin to beet root ratio during storage.

Soybean (Glycine Max)

Soybeans are considered to contain various healthy compounds, such as iso-flavones, saponins, and anthocyanins. Bioactive compounds obtained from spray-dried soybean extract with added colloidal silicon dioxide, maltodextrin, and starch were investigated by Georgette et al. [2008]. These authors showed that the product with colloidal silicon dioxide experienced lower degradation of phenolic content and antioxidant activity than the product containing maltodextrin and starch. This study also indicated that colloidal silicon dioxide samples had high polyphenol content, genistein content, and antioxidant properties compared to those without colloidal silicon dioxide. From this paper, the authors also concluded that spray-dried soybean extract could be used as a functional food ingredient with high antioxidant activity.

Tomato (Lycopesicon Esculentum Mill)

Tomatoes are one of the most popular and nutritious foods and are widely grown in tropical regions. They are rich in carotenoids (lycopene), ascorbic acid (vitamin C), vitamin

E, and dietary fiber [Davis and Hobson, 1981]. Lycopene is susceptible to oxidants, light, and heat due to the presence of unsaturated bonds in its molecular structure [Shu et al, 2006]. Microencapsulation is one of best techniques for protecting lycopene from heat, oxidation, and heat. Lycopene microcapsules were prepared by spray drying containing different ratios of wall material, such as gelatin and sucrose [Shu et al., 2006]. These authors found that although microencapsulated lycopene shows some isomerization, microencapsulated samples have good storage stability compared to control (without microencapsulate). Finally, these authors concluded the optimization of lycopene as follows: ratio of gelatin and sucrose of 3/7, ratio of core and wall material of 1/4, feed temperature of 190^0C, homogenization pressure of 40 MPa, and lycopene purity of not less than 52%.

Pepper (Capsicum Annuum)

Pepper is generally used as a vegetable and spice. The active ingredient in *pepper* is *oleoresin capsicum.* Oleoresin consists of nine major carotenoid pigments [Minguez-Mosquera and Hornero-Mendez, 1993]. Shaikh et al. [2006] investigated the microencapsulation of black pepper oleoresin using gum arabic and modified starch by spray drying. Gum arabic afforded greater protection of oleoresin than modified starch. Red chili oleoresin-in-water emulsions were spray-dried using various ratios of wall materials, such as mesquite gum, maltodextrin, and whey protein concentrate [Perez-Alonso et al., 2008]. Results from this study showed that the best protection against oxidation is provided when the wall to core ratio and water activity are 4:1 and 0.436, respectively. Rodrigue-huezo et al [2004] showed that the stability of carotenoids upon red chili oleoresin degradation remains almost constant in water-in-oil-water microcapsules made from gellan gum, mesquite, and maltodextrin at a water activity of 0 to 0.515 for $M_{35\%, 3.9}$. In a report by Rascon et al. [2010], paprika oleoresin microcapsules prepared with gum arabic and soy protein were isolated by spray drying. These authors found that water activities of 0.274 and 0.710 for gum arabic and soy protein isolate, respectively, provided maximal stability of carotenoids in both yellow and red paprika. Additionally, these authors suggested preparation of paprika oleoresin microcapsules using soy protein isolate for improved nutritional value.

Rosa Mosqueta Oleoresin (Rosa Rubiginosa)

Rosa species belong to the rose family. The main components of Rosa canina and Rosa rugosa species are lycopene and β-carotene [Razungles et al., 1989], whereas rubixanthin is the main component of Rosa pomifera [Maki-Fisher et al., 1983]. Robert et al. [2003] encapsulated Rosa Mosqueta (Rosa rubiginosa) with starch and gelatin by spray drying. It was revealed that encapsulation with gelatin provides greater protection of main carotenoid pigments (*trans*-rubixanthin, *trans*-lycopene, and *trans*-β-carotene) than encapsulation with starch. These authors also showed that the degradation rate of main carotenoid pigments is the same in starch. However, *trans*-β-carotene was found to be more stable in gelatin.

Quercus Resinosa (Pinus Strobus)

Quercus resinosa leaves are used in northern Mexico as a refreshing beverage, due to its high content of polyphenolic compounds [Rocha-Guzman et al., 2010]. Quercus resinosa leaves are rich in phenolic content. Quercus resinosa leaves were encapsulated by spray drying with different proportions (11:4%, 9:6%, and 7:8%) of lactose-sodium caseinate

[Rocha-Guzman et al., 2010]. These authors indicated that the total phenolic content and antioxidant activity are higher at a lower concentration (7:8%) of lactose-sodium caseinate, whereas higher inhibition of deoxy-D-ribose oxidation was observed at a higher concentration (11:4%) of lactose-sodium caseinate.

Mengkudu (Morinda Citrifolia)

Nowadays, medicinal plants are major sources of bioactive compounds. *Morinda citrifolia* or 'mengkudu' is one of the best sources of phenolic compounds [Krishnaiah et al., 2009]. Previously, mengkudu encapsulation powder was prepared using carrageenan at different ratios by spray drying. The highest antioxidant activity and flavonoid and total phenolic contents were observed at a M_{core}/M_{wall} ratio of 1/2. According to these authors, M_{core}/M_{wall} plays an important role in the optimization of antioxidant activity.

Pandan Leaf (Pandanus Amaryllifolius)

Pandan leaf is widely found in tropical countries and contains different antioxidants, such as quercetin, carotenoids, chlorophyll derivatives, and polyphenols [Miean and Mohamed, 2001; Ferruzzi et al., 2002; Nor et al., 2008]. Zn-chlorophyll derivatives powder was produced from pandan leaf using different proportions (10 to 30%) of wall material, such as gum arabic, maltodextrin, and osa-modified starch, by spray drying [Porrarud and Pranee, 2010]. This study revealed that 30% osa-modified starch powder has higher total chlorophyll content and antioxidant activity compared to that encapsulated with gum arabic or maltodextrin powder. Osa-modified starch powder also has a longer half life (462 days) than powder modified with gum arabic or maltodextrin powder.

Yerba Mate (Llex Paraguariensis)

Yerba mate is a kind of tea beverage that contains phenolic compounds [Harris et al., 2010]. Yerba mate has been encapsulated with different concentrations of chitosan and tripolyphosphate pentasodium [Harris et al., 2010] by spray drying. These authors showed that polyphenols are more retained in yerba mate with chitosan microspheres than that without chitosan microspheres. Therefore, these authors concluded that chitosan microspheres can be used to maintain the stability of polyphenols.

Ginger (Zingiber Officinale Roscoe)

Ginger rhizomes contain different phytochemicals such as (η)-gingerol, zingerone, and (η)-shogaol [Balladin and Headley, 1998]. These phytochemicals are used as antioxidants and anti-cancer agents [Yogeshwer and Madhulika, 2007]. Ginger juice was obtained using different concentrations (0 to 10%) of maltodextrin and liquid glucose by spray drying [Phoungchandang and Sertwasana, 2010]. These authors have found that ginger juice with 5% glucose contains the highest 6-gingerol content compared to that with other glucose concentrations. Finally, these authors noted that high quality ginger products can be produced at a 120^0C inlet temperature and with 5% liquid glucose.

Turmeric (Curcuma Longa)

Turmeric is used as a natural colorant in the food industry and is originally from tropical South Asia. Polyphenol curcumin is found in turmeric and can be used to protect against several chronic diseases such as cancer, HIV infection, cardiovascular disease, and skin disease [Paramera et al., 2010]. Curcumin microcapsules were prepared by spray drying using different ratios of gelatin and porous starch (core to wall material 1/20, 1/30, 1/40) [Wang et al., 2009]. Recently, Paramera et al. [2010] encapsulated curcumin with modified starch by spray drying. It was demonstrated that the stability of encapsulated curcumin is higher with modified starch compared to non-encapsulated curcumin. This study also found that encapsulated curcumin can be protected from light and oxidative degradation.

CONCLUSION

Encapsulation techniques might improve the bioactivity of compounds from food materials. Encapsulation has also been used as an effective barrier against light and oxygen. Spray drying is the most commonly used method for encapsulation. In our study, it seems that encapsulating agents are an effective means of maintaining the bioactivity of compounds from plant food materials during processing. This paper is an updated report about various bioactive compounds obtained from different fruits, vegetables, herbs, spices, and legumes using different types of encapsulating agents by spray drying. We tried to elucidate specific wall materials for the spray drying of specific bioactive compounds from food plants. This information could be very helpful for food applications, in particular for nutraceuticals and high value foods. It is also very useful for those trying to incorporate encapsulation agents into different food products. However, there is a lack of information on the encapsulation of bioactive compounds in food and pharmaceutical products during processing and storage. Finally, studies should be carried out on encapsulated food products or pharmaceutical products with the goal of retaining nutritive values as well as sensory properties.

REFERENCES

Agner, A. R.; Bazo, A.P.; Ribeiro, L.R.; Salvadori, D.M.F. DNA damage and aberrant crypt foci as putative biomarkers to evaluate the chemopreventive effect of annatto (*Bixa orellana* L.) in rat colon carcinogenesis. *Mutation Research* 2005, 582, 146–154.

Aherne, S. A.; O'Brien, N. M. Dietary flavonols: Chemistry, food content, and metabolism. *Nutrition* 2002, 18, 75–81.

Ahmed, M.; Akter, M. S.; Eun, J. B. Encapsulation by spray drying of bioactive components, physicochemical and morphological properties from purple sweet potato. *LWT-Food Science and Technology* 2010b, 43, 1307-1312.

Ahmed, M.; Akter, M. S.; Eun, J. B. Impact of α-amylase and maltodextrin on physicochemical, functional and antioxidant capacity of spray-dried sweet potato flour. *Journal of the Science of Food and Agriculture* 2010a, 90, 494-502.

Anderson, R. A.; Evans, L. M.; Ellis, G. R.; Khan, N.; Morrist, K.; Jackson, S. K.; Rees, A.; Lewis, M. J.; Frenneaux, M. P. Prolonged deterioration of endothelial dysfunction in response to postprandial lipaemia is attenuated by vitamin C in type 2 diabetes. *Diabetic Medicine* 2006, 23, 258–264.

Aoki, H.; Kieu, N.T.M.; Kuze, N.; Tomisaka, K.; Chuyen, N.V. Carotenoid pigments in Gac fruit (*Momordica cochinchinensis* Spreng). *Journal of Biotechnology* 2002, 66, 2479-2484.

Assuncao, R. B.; Mercadante, A. Z. Carotenoids and ascorbic acid from cashew apple (Anacardium occidentale L.): variety and geographic effects. *Food Chemistry* 2003, 81, 495–502.

Azeredo, H.M.C.; Santos, A.N.; Souza, A.C.R.; Mendes, K.C.B.; Andrade, M. I. R. Betacyanin Stability During Processing and Storage of a Microencapsulation Red Beetroot Extract. *American journal of Food technology* 2007, 2, 307-312.

Balaswamy, K.; Prabhakara Rao, P.G.; Satyanarayana, A.; Rao, D. G. Stability of bixin in annatto oleoresin and dye powder during storage. *LWT - Food Science and Technology* 2006, 39, 952-956.

Balladin, D. A.; Headley, O. Liquid chromatographic analysis of the main pungent principles of solar dried West Indian ginger (*Zingiber officinale* Roscoe). *Renewable Energy* 1998, 18, 257–61.

Barbosa, M. I. M. J.; Borsarelli, C. D.; Mercadante, A. Z. Light stability of spray-dried bixin encapsulated with different edible polysaccharide preparations. *Food Research International* 2005, 38, 989–994.

Bertolini, A. C.; Siani, A. C.; Grosso, C. R. F. Stability of monoterpenes encapsulated in gum arabic by spray-drying. *Journal of Agriculture and Food Chemistry* 2001, 49, 780−785.

Breinholt, V. Desirable versus harmful levels of intake of flavonoids and phenolic acids. In: Kumpulainen, J., Salonen, J.E. (Eds.), Natural Antioxidants and Anticarcinogens in Nutrition, Health and Disease. The Royal Society of Chemistry 1999, Cambridge, pp. 93–105.

Burin, V. M.; Rossa, P. N.; Ferreira-Lima, N. E.; Hillmann, M. C. R.; Boirdignon-Luiz, M. T. Anthocyanins: optimization of extraction from Cabernet Sauvignon grapes, microcapsulation and stability in soft drink. *International Journal of Food Science and Technology* 2011, 46, 186–193.

Buxiang, S.; Fukuhara, M. Effects of co-administration of butylated hydroxytoluene, butylated hydroxyanisole and flavonoid on the activation of mutagens and drug-metabolizing enzymes in mice. *Toxicology* 1997,122, 61-72.

Bylaite, E.; Nylander, T.; Venskutonis, R.; Jonsson, B. Emulsification of caraway essential oil in water by lecithin and β-lactoglobulin: Emulsion stability and properties of the formed oil aqueous interface. *Colloids and Surface B: Bio interfaces* 2001, 20, 327-340.

Cabrita, L.; Fossen, T.; Andersen, O. M. Colour and stability of the six common anthocyanidin 3-glucosidesin aqueous solutions. *Food Chemistry* 2000, 68,101–107.

Cai, Y. Z.; Sun, M.; Corke, H. Colorant properties and stability of *Amaranthus* betacyanin. *Journal of Agriculture and Food Chemistry*1998, 46, 4491–4495.

Cai, Y.Z.; Corke, H. Production and Properties of spray-dried amaranthus betacyanin pigments. *Journal of Food Science* 2000, 65, 1248-1252.

Cameron, E.; Pauling, L.; Leibovitz, B. Ascorbic Acid and Cancer: A Review. *Cancer Research* 1979, 39, 663-681.

Cavalcante, A. A. M.; Rubensam, G.; Erdtmann, B.; Brendel, M.; Henriques, J. A. P. Cashew (Anacardium occidentale) apple juice lowers mutagenicity of aflatoxin B1 in S. typhimurium TA 102. *Genetics and Molecular Biology* 2005, 28, 328–333.

Chandra, A.; Nair, M. G.; Lezzoni, A. Evaluation and characterization of the anthocyanins pigments in tart cherries (*Prunus cerasus L.). Journal of Agricultural and Food Chemistry* 1992, 40, 967– 969.

Choa, S.Y.; Park, J.W.; Batt, H.P.; Thomas, R. L. Edible films made from membrane processed soy protein concentrates. *LWT–Food Science and Technology* 2007, 40,418–423.

Chopda, C. A.; Barrett, D. M. Optimization of guava juice and powder production. *Journal of Food Processing Preservation* 2001, 25, 411-430

Dangles, O.; Saito, N.; Brouillard, R. Kinetic and thermo-dynamic control of flavylium hydration in the pelargonidin-cin-namic acid complexation. Origin of the extraordinary flower colordiversity of Pharbitisnil. *Journal of the American Chemical Society* 19993, 115, 3125–3132.

Davis, J.N.; Hobson, G.E. The constituents of tomato fruit the influence of environment, nutrition and genotype. *Critical Reviews in Food science and Nutrition* 1981, 15, 205

De Oliveria, M.A.; Maria, G.A.; De Figueiredo, R.W.; De Souza, A.C.R.; De Brito, E.S.; De Azeredo, H.M.C. Additional of cashew tree gum to maltodextrin-based carriers for spray drying of cashew apple juice. *International Journal of Food Science and Technology* 2009, 44, 641-645.

Del Pozo-Insfran, D.; Brenes, C.H.; Talcott, S.T. Phytochemical composition and pigments stability of acai (Euterpe oleracea Mart.). *Journal of Agricultural and Food Chemistry* 2004, 52, 1539-1545.

Desai, K. G. H.; Park, H. J. Encapsulation of vitamin C in tripolyphosphate cross-linked chitosan microspheres by spray drying. *Journal of Microencapsulation* 2005, 22,179-192.

Desobry, S.A.; Netto, F.M.; Labuza, T.P. Influence of maltodextrin systems at an loss during storage equivalent 25DE on encapsulated β-carotene. *Journal of Food Processing Preservation* 1999, 23, 39-55.

Dib taxi, C.M.A.; De menezes, H.C.; Santos, A.B.; Grosso, C.R.F. Study of the micro-encapsulation of camu-camu (Myrciaria dubia) juice. *Journal of Microencapsulation* 2003, 20, 443-448.

Dickinson, E. Hydrocolloids at interfaces and the influence on the properties of dispersed system. *Food Hydrocolloids* 2003, 17, 25-39.

Douglas, R. M.; Chalker, E. B.; Treacy, B. Vitamin C for preventing and treating the common cold. *Cochrane Database System Review* 2000, 2, CD000980.

Duthie, G.G.; Duthie, S.J.; Kyle, J.A.M. Plant polyphenols in cancer and heart disease: implications as nutritional antioxidants. *Nutrition Research Reviews* 2000, 13, 79–106.

Dziedzic, S. Z.; Hudson, B. J. F. Phenolic acids and related compounds as antioxidants for edible oils. *Food Chemistry* 1984, 14, 45-51.

Dziedzic, S. Z.; Hudson, G. J. F. Polyhydrochalcones and flavanones as antioxidants for edible oils. *Food Chemistry* 1983, 12, 205-212.

Edwards, A. J.; Vinyard, B. T.; Wiley, E.R.; Brown, E. D.; Collins, J. K.; Perkins-Veazie, P.; Baker, R. A.; Clevidence, B. A. Consumption of watermelon juice increases plasma concentration of lycopene and β-carotene in humans. *Journal of Nutrition* 2003,133, 1043–1050.

Elliott, J. G. Application of antioxidant vitamins in foods and beverages. *Food Technology* 1999, 53, 46–48.

Enwonwu, C. O.; Meeks, V. I. Bionutrition and oral cancer in humans. *Critical Reviews in Oral Biology and Medicine* 1995, 6, 5–17.

Ersus, S.; Yurdagel, U. Microencapsulation of anthocyanin pigments of black carrot (Daucuscarota L.) by spray drier. *Journal of Food Engineering* 2007, 80, 805-812.

Escribano, J.; Pedreno, M.A.; Garcia-Carmona, F.; Munoz, R. Characterization of the antiradical activity of betalains from Betavulgaris L. roots. *Phytochemical Analysis* 1998, 9, 124–127.

Ferruzzi, M. G.; Böhm, V.; Courtney, P. D.; Schwartz, S. J. Antioxidant and antimutagenic activity of dietary chlorophyll derivatives determined by radical scavenging and bacterial reverse mutagenesis assays. *Journal of Food Science* 2002, 67, 2589–2595.

Fossen, T.; Cabrita, L.; Andersen, O. M. Colour and stability of pure anthocyanins influenced by pH including the alkaline region. *Food Chemistry* 1998, 63,435–440.

Fuhramn, B.; Elis, A.; Aviram, M. Hypocholesterolemic effect of lycopene and β-carotene is related to suppression of cholesterol synthesis and augmentation of LDL receptor activity in macrophage. *Biochemical and Biophysical Research Communications* 1997, 233:658–62.

Gandia-Herrero, F.; Jimenez-Atienzar, M.; Cabanes, J.; Garcia-Carmona, F.; Escribano, J. Stabilization of the Bioactive Pigment of Opuntia Fruits through Maltodextrin Encapsulation. *Journal of Agricultural and Food Chemistry* 2010, 58, 10646–10652.

Gaulejac, N.; Glories, Y.; Vivas, N. Free radical scavenging effect of anthocyanins in red wines. *Food Research International* 1999, 32, 37-333.

Georgetti, S. R.; Casagrande, R.; Souza, C. R. F.; Oliveira, W. P.; Fonseca, M. J. V. Spray drying of the soy bean extract: Effects on chemical properties and anti oxidant activity. *LWT- Food Science and Technology* 2008, 41, 1521–1527.

Giusti, M.M.; Rodriguez-Saona, L. E.; and Wrolstad, R. E. Molar absorptivity and color characteristics of acylated and non-acylated pelargonidin-based anthocyanins. *Journal of Agricultural and Food Chemistry* 1999, 47, 4631–4637.

Glicksman, M.; Schachat, R. E. Industrial Gums, Academic Press, New York, 1959.

Grabowski, J.A.; Truong, V. D.; Daubert, C.R. Nutritional and rheological characterization of spray dried sweet potato powder. *LWT- Food Science and Technology* 2008, 41, 206–216

Grenha, A.; Grainger, C.I.; Dailey, L.A.; Seijo, B.; Martin, G. P.; Remu˜nán-López, C.; Forbes, b. Chitosan nanoparticles are compatible with respiratory epithelial cells in vitro. *European Journal of Pharmaceutical Sciences* 2007, 31(2), 73–84.

Ha, T. J.; Kubo, I. Lipoxygenase inhibitory activity of anacardic acids. *Journal of Agricultural and Food Chemistry* 2005, 53, 4350–4354.

Harris, R.; Lecumberri, E.; Mateos-Aparicio, I.; Mengibar, M.; Heras, A. Chitosan nanoparticles and microspheres for the encapsulation of natural antioxidants extracted from llex paraguariensis. *Carbohydrate polymers* 2010, In Press.

Haslam, E. Nature's palette. Chemistry in Britain 1993, 29, 875–878.

He, Z.; Xia, W. Analysis of phenolic compounds in Chinese olive (Canarium album L.) fruit by RPHPLC–DAD–ESI–MS. *Food Chemistry* 2007, 105 1307–1311.

Hendry, G. A. F.; Houghton J. D. Natural Food Colorants. Blackie Academic and Professional 1996, 40-79.

Hirose, M.; Takesada, Y.; Tanaka, H.; Tamano, S.; Kato, T.; Shirai, T. Carcinogenicity of antioxidants BHA, caffeic acid, sesamol, 4-methoxyphenol and catechol at low doses, either alone or in combination and modulation of their effects in a rat medium-term multi-organ carcinogenesis model. *Carcinogenesis* 1998, 19, 207-212.

Iqbal, K.; Khan, A.; Khattak, M. M. A. K. Biological significance of ascorbic acid (vitamin C) in human health – A review. *Pakistan Journal of Nutrition* 2004, 3, 5–13.

Jayaprakasam, B.; Vareed, S. K.; Olson, L. K.; Nair, M. G. Insulin secretion by bioactive anthocyanins and anthocyanidins present in fruits. *Journal of Agricultural and Food Chemistry* 2005, 53, 28–31.

Johnson E.J. The role of carotenoids in human health. *Nutrition In Clinical care* 2002, 5, 47-49.

Kanner, J.; Harel, S.; Granit, R. Betalains—a new class ofdietary cationized antioxidants. *Journal of Agricultural and Food Chemistry* 2001, 49, 5178–5185.

Kapadia, G. J.; Tokuda, H.; Konoshima, T.; Nishino, H. Chemoprevention of lung and skin cancer by Beta vulgaris (beet)root extract. *Cancer Letters* 1996, 100, 211–214.

Kaur, C.; Kapoor, H. C. Antioxidants in fruits and vegetables–the millennium's health. *International Journal of Food Science and Technology* 2001, 36, 703–725.

Kha, T. C.; Nguyen, M. H.; Roach, P. D. Effects of spray drying conditions on the physicochemical and antioxidant properties of the Gac (*Momordica cochinchinensis*) fruit aril powder. *Journal of Food Engineering* 2010 98, 385–392.

Khachik, F.; Beecher, G. R.; Goli, M. B.; Lusby, W. R. Separation, identification and quantification of carotenoids in fruits vegetables and human plasma by high performance liquid chromatography. *Pure and Applied Chemistry* 1991, 63, 71-80.

Khachik,F.; Bertram,J. S.; Huang,M.-T.; Fahey,J. W.; Talalay, P. Dietary carotenoids and their metabolites as potentially useful chemoprotective agents against cancer. In Packer,L. Hiramatsu,M. and Yoshikawa,T. (eds) *Proceedings of the International Symposium on Antioxidant Food Supplements in Human Health.* Academic Press, 1999, London, pp. *203–229.*

Kim, S. Y.; Jeong, S. M.; Kim, S. J.; Jeon, K. I.; Park, E.; Park, H. R.; Lee, S. C. Effect of heat treatment on the antioxidative and antigenotoxic activity of extracts from persimmon (*Diospyros kaki L.)* peel. *Bioscience Biotech and Biochemistry* 2006, 70, 999–1002.

Kirby, C.J.; Whittle, C.J.; Rigby, N.; Coxon, D.T.; Law, B. A. Stabilization of ascorbic acid by microencapsulation in liposomes. *International Journal of Food Science and Technology* 1991, 26, 437-449.

Kosaraju, S. L.; D'ath, L.; Lawrence, A. Preparation and characterization of chitosan microspheres for antioxidant delivery. *Carbohydrate Polymers* 2006, 64, 163–167.

Kosaraju, S. L.; Labbett, D.; Emin, M. Konczak I.; Lundin, L. Delivering polyphenols for healthy ageing. *Nutrition and Dietetics* 2008, 65, S48–S52.

Krishnaiah, D.; Sarbatly, R.; Hafiz, A. M. M.; Hafeza, A.B.; Rao, S. R. M. Study on Retention of Bioactive Compounds of Morinda Citrifolia L. Using Spray–Drying. *Journal of Applied Sciences* 2009, 17, 3090-3097.

Krishnaiah, D.; Sarbatly, R.; Nithyanandam, R. A review of the antioxidant potential of medicinal plant species. *Food and Bio products Processing* 2010, In Press.

Krishnan, S.; Bhosale, R.; Singhal, R. S. Microencapsulation of cardamom oleoresin: Evaluation of blends of gum arabic, maltodextrin and a modified starch as wall materials. *Carbohydrate Polymers* 2005, 61, 95–102.

Kubo, J.; Lee, J. R.; Kubo, I . Anti-helicobacter pylori agents from the cashew apple. *Journal of Agricultural and Food Chemistry* 1999, 47,533–537.

Landy, P.; druaux, C.; Voilley, A. Retention of aroma compounds by proteins in aqueous solution. *Food Chemistry* 1995, 54, 387-392.

Lansky, E.P.; Newman, R.A. *Punica granatum* (pomegranate) and its potential for prevention and treatment of inflammation and cancer. *Journal of Ethnopharmacology 2007*, 109, 177–206.

Larson, R. A. The antioxidants of higher plants. *Phytochemistry* 1988, *27*, 969-978.

Lee, D.W.; Collins, T. M. Phylogenetic and ontogenetic influence on the distribution of anthocyanins and betacyanins in leaves of tropical plants. *International Journal of Plant Sciences* 2001, 162, 1141–1153.

Liu, L.; Zhao, S. P.; Gao, M.; Zhou, Q. C.; Li, Y. L.; Xia, B. Vitamin C preserves endothelial function in patients with coronary heart disease after a high-fat meal. *Clinical Cardiology* 2002, 25, 219–224.

Liu, X.; Qiu, Z.; Wang, L.; Chen, Y. Quality evaluation of Panax notoginseng extract dried by different drying methods. *Food and Bioproducts processing* 2010, In Press.

Loke, W. M.; Hodgson, J. M.; Proudfoot, J. M.; McKinley, A. J.; Puddey, I. B.; Croft, K.D. 2008. Pure dietary flavonoids, quercetin and (-)-epicatechin augmentnitric oxide products and reduce endothelin-1 acutely in healthy human volunteers. *American Journal of Clinical Nutrition* 2008, 88, 1018–1025.

Majeti, N.V.; Kumar, R. A review of chitin and chitosan applications. *Reactive and Functional Polymers* 2000, 46, 1–27.

Marki-Fisher, E., U. Marti, R. Buchecker, and C.H. Eugster, Das Carotinoid spektrum der Hagebutten von *Rosa pomifera* Nachweis von (5*Z*)-Neurospin; Synthese von (3*R*,15*Z*) Rubixanthin, *Helvetica Chimica Acta 1983,* 66, 495–513.

Mattila, P. Hellstrom, J. Phenolicacids in potatoes, vegetables, and some of their products. *Journal of Food Composition and Analysis* 2007, 20, 152–160.

Mattila, P.; Hellstrom, J.; Torronen, R. Phenolic Acids in Berries, Fruits, and Beverages. Journal *of Agricultural and Food Chemistry* 2006, 54, 71937199.

Mazza, G.; Miniati, E. Anthocyanins in fruits, vegetables, and grains. CRC Press, 1993, Boca Raton, pp. 85-87.

Mercadante A.Z.; Steck, A.; Rodriguez-Amaya D.; Pfander, H.; Britto, G. Isolation of methyl 9'Z-apo-6'lycopenoate from *Bixa orellana*. *Phytochemistry* 1996, 41, 1201–1203.

Mertens-Talcott, S. U.; Jilma-Stohlawetz, P.; Ríos, J.; Hingorani, L.; Derendorf, H. Absorption, metabolism, and antioxidant effects on pomegranate (*Punica granatum* L.) polyphenols after ingestion of a standardized extract in healthy human volunteers. *Journal of Agricultural and Food Chemistry 2006*, 54, 8956–8961.

Meyer, A. S.; Donovan, J. L.; pearson, D. A.; waterhouse, A. I.; Frankel, E.N. Fruit hydroxycinnamic acids inhibit human low-density lipoproteins oxidation in vitro. *Journal of Agricultural Food Chemistry* 1998, 46, 1783-1787.

Miean, K. H.; Mohamed, S. Flavonoid (myricetin, quercetin, kaempferol, luteolin and apigenin) content of edible tropical plants. *Journal of Agricultural and Food Chemistry* 2001, 49: 3106–3112.

Miles, J. (1992). Encapsulation and delivery of perfumes and fragrances. Second workshop on controlled delivery in consumer products. Secaucus, NJ: Controlled Release Society.

Minnguez-Mosquera, M. I.; Hornero-Mendez, D. Separation and quantification of the carotenoid pigments in red peppers (Capsicum annuum L.), paprika and oleoresin by reversed-phase HPLC. *Journal of Agricultural and Food Chemistry* 1993, 41, 1616-1620.

Mistry, T. V.; Cai, Y.; Lilley, T. H.; Haslam, E. Polyphenol interactions. Part 5: anthocyaninco-pigmentation. *Journal of the Chemical Society Perkin Transactions* 1991, 2, 1287–1296.

Moreira, G. E. G.; De Azeredo, H. M. C.; Desouza, A. C. R.; Debrito, E. S.; Demedeiros, M. A. D. Ascorbic acid and anthocyanin retention during spray drying of acerola pomace extract. *Journal of Food Processing and Preservation* 2010, 34, 915-925.

Moser. U.; Bendich, . A. Vitamin C, - *In* Handbook of Vitamins. 2nd Ed. (L, J, .Machlin. ed.) pp. 195-232, Marcel Decker. 1990,New York. NY ISBN O-8247-S351-4.

Namiki, M. Antioxidants/antimutagens in food. Critical review *Food Science Nutrition* 1990, 29, 273–300.

Neill, S.O.; Gould, K. S.; Kilmartin, P. A.; Mitchell, K. A.; Markham, K. R. Antioxidant activities of red versus green leaves in *Elatostema rugosum. Plant, Cell and Environment* 2002, 25, 539–547.

Nor, F.M.; Mohamed, S.; Idris, N.A.; Ismail, R. Antioxidative properties of *Pandanus amaryllifolius* leaf extracts in accelerated oxidation and deep frying studies. *Food Chemistry* 2008, 110, 319–327.

Osawa, T. Novel natural antioxidants for utilization in food and biological systems. In *Postharvest Biochemistry of plant Food-Materials in the Tropics*; Uritani, I.,Garcia, V. V., Mendoza, E. M. Eds.; Japan Scientific Societies Press: 1994,Tokyo, Japan, pp 241-251.

Osorio, C.; Acevedo, B.; Hillebrand, S.; Carriazo, J.; Winterhalter, P.; Morales A. L. Microencapsulation by spray-drying of anthocyanin pigments from corozo (*Bactris guineensis*) Fruit. *Journal of Agricultural and Food Chemistry* 2010a, 58, 6977-6985.

Osorio, C.; Forero, D. P.; Carriazo, J. G. Charaterisation and performance assessment of guava (psidium guajava L.) microencapsulates by spray-drying. *Food Research International* 2010a, In press.

Paramera, E. I.; Konteles, S. J.; Karathanos, V. T. Stability and release properties of curcumin encapsulated in Saccharomyces cerevisiae, b-cyclodextrin and modified starch. *Food Chemistry* 2010, In Press.

Partanen, R.; yoshii, H.; Kallio, H.; Yang, B.; Forsshell, P. Encapsulation of sea buckthorn kernel oil in modified starches. *Journal of the American Oil Chemists Society* 2002, 79, 219-223.

Paśko, P.; Bartoń, H.; Zagrodzki, P.; Chłopicka, J.; Iżewska. A.; Gawlik, M.; Gawlik, M.; Gorinstein, S. Effect of amaranth seeds in diet on oxidative status in plasma and selected tissues of high fructose-fed rats. *Food Chemistry* 2010, In press.

Pedreno, M. A.; Escribano, J. Studying the oxidation and the antiradical activity of betalain from beetroot. *Journal of Biological Education* 2000, 35, 49–51.

Pennington, J. A. T.; Fisher, R. R. Food component profiles for fruit and vegetable subgroups. *Journal of Food Composition and Analysis* 2010, 23, 411-418.

Perez Gutierrez, R.M.; Mitchell, S.; Vargas Solis, R. Psidium guajava: a review of its traditional uses, phytochemistry and pharmacology. *Journal of Enthnopharmacology* 2008, 117, 1-27.

Perez-Alonso, C., Cruz-Olivares, J., Barrera-Pichardo, J. F., Rodriguez-Huezo, M. E. Baez-Gonzalez, J. G., Vernon-Carter, E. J. DSC thermo-oxidative stability of red chili oleoresin microencapsulated in blended biopolymers matrices. *Journal of Food Engineering* 2008, 85, 613-624.

Peterson, J. M. S.; Dwyer, J.; Dsc, R. D. Flavonoids: Dietary occurrence and biochemical activity. *Nutrition Research* 1998, 18, 1995–2018.

Phoungchandang, S.; Sertwasana, A. Spray-drying of ginger juice and physicochemical properties of ginger powders. *Science Asia* 2010, 36, 40–45.

Piattelli, M., Minale, L., Prota, G., 1964. Isolation, structure and absolute configuration of indicaxanthin. *Tetrahedron* 1964, 20, 2325–2329.

Piattelli, M.; Minale, L.; Prota, G. Isolation, structure and absolute configuration of indicaxanthin. *Tetrahedron* 1964, 20, 2325–2329.

Pitalua, E.; Jimenez, M.; Vernon-Carter, E. J.; Beristain, C. I. Antioxidative activity of microcapsules with beet root juice using gum Arabic as wall material. *Food and Bio products Processing* 2010, 88 253–258.

Porrarud, S.; Pranee, A. Microencapsulation of Zn-chlorophyll pigment from Pandan leaf by spray drying and its characteristic. *International Food Research Journal* 2010, 17, 1031-1042.

Prior, R. L.; Wu, X. Anthocyanins: Structural characteristics that result in unique metabolic patterns and biological activities, *Free Radical Research* 2006, 40, 1014–1028.

Qi, Z. H.; Xu, A. Starch based ingredients for flow encapsulation. *Cereal Food World* 1999, 44, 460–465.

Quek, S.Y.; Chok, N. K.; Swedlund, P. The physicochemical properties of spray-dried water melon powders. *Chemical Engineering and Processing* 2007, 46, 386–392.

Rao, A.V.; Rao, L.G. Carotenoids and human health. *Pharmacological Research* 2007, 55, 207–216.

Rascón, M. P.; Beristain, C. I.; García, H. S.; Salgado, M. A. Carotenoid retention and storage stability of spray-dried encapsulated paprika oleoresin using gum Arabic and Soy protein isolate as wall materials. *LWT-Food Science and Technology* 2010, In Press.

Razungles, A., J. Oszmianski, and J. Sapis, Determination of Carotenoids in Fruits of *Rosa* sp. (*Rosa canina and rugosa*) and Chokeberry (*Aronia melanocarpa*), *Journal of Food Science* 1989, *54*, 774–775.

Regan O. J.; Mulvihill D. M. Preparation, characterisation, and selected functional properties of sodium caseinate-maltodextrin conjugates. *Food Chemistry* 2009, 115, 1257-1267.

Righetto, A. M.; netto, F. M. Vitamin C stability in encapsulated green West Indian cherry juice and in encapsulated synthetic ascorbic acid. *Journal of the Science of Food and Agriculture* 2006, 86, 1202-1208.

Rinaudo, M.; Chitin and chitosan: Properties and applications, *Progress in Polymer Science* 2006, 31, 603–632.

Robert, P.; Carlsson, R. M.; Romero, N.; Masson, L. Stability of spray-dried encapsulated carotenoid pigments from rosa mosqueta (*Rosa rubiginosa). Journal of the American Oil Chemists Society* 2003, 80, 1115-1120.

Robert, P.; Gorena, T.; Romero, N.; Sepulveda, E.; Chavez, J.; Saenz, C. Encapsulation of polyphenols and anthocyanins from pomegranate (Punica granatum) by spray drying, *International Journal of Food Science and Technology* 2010, 45, 1386–1394.

Rocha-Guzman, N. E.; Gallegos-Infante, J. A.; Gonzalez-Laredo, R. F.; Harte, F.; Medina-Torres, L.; Ochoa-Martinez, L. A.; Soto-Garcia, M. Effect of High-Pressure Homogenization on the Physical and Antioxidant Properties of Quercus resinosa Infusions Encapsulated by spray-Drying. *Journal of Food Science* 2010, 75, N57-N61.

Rodriguez-Hernandez, G. R.; Gonzalez-Garcia, R.; Grajales-Lagunes, A.; Ruiz-Cabrera, M. A. Spray-drying of cactus pear juice (*Opuntia streptacantha*): effect on the physicochemical properties of powder and reconstituted product. *Drying Technology* 2005, 23, 955-973.

Rodriguez-huezo, M. E.; Pedroza-islas, R.; Prado-barragan, L. A.; Beristain, C. I.; Vernon-carter, E. J. Microencapsulation by Spray Drying of Multiple Emulsions Containing Carotenoids. *Journal of Food Science* 2004, 69, E351-E359.

Saenz, C. Processing technologies: An alternative for cactus pear (Opuntia spp.) fruits and cladodes. *Journal of Arid Environments* 2000, 46, 209-225.

Saenz, C.; Tapia, S.; Chavez, J.; Robert, P. Microencapsulation by spray drying of bioactive compounds from cactus pear (*Opuntia ficus-indica*). *Food chemistry* 2009, 114, 616-622.

Saint-Cricq de Gaulejac, N.; Provost, C.; Vivas, N. Comparative study of polyphenol scavenging activities assessed by different methods. *Journal of Agriculture and Food Chemistry* 1999, 47, 425–431.

Salas, M.P.; Celiz, G.; Geronazzo, H.; Daz, M.; Resnik, S. L. Antifungal activity of natural and enzymatically-modified flavonoids isolated from citrus species. *Food Chemistry* 2011,124, 1411–1415

Sari, H. Flavonols and phenolic acids in berries and berry products. PhD thesis 2000, 16-17, Kuopio University.

Schwartz, S. J.; Von Elbe, J. H. Identification of betanin degradation products. European *Food Research and Tecxhnology* 1983,176, 448–453.

Shahidi, F.; Han, X.-Q. Encapsulation of food ingredients. *Critical Reviews in Food Science and Human Nutrition* 1993, 33, 501-547.

Shaikh,j.; Bhosale, R.; Singhal, R. Microencapsulation of black pepper oleoresin. *Food Chemistry* 2006, 94, 105-110.

Shu, B.; Yu, W.; Zhao, Y.; Liu, X. Study on Microencapsulation of lycopene by spray-drying. *Journal of Food Engineering* 2006, 76, 664-669.

Silva, C. R.; Antunes, L. M. G.; Bianchi, M. L. Antioxidant action of bixin against cisplatin-induced chromosome aberrations and lipid peroxidation in rats. *Pharmacological Research* 2001, 43, 561–566.

Siva, R.; Mathew, G. J.; Venkat, A.; Dhawan, C. An alternative tracking dye for gel electrophoresis. *Current Science* 2008, 94, 765–767.

Stevens, C. V.; Meriggi, A.; Booten, K. Chemical modification of inulin, a valuable renewable resource, and its industrial applications. *Biomacromolecules* 2001, 2, 1-16.

Stintzing, F. C.; Carle, R. Functional properties of anthocyanins and betalains in plants, food, and in human nutrition. *Trends in Food Science and Technology* 2004, 15, 1, 1-38.

Sultana, B.; Anwar, F. Flavonols (kaempferol, quercetin, myricetin) contents of selected fruits, vegetables and medicinal plants. *Food Chemistry* 2008, 108, 879–884.

Tharanathan, R. N.; Kittur, F. S. Chitin-The undisputed biomolecule of great potential. *Critical Reviews in Food Science and Nutrition* 2003, 43, 61–87.

Thornhill, S. M.; Kelly, A. M. Natural treatment of perennial allergic rhinitis. *Alternative Medicine Review* 2000, 5, 448–454.

Thresiamma, K. C.; George, J.; Kuttan, R. R. Protective effect of curcumim, ellagic acid and bixin on radiation induced genotoxicity. *Journal of Experimental and Clinical Cancer Research* 1998, 17 (1998) 431–434.

Tonon, V. R.; Brabet, C.; Hubinger, M. D. Anthocyanin stability and antioxidant activity of spray-dried acai (Euterpe oleracea Mart.) juice produced with different carrier agents. *Food Research International* 2010, 43, 907-914.

Tonon, V. R.; Brabet, C.; Hubinger, M. D. Influence of process conditions on the physicochemical properties of acai (*Euterpe oleraceae Mart*) powder produced by spray drying. *Journal of Food Engineering* 2008, 88, 411-418.

Tripoli, E.; Guardia, M. L.; Giammanco, S.; Majo, D. D.; Giammanco, M. Citrus flavonoids: Molecular structure, biological activity and nutritional properties: A review. *Food Chemistry* 2007,104, 466–479.

Van Hal, M.V. Quality of sweet potato flour during processing and storage. *Journal of Food Reviews International* 2000, 16,1-37.

Wagner, L. A.; Warthesen, J. J. Stability of Spray-Dried Encapsulated Carrot Carotenes. (1995). *Journal of Food Science* 1995, 60, 1048-1053.

Wang, H.; Nair, M. G.; Strasburg, B. M.; Chang, Y. C.; Booren, A. M. Antioxidant and anti-inflammatory activities of anthocyanins and their aglycones, cyanidin, from tart cherries. *Journal of Natural Products* 1999, 62, 294– 296.

Wang, y.; Lu, Z.; Lv, F.; Bie, X. Study on microencapsulation of curcumin pigments by spray drying. *European Food Research Technology* 2009, 229, 391-396.

Wyler, H.; Mabry, T. J.; Dreiding, A. S. Uber die Konstitution des Randenfarbstoffes Betanin: Zur Struktur des Betanidins. *Helvetica Chimica Acta* 1963, 46, 1745–1748.

Yang, J.; Gadi R. L. Effects of steaming and dehydrationon anthocyanins, antioxidantactivity, total phenols and color characteristics of purple-fleshed sweet potatoes (Ipomoea batatas). *American Journal of Food Technology* 2008, 3,224–234.

Yogeshwer, S.; Madhulika S. Cancer preventive properties of ginger: A brief review. *Journal of Food Science* 2007, 45, 683–90.

Zhang, L. X.; Cooney, R. V.; Bertram, J. S. Carotenoids enhance gap functional communication and inhibit lipid peroxidation in C3H/10T1/2 cells; relationship to their chemopreventative action. *Carcinogenesis* 1991, 12, 2109-2114.

Zhang, L.; Mou D.; Du, Y. Procyanidins: extraction and micro-encapsulation, *Journal of Agricultural and Food Chemistry* 2007, 87, 2192–2197.

In: Sprays: Types, Technology and Modeling
Editor: Maria C. Vella, pp. 135-180
ISBN 978-1-61324-345-9

Chapter 3

DROP FORMATION OF PRESSURE ATOMIZERS IN A LOW PRESSURE ENVIRONMENT

***Frank S. K. Warnakulasuriya*[1] *and William M. Worek*[2]**
[1]Department of Marine Engineering Technology,
Texas A and M University at Galveston,
Galveston, Texas, USA
[2]Department of Mechanical and Industrial Engineering,
University of Illinois at Chicago,
Chicago, Illinois, USA

ABSTRACT

In this paper, the effect of variable density, viscosity and surface tension of the fluid on drop characteristics (i.e. drop size, drop velocity, number density and drop shape) and spray formation characteristics in terms of spray angle, drop size distribution, spray shape and penetration (mean resident time) due to variations of fluid temperature, differential pressure of fluid across nozzle orifice and fluid flow rate of highly temperature dependent high viscous salt solution was studied. This work was mainly conducted to study and validate the drop absorption concept developed to be utilized in absorption refrigeration cycles. This study investigated the characteristic of drops and sprays for two sets of pressure atomizers (i.e., swirl jet atomizers and full jet atomizers). In swirl jet atomizer testing, four different models (sizes) and in full jet atomizer testing, three different models (sizes) were tested. For each selected nozzle, by changing the fluid temperature and the nozzle pressure, an average of 30 different spray flow trials were conducted, which totaled to 200 tests.

In the process of designing and setting up the experimental test rig, several factors needed to be considered. This include building of a single nozzle spray absorber chamber that is able to attain a low pressure of 1.23 kPa, development of solution tempering system that is able to vary the solution temperature in the range of 60°C to 90°C, setting up of leak proof high vacuum system with leakage rate less than 4.67×10^{-6} scc/sec, assembling a Phase Doppler Particle Analyzer (PDPA) and high speed digital imaging system to measure drop and spray characteristics, and development of complete data acquisition and control system to control the process and collect all of the data. In addition, for comparison purpose, a spreadsheet program based on existing drop formation models was developed.

In this study, the ranges of input parameter variation were controlled by the development and expansion of the spray due to nozzle geometry of each individual nozzle, variation of properties of sprayed fluid and changing of the flow condition inside the vacuum chamber that limited by the diameter and length of the chamber. During the testing, input parameters included fluid density, viscosity ratio i.e. k-factor of dispersed to continuous phase flow and surface tension of fluid were varied in the ranges of 2502 kg/m^3 – 2524 kg/m^3, 1015 – 1708 and 0.2073 kg/s^2 – 0.2575 kg/s^2, respectively. The input parameters of mass flow rate and differential pressure across the nozzle related to different flow conditions were varied from 0.011 kg/sec to 0.045 kg/sec and 21.31 kPa to 519.33 kPa, respectively. The input parameters related to nozzle geometry that is nozzle constants and flow numbers were varied from 0.69 to 6.9 and 3.6×10^{-7} to 2.3×10^{-5}, respectively.

All the tested sprays showed the breaking of drops is mainly caused by viscous effect. The range of the droplet Reynolds number (Re) observed was between 2.38×10^2 and 2.14×10^3 and the range of Weber number (We) was between 1.22×10^{-5} and 3.05. The ranges of both Bond number (Bo) and Morton number (M) that define the shape of the droplets were between 0.0037 – 0.043 and 6.14×10^{-12} – 3.16×10^{-23}. The drops generated in the vacuum chamber were largely spherical drop in creeping and high-speed flows and drops with larger Re number and Bo number showed some "wobbling" effects. The mean volumetric diameter (MVD) of the drops that generated from nozzles were varied from 200 to 500 micrometers and the drop velocities were varied from 4.5 m/sec to 16.75 m/sec. The distribution factor Q for drops varied from 4.6 to 6.8. Since the pressure atomizers used were limited to swirl jet and full jet atomizers due to the overall requirement of generating smallest possible droplets, the spray shapes were limited only to "full cone" and "hollow cone" spray patterns. The range of spray angles that were generated by selected nozzles varied from 32 degrees to 98 degrees. The drop mean resident times that depended on drop velocities and length of the absorber were range from 0.12 sec to 0.33 sec.

The experimental results agreed well with the spreadsheet program developed based on published drop formation models. The values were within the percentages of 79% to 94% and the experimental scatter was in the range of the error margins of the equipments.

Keywords: Absorption refrigeration, drop formation, spray in low-pressure environment, drop model, drop measurement, spray measurement, spray modeling, experimental spray studies

NOMENCLATURE

a = radius of drop, m

D = diameter of the drop, m

k = k-Factor; viscosity ratio = viscosity of disperse phase/viscosity of continuous phase

$\dot{m}$ = solution mass flow rate, kg/s

P = Pressure, kPa

ΔP = differential pressure across the nozzle, kPa

We = Weber number

G = Gravitational constant, m/s^2
Re = Reynolds number
t = Thickness of the film (sheet), mm
U = speed of the drop, m/s
U_R = Relative velocity, m/s
C_D = Discharge coefficient
Eo = Eotvos number
Bo = Bond number
FN = Flow number
A = Area, m^2
d = diameter, m
K = Nozzle constant
V_o = Representative velocity, m/s
Q = The fraction of the spray flow contained in drops with a diameter less than "d^*"
d^* = the independent variable
q = Distribution factor
X^* = the characteristic diameter of the distribution
N = Number of drops
MVD = Mean volumetric diameter, m
SMD = Sauter mean diameter, m
$A, B, \ldots$ = Constant coefficients of power-law equations
$\left(\frac{\partial \rho}{\partial \Delta}\right)$ = effect of the error on the ultimate calculation, in this thesis, the effect of the error on the absorption ratio.

Greek Symbols

θ_m = Spray angle
Δ = error of some particular reading
ρ = density of a substance
ν = kinematic viscosity
μ = dynamic viscosity
σ = surface tension
$\Delta\rho$ = Difference of density between dispersed and continuous phases

Subscripts/Superscripts

L = liquid
A = Ambient air
G = Gas
R = Relative

Crit= Critical value
Norm = Normalized parameter
d = Dispersed phase
c = Continuous phase
IP = Inlet port
S = Swirl chamber
O = Orifix
a, b,... = Coefficients of power of power-law equations

INTRODUCTION

Absorption cooling systems have been built over the last 100 years. System designs range from using ammonia/water, sulfuric acid/water, various organic absorption pairs and lithium bromide/water. Large capacity lithium bromide/water chillers are the most commonly produced absorption-cooling machine today (ASHRAE, 1989). These absorption machine consist of the generator where water, the refrigerant is boiled from a weak aqueous lithium bromide solution, the condenser where the water refrigerant is liquefied, the evaporator where liquid water refrigerant is boiled at very low pressure, the solution heat exchanger which recuperates heat from the strong solution leaving the generator, and absorber, where the water is reintroduced back into the lithium bromide solution. The system schematic is shown in the Figure 1. All of the processes involved are well understood except for the absorption process in the absorber. This is the one process that is peculiar to an absorption-cooling machine.

Today's technologies have developed from single-effect cycles, described above, to multi-effect cycles. In multi-effect high performance absorption cycle chillers, the high temperature loops operate with highly concentrated salt solutions that are relatively viscous and corrosive. Consequently, the design of the corresponding apparatus, especially the absorber, presents engineering and material challenges. Currently, this system component is associated with an undesirable cost premium and reduction of effectiveness due to the size of the absorber.

The design that has been used in the absorber of conventional large commercial absorption chillers is a falling-film heat exchanger as shown in Figures 2 and 3. In an absorber, the rate which the refrigerant (i.e., water in a LiBr-H_2O system) in the form of vapor is absorbed by the concentrated salt solution is very important. It has long been recognized that if this rate could be increased beyond that which occur in the falling film absorbers, the size of the absorber could be reduced in size and its performance could be enhanced.

A variety of absorption processes have been modeled in the past. These include absorption in falling films [1 - 5], absorption in laminar flow on vertical tubes [6], absorption in turbulent flow in vertical tubes [7], and absorption in flow over horizontal tubes [8, 9]. In addition, sprays that can produce such small droplets in absorption industry have been studied by several investigators [10 - 13].

It is apparent that if the total absorption area of the salt solution could be increased, the rate of absorption of the water vapor by the salt solution would also increase. One obvious way of doing this is to introduce the absorption fluid in the form of the fine droplets, which could increase the rate of the absorption by an order of magnitude. It is known that the size of these droplets decreases, the total area exposed to the vapor increases, and the rate of absorption increases accordingly. Numerous studies [10 - 14] have confirmed the above effect demonstrating the improvement of the absorption rate with experiments, analytical and numerical calculations. As shown in Figures 2 and 3, a conventional absorber consists of a bundle of tubes covered with the absorbing solution which are surrounded by the water vapor to be absorbed. The rate of the absorption is directly proportional to the total tube area. If all the solution on these tubes can be converted into the drops with an average of diameter of 300 microns, the total area would be increased by more than 200 times. Therefore, the rate of absorption could theoretically be increased by a similar factor.

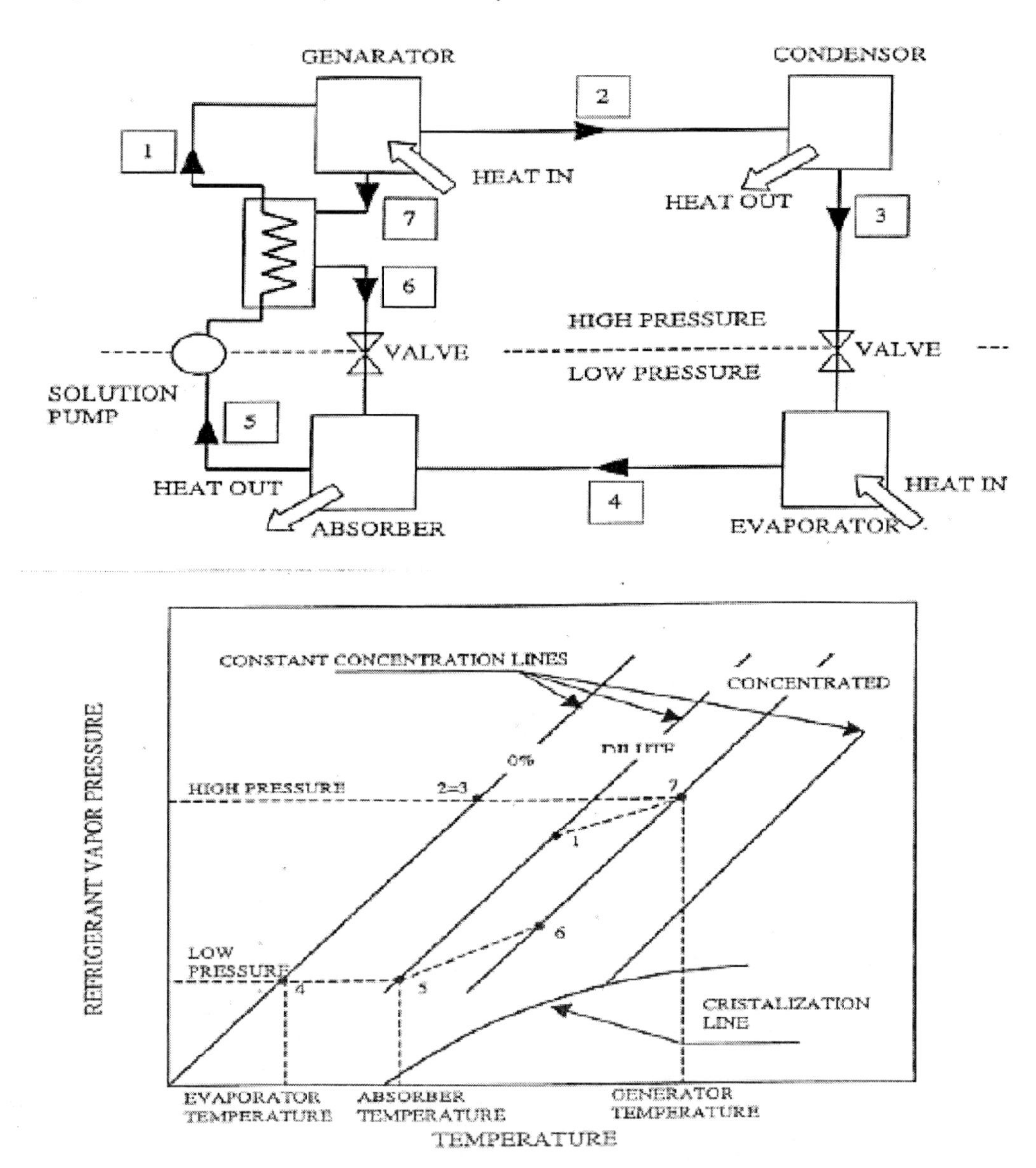

Figure 1. Absorption cooling cycle.

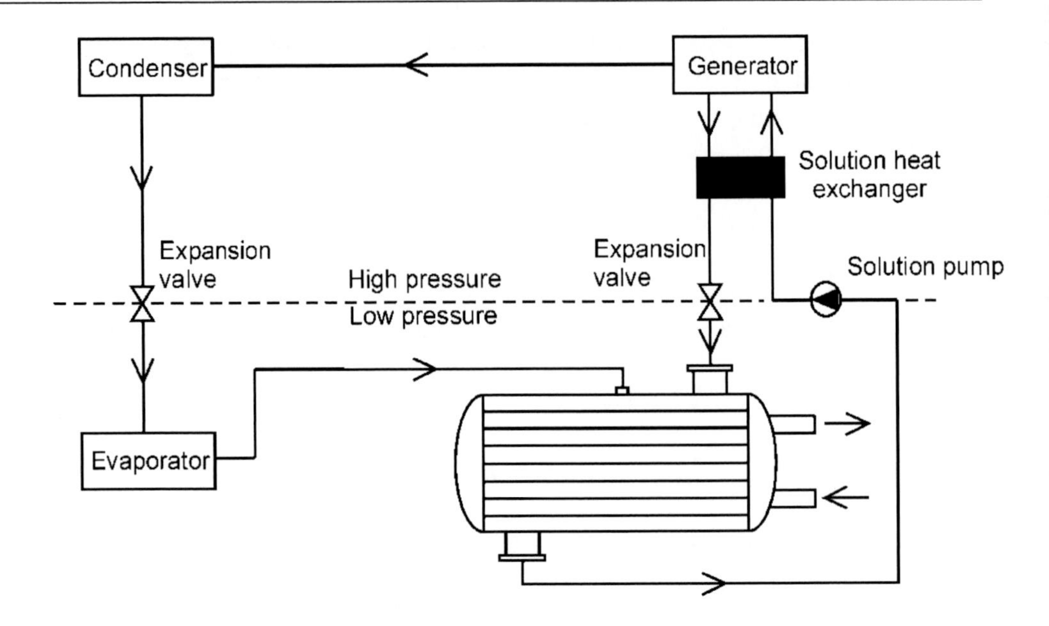

Figure 2. A conventional absorption chiller.

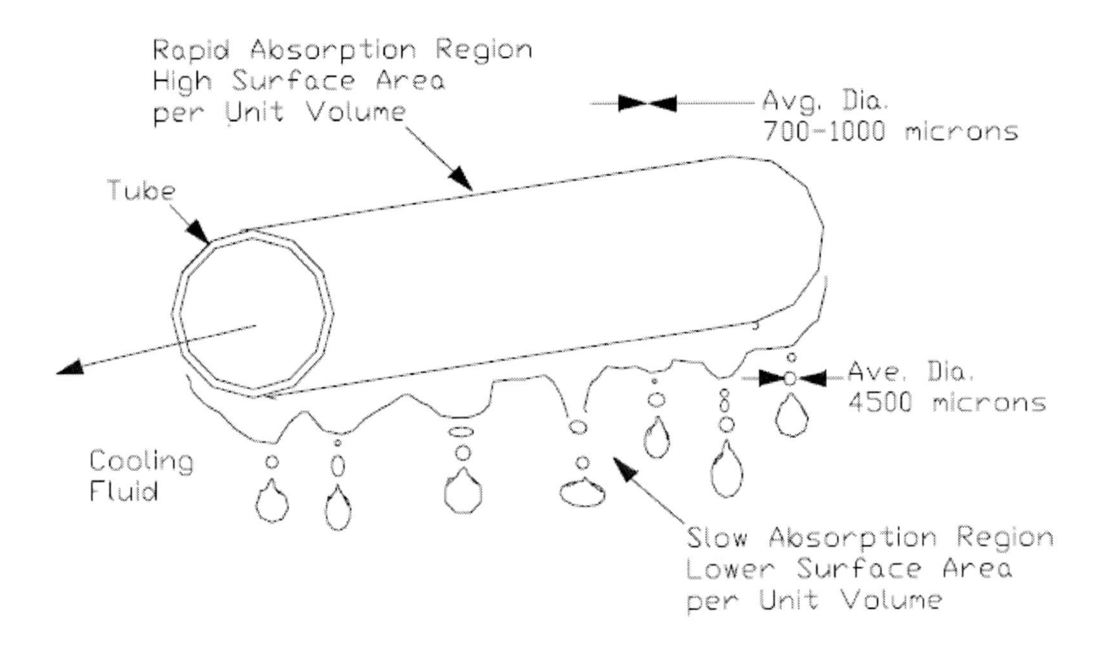

Figure 3. Solution distribution of present absorber designs.

CONCEPT

The spray absorption concept shown in Figure 4 could be advantageously applied to any absorption cooling system. However, its benefits are most visible in applying it to multi-effect high-efficiency absorption chillers, and it allows for simplifications of those systems, a long held ambition of the absorption industry.

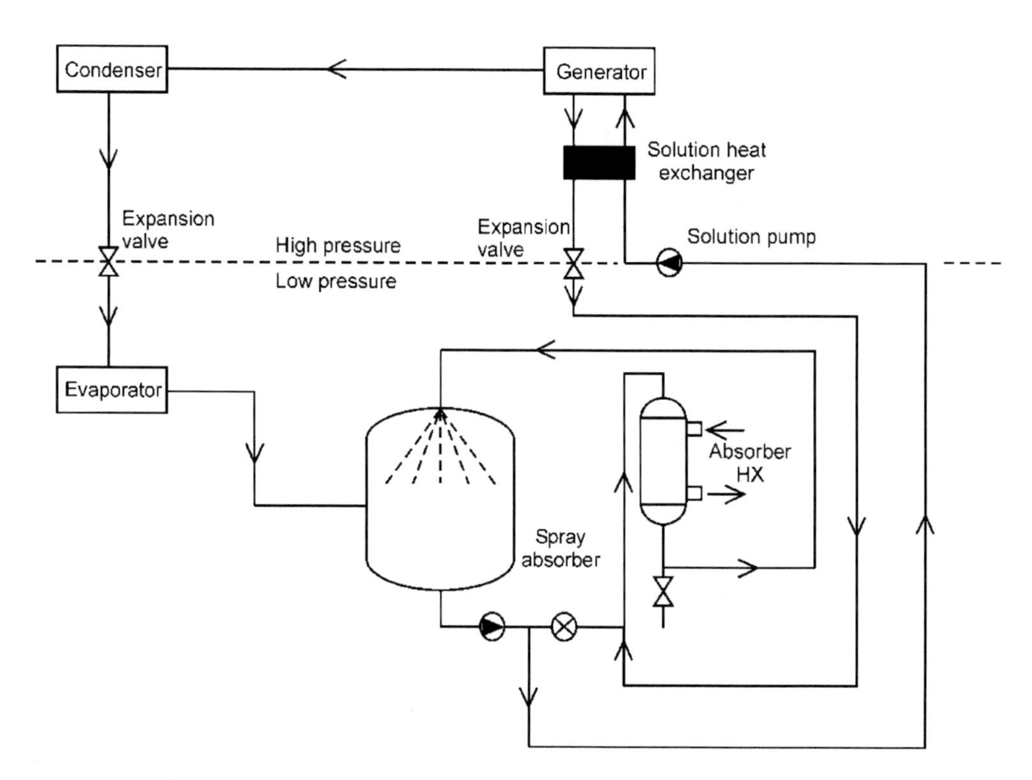

Figure 4. Spray absorber with solution sub-cooler.

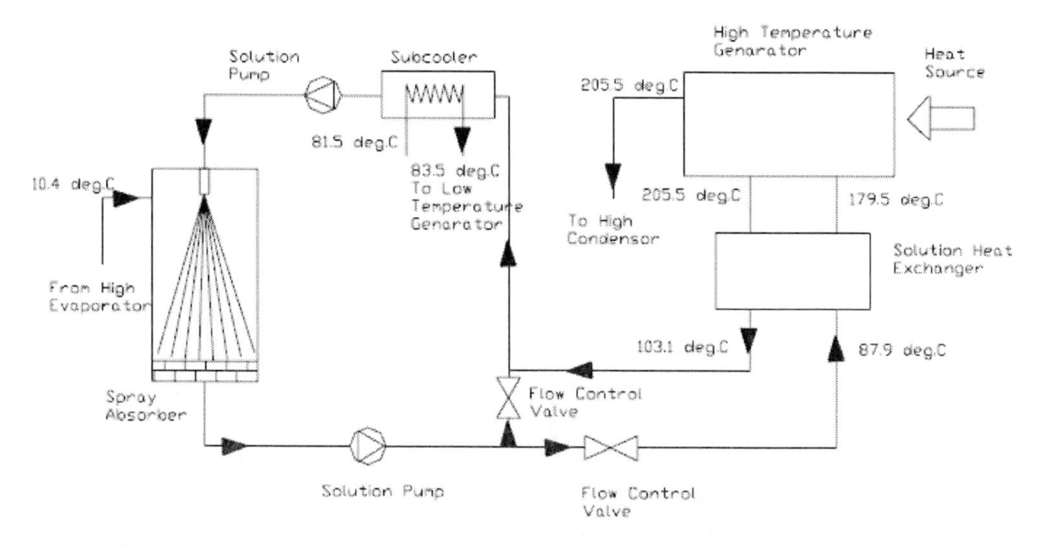

Figure 5. High Temperature Loop of triple-effect Spray Absorption System.

As mentioned earlier, the absorber and the evaporator of conventional large chillers basically rely upon falling film heat exchangers and this approach requires that the cooling system reject heat through a cooling tower. This is true for both single and double-effect systems using lithium bromide-water solutions where the solution concentration is around

62%. These machines have efficiencies defined by the coefficient of performance (COP) that vary from values of about 0.6-0.7 for single-effect systems to values of about 1.0-1.2 for double-effect systems.

Triple-effect concepts, in which the high temperature loop shown in the Figure 5, are been considered that could extend the COP values to 1.5 by using a proprietary solution that varies in salt concentration in the 85 to 90% range. Concomitant with the higher salt concentration, other solution properties pose technical question that must be answered. These include the viscosity that affects pumping requirements as well as the mass and heat transfer properties. Of particular importance is the rate that the refrigerant (water) in the form of vapor is absorbed by the concentrated salt solution in the absorber section of the chillers. It has long been recognized that if this rate could be increased beyond that occurring in the falling film type of absorber, this component could be reduced in size and, because this component is usually the most expensive part of the total system, cost reduction would be possible.

The advantages of the spray absorption approach are considerable. They are:

- Reducing the absorber surfaces reduces the size and weight of absorber.
- Since the operating temperature of the system is high, evaporator pressures can be as high as 1200 to 1350 Pa, as opposed to less than 800 Pa on conventional machines. This raises the temperature in the absorber about 3°C and improves system efficiency.
- Due to the elevated absorber temperatures, this cycle can be air-cooled and since the absorber and condenser reject heat directly to the outdoor air in small system, no cooling tower is required.
- For large system, heat can be rejected in a liquid-liquid heat exchanger and intensifying heat transfer reduces the size of the heat exchanger.

At present, absorption systems sell at a considerable premium in price compared to electrically driven vapor-compression systems. In general, this is due to the larger size and weight of these systems. The weight problem can be attributed to both the large heat exchangers and the heavy containment vessels needed to allow these systems to operate at high vacuums. The size of these vessels is dictated by the heat and mass transfer processes they contain.

Any improvement in the transport process rates would reduce the weight and the bulk of the interior components and similarly reduce the size of the required containment vessels. For any significant improvement in the cost of these cycles, process enhancement are the key.

The cycle benefits would be derived from the two following operating characteristics:

- Absorption heat release rates from 10 to 20 kW/m^2 are common from the falling films in existing absorbers. On the other hand, with a 10°C temperature difference across the sub-cooler, forced convection liquid-to-liquid heat exchange rates can be in excess of 100 kW/m^2. The surface area of the absorber cooler required for the system shown in Figure 4 can thus be smaller than that of the cooling tubes in the falling film absorber in Figure 2.

- Given equal absorber volumes an ideally dispensed spray would present nearly 10 times the solution surface available from a falling film absorber. In practice, an ideal well-dispersed drop is not attainable and actual area enhancement would be less.

If the spray absorption flux is comparable to that of the falling film, sprays have considerable advantage in size and material intensity over a conventional absorber. There is indication that the absorption rates in sprays are actually higher, bringing added benefits to the concept.

Consequently, it is necessary to confirm the previous results and demonstrate that the benefits of the spray absorption concept can be successfully achieved, particularly with the highly viscous solutions of multi-effect cycle machines. Beyond our recognition, this has been recognized by a major absorption chiller manufacturer who believes that this research is a necessary prerequisite for implementing the concept in their triple-effect chiller designs.

The primary goal of this work is to provide empirical evidence that a proper spray can be generated using a viscous solution that is required to be used in triple-effect absorption chillers. In this work, as a first step of investigation of the performance of the spray absorbers, the feasibility of creating the optimum drop sizes defined by the previously mentioned investigators [10 - 13] with a new absorbent solution was studied. Based on previous studies conducted along atomization of high viscous solution [15 - 18], the most appropriate atomization for the applications can be considered as the liquid cone and sheet disintegration. Even with high viscous solutions, these atomization patterns form a conical shaped liquid sheet or liquid cone, which gives higher flow rates and smaller droplets. These types of disintegrations can be effectively achieved with pressure-swirl and pressure-jet atomizers.

In order to generate the smallest possible droplet sizes keeping higher flow rates, when selecting pressure-swirl and pressure-jet atomizing nozzles to conduct the experiments, recommendations given by all the above investigators, specifications given by the manufacturer and results obtained from the previous research carried out by authors for many different kinds of pressure nozzles for similar operating conditions were used.

Drop Formation Theory

The formation of a bulk liquid into a large number of fine particles by engineering devices is called atomization. A stream of a large number of such particles is called a spray, and such engineering devices are called atomizers or nozzles. There are several basic processes associated with all method of atomization, such as the hydrodynamics of the flow within the atomizer which govern the turbulence properties of the emerging liquid stream. The development of the jet or sheet and the growth of the small disturbances which eventually leads the disintegration into ligaments and drops, are also of primary importance in determining the shape and penetration of the resulting sprays, as well as its details characteristic of number density, drop velocity, and drop size distribution as function of time and space.

All above characteristics are markedly affected by:

- Nozzle size and geometry
- Physical properties of the surrounding medium
- Physical properties of the sprayed liquid

The basic influences of above factors for drop formation can be indicated as,

$$Mean\ Drop\ Size \propto \sqrt{Jet\ Diameter} \tag{1}$$

$$Mean\ Drop\ Size \propto \sqrt{Sheet\ Thickness} \tag{2}$$

$$Mean\ Drop\ Size \propto \Delta P_L^A \tag{3}$$

$$Mean\ Drop\ Size \propto \rho_G^B \tag{4}$$

The flow and spray characteristics of most atomizers are strongly influenced by the liquid properties of density, viscosity, and surface tension.

$$Mass\ Flow\ Rate \propto \sqrt{\rho_L} \tag{5}$$

$$\begin{pmatrix} Minimum\ energy\ required \\ for\ atomization \end{pmatrix} = \begin{pmatrix} Surface \\ tension \end{pmatrix} \times \begin{pmatrix} Increase\ of\ liquid\ surface \\ area\ due\ to\ atomization \end{pmatrix} \tag{6}$$

For correlation of drop size data, whenever the surface tension is important (drop breakup in turbulent flow fields), the Weber number, is a useful dimensionless parameter. The Weber number for drop form by above mechanism can be defined as:

$$Weber\ Number(We) = \frac{Inertial\ Force}{Surface\ Force} = \frac{\rho_G U_R^2 D}{\sigma} \tag{7}$$

$$We_{crit} = \frac{8}{C_D} \tag{8}$$

For correlation of drop size data, whenever the viscosity of liquid is important (drop breakup in viscous flow fields), the Reynolds number, is a useful dimensionless parameter.

$$Reynolds Number(\mathrm{Re}) = \frac{InertialForce}{ViscousForce} \tag{9}$$

$$Re_d = \frac{UD}{\nu_d} \tag{10}$$

$$Re_c = \frac{UD}{\nu_c} \tag{11}$$

In many respects, viscosity is the most important liquid property. Although, in an absolute sense, its influence on atomization is no greater than that of surface tension, its importance stems from the fact that it affects not only the drop size distribution in the spray, but also the nozzle flow rate and spray pattern. An increasing viscosity lowers the Reynolds number and hinders the development of any natural instabilities in the jet or sheet. The combined effect is to delay disintegration and increase the size of the drops in the spray.

Drop Shape

The combined influence of the above three factors on the drop shape can be expressed by the Eotvos number (sometimes referred to as the Bond number), which relates diameter and density to the surface tension. Drops with low Eotvos number tend to remain spherical for very high continuous-phase Reynolds number.

$$Eo = \frac{g\Delta\rho D^2}{\sigma} \tag{12}$$

The Morton number relates the viscosity of the continuous phase to surface tension on droplet surface. (In the literature, this group is often simply referred to as the M-group or property group).

$$M = \frac{g\,\mu^4\,\Delta\rho}{\rho^2\,\sigma^3} \tag{13}$$

The stream emerging from the nozzle could be categorized as a jet or liquid sheet, according to pressure atomizer geometry. During the primary atomization through the nozzle, a jet or sheet emerging is broken down into drops. According to the previous authors, the most appropriate atomization for our kind of application is the liquid sheet or jet disintegration. Many atomizers do not form jets of liquid, but rather form flat or conical sheets.

Atomization Mechanism

When the sheet or jet of liquid emerges from a nozzle, its subsequent development is influenced mainly by its initial velocity and the physical properties of the liquid and the ambient gas. To expand the sheet or jet against the contracting surface tension force, minimum sheet or jet velocity is required, which is provided by pressure, aerodynamic drag, or centrifugal force, depending on the type of atomizer.

Fraser and Eisenklam [19] define three modes of sheet disintegration. They are rim, wave, and perforated-sheet disintegration. In the rim mode, forces created by surface tension cause the free edge of a liquid sheet to contract into the thick rim, which then breaks up by the mechanism corresponding to the disintegration of free jets. When this occurs, the resulting drop continues to move in the original flow direction, but remains attached to the receding surface by thin threads that also rapidly break up in to rows of drops. This mode of disintegration is most prominent where the viscosity and surface tension of the liquid are both high. It tends to produce large drops, together with numerous small satellite droplets

From a large number of tests on a wide variety of liquids, Dombrowski and Fraser [20] concluded that liquid sheet with high surface tension and viscosity are the most resistant to disruption, and the effect of liquid density on sheet disintegration is negligibly small.

At the boundary between a liquid and the ambient air the balance of forces may be expressed by the equation

$$P_L - P_A = -\sigma \frac{d^2 h}{dx^2} \tag{14}$$

Atomization is often accomplished by discharging the liquid at high velocity into a relatively slow moving stream of air or gas. An ideal atomizer would possess the ability to provide the required atomization over a wide range of liquid flow rates, rapid response to changes in liquid flow rate, freedom from flow instabilities, low power requirements, capability for scaling, to provide design flexibility, low cost, light weight, ease of maintenance, and ease of removal for maintenance.

Atomizers can be divided to the basic groups such as pressure atomizers, rotary atomizers, air-assist atomizers, air-blast atomizers, effervescent atomizers, electrostatic atomizers, ultrasonic atomizers, and whistle atomizers. During the selection of the nozzle, in addition to the above recommendation for achieving smaller droplets and a well-dispersed spray, the manufacture's recommendations were considered.

Pressure atomizers rely on the conversion of pressure into kinetic energy to achieve high relative velocity between the liquid and the surrounding gas. They include plain-orifice, simplex, variable geometry, duplex and duel-orifice nozzles.

In pressure swirl nozzle, the internal flow characteristics are of primary importance, because they govern the thickness and uniformity of the annular liquid film formed in the final discharge orifice as well as the relative magnitude of the axial and tangential component of velocity of the film. Despite the geometric simplicity of the pressure swirl nozzle, the hydrodynamic processes occurring within the nozzle are highly complex. Nevertheless, the early theories based on the assumption of frictionless flow soon led to the formulation of a quantitative relationship between the main atomizer dimensions and various flow parameters, such as nozzle flow number, discharge coefficient, film thickness and initial spray cone angle.

Spray Characteristics

Due to the complexity of hydrodynamic processes occurring in the nozzle and lack of any quantitative theory for flow in orifices over wide ranges of Reynolds number has led to a number of empirical correlation for predicting film thickness, cone angle, and coefficient of discharge for viscous flow field. Out of large numbers of empirical correlation equations for the above parameters, one equation for each was selected using the ranges of Reynolds numbers in Table 1.

Table 1. Numerical Coefficient for Differing Drop Sizes

Drop size (Micro meters)	Stokes Terminal Velocity (m/s)	Theoretical Average Droplet Velocity (m/s)	Vapor Reynolds Number	Droplet Reynolds Number	Droplet Bond Number	Droplet Weber Number	Droplet Morton Number
10	2.90E-03	1.45E-03	1.27E-04	1.51E-03	9.15E-06	1.28E-12	5.99E-14
100	2.90E-01	1.45E-01	1.27E-01	1.51E+00	9.15E-04	1.28E-07	5.99E-14
150	6.52E-01	3.26E-01	4.29E-01	5.10E+00	2.06E-03	9.74E-07	5.99E-14
200	1.16E+00	5.80E-01	1.02E+00	1.21E+01	3.66E-03	4.10E-06	5.99E-14
300	2.61E+00	1.30E+00	3.43E+00	4.08E+01	8.23E-03	3.12E-05	5.99E-14
350	3.55E+00	1.78E+00	5.45E+00	6.48E+01	1.12E-02	6.74E-05	5.99E-14
375	4.08E+00	2.04E+00	6.71E+00	7.97E+01	1.29E-02	9.51E-05	5.99E-14
400	4.64E+00	2.32E+00	8.14E+00	9.67E+01	1.46E-02	1.31E-04	5.99E-14
425	5.24E+00	2.62E+00	9.76E+00	1.16E+02	1.66E-02	1.78E-04	5.99E-14
450	5.87E+00	2.93E+00	1.16E+01	1.38E+02	1.85E-02	2.37E-04	5.99E-14
475	6.54E+00	3.27E+00	1.36E+01	1.62E+02	2.06E-02	3.10E-04	5.99E-14
500	7.25E+00	3.62E+00	1.59E+01	1.89E+02	2.29E-02	4.01E-04	5.99E-14
1000	2.90E+01	1.45E+01	1.27E+02	1.51E+03	9.15E-02	1.28E-02	5.99E-14
2000	1.16E+02	5.80E+01	1.02E+03	1.21E+04	3.66E-01	4.10E-01	5.99E-14
3000	2.61E+02	1.30E+02	3.43E+03	4.08E+04	8.23E-01	3.12E+00	5.99E-14

The effective flow area of a pressure atomizer, usually described in terms of flow number, is expressed as:

$$FN = \frac{\dot{m}_L}{(\Delta P_L)^{0.5} \times \rho_L^{0.5}} \tag{15}$$

After the follow up work of Eisenklam [21], and Dombrowski and Hassan's [22] a dimensionally correct equation to correlate their experimental data on discharge coefficients, Rizk and Lefebvre [23] derived the following relationship for C_D:

$$C_D = 0.35\left(\frac{A_{IP}}{D_S \times d_0}\right)^{0.5}\left(\frac{D_S}{d_0}\right)^{0.25} \tag{16}$$

Rizk and Lefebvre [23] have suggested that for $t/d_0 << 1$, the original equation developed by Geffen and Muraszew [24] can be written more succinctly, while still retaining all its essential features, as:

$$t = 2.7\left[\frac{d_0\ FN\ \mu_L}{\left(\Delta P_L\ \rho_L\right)^{0.5}}\right]^{0.25} \tag{17}$$

Based on numerous calculations of spray angle using equations previously developed, Rizk and Leferbvre [25] derived the following dimensionally correct equation for spray angle:

$$2\theta_m = 6K^{-0.15}\left(\frac{\Delta P_L\ d_0^2\ \rho_L}{\mu_L^2}\right)^{0.11} \tag{18}$$

$$K = \frac{A_P}{D_S\ d_0} \tag{19}$$

The spray properties of importance include mean drop size, drop size distribution, droplet number density, cone angle, and penetration. The quality of the fineness of an atomization process is usually described in terms of mean drop size. Unfortunately, the physical processes involved in atomization are not yet sufficiently well understood for mean diameters to be expressed in terms of equations derived from basic principles. The simplest case of the breakup of a liquid jet has been studied theoretically for more than a hundred years (Lord Rayleigh, 1878), but the results of these studies have failed to predict the spray characteristics to a satisfactory level of accuracy.

Drop Characteristics

The absence of any general theoretical treatment of the atomization process has led to the evolution of empirical equations to express the relationship between the mean drop size in the spray and the variables of liquid properties, gas properties, flow conditions, and atomizer dimensions. Owing to the complexity of the various physical phenomena involved in pressure swirl nozzles, the study of atomization has been pursued principally by empirical methods, yielding correlation for mean drop size of the form:

$$SMD \propto \sigma^a\, \nu^b\, \dot{m}_L^c\, \Delta P_L^d \tag{20}$$

Considering liquid properties, ambient condition, and nozzle geometry, Babu [26] used regression analysis to determine the following equation for the drop size for low Reynolds number flow where $\Delta P_L << 2.8$ MPa:

$$SMD = 133 \frac{FN^{0.64291}}{\Delta P_L^{0.2265} \rho_L^{0.3215}} \tag{21}$$

For similar conditions, analysis of the experimental data of Lefebvre [27] yielded the following equation for mean drop size:

$$SMD = 2.25 \sigma^{2.25} \mu_L^{0.25} m_L^{0.25} \Delta P_L^{-0.5} \rho_A^{-0.25} \tag{22}$$

Stokes [28] analysis for steady flow of a real fluid past a solid sphere showed that the drag coefficient on a sphere can be expressed as

$$C_D = \frac{24}{\text{Re}} \tag{23}$$

and hence the drop velocity can be expressed as

$$U_{st} = \frac{g D^2 \Delta\rho}{18 \mu_L} \tag{24}$$

Stokes law holds only for the flows that are entirely dominated by viscous forces, that is, for flow characterized by low Reynolds numbers. But for the larger drops with higher Re numbers, Rybczynski-Hadamard velocity of a falling drop can be related to Stokes velocity by:

$$U_{HR} = \frac{3U_{st}(1+k)}{(2+3k)} \tag{25}$$

For $k = 1000$ to 1700, the Stokes velocity can be used with little loss of accuracy and for the droplets less than 400 microns, half of the Stokes velocity can be used as an average velocity with little loss of accuracy, as shown in the Table 1.

Laminar circulation within a spherical drop occurs in the form of a Hill's vortex [29, 30]. This is a toroidal-shaped circulation pattern that typically occurs in spherical drops at high Re and low values of k. Besides the high value of k for salt solution in an atmosphere of low-pressure water vapor, the slow internal circulation within the drop can occur due to the smaller diameters [12, 13].

This motion can be expressed with a representative velocity of fluid on the drop surface at 90 degree to the surrounding vapor flow by [29, 31]:

$$V_o = \frac{U}{2(1+k)} \tag{26}$$

Due to the heterogeneous nature of the atomization process, the ligaments formed by the various mechanism of sheet disintegration vary widely in diameter and the resulting main drops and satellite drops vary in size correspondingly. Therefore, practical nozzles do not produce sprays of uniform drop size at any given operating condition. Instead, the spray can be regarded as a spectrum of drop size distributed about some arbitrarily defined mean value. Thus, in addition to the mean drop size, another parameter of importance in the definition of the spray is the distribution of drop sizes it contains.

An instructive picture of drop size distribution may be obtained by plotting a histogram of drop size, each ordinate representing the number of drops whose dimension fall between the limits D - ΔD/2 and D + ΔD/2. As ΔD is made smaller, the histogram assumes the form of a frequency curve that may be regarded as a characteristic of the spray, provided it is based on sufficiently large samples. Such a curve is usually referred to as a frequency distribution curve.

Because the graphical representation of drop size distributions is laborious and not easily related to experimental results, many workers developed mathematical models to obtain parameters of drop distribution from a limited number of drop size measurements. The Rossin-Rammler distribution used in this experiment is the most commonly accepted drop distribution function [18]. This distribution fits the spray distribution around two factors, X^* and q and is of the form in:

$$1 - Q = \exp\left\{ -\left(\frac{d^*}{X^*} \right)^q \right\} \tag{27}$$

As q goes to infinity, the spray becomes perfectly uniform in drop size. Practical sprays have values of q between 1 and 7. That is, 63.2 % of the solution is in drops of smaller diameter than X^*. Although X^* is not a mean diameter of the distribution, it can be calculated directly from the mean diameters.

Three mean diameters were calculated in these drop measurements. The following simplified nomenclature and definitions from Lefebvre [18] are used.

MLD is the mean diameter of the drops that defined as

$$MLD = D_{10} = \frac{\int_0^\infty N\, D\, dD}{\int_0^\infty N\, dD} \tag{28}$$

Half of the volume of the spray is in drops smaller than this diameter. Ryan [13] used this diameter to calculate the absorption rates. It is also called the mean mass diameter by some authors.

$$MVD = D_{30} = \left[\frac{\int_0^\infty N\, D^3\, dD}{\int_0^\infty N\, dD} \right]^{\frac{1}{3}} \tag{29}$$

Sauter Mean Diameter (*SMD*) is the diameter of the drop whose ratio of volume to surface area is the same as entire spray. Most of the empirical correlating equations that appear in the literature are based on this diameter.

$$SMD = D_{32} = \frac{\int_0^\infty N\, D^3\, dD}{\int_0^\infty N\, D^2\, dD} \tag{30}$$

EXPERIMENT

The experimental setup developed for this experiment was intended to produce a reproducible spray using experimental nozzles manufactured by Spraying System Incorporated, and provide good visibility of the spray process for general observation, high speed and video imaging of the spray and laser measurement of drop sizes and speeds. The design and construction of the experimental setup considered the required operating conditions, characteristics of nozzles and behavior of sprays at least under normal operating (atmospheric pressure) condition, as given in the Table 2 and 3.

Table 2. Ranges of Operating Parameters

Operating parameter	Required range
Required concentration of salt solution at nozzle	84%
Required nozzle pressure range	0-5.2x105 Pa
Required absorber pool temperature	92.2 oC
Required solution temperature range at nozzle	65-85 oC
Required absorber pressure	1.3x103 Pa

The nozzle pressure of any particular test depended on the development of the spray for that specific nozzle and the maximum spray angle that any spray can reach before contact with the absorber chamber wall, due to the limited diameter of the chamber. Whenever it was possible, the fully-developed spray from each nozzle was obtained and if it was not possible, the nozzle pressure was deviated from the range given in the Table 2. The selection of two nozzles types listed in the Table 3 for this experiment was based on the previous studies performed by industry, nozzle manufacturer and previous experiments conducted by this investigators on the characterization of nozzles. The nozzle specifications given in Table 3 was based on the spray of water at 6.89×10^4 Pa nozzle pressure under atmospheric condition supplied by the manufacturer. Even though the spray characteristic given in the table 3 seems very appropriate for absorption application, due to the high viscous nature of the salt solution, dramatic changes in nozzle performance can be expected when running the new absorber solution.

Table 3. Nozzle Specifications

Type	Model	Nomen-cleture	Nozzle Constant (K)	Drop Size (Water) $MVD \times 10^6$ (m)	Capacity (Water) $Q \times 10^{-6}$ (m^3/s)	Spray Angle (Degree)
Whirl Jet	1/8BXSS1	SJ1	0.47	252	6.3	52
Whirl Jet	1/8BXSS2	SJ2	0.32	263	1.2	54
Whirl Jet	1/8BXSS2W	SJ2W	0.25	268	1.6	114
Whirl Jet	1/8BXSS3	SJ3	0.22	275	1.9	56
Full Jet	1/8GGSS2	FJ2	1.65	325	1.2	43
Full Jet	1/8GGSS3	FJ3	1.03	346	1.8	58
Full Jet	1/8GGSS2.8W	FJ3W	0.5	338	1.8	120

Experimental Test Rig

In order to achieve the operating conditions and conduct the data collection process using above mentioned laser and imaging system, the developed experimental setup was consists of five major systems as described below as well as shown in the Figure 6.

Vacuum Spray Absorber Chamber: The spray chamber was a 304 stainless steel cylindrical welded tank with a flat top and bottom. The diameter of the tank was 60 cm and the height was 120 cm. the chamber consists of eight MDC glass viewports having a 20 cm diameter and a 12.5 cm view diameter and one quick access MDC viewport door. The viewports were placed around the cylinder such that the laser instrument could be aligned to take the required measurements. The top surface of the tank contained five NPT connectors with a center one was reserved to supply solution to the nozzle. The other connectors were mainly use for electrical and sensor wirings. In order to control the temperature inside the chamber as well as the temperature of solution pool, a heater with two T-type thermocouples were inserted into the chamber. The pressure of the chamber was controlled using attached vacuum system which was controlled by the chamber pressure monitor and controller. The level controller which controlled the pool level sustained the vacuum while returns the solution back to the storage.

Solution Tempering System: The solution in the storage tank as can be seen in the Figure 6 was pumped to pressurize the nozzle using the Liquiflo gear pump with a sealed magnetically coupled drive that can operate under very low net positive suction head. The valves that controlled the flow rates as well as liquid pressure in the system was Nupro bellow valves. The solution tempering system consisted of Parker in-line heater, Series N93 Vulcan 1500 W inline cartridge heater and a water pump as shown in the Figure 6.

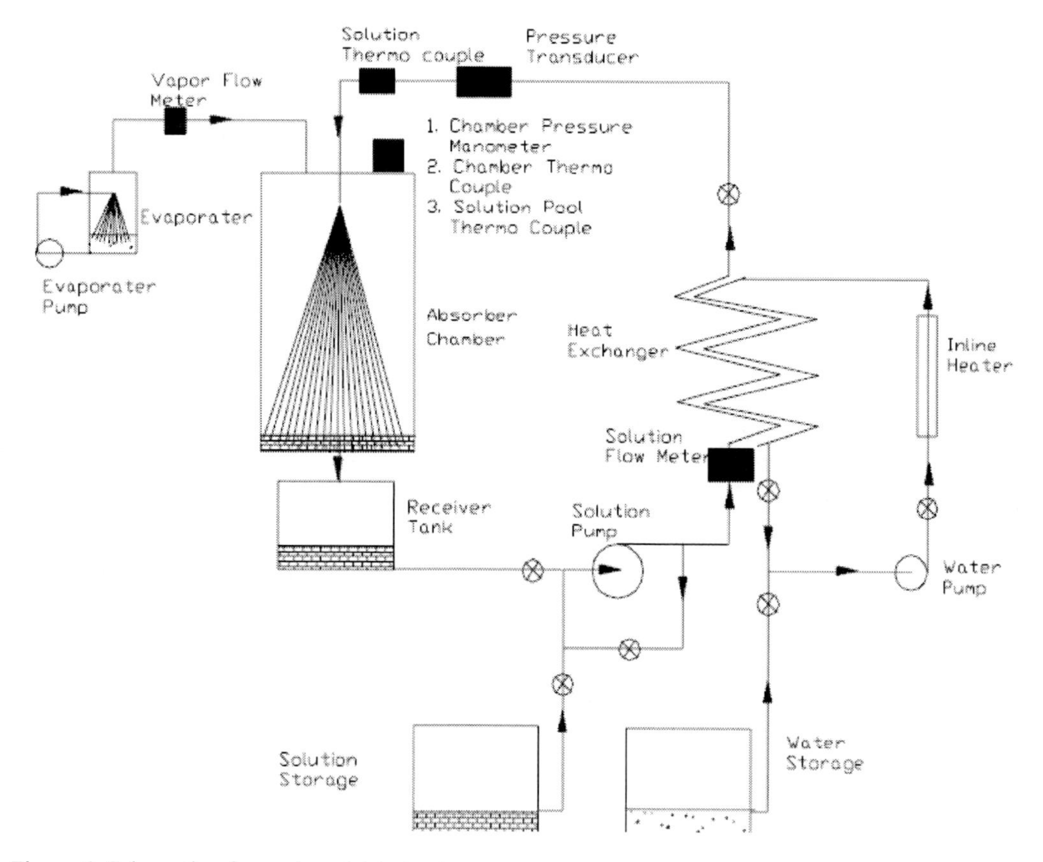

Figure 6. Schematic of experimental test setup.

High Vacuum System: Vacuuming was one of the most important aspect of this experiment, as the chamber should have pure water vapor environment. To attain a quick vacuum, two vacuum pumps, Welsh DuoSeal model 11402 and Edwards high vacuum pump model ED 200 A was used. Both pumps were able to produce a vacuum less than 8 Pa. To minimize the degassing and avoid water vapor mixing with pump oil, which reduces the pump efficiency and hence degrade the vacuum, a dry ice fore-line trap was used. The vacuum system was controlled and broken by manually controlled two MDC type KVA-100 vacuum valves. Importantly, during the testing of high vacuum, a Varian 938-41 Porta-Test helium mass spectrometer leak detector was used to check the leak rate and vacuum integrity.

Particle Measurement and Imaging System: The Aerometric Phase Doppler Particle Analyzer (PDPA) is a unique and sophisticated instrument which utilizes the light scattered by spherical particles to obtain simultaneous size and velocity measurement. This instrument was capable of making in-situ and nonintrusive measurements which are important when measuring easily deformable particles or probing sensitive flow fields. The PDPA which comprised of 5 major components consisting of a transmitter, a receiver, a signal processor, a motor controlled box and a computer can be seen in the Figure 7. The model XMT-1100-4S transmitter was design to generate tow equal intensity laser beams and focus them to an intersection point, which forms a measurement volume or fringe volume. The transmitter also contains a Spectra-Physics model SP-106 10 mW polarized helium-neon laser, with a wavelength of 623 micrometers and an output beam of diameter 0.68 mm. The receiver model RCV-2100 is a scattered light collection system designed to accurately provide the single-phase shifts necessary for drop size determination. The package consists of a highly efficient lens system for light collection, a special filter (a slit of 100 microns width and a 1 mm length) for exact probe volume definition, a light partitioning prism assembly for collecting the discrete light signal, and multiple detectors complete with preamplifiers, and a Photomultiplier tube.

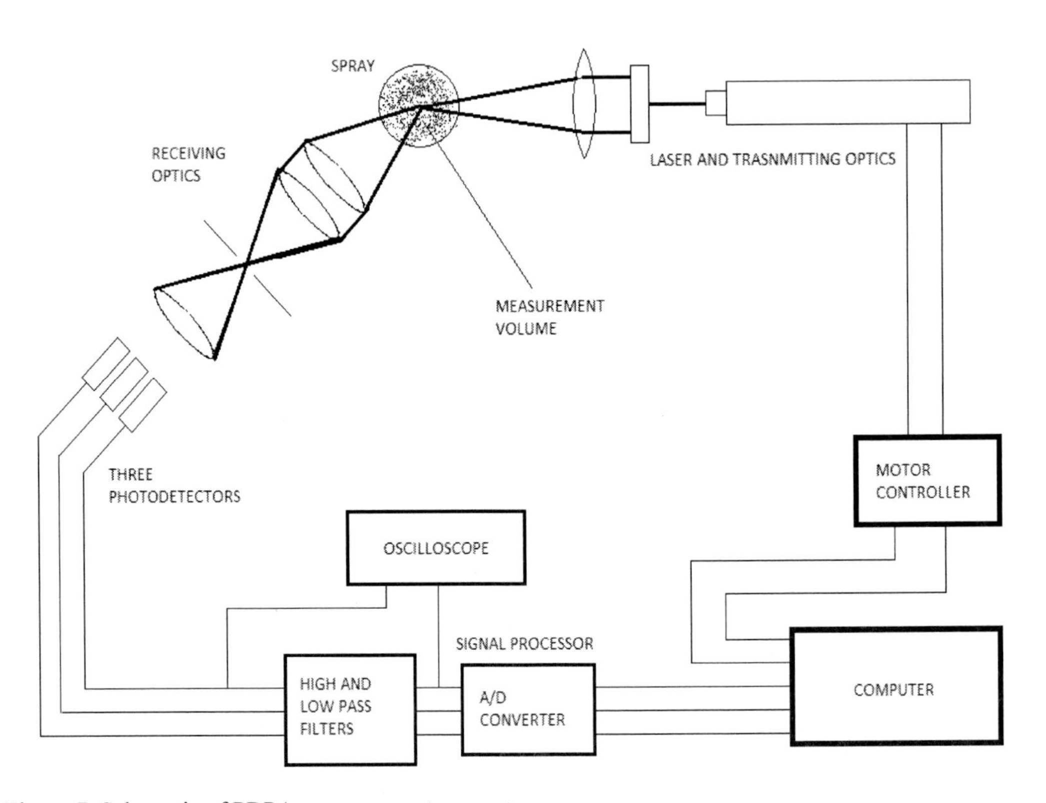

Figure 7. Schematic of PDPA measurement apparatus.

The signal processor model PDP 3100 is a high-speed analogue/digital signal processor that accepts three amplified photomultiplier signals as inputs. The model MCB-7100-1 motor controller is designed to monitor and control both the frequency shift and the track select located in the transmitter.

In addition to using the PDPA system for drop measurements, a digital camcorder and a motion analyzer developed by Kodak, together with plotting techniques were used to measure the spray characteristics of the spray angle and lamella developments.

Data Acquisition and Control System: The control system consisted of a well-tune PID loop programmable temperature controllers to control the nozzle, pool and chamber temperature. The flow rate of solution was measured by a calibrated Rosemount Micromotion D6 corolius-type flow meter with a remote elite model RFT9739 field mount transmitter while the nozzle pressure was measured by an Omega PX 203 series pressure transducer. The Edwards model 655 Barocel capacitance manometer with a measuring range of 0 – 10 millibar was used to measure and controlled the chamber pressure. The data acquisition system consisted of a Hewlett-Packard model HP3852A data acquisition unit with two 20-channel relay multiplexer boards attached to a computer.

Experimental Procedure

The most difficult task of the experiment was aligning the laser system. After selecting the optical configuration that will be used for testing, the beam cross section was focused on an alignment tool made with tape attached to a small piece of mirror, as this suppose to be a constant position weak reflector. After the laser was turned on, the transmitter was set on the greatest spacing. When aligning the transmitter, caution was taken to make the transmission beam perpendicular to the chamber view port glass. Using the 3-D traversing mechanism built onsite, the receiver was aligned to the orientation of transmitter. By looking into the slit through receiver viewing port, the receiver was shifted until the alignment spot from the weak reflector was at the center of the slit. Using the traversing mechanism attached to the receiver, the focus of the spot was adjusted as sharp as possible. Then the both transmitter and receiver were firmly fixed. In order to catch the reflected beam from the solution droplets, both the transmitter and the receiver has to be in the same horizontal plane. As per manufacturer's recommendation for fairly large particles, the 90-degree collection angle was used. After initiating the solution flow, the precise alignment was performed by monitoring signals using an oscilloscope and real-time data histograms.

During the testing period, the whole system was kept at minimum required temperature to avoid crystallization of the high viscous salt solution. The overall experiment was performed considering the temperature as primary variable since the viscosity and density of this salt solution was highly dependent on temperature. For each nozzle tested, 5 to 7 temperature setting were used varying temperature of the solution from 65°C to 85°C. Under each temperature setting for each nozzle, 4 to 7 different flow rate were considered to collect the data depending on the development of the spray inside the vacuum chamber. Using the configured data collection system which was connected to experimental setup, the solution temperature, chamber temperature, pool temperature, nozzle pressure, chamber pressure and solution flow rates were recorded and saved for data analysis. When conducting the experiment, the solution temperature was carefully set to obtain the correct density and viscosity values. When collecting data pertaining to drops (i.e. drop size, drop velocity and number density), the flow field was discretized and the transmitter and receiver were moved along the trajectories to obtain the best representation of the spray flow field as shown in the Figure 8. For each experimental run, the droplet data was collected from 5 to 25 points for each horizontal setting, depending upon the visibility of the spray through sight glasses, for 5-10 different horizontal settings for each spray. At each data collection point, the drop data was acquired for up to 500 drops or after an allocated time of 50 seconds was reached. Further, during each spray flow, the digital camcorder and the Kodak motion analyzer was utilized to capture and store images of sprays flow closer to nozzle to measure the spray angle and lamella development.

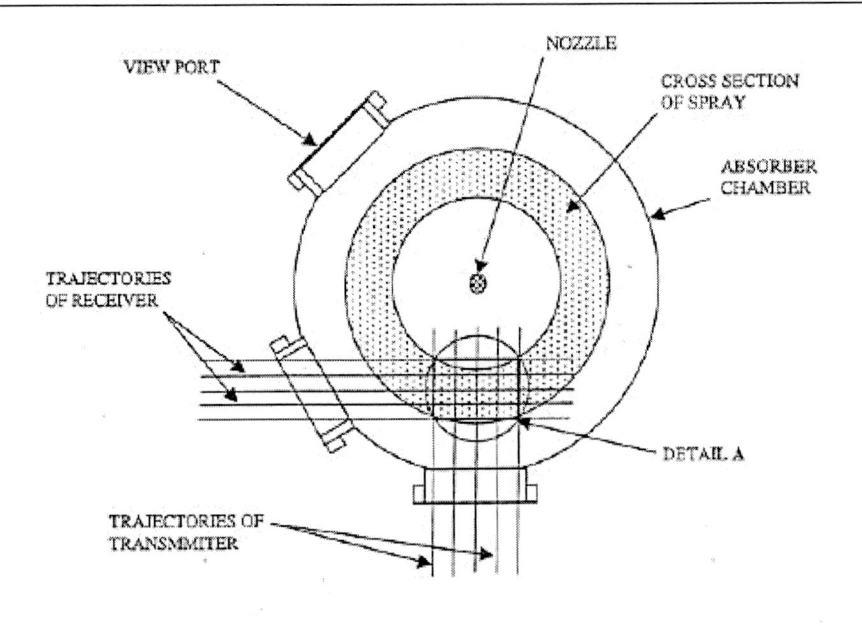

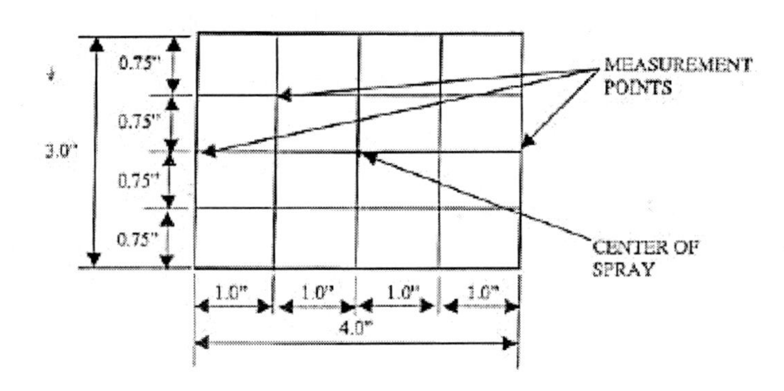

Figure 8. Schematic diagram of transmitter and receiver trajectories of laser measurement system (above) and example measuring field shown as Detail A (below).

UNCERTAINTY ANALYSIS

The uncertainties associated with the results of this experiment are primarily caused by instrumentation and operating error. The instrumentation error, which is due to the accuracy, repeatability and calibration of instruments, affected the experimental results. The operating error caused by the variation of parameters such as properties of the solution and the operating conditions also introduced error into the experimental results.

The error analysis used the root mean square method of the form:

$$RMS\ Error = \sqrt{\sum \left[\Delta \left(\frac{\partial \rho}{\partial \Delta} \right) \right]^2} \tag{31}$$

The *RMS* errors calculated on experimental results of the drop size related to each Swirl Jet (SJ) nozzle tested are given in the Table 4 and that for Full Jet (FJ) are given in Table 5.

Table 4. RMS total of error limits on experimental results for SJ's

Error source	Error	Percentage effect on the results			
		SJ1	SJ2	SJ2W	SJ3
		[±]	[±]	[±]	[±]
Instrument error					
Nozzle temperature	0.45oF	1.2	0.9	1.4	0.8
Nozzle pressure	0.25%	1.3	0.75	1.45	0.5
Chamber pressure	0.1 mbar	1.35	0.8	1.35	0.65
Solution flow rate	0.2% flow	1.75	1.25	2.05	0.9
Operating error					
Solution concentration	0.1%	2.3	1.6	2.5	1.2
Solution density	0.2%	2.5	1.8	2.7	1.6
Solution viscosity	0.3%	2.4	2.1	2.6	1.7
RMS total of all errors		2.3	1.9	2.5	1.5

Table 5. RMS total of error limits on experimental results for FJ's

Error source	Error	Percentage effect on the results		
		FJ2	FJ3	FJ3W
		[±]	[±]	[±]
Instrument error				
Nozzle temperature	0.45oF	1.4	1.3	1.8
Nozzle pressure	0.25%	1.5	0.95	1.65
Chamber pressure	0.1 mbar	1.45	0.9	1.75
Solution flow rate	0.2% flow	1.95	1.35	2.15
Operating error				
Solution concentration	0.1%	2.5	1.7	2.6
Solution density	0.2%	2.6	1.9	2.8
Solution viscosity	0.3%	2.6	2.3	2.7
RMS total of all errors		2.6	2.2	3.2

Results and Discussion

From the experimental data obtained for the nozzle flows, for all the tested swirl jet nozzles of 1/8BXSS1 (SJ1), 1/8BXSS2 (SJ2), 1/8BXSS2W (SJ2W) and 1/8BXSS3 (SJ3) and all the tested full jet nozzles of 1/8GGSS2 (FJ2), 1/8GGSS3(FJ3) and 1/8GGSS2.8W (FJ3W) under fully-developed and controlled-flow conditions with the new salt solution, the data were analyzed and ranges of performance were tabulated and are shown in Table 6. These ranges of performance of all the nozzles tested shows some significant deviation of performance of nozzles compared to nozzle specifications supplied by the nozzle manufacturer (which were based on flow of water at atmospheric condition) as predicted based on the empirical correlation developed by the previous researchers [21 - 26]. In general, the mean drop diameters that were obtained in this low pressure environment are as large as 150% for comparable testing parameters under atmospheric condition. The comparison of spray angles between both environments shows average deviation of around 25% as can be seen in Table 3 and Table 6. The two major variables that affect on this deviation are 1) pressure difference across the nozzle (ΔP) due to the ultra low pressure ambient condition and 2) the large viscosity ratio of disperse to continuous phase (k-Factor) due the solution of spray was high viscous salt solution.

Table 6. Nozzle performance with "new absorbent" salt solution

Nozzle	Gage Pressure Range (kPa)	Flow Rate Range (kg/s)	Drop Diameter MVDx10^6 (m)	Drop Velocity Range (m/s)	Spray Angle Range (Degree)
1/8BXSS1 (SJ1)	14.8-90	0.018-0.025	373-411	8.85-11.74	40.5-52
1/8BXSS2 (SJ2)	3.5-73	0.027-0.034	387-420	9.81-12.52	40.8-53.4
1/8BXSS2W (SJ2W)	(78.6)-(57.5)	0.019-0.024	383-469	5.21-11.06	84.0-98.0
1/8BXSS3 (SJ3)	(47.2)-16.4	0.031-0.043	384-423	9.85-12.93	44.6-54.1
1/8GGSS2 (FJ2)	139.3-419.65	0.013-0.019	417-485	12.43-16.44	40.5-50.6
1/8GGSS3(FJ3)	105.0-383.5	0.016-0.027	451-492	14.95-16.18	45.2-55.8
1/8GGSS2.8W (FJ3W)	0.76-33.5	0.025-0.029	459-480	10.70-12.37	68.0-84.8

As discussed in the theory above, the nozzle constant, K , represents the geometry and size of the atomizer (nozzle) while the flow number, FN, represents the effective flow area or mass flow possibility (or effectiveness) of the nozzle. For any particular type of nozzle, the relationship of mass flow rate through the nozzle pertaining to any ΔP should therefore be correlated through the value of K in order to coincide the data for all the nozzle tested in that particular type. For this case, it will be either the all the swirl jet nozzles (SJ's) or all the full jet nozzles (FJ's). The power-law relationship, which is established based on the plots in Figure 9 for swirl jet nozzle can be given as:

$$\dot{m}/K = C\left(\Delta P\right)^{0.45} \tag{32}$$

where:

k-Factor	1000	1100	1200	1300	1400	1500	1700
C	0.0017	0.0015	0.0013	0.0011	0.0009	0.0007	0.0005

All the tested models of SJ1, SJ2, SJ2W and SJ3 agreed with above relation within ± 2 % of error as indicated using the 2% error bars on the Figure 9. However, as can be seen further in the figure as well as discussed above, the viscosity has played a significant role not only on drop formation but also on controlling the fluid flow aspects. Therefore, the plots that represent the data in the figure has to be separated based on the viscosity ratio, i.e. *k*-Factor. As it can be observed, when *k*-factor increases, mass flow rate of the nozzle decreases for a same pressure and visa-versa.

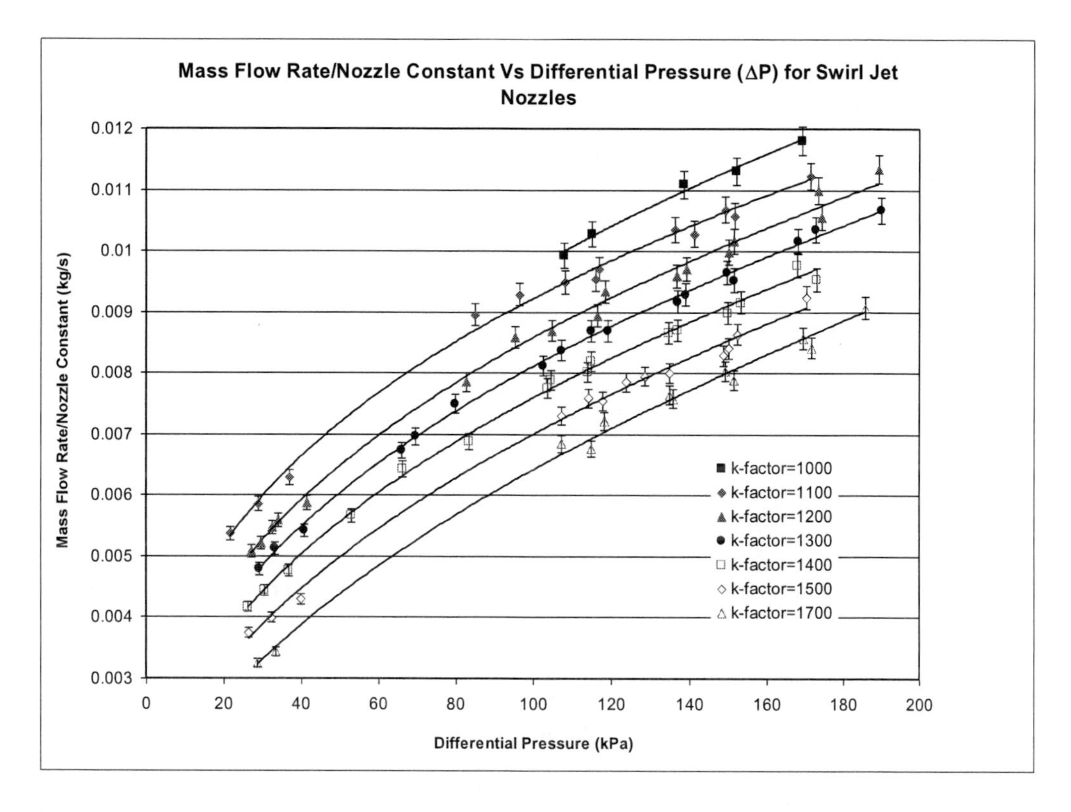

Figure 9. Normalized mass flow rate with respect to nozzle constant (K) to pressure gradient (ΔP) across the nozzle for swirl jet nozzles.

The above property was also exhibited by all the tested full jet nozzles and for that case, the relationship of normalized mass flow rate to differential pressure across the nozzle, ΔP was as plotted on Figure 10.

$$\dot{m}/K = C(\Delta P)^{0.55} \tag{33}$$

where:

k-Factor	1000	1100	1200	1300	1400	1500	1700
C	0.0013	0.0012	0.0011	0.0010	0.0007	0.0006	0.0004

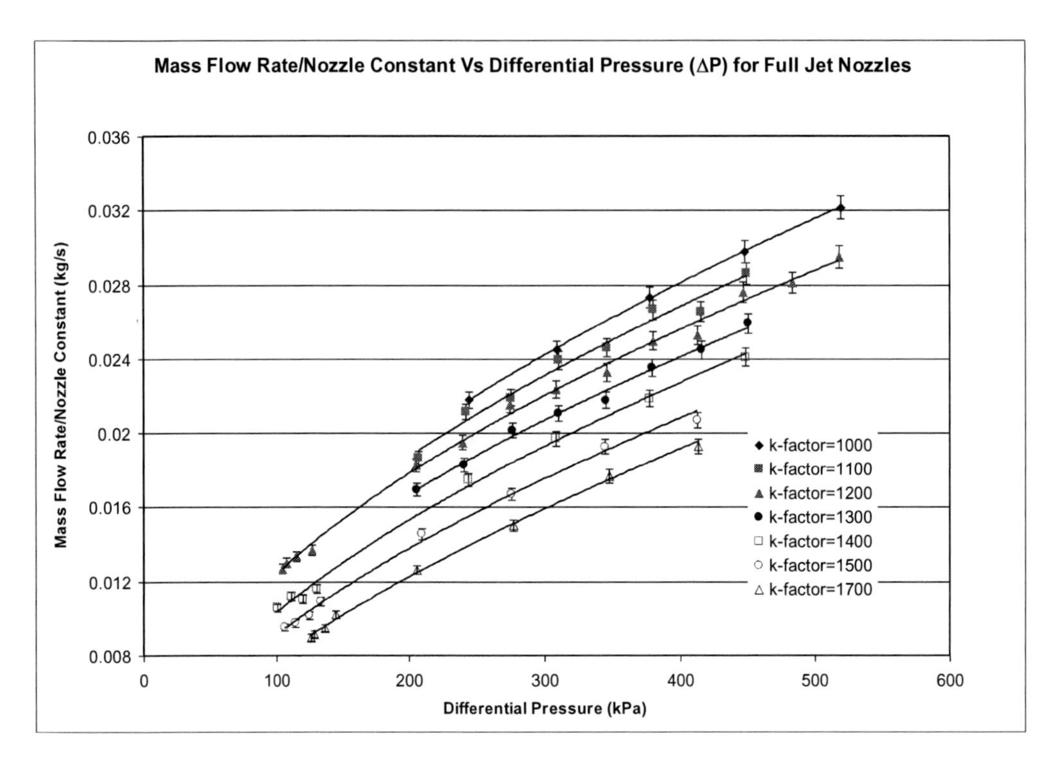

Figure 10. Normalized mass flow rate with respect to nozzle constant (K) to pressure gradient (*ΔP*) across the nozzle for full jet nozzles.

The flow data observed for tested full jet nozzles of FJ2, FJ3 and FJ3W also was predicted using the above relationship within $\pm$ 2%, as indicated with error bars.

However, when the Flow Number (*FN*) was considered for normalizing the mass flow rate for different sizes (or models) of the same type of nozzle, the data points of flow rate relevant to all the sizes of nozzles were aligned regardless of viscosity ratio, the *k*-factor, as can be seen in Figure 11. The relationship between normalized mass flow rate to ΔP is given in the below Equation (34).

$$\frac{\dot{m}}{FN} = 50.22\left(\Delta P\right)^{0.5} \tag{34}$$

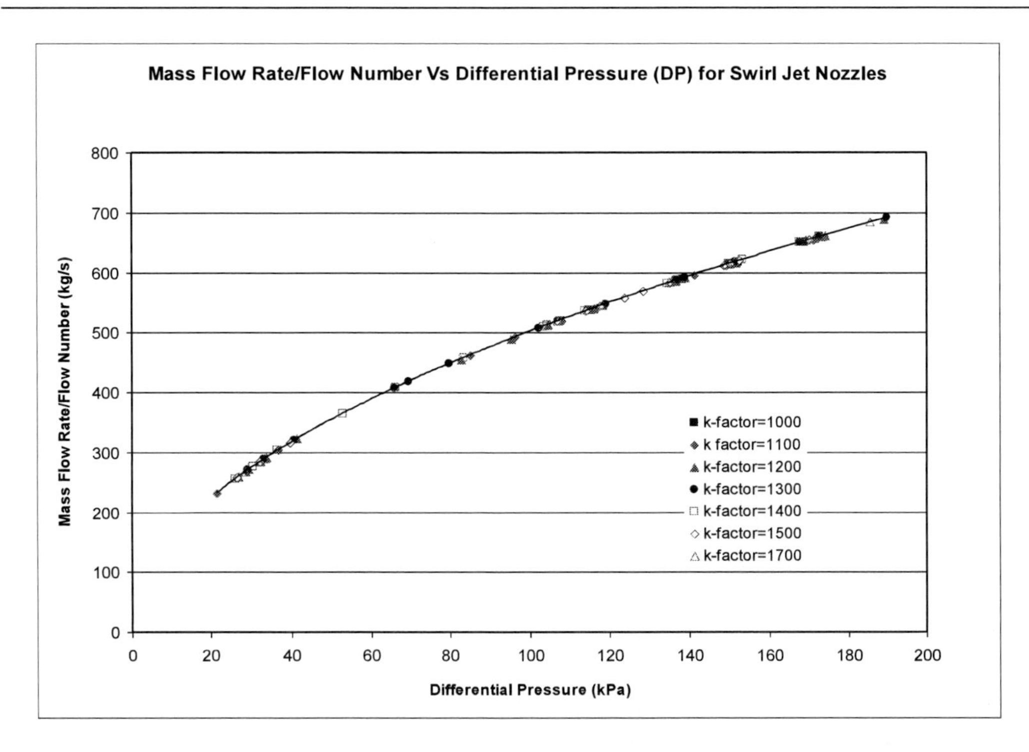

Figure 11. Normalized mass flow rate with respect to flow number (FN) to pressure gradient (delta P) across the nozzle for swirl jet nozzles.

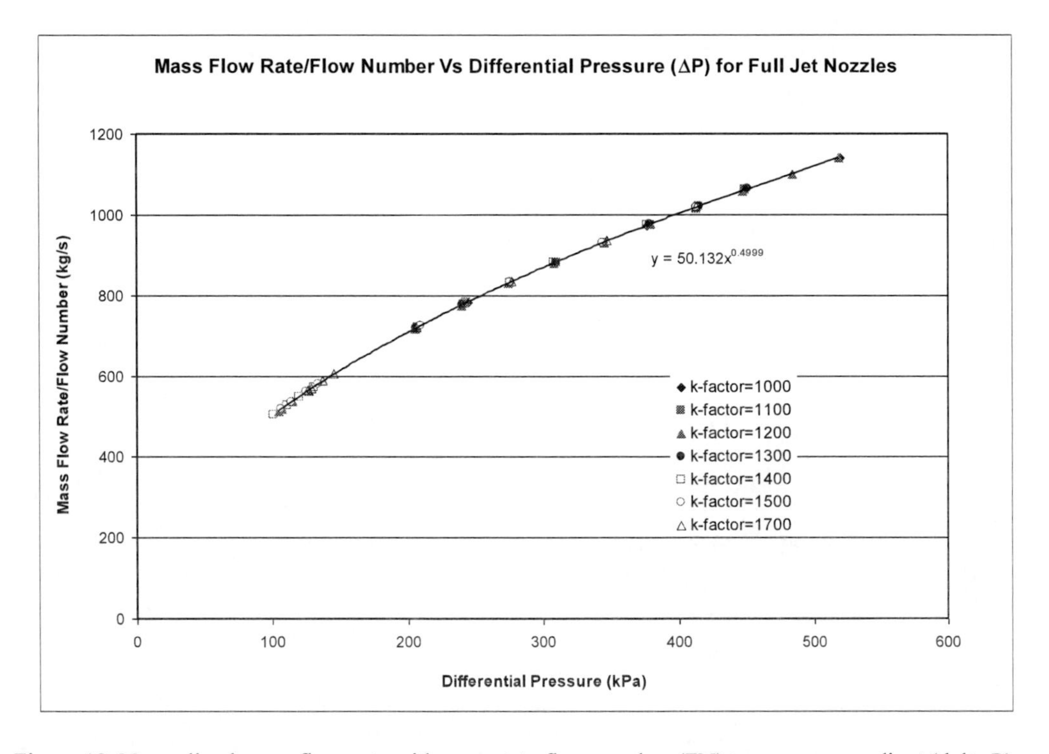

Figure 12. Normalized mass flow rate with respect to flow number (FN) to pressure gradient (delta P) across the nozzle for full jet nozzles.

The reason for this behavior was when normalizing the mass flow rate by dividing with *FN*, the resultant will be only a function of ΔP and density of liquid, as was discussed previously. However, this further proves that the dependency of mass flow rate to ΔP across the nozzle is independent from the density variation of the fluid. Similar clarification for the full jet nozzles is valid as can be seen in the Figure 12. The above power-law relationship for FJ's is:

$$\frac{\dot{m}}{FN} = 50.13(\Delta P)^{0.5} \tag{35}$$

The plots shown in Figures 13 through 28, mainly represent the variation of drop parameters of drop size, drop velocity, drop distribution and spray angle with respect to two dominant primary variables of pressure difference across the nozzle, i.e. ΔP and viscosity ratio of dispersed to continuous phase i.e. *k*-Factor for both swirl jet and full jet nozzles, as discussed above.

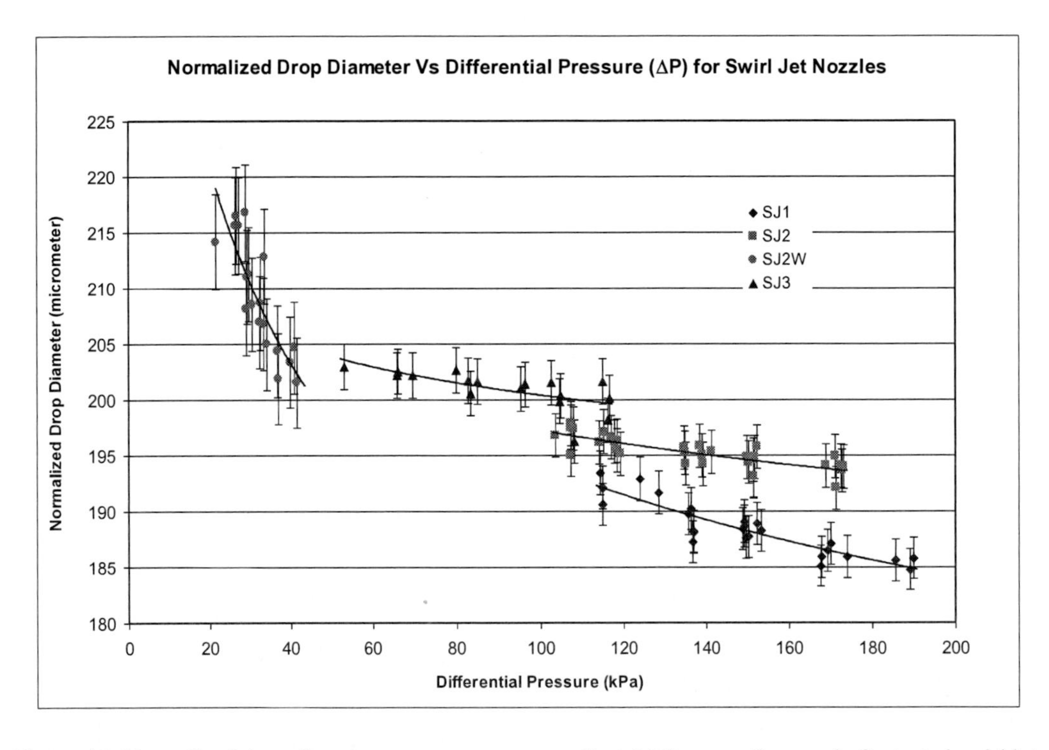

Figure 13. Normalized drop diameter versus pressure gradient (ΔP) across the nozzle for tested swirl jet nozzles (SJ's) with prospective error bars.

Drop Size

The plots shown on the Figures 13 and 14 represent the variation of drop size with respect to the ΔP for both SJ's and FJ's. These results show that the drop size decreases when ΔP increases. This fact is true regardless of whether the pressure inside the chamber is negative (vacuum) or positive (high pressure) with relative to atmospheric pressure. The relationship between drop diameter (D) and the differential pressure (ΔP) can be established as a power law equation, $D = A(\Delta P)^a$. However, because experiments were performed at different temperatures that led to present different density and viscosity of salt solution, it was necessary to alleviate the dependency of drop size on those two parameter to obtain a direct relationship between ΔP and drop size. In order to address this issue, the drop size was normalized dividing by arbitrary power terms of density and viscosity of the liquid. Then, using a curve-fitting technique, it is found that those power terms for both viscosity and density were equal. This agrees with published literature which is 0.18 for swirl jet nozzles and 0.17 for full jet nozzles [27]. Therefore, the power-law equation established for normalized diameter for swirl jet nozzles can be given as:

$$D_{Norm} = \frac{D}{(\mu_L \rho_L)^{0.18}} = A(\Delta P)^a \tag{36}$$

and that for the full jet nozzles can be given as:

$$D_{Norm} = \frac{D}{(\mu_L \rho_L)^{0.17}} = A(\Delta P)^a \tag{37}$$

where:

Nozzle	SJ1	SJ2	SJ2W	SJ3	FJ2	FJ3	FJ3W
A	276.0	230.6	316.6	224.5	300.4	303.9	440.0
"a"	-0.0764	-0.0339	-0.1352	-0.0247	-0.0531	-0.0491	-0.1295

All the data points obtained for all seven nozzles tested were agreed with respective RMS error limit for each nozzle as presented in both figures.

When considering the effect of viscosity ratio (*k*-Factor) on the drop diameter as can be seen in Figures 15, the swirl jets shows a uniform variation as the drop size increases with increasing *k*-Factor regardless of whether the nozzle is a normal or wide angle nozzle. However, for the case of full jet nozzles, the drop size increment for wide angel nozzle is much slower than that for the normal nozzle, as can be seen in Figure 16. This is a much desired property when it comes to absorption applications as nozzle requirements for commercial application should be independent of the mass flow rate increment of the nozzle. The relationships between drop diameter and *k*-Factor developed for both cases needed normalizing as in this case, the density of the fluid and pressure different across the nozzles acted as secondary variables due to the variation of flow condition based on temperature variations. Using the relationship published in previous article [27] and curve fitting, once again two power-law equations were established for both the types of nozzle tested as can be seen in Equations (38) and (39).

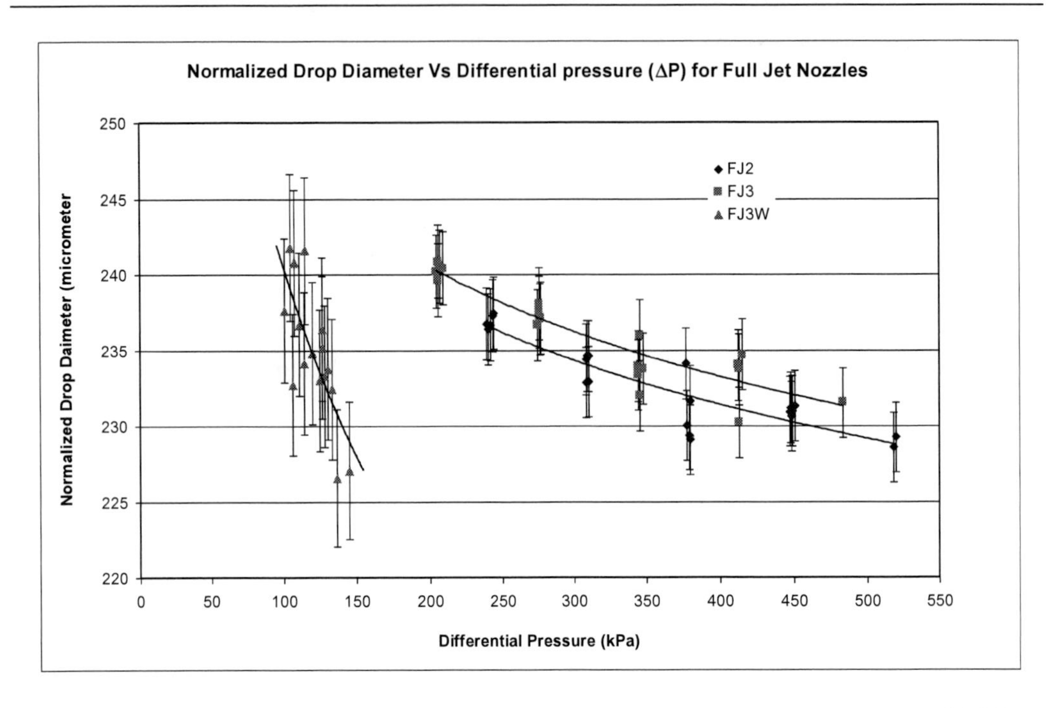

Figure 14. Normalized drop diameter versus pressure gradient (ΔP) across the nozzle for tested full jet nozzles (FJ's) with prospective error bars.

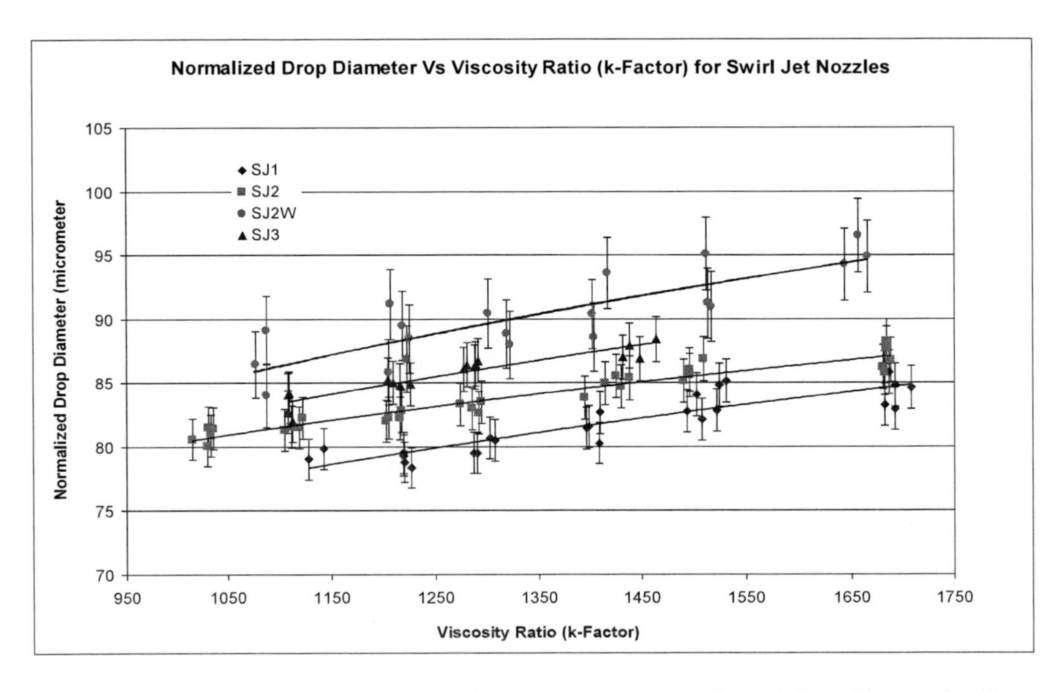

Figure 15. Normalized drop diameter versus viscosity ratio (*k*-Factor) for tested swirl jet nozzles (SJ's) with prospective error bars.

For swirl jet nozzles, the relationship is:

$$D_{Norm} = \frac{D}{\rho_L^{\ 0.2} (\Delta P)^{0.12}} = B(\Delta P)^b \tag{38}$$

and for full jet nozzle, the relationship is:

$$D_{Norm} = \frac{D}{\rho_L^{\ 0.2} (\Delta P)^{0.16}} = B(\Delta P)^b \tag{39}$$

where:

Nozzle	SJ1	SJ2	SJ2W	SJ3	FJ2	FJ3	FJ3W
B	20.2	27.5	21.6	18.4	35.5	35.0	78.3
"b"	0.1931	0.155	0.2019	0.1931	0.1383	0.1425	0.0312

Once again, all the data points obtained for all seven nozzle tested agreed with respective RMS error limit for each nozzle as presented in both figures.

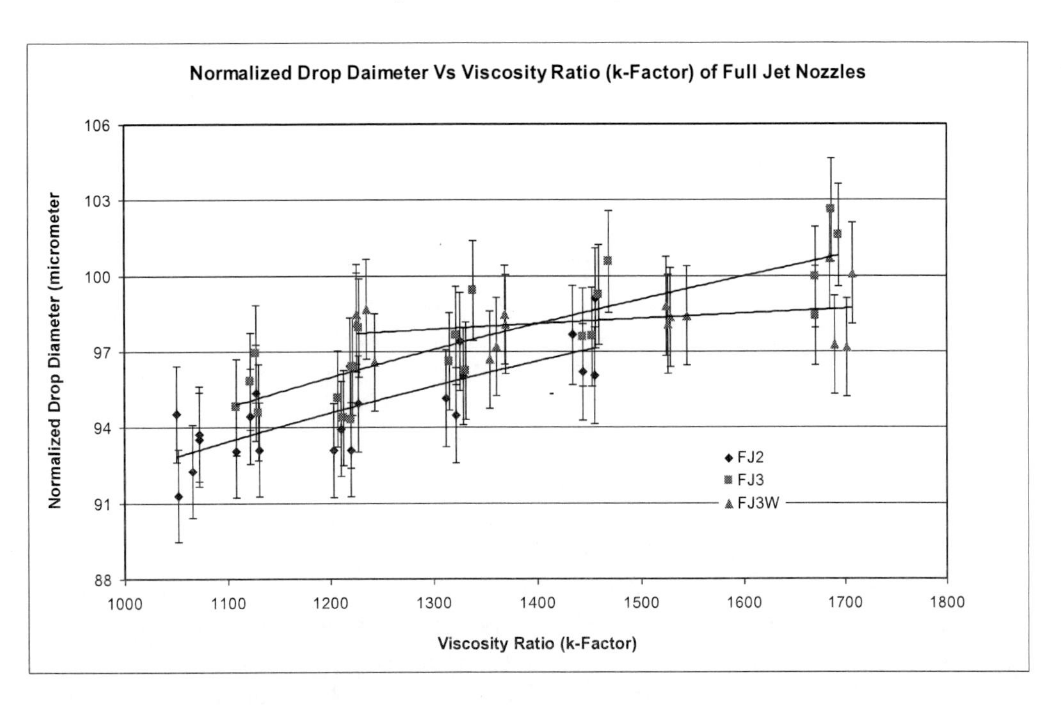

Figure 16. Normalized drop diameter versus viscosity ratio (k-Factor) for tested full jet nozzles (FJ's) with prospective error bars.

Drop Velocity

According to published data pertaining to spray flows [28, 29], the drop velocity also is a function of many variables of the fluid, the nozzle geometry and surrounding conditions. From the above variables, the major factors that affect the drop velocity are the drop diameter, the density of both the drop and the surrounding environment, the pressure difference across the nozzle and the viscosity ratio of the disperse to the continuous phase. However, when it comes to moderately high velocity flows, the dependency of velocity on drop size diminishes. The point that was given above is able to be observed in nozzle flows shown in Figure 17 for swirl jet nozzles and Figure 18 for full jet nozzles. In addition, the drop flow from the full jet nozzle showed a clear independency of drop velocity on difference in density as given in Figure 18. The power-law equations established based on the data shown in Figures 17 and 18 correlate the drop velocity to *k*-Factor for swirl jet nozzles and full jet nozzles and are given in Equation (40) and (41).

For swirl jet nozzles:

$$U_{Norm} = \frac{U}{(\Delta\rho)^{0.24}(\Delta P)^{-0.42}} = C(k)^{c} \tag{40}$$

and for full jet nozzles:

$$U_{Norm} = \frac{U}{(\Delta\rho)^{0.004}(\Delta P)^{-0.05}} = C(k)^{c} \tag{41}$$

where:

Nozzle	SJ1	SJ2	SJ2W	SJ3	FJ2	FJ3	FJ3W
C	95.2	48.4	2451.6	38.6	35.8	16.4	139.1
"c"	-0.5949	-0.4889	-0.9624	-0.4370	-0.1650	-0.0467	-0.3795

Once again, all the data points obtained for all seven nozzle tested were agreed with respective RMS error limit for each nozzle as presented in both figures.

In order to find the dependency of drop velocity to pressure difference across the nozzle, ΔP, Figure 19 for swirl jet nozzles and Figure 20 for full jet nozzles were developed. Comparison of those figures shows that while the drop velocity of swirl jet is somewhat depended on ΔP, the velocity of drop flow of full jet nozzles are fairly independent of the pressure difference across the nozzle. The power-law correlations that were developed based on the plots for the swirl jet nozzle is given in Equation (42) and that for the full jet nozzles is given by Equation (43). In general, observation of the results obtained through this series of experiments for drop velocity, as can be seen in Figures 17, 18, 19 and 20, shows a very strong relationship between the drop velocity and the viscosity ratio, and specifically to the viscosity of the sprayed solution, as can be seen in Equations 42 and 43.

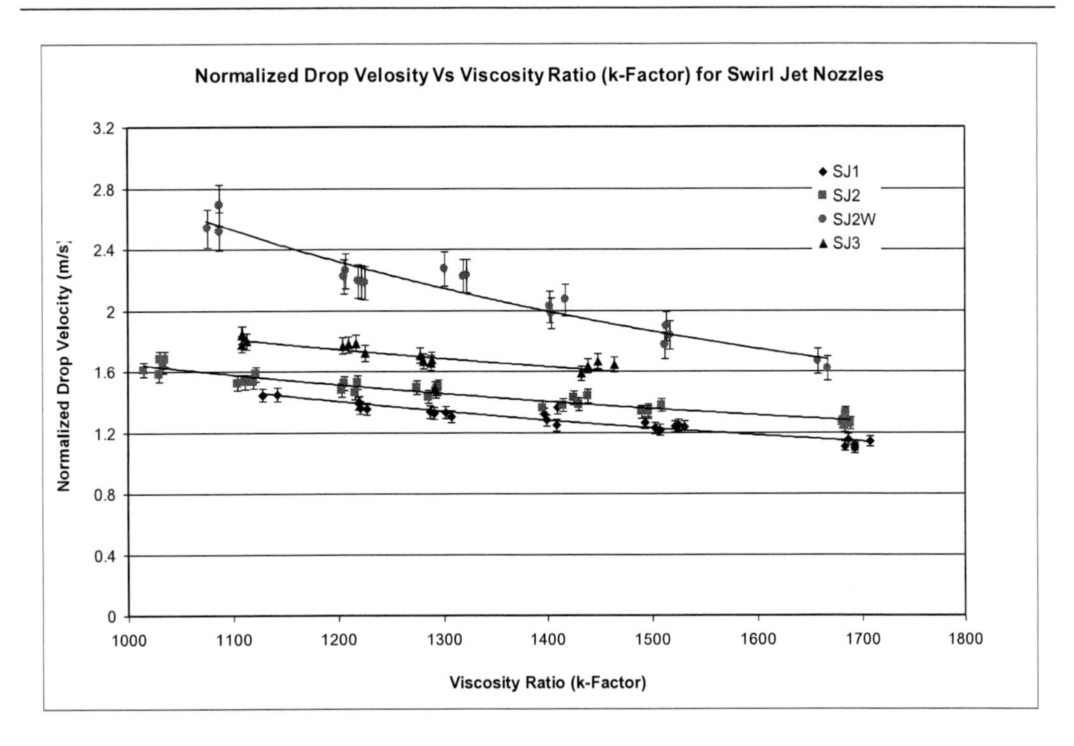

Figure 17. Normalized drop velocity versus viscosity ratio (k-Factor) for tested swirl jet nozzles (SJ's) with prospective error bars.

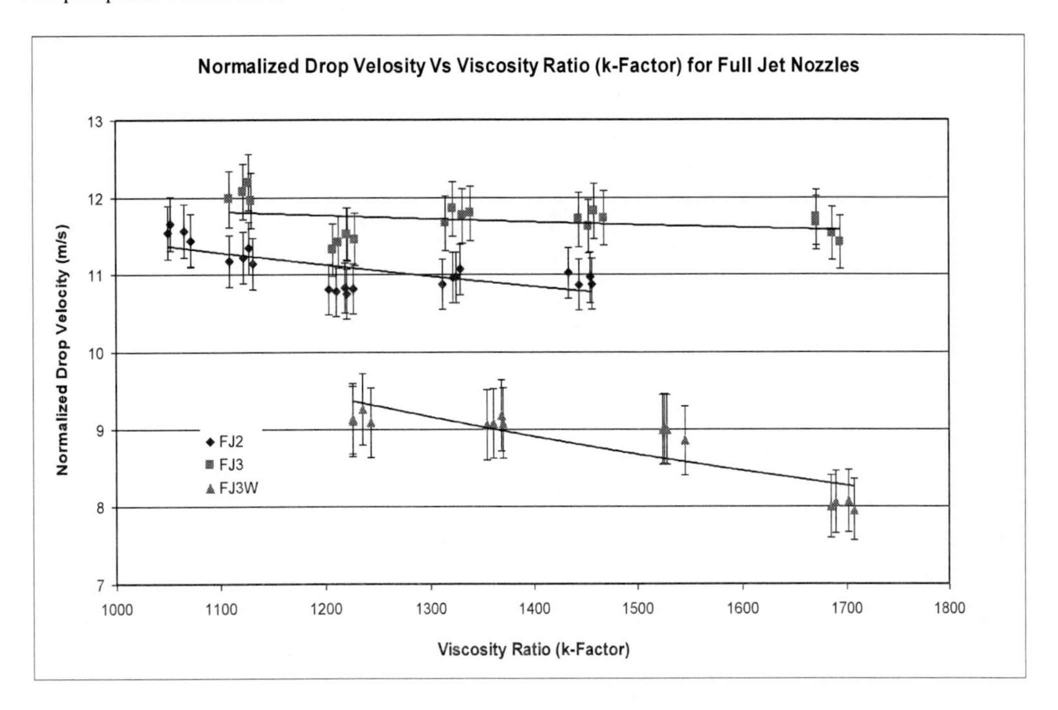

Figure 18. Normalized drop velocity versus viscosity ratio (k-Factor) for tested swirl jet nozzles (SJ's) with prospective error bars.

Power-law equation for swirl jet nozzles is:

$$U_{Norm} = \frac{U}{(\mu_L)^{-0.6}(\Delta\rho)^{0.05}} = D(\Delta P)^d \tag{42}$$

and that for the full jet nozzles is:

$$U_{Norm} = \frac{U}{(\mu_L)^{-0.3}(\Delta\rho)^{0.05}} = D(\Delta P)^d \tag{43}$$

where:

Nozzle	SJ1	SJ2	SJ2W	SJ3	FJ2	FJ3	FJ3W
D	0.24	0.26	0.12	0.88	5.39	6.42	4.51
"d"	0.3009	0.2986	0.6005	0.0662	0.0339	0.0225	0.0583

Once again, all the data points obtained for all seven nozzle tested were agreed with respective RMS error limit for each nozzle as presented in both figures.

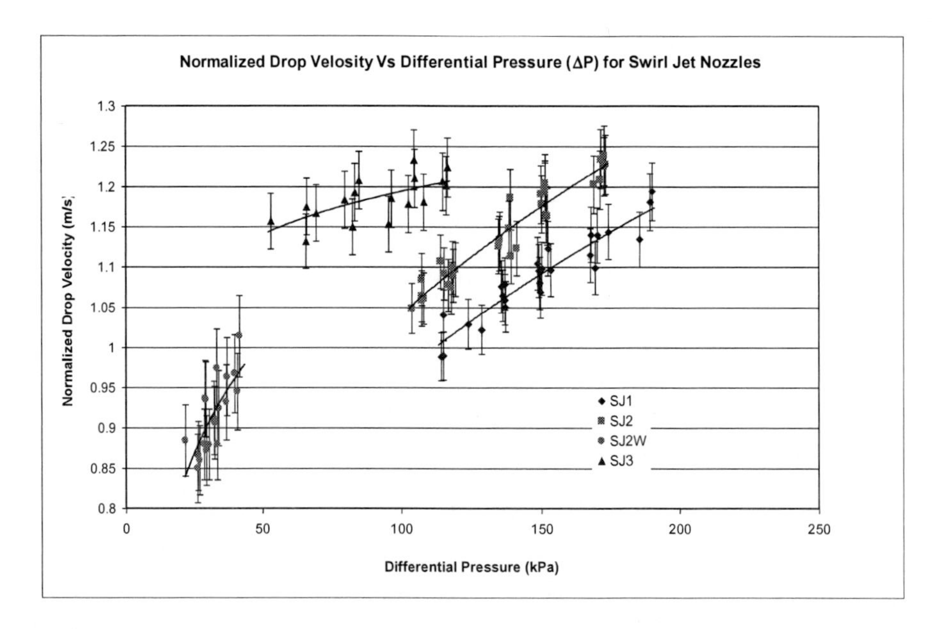

Figure 19. Normalized drop velocity versus differential pressure (ΔP) for tested swirl jet nozzles (SJ's) with prospective error bars.

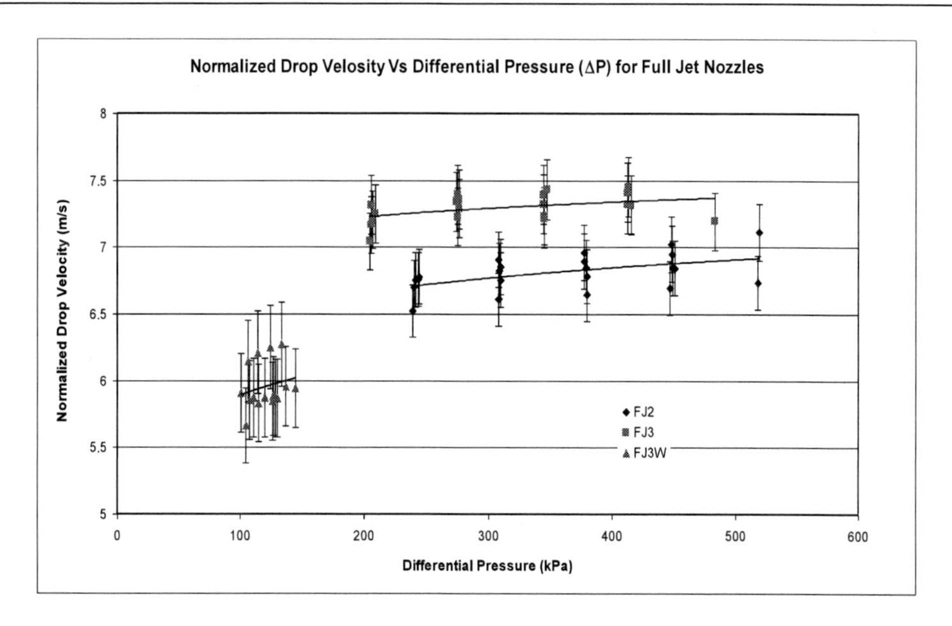

Figure 20. Normalized drop velocity versus differential pressure (ΔP) for tested full jet nozzles (FJ's) with prospective error bars.

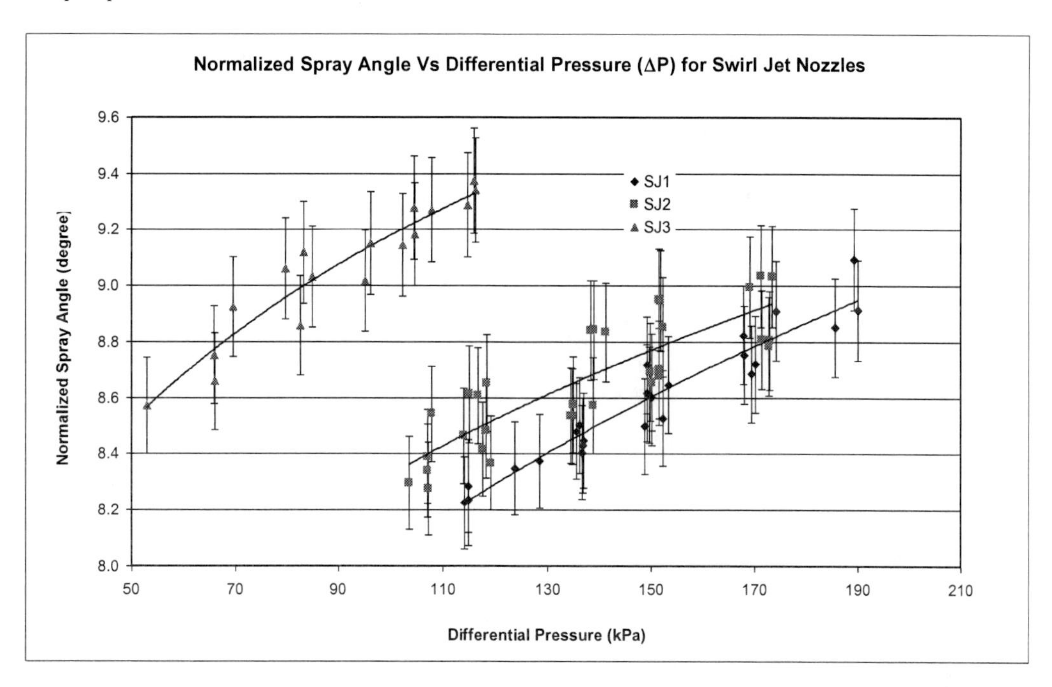

Figure 21. Normalized spray angle versus differential pressure (ΔP) for tested swirl jet nozzles (SJ's) with prospective error bars.

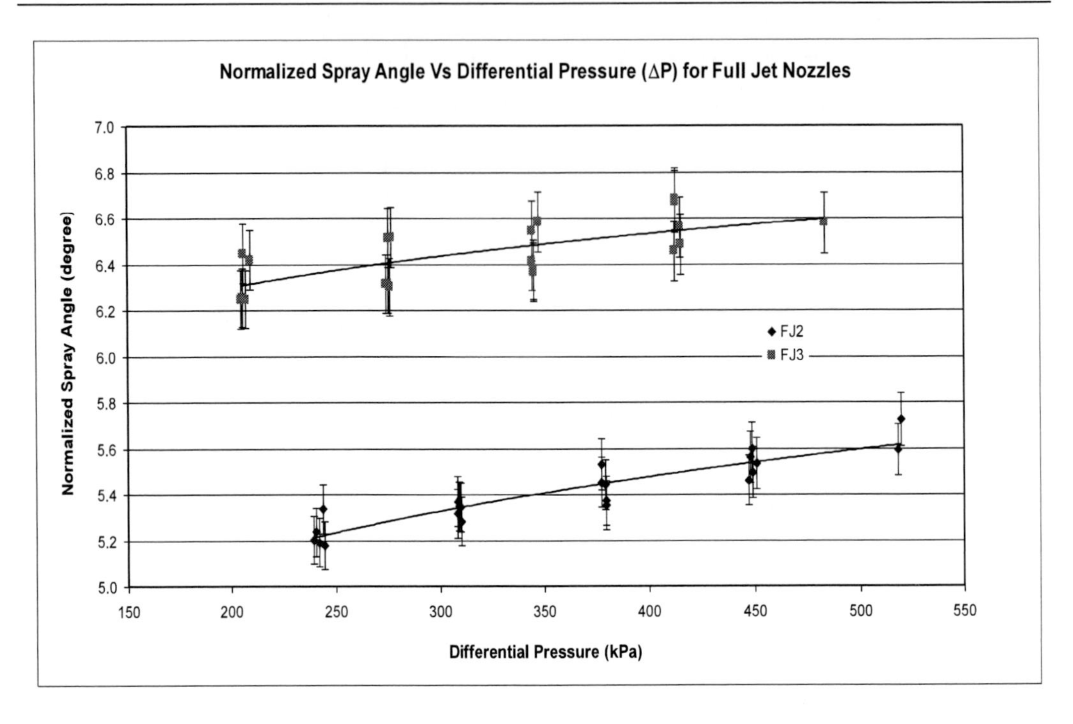

Figure 22. Normalized spray angle versus differential pressure (ΔP) for tested full jet nozzles (FJ's) with prospective error bars.

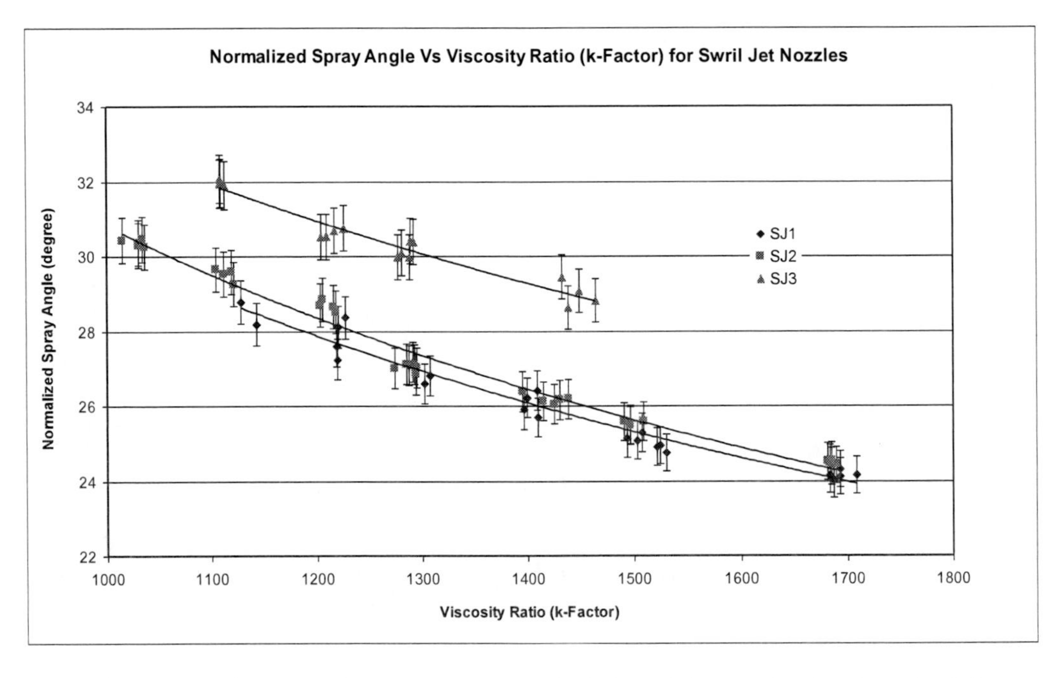

Figure 23. Normalized spray angle versus viscosity ratio (k-Factor) for tested swirl jet nozzles (SJ's) with prospective error bars.

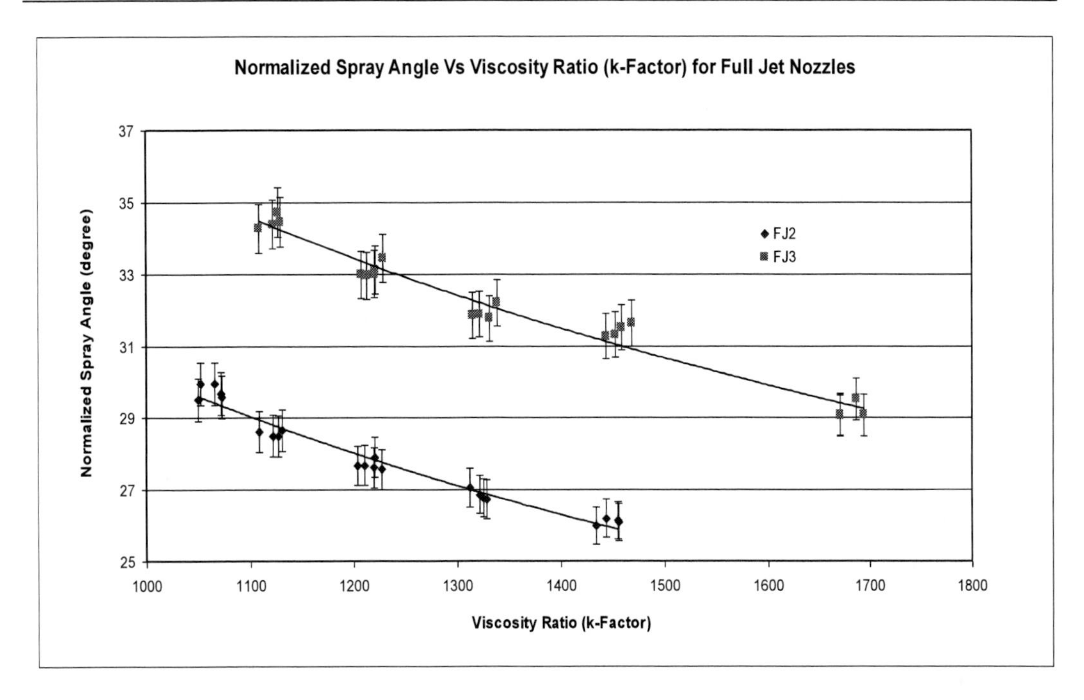

Figure 24. Normalized spray angle versus viscosity ratio (*k*-Factor) for tested full jet nozzles (FJ's) with prospective error bars.

Spray Angle

In addition to considering the characteristic of the drops in the spray flow, it is also important to study the characteristic of spray such as spray angle and the drop distribution factor to understand the drop flow from any specific atomizer or nozzle to match the performance of the nozzle to any specific application. Considering this fact and using the imaging techniques described in the experimental section, first, the spray angle of all the generated spray flows were measured and the data were recorded. Using recorded data, the plots necessary to show the dependency of spray angle on the differential pressure across the nozzle (ΔP) for each nozzle tested were developed as shown in Figures 21 and 22. These results shown in the figures also were dependent on the density and viscosity of the spray solution and therefore it was necessary to normalize the spray angle with respect to those two parameters as indicated in Equations (44) and (45) which were developed based on the data given in those two figures.

Power-law equation developed for swirl jet nozzles is:

$$\theta_{Norm} = \frac{\theta}{(\mu_L)^{-0.44}(\rho_L)^{0.08}} = E(\Delta P)^e \tag{44}$$

and that for the full jet nozzles is:

$$\theta_{Norm} = \frac{\theta}{(\mu_L)^{-0.54}(\rho_L)^{0.08}} = E(\Delta P)^e \tag{45}$$

where:

Nozzle	SJ1	SJ2	SJ3	FJ2	FJ3
E	3.75	4.60	5.57	3.07	4.77
"e"	0.1659	0.1268	0.1086	0.0965	0.0526

Once again, all the data points obtained for all six nozzle tested were agreed with respective RMS error limit for each nozzle as presented in both figures.

Secondly, using the collected data for the spray angle for all the experimented sprays, the dependency of spray angle on the viscosity ratio (*k*-Factor) was analyzed. Based on this analysis, the plots that were developed for both swirl jet nozzles and full jet nozzles and are shown in Figures 23 and 24, respectively. The established power-law relation between normalized spray angle and the *k*-Factor is given in Equations (46) and (47), below. This analysis also shows the spray angle's strong relationship on ΔP and the weak relationship to the density of the sprayed solution.

The power-law equation that represents the above relation for swirl jet nozzles is:

$$\theta_{Norm} = \frac{\theta}{(\Delta P)^{0.11}(\rho_L)^{0.02}} = F(\mu_L)^f \tag{46}$$

and that for the full jet nozzles is:

$$\theta_{Norm} = \frac{\theta}{(\Delta P)^{0.08}(\rho_L)^{0.01}} = F(\mu_L)^f \tag{47}$$

where:

Nozzle	SJ1	SJ2	SJ3	FJ2	FJ3
F	598.5	711.4	398.0	515.1	524.8
"f"	-0.4325	-0.4545	-0.3603	-0.4107	-0.3884

Once again, all the data points obtained for all six nozzle tested were agreed with respective RMS error limit for each nozzle as presented in both figures.

Drop Distribution Factor

The second spray characteristic that was considered was the drop distribution factor, q, which represents uniformity of the spray in terms of drop size as explained in the theory section. This was the most difficult parameter to determine its relationship between the viscosity ratio of disperse to continuous phase (*k*-factor), and pressure difference across the nozzle, ΔP. The reason for this difficulty was its weak dependence on all the primary drop flow parameters as can be seen by the scattered data in Figures 25 and 26. The power-law equations developed based on these plots for both swirl jet nozzles and full jet nozzles are direct relationships between the parameters and can be represented with a single equation as given in the Equations (48). However one very prominent observation that can be observed from these plots are, when ΔP increases, the drop distribution factor, q, for swirl jet nozzles increases, and contrary to that it decreases for full jet nozzles.

$$q = H\left(\Delta P\right)^{h} \tag{48}$$

where:

Nozzle	SJ1	SJ2	SJ2W	SJ3	FJ2	FJ3	FJ3W
H	2.02	2.56	4.23	3.36	6.17	5.39	5.56
"h"	0.1554	0.1086	0.0554	0.0529	-0.0504	-0.0182	-0.0334

Error bars displayed in the plots of below figures related to distribution factor, q, is developed based on the visual inspection and are ± 3% for normal swirl and full jet nozzles and ± 5% for wide angle nozzles.

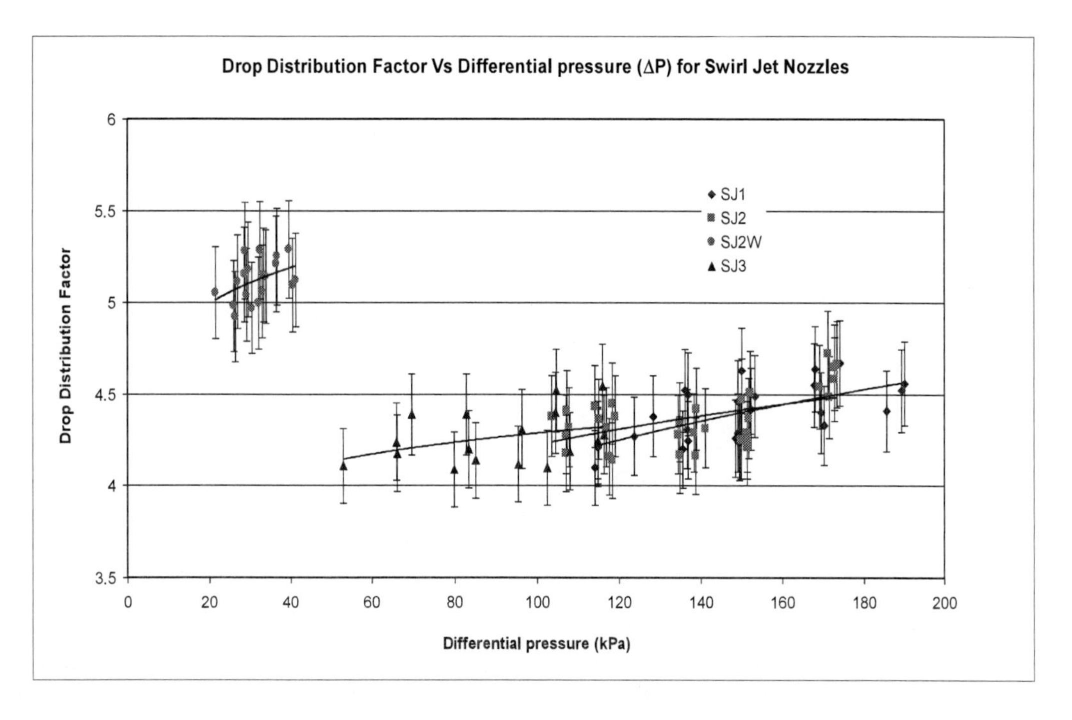

Figure 25. Drop distribution factor (q) versus differential pressure (ΔP) for tested swirl jet nozzles (SJ's) with prospective error bars.

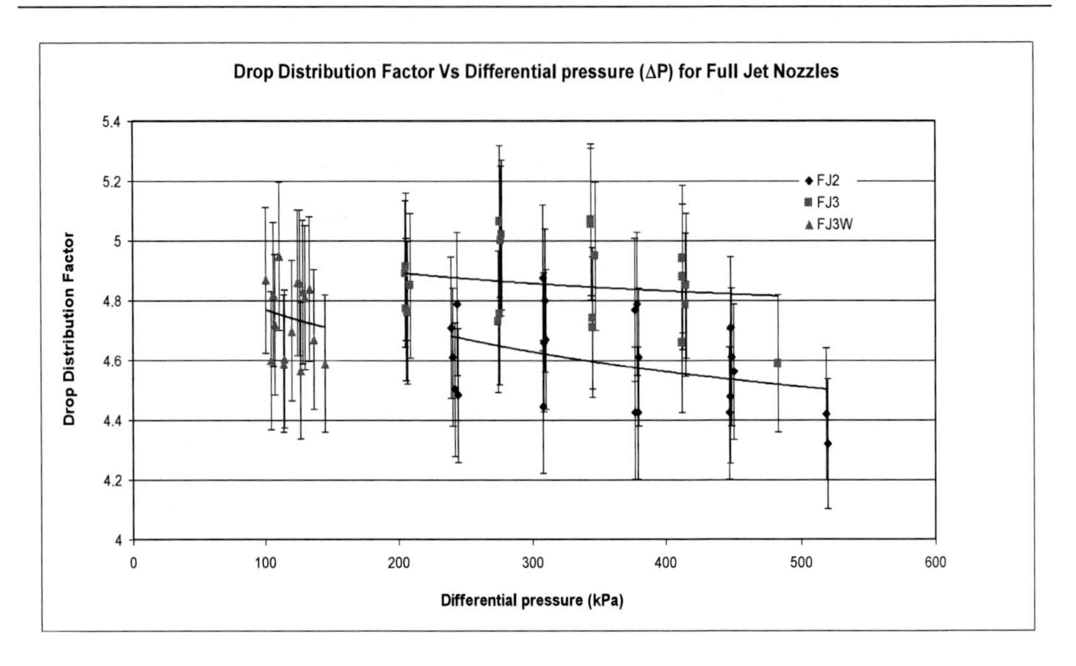

Figure 26. Drop distribution factor (q) vs. differential pressure (ΔP) for tested full jet nozzles (FJ's) with prospective error bars.

The variation of drop distribution factor with respect to viscosity ratio (k-Factor) shows the similar independence of q from all the drop and spray flow parameters. Further, as was exhibited in the previous case, the trend of the variation of q with respect to variation of the k-Factor for swirl jet and full jet nozzles are opposite to each other. However, as can be seen in Figures 27 and 28, for swirl jet nozzles when k-Factor increases, the drop distribution factor, q, decrease and for full jet nozzle, it increases. The power-law equation that was established for these cases also are the same as the previous cases and is given by Equation (49).

$$q = J\left(k\right)^{j} \tag{49}$$

where:

Nozzle	SJ1	SJ2	SJ2W	SJ3	FJ2	FJ3	FJ3W
J	13.16	7.49	7.80	6.84	1.09	1.87	3.15
"j"	-0.1514	-0.0752	-0.0846	-0.0401	0.2020	0.1322	0.0561

For this case also, the error bars displayed in the plots of below figures related to distribution factor, q, is developed based on the visual inspection and are $\pm 3\%$ for normal swirl and full jet nozzles and $\pm 5\%$ for wide angle nozzles.

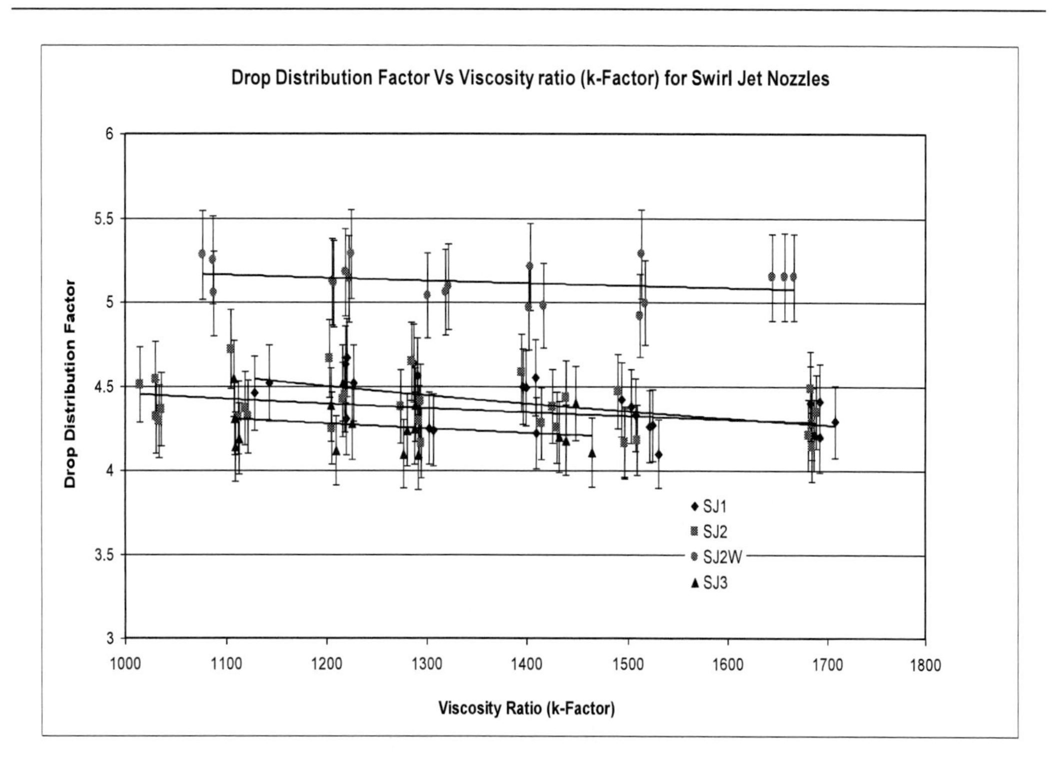

Figure 27. Drop distribution factor (q) vs. viscosity ratio (k-Factor) for tested swirl jet nozzles (SJ's) with prospective error bars.

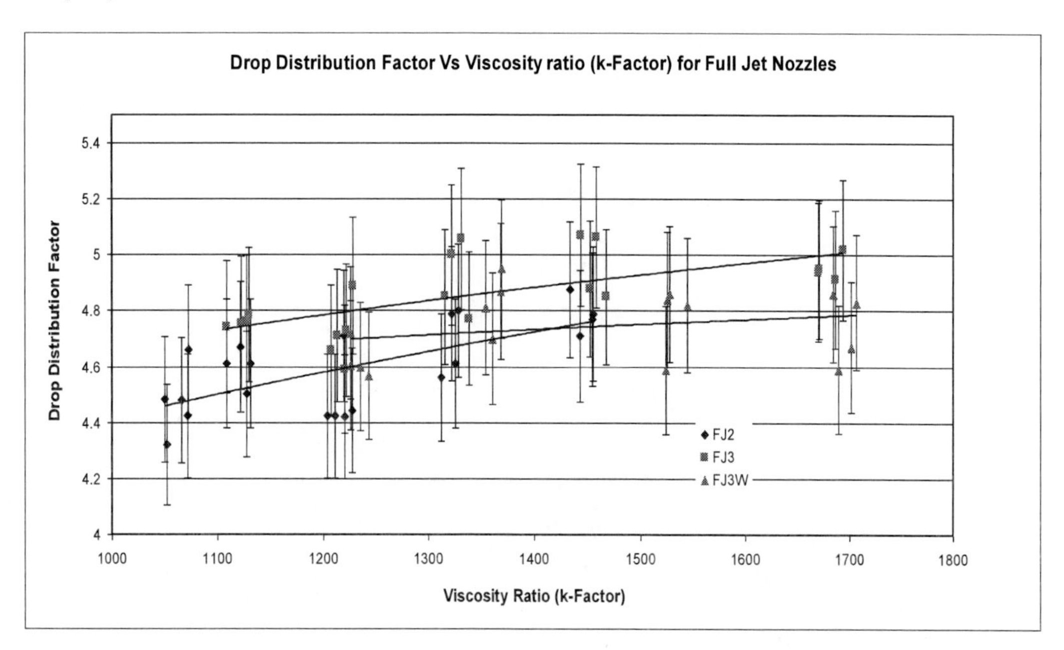

Figure 28. Drop distribution factor (q) vs. viscosity ratio (k-Factor) for tested full jet nozzles (FJ's) with prospective error bars.

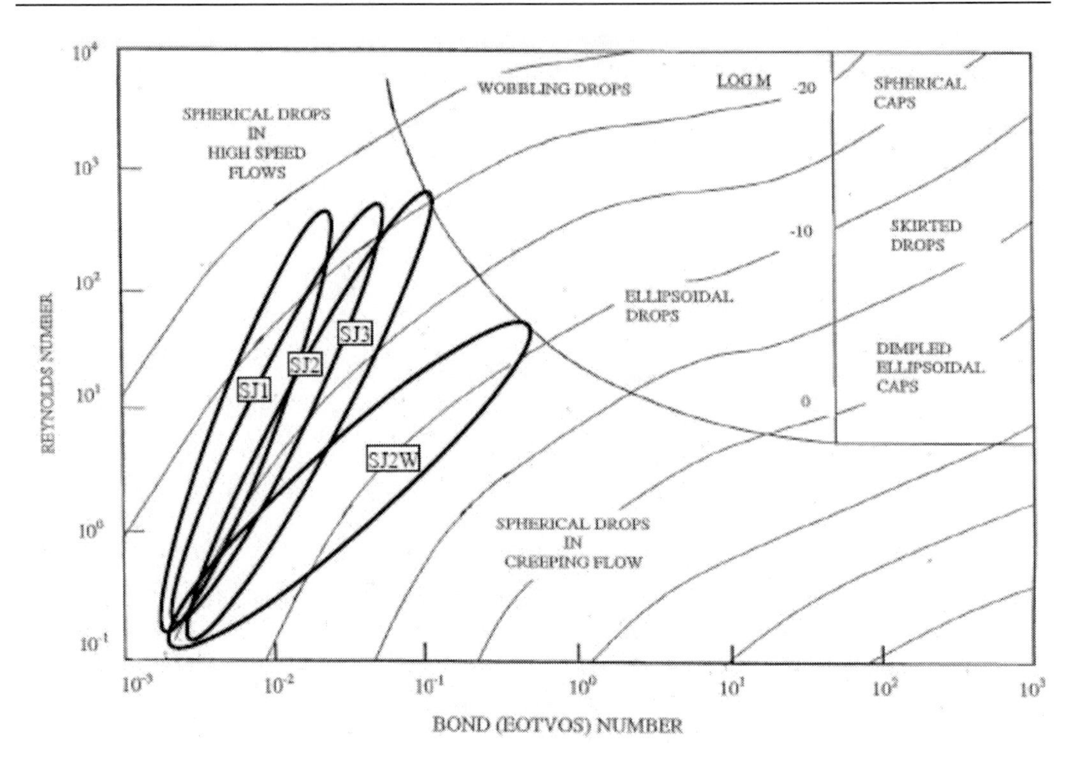

Figure 29. The drop regimes of experimented nozzle sprays represented for all four models of swirl jet nozzles on drop shape chart.

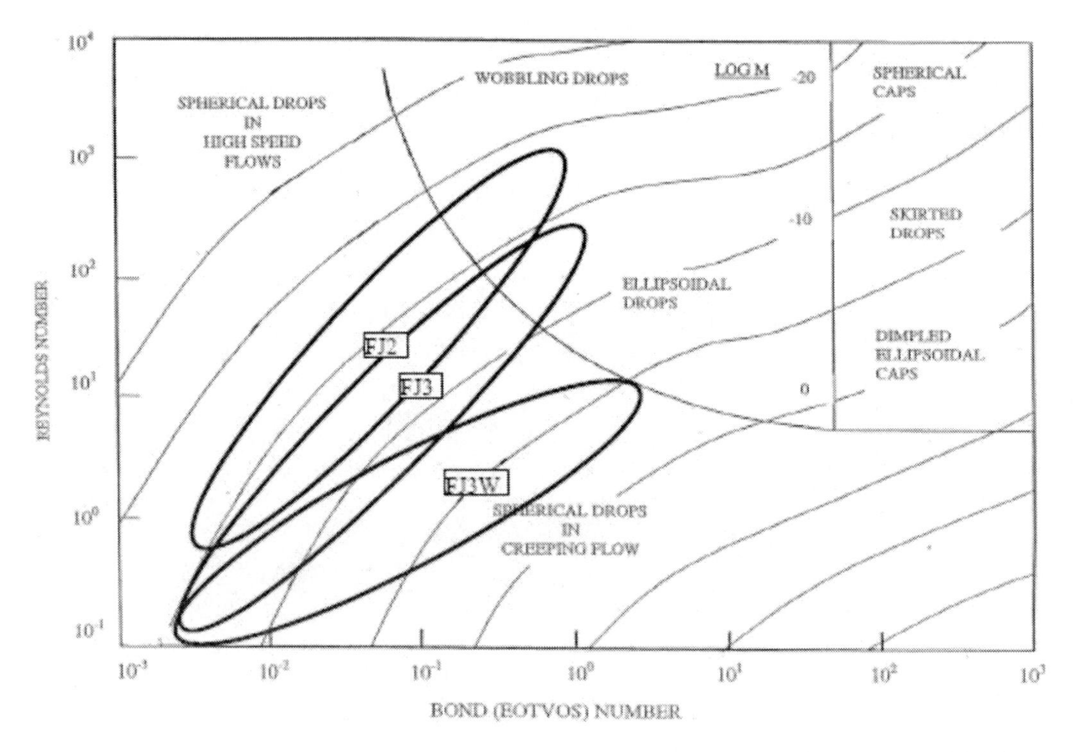

Figure 30. The drop regimes of experimented nozzle sprays represented for all three models of full jet nozzles on drop shape chart.

Drop Shape Chart

In the process of experimenting, studying and analyzing drops and drop flows such as sprays, one of the primary aspect that any researcher should pay attention to is the drop shape, change of shape and behavior of the drop in its flight. The main reason that above factors, especially shape of the drops and their variation is important is, the most of the earlier studies and their findings have done based on the assumption of the shape of the drop. For an example, if someone consider the equation for drag force derived for spherical drop when his drops are more of skirted shape, his whole analysis could be invalid. Considering the above fact, the first step of the analysis portion of this research was to study and define the shape of the drops generated in this experiments based on the drop shape charts shown in Figures 29 and 30. As the drop shape charts are mainly based on the droplets Reynolds and Bond (Eotvos) numbers, when defining the drop shapes as per below figures, first, those numbers were calculated for droplets from each nozzle for each experiment conducted using the smallest, mean volumetric and largest diameter of the each spray. Then, considering both Reynolds and Bond number for each drop size consider with respective drop velocity, the drop regimes shown in below figures were developed. Figure 29 as shown below represent the drop shape regimes for all the four tested swirl jet nozzles of SJ1, SJ2, SJ2W and SJ3, whereas Figure 30 below represent the drop shape regimes of all the full jet nozzle of FJ2, FJ3 and FJ3W, which were tested in this experiment.

When analyzing the shapes of the drop generated from swirl jet nozzles as per Figure 29, is can be observed that generic swirl jet nozzles of SJ1, SJ2, and SJ3 tended to generate spherical drops however, due to comparatively high velocity and gravitational force overcoming the surface tension effects, it stays more towards the high velocity regime. In the case of wide angle nozzle of SJ2W, due to low velocities, drops shows somewhat creeping flows. When consider the drop regime of the full jet nozzles, it is also showed similar behavior as comparable swirl jet nozzles, however the large droplets generated from FJ2 and FJ3 due to the dominant effect of surface tension shows some "wobbling" effects as can be seen in the Figure 30.

CONCLUSIONS

In the process of formation the drops, the nozzle geometry, properties of sprayed solution and properties of surrounding medium played equally important, comparable and very well interconnected rolls. However, the level of effects of certain parameters belongs to above each factor may affect more than other parameters, depending once again on which of the above factors dominate the entire drop formation process. The major parameters including nozzle pressure, ambient pressure, flow rate, density, viscosity and surface tension of the fluid, density and viscosity of surrounding medium, in this case water vapor, and nozzle parameters of nozzle constant and flow number, considered for this analysis were selected based on the previous experiments of these authors and other previous publications [15 - 26].

For a spray nozzle, the flow rate of the nozzle is directly proportional to the nozzle pressure. Therefore, any change in nozzle pressure directly affects the flow rate and the nozzle performance. Increasing the pressure difference across the nozzle decreases the drop sizes which increases the drop quality due to the increase of drop velocity with higher discharge pressures. However, as the viscosity of the used solution was fairly high, the above relationship between drop size and nozzle pressure was also considerably affected by the viscosity of the solution as increasing of viscosity ratio i.e. *k*-Factor has always increase the drop size regardless of the nozzle pressure. Even though, the surface tension of the liquid sprayed is very important factor that affect the drop breakups, it was not the case for this set of experiments as drop breakup of most of the sprays took place due to inertia effect than turbulent effects. This can be clearly observed from Weber numbers calculated related to these drop flows is comparatively smaller compared to the critical Weber number as presented in Table 1.

Tested swirl jet nozzles were able to create higher swirl with high angular velocity flows that lead to smaller drop size being generated despite having a solution with high viscosity and high surface tensions. The main effect of increasing the viscosity of the solution was not on the droplet size but on the spray angle, since the variation of spray angle is larger than the variation of drop size with respect to viscosity ratio. Since the spray angle was highly depended on the viscosity of the solution, it was observed that reduction of the angle of the spray increases the drop speed. However as can be seen from above figures, the effect of drop size is really negligible on drop speed for fairly high speed flows. Hence, it is difficult to define a direct relationship between drop size and the drop speed, and defining the dependency of drop size on combined effects of all the above considered parameters.

REFERENCES

[1] Grigor'eva N. I.; Nakoryakov V. E. Combined Heat and Mass Transfer Dhuring Absorption in Drops and Films, *Journal of Engineering Physics*, 1976, volume 32, 243-247.

[2] Grigor'eva N. I.; Nakoryakov V. E. Exact Solution of Combined Heat and Mass Transfer Dhuring Film Absorption, *Journal of Engineering Physics*, 1977, volume 32, 1349-1353.

[3] Nakoryakov V. E.; Grigor'eva N. I. Joint Heat and Mass Transfer Dhuring Film Absorption on Drops and on Films, *IFZh*, 1977, volume 32(3), 399-405.

[4] Nakoryakov V. E.; Grigor'eva N. I. Calculation of Non-isothermal Absorption of the Initial Portion of a Down Flowing Films, Oak Ridge National Laboratories, Central Research Library, 1978, volume 14(4), 483-488.

[5] Nakoryakov V. E.; Bardukov, A. P.; Bufetov P. F.; Grigor'eva N. I.; Dorokhov A. R.; Experimental Study of Non-isothermal Absorption by a Falling Liquid Film, Oak Ridge National Laboratories, Central Research Library, 1980, volume 14(5), 755-758.

[6] Grossman G. Simultaneous Heat and Mass Transfer in Film Absorption Under Laminar Flow, *International Journal of Heat and Mass Transfer*, 1983, volume 26(3), 357-371.

[7] Grossman G; Heath M. T. Simultaneous Heat and Mass Transfer in Absorption of Gases in Turbulent Liquid Films, *International Journal of Heat and Mass Transfer*, 1984, volume 27(12), 2365-2376.

[8] Andberg J. W.; Vleit V. C. Non-isothermal Absorption of Gases into Falling Liquid Films, Proceeding of the ASME-JSME Joint Conference, 1987, 423-431.

[9] Andberg J. W.; Vleit V. C. Absorption of Vapor into Liquid Film Flowing Over Cooled Horizontal Tubes, Author's Manuscript, 1990.

[10] Benbrahim A.; Prevost M.; Bugarel R. Performance of a Composite Absorber, Spraying and Falling Film, Proceeding of the International Workshop on Advanced Heat Pumps, 1986, 248-256.

[11] Summerer F.; Flamensbeck M.; Riesch P.; Ziegler F.; Alefeld G. A Cost Effective Absorption Chiller with Plate Heat Exchanger using Water and Hydroxides, Applied Thermal Engineering, 1989, volume 18, 413-425.

[12] Morioka I.; Kiyota M.; Ousaka A.; Kobayashi T. Analysis of Steam Absorption by a Sub-cooled Droplet of Aqueous Solution of LiBr, *JSME International Journal*, Series II-35, 1992 458-464.

[13] Ryan W. Water Absorption in an adiabatic spray of aqueous lithium bromide solution, Ph.D. Thesis, Illinois Institute of Technology, Chicago, IL USA, 1995.

[14] Warnakulasuriya F. Heat and Mass Transfer and Water Absorption Properties of New Absorbent Droplets, Ph.D. thesis, University of Illinois at Chicago, Chicago, IL, 1999.

[15] DeCorso S. M. Effect of Ambient and Fuel Pressure on Spray Drop Size, *ASME J. Eng. Power,* issue 82:10, 35-38.

[16] Jones A. R. Design Optimization of a Large Pressure-jet Atomizer for Power Plant, Proceedings of the 2nd International Conference on Liquid Atomization and Sprays, 1992, 181-185.

[17] Wang X. F.; Lefebvre A. H. Mean Drop Sizes from Pressure-swirl Nozzles, AIAA *Journal of Propulsion Power*, 1985, volume 1(3), 200-204.

[18] Lefebvre, A.H. Atomization and Sprays, Hemisphere Publishing Corporation: New York, NY, 1989.

[19] Fraser R. P.; Eisenklam P. Research into the Performance of Atomizers for Liquids, *Imperial Collage Chemical Engineering Society Journal*, 1953, volume 7, 52-68.

[20] Dombrowski N.; Fraser R. P. A Photographic Investigation into the Disintegration of Liquid Sheets, *Journal of Mathematical and Physical Sciences*, 1954, volume 247, 101-130.

[21] Eisenklam P. Atomization of Liquid Fuel for Combustion, *Journal of Institute of Fuel,* 1961, volume 34, 130-143.

[22] Dombrowski N.; Hassan D. The Flow Characteristics of Swirl Centrifugal Spray Pressure Nozzles with Low Viscosity Liquids, *AIChE. Journal*, 1969, volume 15, 604-612.

[23] Rizk N. K.; Lefebvre A. H. Internal Flow Characteristics of Simplex Swirl Atomizers, *AIAA Journal of Propulsion Power,* 1985, volume 1, 193-199.

[24] Giffen E.; Muraszew A. Atomization of Liquid Fuel, Chapman and Hall, London, 1953.

[25] Rizk N. K.; Lefebvre A. H. Prediction of Velocity Coefficient and Spray Cone Angle for Simplex Swirl Atomizers, Proceedings of 3rd International Conference on Liquid Atomization and Spray Systems, 1985, section 111C/2, 1-16.

[26] Babu K. R.; Narasingham N. V.; Narayanswamy K. Prediction of Mean Drop Size of Fuel Sprays from Swirl Spray Atomizer, Proceedings of 2nd International Conference on Liquid Atomization and Spray Systems, 1982, 99-106.

[27] Lefebvre A. H. Gas Turbine Combustion, Hemisphere, Washington, D.C., 1983.

[28] Stokes G. G. Scientific Papers, University Press, Cambridge, 1901.

[29] Clift R.; Grace J.; Weber M. Bubbles, Drops and Particles, Academic Press, Harcourt, Brace, Jovanovich, 1978.
[30] Panton R. L. Incompressible Flow, John Wiley and Sons, New York, NY, 1984.
[31] Levich V. G. Physicochemical Hydrodynamics, Prentice-Hall, Inc., Englewood Cliffs, NJ, 1962.

In: Sprays: Types, Technology and Modeling
Editor: Maria C. Vella, pp. 181-208
ISBN 978-1-61324-345-9

Chapter 4

SPRAY DRYING: THE SYNTHESIS OF ADVANCED CERAMICS

Rigoberto López-Juárez,[1] Simón Y. Reyes-López,[2] José Ortiz-Landeros,[1,3] and Federico González-García[4]

[1]Instituto de Investigaciones en Materiales,
Universidad Nacional Autónoma de México, Ciudad Universitaria,
México D.F., México

[2]Departamento de Ingeniería Química,
División de Ciencias e Ingenierías, Campus León,
Universidad de Guanajuato, León, Guanajuato, México

[3]Departamento de Ingeniería Metalúrgica,
Escuela Superior de Ingeniería Química e Indústrias Extractivas,
Instituto Politécnico Nacional, México DF, México

[4]Departamento de Ingeniería de Procesos e Hidráulica,
Universidad Autónoma Metropolitana-Iztapalapa,
México D.F., México

ABSTRACT

Spray-drying is a well-known processing technique commonly used for granulating powder materials. In fact, spray drying has been developed for many industrial applications due to its ability to produce high volumes of uniform particles with identical characteristics. An example of this is its extensive use in the ceramic industry coupled with other processing stages like die-pressing.

In this report, we discuss the advantages of using the Spray Drying method (SD) in the synthesis and processing of advanced ceramics, through some cases of study which include the preparation of yttrium aluminum garnet -$Y_3Al_5O_{12}$- (YAG), α-alumina, $K_{0.5}Na_{0.5}NbO_3$ (KNN), $(K_{0.48}Na_{0.52})_{0.96}Li_{0.04}Nb_{0.85}Ta_{0.15}O_3$ (KNLNT), and Cu/Mg/Al mixed oxides. The applications of these ceramics include optical, piezoelectric, ferroelectric, structural, ion exchange/adsorption, pharmaceutical, photochemical, and electrochemical.

The synthesis of different advanced materials was performed by SD of aqueous solutions and organic compounds. For YAG, KNN, and KNLNT preparation, citric acid was used as a chelating agent, while the synthesis of α-alumina was achieved by using aluminum formate ($Al(O_2CH)_3$) as a metal-organic precursor. Besides, macroporous Cu/Mg/Al mixed oxides were fabricated by using layered double hydroxides as building blocks and submicrometric polystyrene spheres as pore former agents.

The results demonstrate that spherical agglomerates can be obtained in most cases after being calcined, depending upon the equipment operation conditions, composition, concentration, and organic compound content. Furthermore, the agglomerates with narrow size distribution formed by nanometric particles can be produced. This route shows that the calcination temperatures can be reduced considerably compared with the conventional ceramic method. These features demonstrate the feasibility of this method for the synthesis of advanced ceramics.

1. Introduction

In the field of ceramic processing, and particularly in the synthesis of advanced ceramics, a set of well-known chemical methods are used. Among them, sol-gel, precipitation, combustion, and polymerization have become popular. Roughly speaking, the materials synthesized by the chemical route, regardless of the method used, are fine grained, chemically homogeneous, and show low synthesis temperature powders. Nevertheless, parameters such as temperature, pH, and other variables must be strictly controlled to avoid formation of undesirable phases. In this context, to synthesize a "new material", it is necessary to explore several methods in order to find the most convenient, because every system has its own complexity.

Chemical methods have advantages and disadvantages. An advantage of the chemical methods is that, homogeneous, fine grained powders are obtained with controllable particle size distribution; although, the majority of reagents are quite expensive and some techniques are time consuming. For instance, sol-gel demands many days for producing the final product. This is because low drying rates are essential to avoid defects in consolidated ceramic bodies. In chemical precipitation of complex oxides, different precipitation rates yield powders with inhomogeneous compositions that can produce secondary phases.

In this context, spray drying represents a practical an economical method. In fact, in the ceramic industry, one common stage employed to produce granulated powders is the spray drying (SD). Granulation is necessary for further processing of bulk products, especially when pressed samples are produced. This drying method is also used in the pharmaceutical and food industries; for example in the production of drug delivery powders [1-5], milk, and fruit juices [6-9]. Despite its versatility for materials processing, spray drying itself is not a common synthesis method of fine ceramics. It has not been exploited in its whole potential, and only some reports about the synthesis of ferroelectrics, superconductors, catalysts, pigments, optical ceramics, cathode ceramics, and other materials are found in related literature [10-24].

In this chapter we present both, the synthesis and processing of some advanced ceramics via spray drying. Throughout the text, some of the main features related to spray drying techniques are reviewed. Different cases are presented, showing the feasibility of this technique to produce chemically homogeneous spherical agglomerates that are desirable for further processing stages such as dry pressing, and how the agglomerate size distribution can be controlled by changing some parameters.

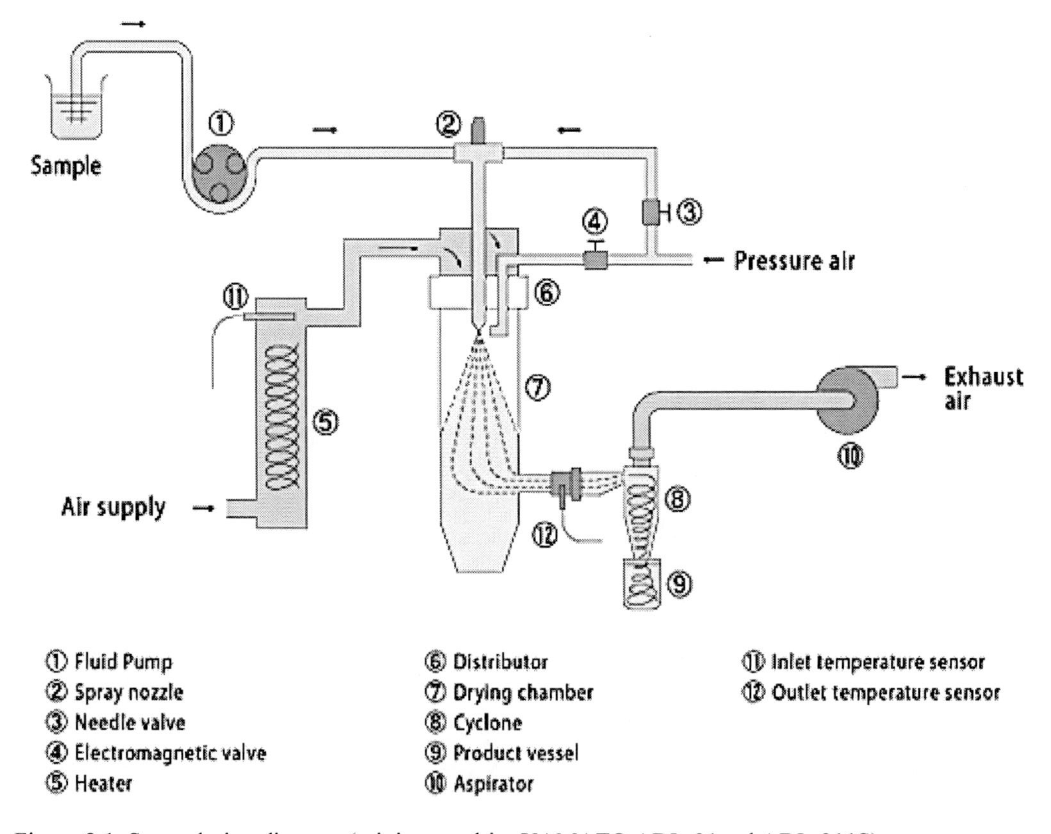

Figure 2.1. Spray drying diagram (mini spray drier YAMATO ADL-31and ADL-311S).

2. Spray Drying Equipment Description

As it was stated above, the spray drying is quite important in the industry. The industrial applications demand huge units that must process several tons per day of products. In contrast, the laboratory equipment is smaller, and the characteristics are somewhat different.

Whatever the size of the unit, the apparatus includes an atomizer, the drying chamber where the solvent is vaporized, and the cyclone for powder collection. The atomizer can be a pressure nozzle, a two fluid nozzle, or a rotating disc. Conventionally, in lab units, the two fluid nozzles are most desirable since it produces smaller drops with narrow size distribution. In Figure 2.1, a spray dryer diagram is shown. The gas for drying in most cases is hot air because of its availability, but nitrogen or other gases can be used in cases where oxidation must be avoided.

For a water-based feed, the inlet air temperature can be used between 160 to 200°C. The outlet temperature is commonly lower (70-120°C). For the synthesis of fine ceramics, the drying temperature is not high enough to produce crystalline powders, which is achieved by calcination of the dried powders. Frequently, this process is carried out in a high temperature furnace. For each composition, a study should be performed to establish the optimum calcination temperature. Typically, relatively low temperatures are used (less than 900°C). It is worth mentioning that the crystallization temperatures of finae ceramics are reduced by using SD.

Regarding the feed characteristics, it can be a slurry or solution; nevertheless for the synthesis of fine ceramics, solutions are preferred. The starting reagents are highly soluble salts i.e. nitrates, sulfates, carbonates of the appropriate cations. The only restriction is that they must not react with the equipment parts. In fact, water is the most popular solvent because it's environmentally friendly and has a low cost, but organic solvents are used in the event that water produces difficulties. In the case of slurries, they need to have small particles with the suitable additives for maintaining high dispersal and to prevent the trapping of the nozzle. It is easier to use a rotating disc atomizer which is ideal for drying slurries.

In regards to the powder morphology, it depends upon drying conditions (temperature, atomization pressure, and solid concentration in solution), but also on reagent properties. Therefore, spheres, hollow spheres, and doughnuts are some of the frequently produced powders. In ceramic application, regular solids, such as spheres, are more desirable for compaction [25], but for drug delivery, hollow spheres are highly recommended [1]. The two phenomena that control the drying and particle morphology are mass and heat transfer rates. The mass transfer rate for ideal solid spheres can be expressed as the following [25]:

$$J = 4\pi R_d^2 K_c \left(\frac{P_1^S}{R_g T_S} - \frac{P_1^B}{R_g T_B} \right) \tag{1}$$

where J is the mass transfer rate, R_d the droplet radius, K_c the mass transfer coefficient, P_1^S is the solvent partial pressure at the drop surface, P_1^B is the solvent partial pressure in the drying gas, T_S is the surface temperature, T_B is the drying gas temperature, and R_g is the universal gas constant. The mass transfer coefficient is calculated with the aid of the Sherwood, Reynolds, and Schmidt numbers [25].

The heat transfer rate is expressed by:

$$Q = 4\pi R_d^2 h (T_B - T_S) \tag{2}$$

where h is the heat transfer coefficient (h depends of Nusselt and Prandt numbers [25]).

For the formation of spherical particles (dried non-hollow agglomerates), the mean diameter of them (D_p) can be approximated by:

$$D_p = D_d \left(\frac{\rho_d C_d}{\rho_p C_p} \right)^{1/3} \tag{3}$$

where D_d is the droplet mean diameter, C_d and C_p are the solid weight solid fractions in the droplet and the dry agglomerate, respectively, and ρ_d and ρ_p are the densities of droplet and particle.

These properties are related to the feed characteristics and drying conditions, thus, one can modify them and study the effects on particle morphology and humidity content in the final products. The mean diameter equation stated above for agglomerate size calculation is just an approximation, and depending on the atomizer type, there is a specific equation that predicts (also approximately) the droplet size (D_d) for a most accurate D_p estimation [26]. In our case, we will use a two fluid atomizer (pneumatic) for which D_p is given by:

$$D_d = \frac{535 \times 10^3 \sqrt{\rho}}{V_{REL}\sqrt{\rho}} + \left[597 \left(\frac{\mu}{\sqrt{\sigma\rho}}\right)^{0.45} \times \left(\frac{1000 V_{FL}}{V_{AIR}}\right)\right] \tag{4}$$

where σ, ρ, and μ are the fluid surface tension, density, and viscosity, respectively, and V_{FL} and V_{AIR} are volumetric flow rates of the feed and air, respectively, and V_{REL} may be the outlet velocity of air.

Despite the agglomerate size and the fact that the heating and mass transfer rates can be approximated by the above equations, the modeling of spray drying continues to be one challenge for engineers, and some studies are reported mainly for food production [27-30]. In the industry, this is even more complicated due to feeds with higher solid concentrations that are non-Newtonian and viscous. This could be the issue for another discussion; this chapter only deals with the processing of low concentration solutions in a lab spray dryer, with two fluid nozzles for the synthesis of advanced ceramics.

3. Overview of Some Ceramic Systems Synthesized by Spray Drying

In this section we will introduce some important features, applications, and the efforts made until now for the synthesis of some materials.

3.1. Macroporous Cu-Mg-Al Mixed Oxides

Hydrotalcite-like compounds, also referred to as LDHs, constitute a family of isomorphous and isostructural compounds which posses a characteristic layered structure formed by the stacking of brucite-like sheets, where partial substitution of divalent cations by trivalent cations promotes the presence of residual positive charges which are compensated by anionic species and water molecules located in the interlayer space [31]. The LDHs can be represented by the general formula:

$$[M^{2+}{}_{1-x} M^{3+}{}_{x} (OH)_2]^{x+} [(A^{n-})_{x/n}\, mH_2O] \tag{5}$$

where M^{2+} and M^{3+} are the divalent and trivalent metallic ions respectively, A^{n-} is the compensating charge anion, and x is the molar ratio $M^{3+}/(M^{2+} + M^{3+})$, that is generally found in the range of 0.2-0.33 [30-38].

Nowadays, the increasing attention on these sorts of materials is due to rising facilities to incorporate a broad variety of both metals and interlayer species, including organic and biological molecules; therefore, LDHs posses a wide range of potential applications in several fields such as molecular sieves, ion exchange/adsorption, pharmaceutics, and electrochemistry, among others [33-38]. In addition, since thermal decomposition of LDHs gives place to the formation of stable and non- stoichiometric mixed metal oxides with homogenous metal distribution and high specific surface area and strong basic properties [39-42], then they are suitable to be used in the field of heterogeneous catalysis including organic chemistry and environmental catalysis, as both catalyst precursors and catalyst supports. In fact, recently, mixed oxide catalysts obtained by thermal decomposition of Cu-Mg-Al LDHs have been widely investigated due to their interesting properties for environmental applications such as NO_x reduction and SO_x traps [43-46], catalytic reduction of nitrates in water [47-48], and others, such as the oxidation of phenol aqueous solutions [49-50].

In recent years, template-based synthesis of mesoporous and macroporous materials using supramolecular arrays of surfactants [51] and submicrometric polymer spheres as templates, respectively [52], have been proposed as feasible methods to prepare ordered porous frameworks of amorphous or crystalline inorganic oxides and several others materials including LDHs [53-55]. The potential applications for such kinds of materials includes the fabrication of membranes, catalyst, and catalysis supports, where the principal advantages related to this field are the good textural properties such as uniform open porosity, high surface area, and high pore volume.

In addition, due to the versatility of the synthesis method, it has been successfully coupled to several processing techniques such as dip coating [56-57], slip casting [58], and spray drying processing [59]. In the last case, homogeneous solutions or suspensions containing an adequate composition of both precursors and templates are sprayed by the atomizer; the novel feature is the fact that the self assemblage between precursor material and template can be attained easily during the drying stage. Further removal of the template gives place to the formation of a porous microstructure.

3.2. Lead-Free Ferroelectric Ceramics

Lead zirconate-titanate (PZT) piezoelectric (ferroelectric) ceramics have a great number of applications due to their outstanding piezoelectric properties. In recent years, intense investigation has been performed over an important number of lead-free materials in order to replace PZT-based piezoelectric ceramics. The environmental protection has been the driving force behind these efforts because of the great amount of disposed electronic products containing PZT. In the last decade, Saito and coworkers [60] reported the synthesis and properties of lead-free materials based on niobium and alkaline cations. Since then, the research of bismuth–sodium titanates and potassium–sodium niobates has increased dramatically. Most of these reports are focused in the evaluation of piezoelectric, dielectric, and ferroelectric properties. Only a few of the researches have tried to explore new chemical methods for the synthesis of these ceramics [61-62]. It is important to emphasize that in the majority of scientific articles on this subject, the authors have employed the solid state reaction method for the synthesis of the powders. This method demands high temperature heat treatment to achieve the perovskite phase, and secondary phases are quite frequent. These undesirable phases reduce the performance of piezoelectric materials. The presence of these phases are related to the high volatility of alkaline elements (K^+, Na^+, Li^+), and soft chemical methods need to be developed for the production of homogeneous lead-free ferroelectric powders. The spray drying could be an option for this purpose, with the aid of additional chemical steps to stabilize niobium and tantalum in aqueous solutions [12-13].

3.3. Yttrium Aluminum Garnet (YAG)

Yttrium Aluminum Garnet (YAG) has a cubic structure that belongs to spatial group *Ia*3*d* with a lattice parameter of 12.01 Å. This compound has several applications, either as pure phase or doped with elements such as Nd^{3+}, Eu^{3+}, Tb^{3+}, Cr^{3+}, and Ce^{3+} in laser system production and the coating of electronic devices, including the phosphor for cathode ray tubes [63–67]. These applications are intimately related to the optical properties, chemical stability at high temperatures, excellent corrosion resistance, and good mechanical properties. There are several methods reported in the literature for the synthesis of YAG powders. The conventional solid state reaction method [68] for the synthesis of YAG requires high temperatures (1600°C) and long heat treatment. Several wet methods have been developed to diminish the crystallization temperature and increase the purity of the resulting YAG materials [69–75]. The presence of intermediate phases such as metastable $YAlO_3$, $Y_4A_2O_9$, and Y_2O_3 has been observed in several synthesis methods [72-73]. Li *et al.* [76] has recently succeeded in reducing the crystallization temperature, obtaining the YAG crystalline phase approximately at 300°C for 2 hours, using a pressurized reactor. Different methods to prepare doped and undoped YAG and Al_2O_3–Y_2O_3 fibers have alo been reported in the related literature such as sol–gel [77-78].

3.4. α-Alumina (α-Al_2O_3)

Alumina (aluminium oxide, Al_2O_3), in various forms, has been intimately involved in our daily life. Alumina is an important technical ceramic, extensively used in microelectronics, catalysis, refractories, abrasives, and structural applications. Specifically, high quality corundum polycrystalline materials are used as electronic substrates and as bearings in watches and in various high precision devices. Therefore, continuous efforts are being made to develop new processing routes for high quality alumina based ceramics. These efforts are responsible for recent developments in chemical processing (e.g. sol-gel and polymer precursor) approaches, especially for advanced ceramic applications [24].

Some methods for preparing α-alumina are: Calcination of the aluminum hydroxides, transition aluminas, aluminum salts, and solidification from melts. Less common methods are: hydrothermal synthesis at high pressure and vapor-phase transition. Aluminium hydroxides are typical starting materials to produce transition aluminas and α-alumina via calcination. There are two crystalline forms of aluminium monohydroxides: diaspore (α-AlOOH) and boehmite (γ-AlOOH), and three crystalline forms of aluminium trihydroxides: bayerite (α-$Al(OH)_3$), gibbsite (γ -$Al(OH)_3$), and nordstrandite (Al(OH)).

The structure of the starting aluminium hydroxide determines the sequence of the alumina phases and the temperature range of their existence. Depending on the processing methods, the temperature range of the alumina phases may vary. Hydrous oxides and hydroxides form metastable transition alumina phases which transform to stable α-alumina at about 1,200°C. The only hydrated oxide that converts directly to α-alumina is diaspore which forms α-alumina at temperatures as low as 500°C. The direct topotactic conversion of diaspore into α-alumina is due to the structural similarity of these phases, both having a hexagonal close packed anion sublattices. Attempts to synthesize diaspore as a precursor for alpha alumina have been unsuccessful and hence there is an interest in developing techniques to obtain α-alumina at low temperatures. Aluminum formate precursor was readily produced via spray drying and calcined at 1,100°C yielding pure α-alumina powders [24].

4. EXPERIMENTAL

Here we describe the reagents, parameters, and conditions under which the syntheses of the proposed ceramics systems were performed.

4.1. Cu-Mg-Al Mixed Oxides

4.1.1. Synthesis of Latex Template

The monodisperse polystyrene spheres (PS) were prepared by emulsifier-free emulsion polymerization, using ammonium persulphate ($(NH_4)S_2O_8$ 99+%, Sigma-Aldrich) as initiator and sodium bicarbonate (Na_2CO_3, 99+%, Aldrich) as a buffer. The system was purged with nitrogen to eliminate oxygen inhibitor effects. Monodisperse polystyrene spheres with an average diameter (estimated by scanning electron microscopy) of 850 ±20 nm were obtained after reactions for 7h, which were kept at 60 ±2 °C with mechanical stirring at 350 rpm and a monomer/water ratio of 0.125.

4.1.2. Preparation of Macroporous Spray Dried Powders

a) LDHs Precursors Preparation

Cu-Mg-Al hydrotalcite-like compound powders were synthesized with Cu:Mg:Al molar ratios of 0.22:2:1 by conventional coprecipitation method as follows. An aqueous solution of inorganic precursors were prepared with stoichiometric amounts of $Cu(NO_3)_2{\cdot}2.5H_2O$ and $Mg(NO_3)_2{\cdot}6H_2O$, $Al(NO_3)_3{\cdot}9H_2O$ (Aldrich Chemical Company 99.0+%). The solution was prepared at PH=10, with additions of NaOH (1.0 M) and Na_2CO_3 (1.0 M) at room temperature under vigorous mechanical stirrings. The resulting precipitate was filtered and washed several times.

b) Macroporous Mixed Oxides Preparation

Aqueous suspension (with 5.0 % in solids) and pH 6 were prepared using the obtained hydrotalcite precipitate powders. Before the spray drying process, the suspension was stirred for 1 hour in order to achieve a well dispersion of the particles. Then, the slurry was fed into a mini-spray dryer YAMATO-ADL31, at an atomization pressure of 2 kg/cm^2 and at an outlet temperature of 95°C, to fabricate the porous aggregates. On the other hand, when the spray dryer template-assisted method was used, spherical self assembles of hydrotalcite and PS templates were fabricated using a second aqueous suspension consisting of both hydrotalcite particles and PS. This second suspension was prepared maintaining the solid content of 5% and a volume ratio of LDHs:PS, which was estimated as 60:40. The slurry was fed into the spray dryer system using the above-mentioned operation conditions. In this case, the stabilized suspension was obtained using sodium polyethacrylate (NaPM) as dispersant and pH was adjusted to 10 based on zeta potential measurements. Finally, powder precursors were thermal treated at 500 °C for 8 hours in order to obtain the porous mixed oxides.

4.1.3. Powders Characterization

Cu-Mg-Al LDHs were studied by X-ray powder diffraction techniques (XRD). The XRD patterns were recorded using a Siemens D5000 diffractometer with CuK_α radiation in the 2θ range of 5-70°. Lattice parameters of hydrotalcite crystals were determined from diffraction peaks using the R3m space group. Crystallite size was obtained from broadening of XRD peaks using the classical Debye-Scherrer equation. Zeta potential measurements were performed by using a Zeta meter 3.0+ instrument. The electro-kinetical analysis was carried out in the pH range from 4 to 10, adjusting the pH suspension with hydrochloric acid and/or ammonium hydroxide dilute solutions. The morphology and macroporous microstructure of the powders were examined using a scanning electron microscope Jeol model JSM-6400. The specific surface area of calcined samples was determined by nitrogen sorption according to single point BET method using a Quantacrome apparatus.

4.2. Lead-Free Ferroelectric Ceramics

4.2.1. The Powders Synthesis

Two alkali niobate based compositions were synthesized ($K_{0.5}Na_{0.5}NbO_3$ (KNN) and $(K_{0.48}Na_{0.52})_{0.96}Li_{0.04}Nb_{0.85}Ta_{0.15}O_3$ (KNLNT)). Details of the synthesis can be found elsewhere [12-13]. For the synthesis of niobates or tantalates by a soft chemical method, it is crucial to stabilize the niobium ion since it reacts readily with moisture forming niobium oxide. Once the stable niobium/tantalum solution is prepared, the final feed solution with all ions including potassium, sodium, and/or lithium is obtained and dried in a spray drying equipment (as described for LDHs). The obtained dried powders were calcined at 800°C for 1 hour.

4.2.2. Characterization of Synthesized Powders

The powders were characterized by (XRD) at room temperature with Cu Kα radiation using a Bruker Advanced D-8 diffractometer. Scanning electron microscopy (SEM) images were obtained by a Leica Cambridge Stereoscan 440 microscope at 20 kV. Also, the synthesized powders were observed by transmission electron microscopy (TEM) with a Phillips TECNAI F20 super Twin at 200 kV.

4.3. Synthesis and Characterization of Yttrium Aluminum Garnet (YAG) Powders

Aluminum nitrate (99.9% J.T. Baker), yttrium nitrate (99.9% Aldrich), citric acid (J.T. Baker), ethylene glycol (99.9% J.T. Baker), and deionized water were used as raw materials for the synthesis. The procedure for the precursor solution is found elsewhere [19-20]. The solution was fed into a spray dryer to eliminate the solvent and to produce precursor powders. These powders were thereafter calcined from 750 to 1,200°C to promote the crystallization of the YAG phase. Heat treated powders were characterized by XRD using a SIEMENS D5000 diffractometer with Cu Kα radiation operated at 20 kV. Several micrographs of the samples were obtained by SEM (as for LDHs) to estimate the characteristics of agglomerates as well as by TEM (Phillips TECNAI F20 super Twin) with an acceleration voltage of 200 kV.

4.4. Synthesis and Characterization of α-Alumina Powders by Metal-Organic Precursor

Aluminum formiate $Al(O_2CH)_3$ was synthesized by chemical synthesis described by Reyes, Serrato, and Sugita [24]. The method of processing involved a mixture of aluminum metal, with formic acid using as catalyst mercuric chloride, to produce aluminum formate solution which was spray dried to produce fine granulated metal-organic precursor. The granular powder was calcined in dry air at a rate of 5°C/min up to 1,100°C. The grain size distribution was obtained by the AcoustoSizer IIs™ technique (Colloidal Dynamics Inc.). Phases were defined by thermal analysis, infrared spectroscopy (IR), nuclear magnetic resonance (NMR), XRD, and SEM. A 10 mg of sample was run on a thermal analysis instrument (Model Q600 simultaneous differential scanning calorimetry/thermogravimetric analysis (DSC/TGA), TA Instruments). The heating rate was 10°C/min to 1,400°C, with a nitrogen-gas flow of 100 cm^3/min. Attenuated Total Reflection (ATR) techniques were used and fitted with a Fourier transform (The TENSOR™ 27 series FT-IR spectrometer, ZnSe crystal, Bruker Optics Inc.). NMR ^1H, ^{13}C spectra of $[Al(O_2CH)_3]$, were obtained on a Varian Gemini 200 NMR spectrometer by using 30 mg sample dissolved in D_2O. Crystalline products, as a function of processing temperature, were followed by XRD. The microstructure was studied by SEM.

5. Review of Most Prominent Results of the Investigated Ceramics

5.1. Cu-Mg-Al Mixed Oxides

Zeta potential values of different colloidal particles in suspension as a function of pH and NaPM addition are presented in Figure 5.1. As it was expected, while LDHs particles present positive surface charges, PS expose high negative charge values in the range of pH studied. Then, in order to avoid the heterocoagulation phenomena and to guarantee a homogeneous distribution of the two components in the atomized droplets during spray dryer process, surface potential of the LDHs was modified adding 0.35 mL/g of NaPM. As result, modified particles showed the isoelectrical point at pH ~8 and negative zeta potential above this pH value. Therefore, pH 10 was established as the best condition to obtain a colloidal stability of two components in the aqueous suspension.

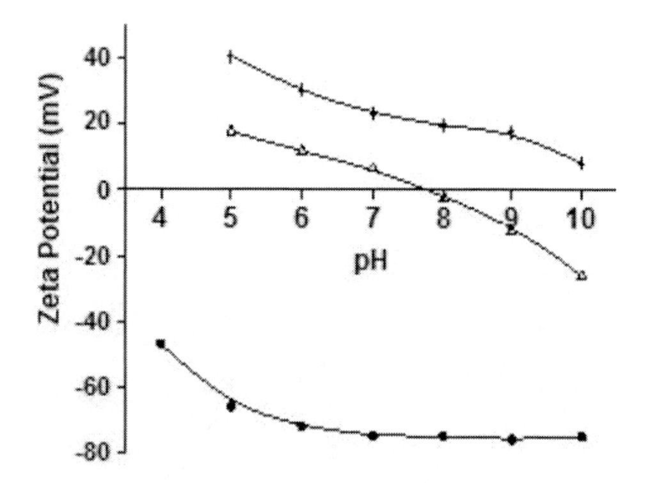

Figure 5.1. Zeta potential as a function of pH for LDHs (+), NaPM on LDHs (Δ), and for polystyrene spheres template (•).

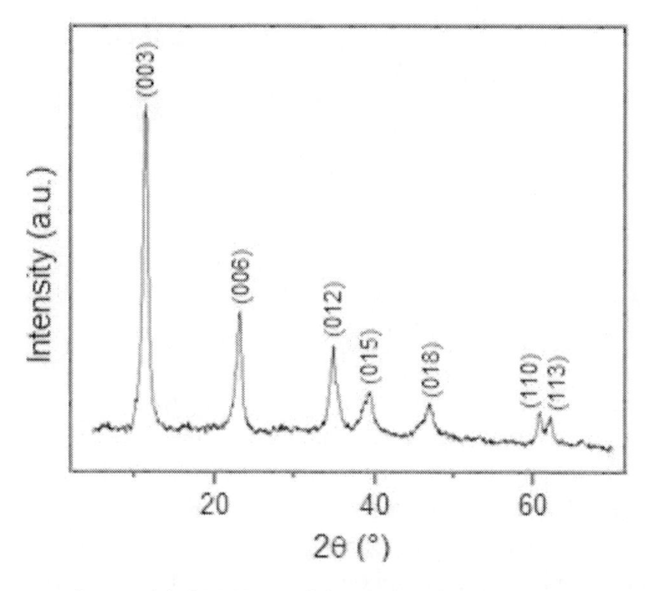

Figure 5.2. XRD pattern of spray dried LDHs particles, indexed pattern correspond to pure phase.

The XRD pattern of spray dried LDHs samples at 95°C (Figure 5.2) shows the presence of a well crystallized hydrotalcite phase, and no other crystalline compounds were identified. The calculated crystallographic parameters *a* y *c,* were 3.055 Å and 23.031 Å, respectively. It is reasonable to presume that Cu and Mg are stabilized in brucite-like sheets [44]. This information suggests the incorporation of Cu^{2+} ions in the LDHs structure, which results in structural disorder and leads to an increase of lattice parameters [39,44]. The crystallite size of hydrotalcite precursor calculated from (003) and (110) diffraction peaks were 75 Å and 140 Å, respectively, showing the nanocrystalline nature of the sample. In addition, assuming the hydrotalcite crystal like-prisms with height equal to crystalline size along (003) planes and a length of prism corresponding to crystalline size along (110) plane, the calculated values indicate a geometry close to plate-like crystals [39].

The spray dried LDHs morphology is shown in Figure 5.3. The micrographs revealed that particles tend to form the typical spherical morphology of a spray drying powder. Then, the obtained granules can be categorized as solid and non-macroporous spheres with a size distribution between 1 and 5 μm. Furthermore, spray drying defects such as hollow or doughnut granule shapes were not observed, indicating an adequate slurry formulation to achieve a well granulation of the powders.

Figure 5.3. LDHs spherical granules exposing the spherical morphology derived from spray drying processing.

Figures 5.4a and 5.4b show the SEM micrographs for the samples prepared by spray drying technique without and with PS as a template, respectively. In the former sample, after calcination, spherical granules were loose; SEM revealed the presence of submicrometric aggregates of primary individual particles, with a diameter of ~ 200 nm. On the other hand, when samples were processed incorporating PS, powders present an open porous microstructure with pore diameter between 500 and 700 nm, and particle size in the range of 4 to10 μm. SEM images reveal a interconnected porosity (Figure 5.5b), not forgetting that this synthesis route represents a practical and economical method to produce high volumes of uniform particles possessing repeatable characteristics [79].

In both cases, calcination produces mixed oxides; XRD analyses confirmed the formation of CuO (JCPDS 03-0884), MgO (JCPDS 03-0998) and spinel-like phases (JCPDS 03-0897). These results are in agreement with previous papers on the Cu/Mg/Al system [44, 79-82]. Finally, the calculated surface areas were 114 and 103 m^2/g for templated and non-templated calcined samples, respectively. It is noteworthy that the high specific surface area cannot be only attributed to the macroporosity, but it also depends on the mesoporosity and the inherent characteristics of the LDHs and their calcination products [39,44]. Consequently, the differences in the surface area can be correlated with the major accessibility that macroporus framework provides to the mesoporous walls in the material. In other words, surface area values reflect the fact that mesopores and macropores are present, simultaneously, in a hierarchical pore microstructure.

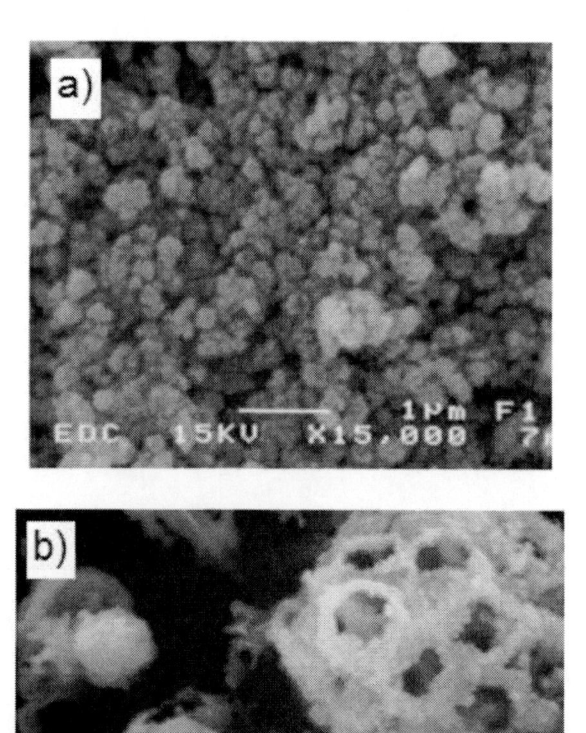

Figure 5.4. Oxide particles obtained by spray drying, a) without template addition, b) without template addition.

5.2. Lead-Free Ferroelectric Ceramics

The SEM and TEM images of KNN and KNLNT powders calcined at 800°C can be seen in Figure 5.5 and 5.6, respectively. It is possible to observe that powders consist of agglomerates that have a particle size smaller than 1 μm (average grain size ~ 300 nm for KNN and ~ 100 nm for KNLNT measured on bright field TEM images). The crystals have a cubic-like shape characteristic of KNN perovskite phase.

Figure 5.5. SEM (a) and bright field TEM (b) images of KNN powder annealed for 1 hour at 800°C.

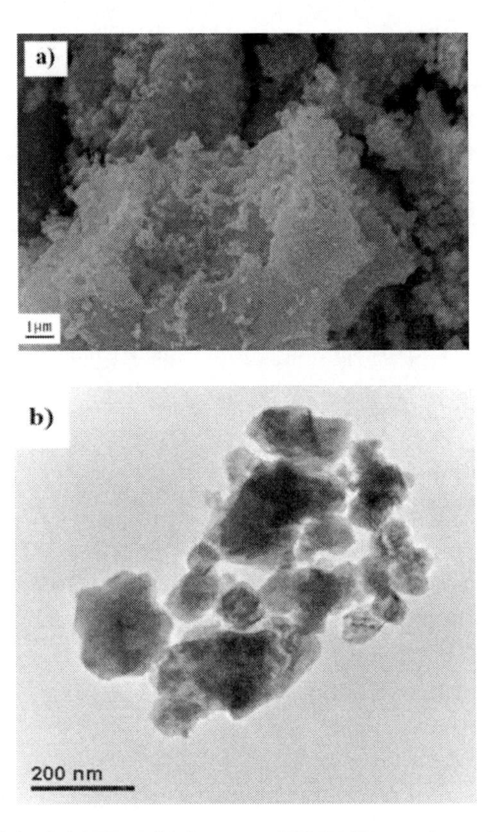

Figure 5.6. SEM (a) and bright field TEM (b) images of KNLNT powders calcined at 800°C for 1 hour.

The small crystal size and the purity of the synthesized powders are crucial issues for the sintering of high density samples. In fact, in the field of ferroelectric ceramics for piezoelectric applications, the density of sintered samples should be higher than 90% of the theoretical value, at which fine powders are desirable. As it is seen in the SEM images, the powders are agglomerated with no spherical morphology; this is due to the considerable amount of citric acid used during the synthesis. The organics produced large quantities of gases at the calcination step, so the breaking of the spheres was produced. Despite the fact that the powders are not spherical, they produced high density pellets with good ferroelectric and piezoelectric properties as reported in previous publications [12-13].

Another feature is that the heat treatment has been reduced to one hour for the precursor powders by spray drying compared with the several hours required with the conventional mixed oxide or the ceramic method (CM). The above is important from the point of view that the short calcinations time avoids considerablely the volatilization of potassium and sodium that takes place when the conventional mixed-oxide route is used [82-84]. Actually the CM also requires milling before and after the calcination step, usually for 12 to 24 hours for each one, and this is not necessary in our methodology. The corresponding XRD patterns of the calcined powders at 800°C are shown in Figure 5.7. It can be seen that the calcined powders have a pure perovskite phase.

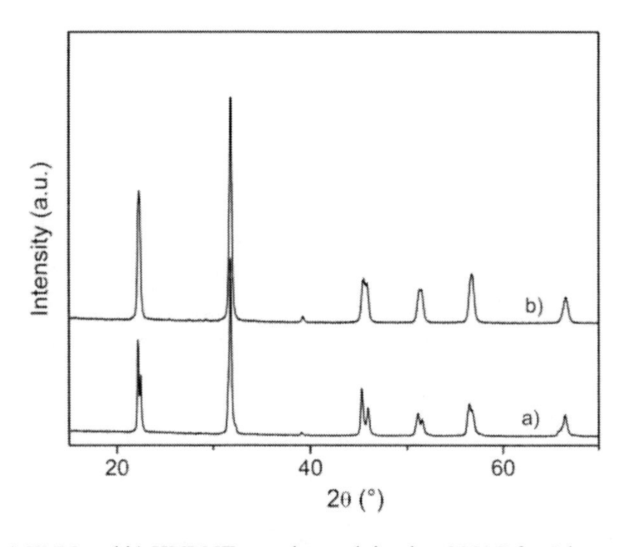

Figure 5.7. XRD of a) KNN and b) KNLNT powders calcined at 800°C for 1 hour.

5.3. Yttrium Aluminum Garnet (YAG) Powders

The synthesized YAG powders after being calcined for 1 hour have spherical morphology, characteristic of spray drying as shown in Figure 5.8. The agglomerates have narrow size distribution; this is the result of using a two fluid nozzle configuration since it produces the finest droplets, among all types of atomizers including one fluid nozzle and rotating discs [25]. As is observed, the spheres are between 0.3 and 2 μm.

Figure 5.8. Calcined YAG powders at: a) 800°C and b) 900°C for 1 hour.

Details of the microstructure of the agglomerates are seen by TEM (Figure 5.9), and the micrographs show the presence of nanometric crystals (Figure 5.9c). As was established in Section 2, the mean size distribution when we consider spherical agglomerates has a direct relation with the mean droplet size produced in the nozzle. Also, the concentration of solids in the feed is also important. Experimentally, we used a dilute feed (~10% wt) and, simultaneously with the conditions already mentioned, made it possible to obtain a narrow size distribution in the YAG ceramics. Normally, in the realm of fine ceramic processing, dilute feed solutions are preferred.

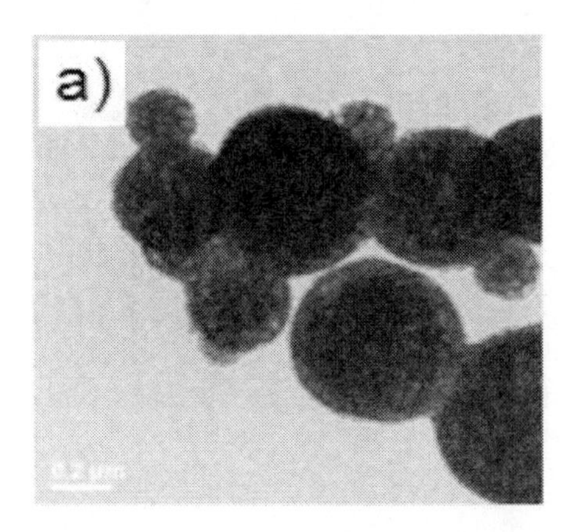

Figure 5.9. (Continued).

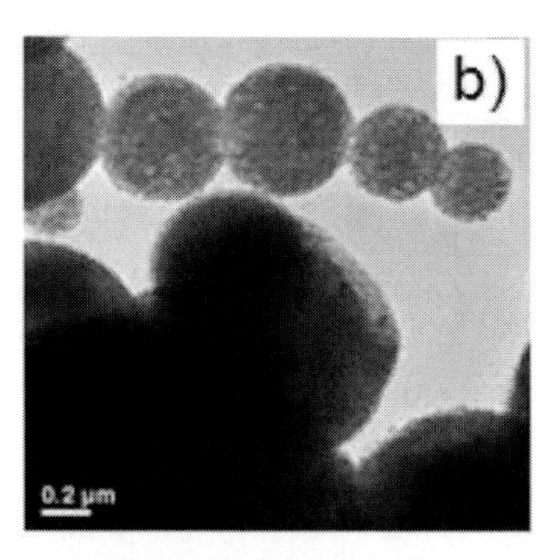

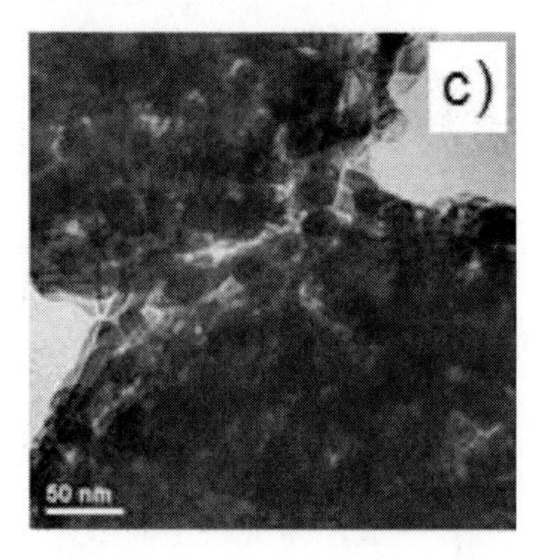

Figure 5.9. Bright field TEM images of YAG powders calcined at 900°C for 1 hour: a) and b) two different general areas, c) nanocrystals between two spheres.

The synthesis temperature for YAG powders is as critical as it is for lead-free ferroelectric ceramics. As mentioned above, when the ceramic route is used, the formation of secondary phases is common (e.g. yttrium-aluminum pyrochlore-YAP); in addition, the calcination step is performed at high temperatures (greater than 1,400°C) and grain growth is unavoidable. Therefore, obtaining low-temperature crystalline ceramics is of great importance. Chemical methods, including spray drying, reduce diffusion distances and the energy barrier for mass transfer, and the synthesis can be achieved below 1,000°C. The XRD results of annealed powders from 750 to 1,200°C are shown in Figure 5.10. It is observed that the crystallization is achieved at 850°C, which is a considerably lower temperature than that required by the ceramic method. The sprayed powders are amorphous since these are a mixture of nitrates, citric acid, and ethylenglycol, that need to be calcined for producing crystallization. The powders remain amorphous below 850°C, and at this temperature Bragg reflections appear, corresponding to the cubic YAG phase. The intimate mixing of the reagents reduces the energy and avoids the appearance of undesirable compounds in the product.

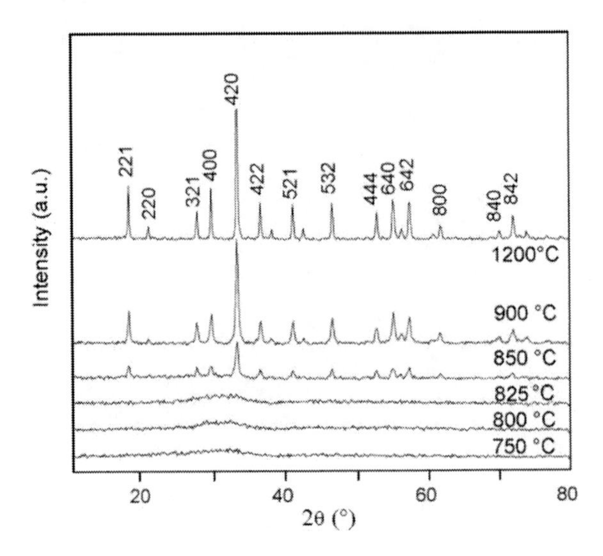

Figure 5.10. XRD patters of annealed YAG powders.

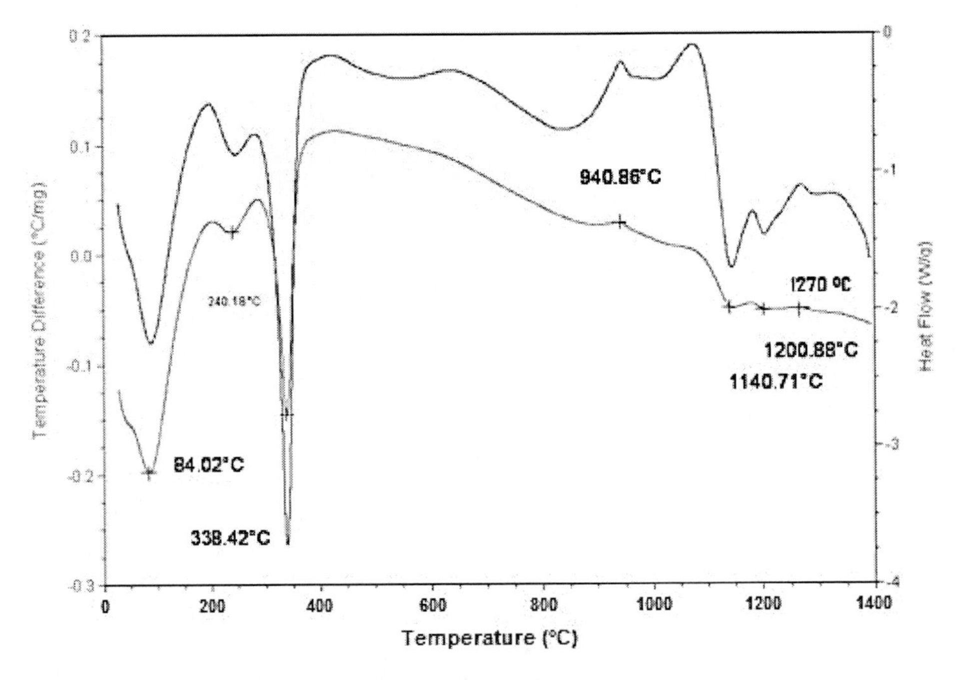

Figure 5.11. DTA and DSC of aluminum formate $Al(O_2CH)_3$.

5.3. α-Alumina Powders by Metal-Organic Precursor

Alpha alumina is the thermodynamically stable phase of aluminum oxides at normal pressure and temperature conditions. The precursors for the synthesis of α-alumina are divided mainly into two groups: those derived of inorganic and those derived of metal-organic salts. Inorganic salts lead to high temperature formation of α-alumina (1,200°C) and metal-organic salt precursors reduce the temperature formation at around 1,100°C.

The application of chemical principles in the solution synthesis and processing of ceramic particles by spray drying has been summarized in a recent paper [24]. This paper reported the synthesis of α-alumina powders with high porosity, using only the aluminium formate $Al(O_2CH)_3$, a metal-organic precursor and the spray drying processing route.

TGA profiles for aluminum formate decomposition show a total ceramic yield of 31.50 wt%, which is similar to the theoretical yield for $Al(O_2CH)_3$. DTA and DSC in the Figure 5.11 reveal endotherms with maxima at 85, 240, and 340°C, corresponding to reactions (1), (2), and (3). The 940°C exothermic peak corresponds to the crystallization of η-alumina; the endothermic peak at 1,140°C shows the formation of α-alumina, and finally the exothermic signal at 1,270°C corresponds to the crystallization of α-alumina, the characteristic enthalpies are summarized in the table 5.1.

$$Al_2(O_2CH)_3 \xrightarrow{85^0 C} Al(OH)(O_2CH)_2 + CO$$

(reac.1)

$$2Al(OH)(O_2CH)_2 \xrightarrow{240^0 C} Al_2O(O_2CH)_4 + H_2O$$

(reac.2)

$$Al_2O(O_2CH)_4 \xrightarrow{340^0 C} Al_2O_3 + 4CO + 2H_2O$$

(reac.3)

Typical morphology of a spray dried powder is observed in the SEM micrographs of Figure 5.12. The granular shape of α-alumina powders are spherical with a particle size distribution $D_p \sim$ 0.80 μm shown in the Figure 5.13. The SEM image in the Figure 5.14 revealed the presence of submicrometric aggregates, of primary individual particles, with a diameter of around 150 nm.

Table 5.1. Characteristic formation enthalpies of α-alumina from $Al(O_2CH)_3$ precursor

Phase	T $_0$C	ΔH^t (J/g)
$AlOH(O_2CH)_2$	84	553
$Al_2O(O_2CH)_4$	240	65
Al_2O_3	340	582
η-Al_2O_3	940	635
α-Al_2O_3	1100	282

A variable, which is known to have an influence on the concentration of the phases of alumina, is the surface area of the powders. The samples derived from aluminum formate precursors, show higher surface area. Figure 5.15 shows the variation of surface area as a function of calcination temperature for alumina powders. The sample calcined at 600°C has the highest surface area (577 m^2/g). As the calcination temperature is increased at 1,000°C, the surface area is significantly reduced to 180 m^2/g, then a gradual decrease to 101 m^2/g at 1,100°C is observed.

Figure 5.12. SEM micrograph of α-alumina powders calcined at 1,100°C.

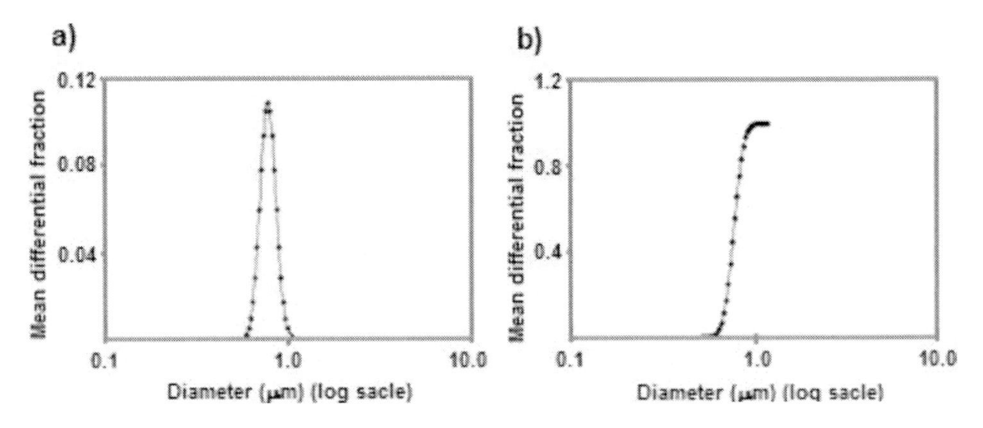

Figure 5.13. Average particle size distribution of α-alumina powders, a) Differential and b) Cumulative.

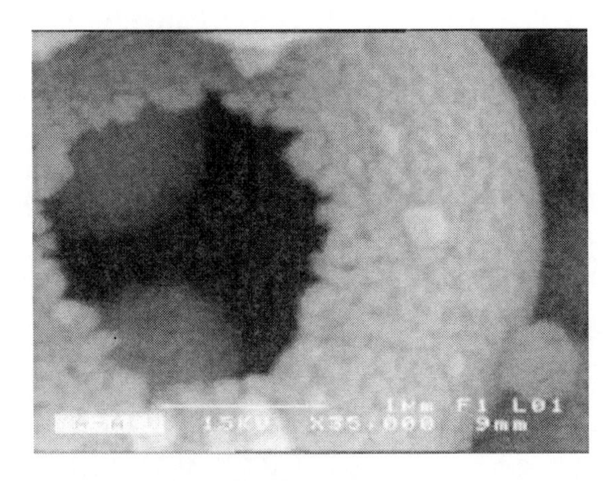

Figure 5.14. SEM micrograph of α-alumina agglomerate sowing nanoparticles of about 150 nm.

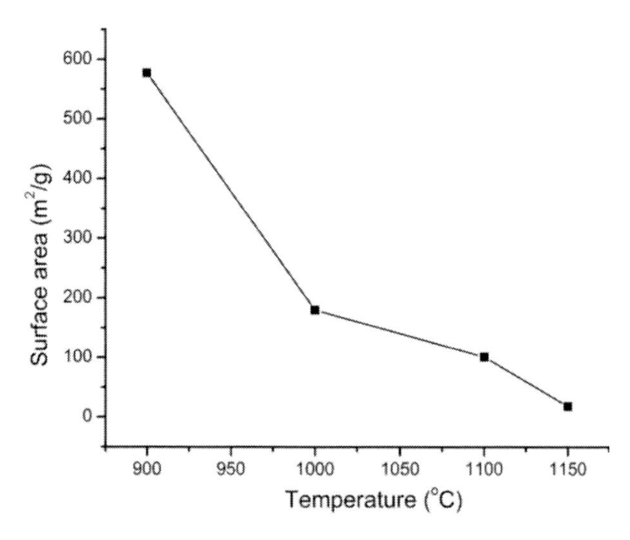

Figure 5.15. Variation of surface area of metal-organic precursor as a function of calcination temperature.

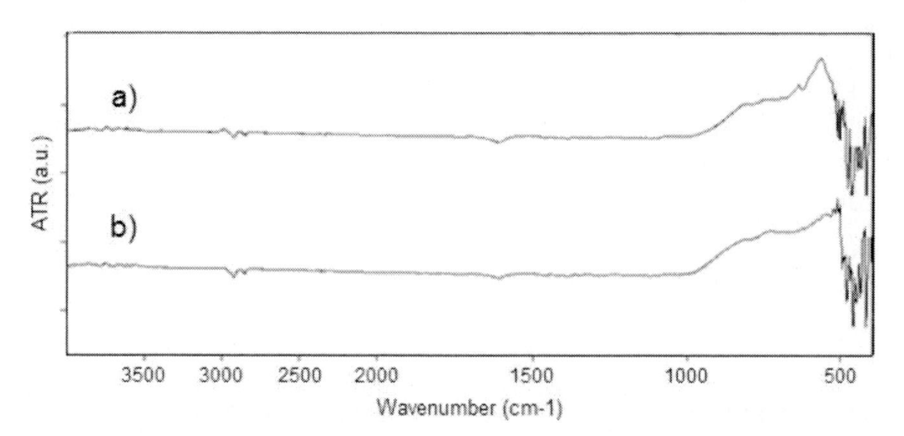

Figure 5.16. IR spectra of $Al(O_2CH_3)$ calcined at various temperatures: a) α-Al_2O_3 at 1,100°C and b) η-Al_2O_3 at 1,000°C.

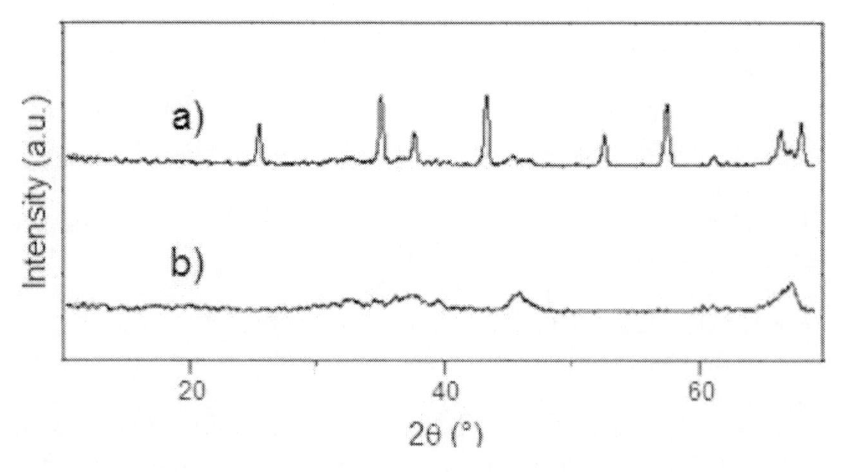

Figure 5.17. DRX of $Al(O_2CH_3)$ calcined at various temperatures: a) α-alumina at 1,100°C and b) η-alumina at 1,000°C.

The sample calcined at 1,150°C possesses a surface area of 11 m^2/g m and it is found to consist of 100% α-Al_2O_3. The surface area of the alumina powders depend, to a great extent, on the nature of the precursor and the calcination temperature. As the calcination temperature increases from 1,000 to 1,100°C, the percentage of α-Al O increases.

In Figure 5.16 and 5.17, the IR spectra and XRD are shown corresponding to the formation of η-alumina at 1,000 C and α-alumina at 1,100 C.

CONCLUSION

Spray drying is a versatile technique for the production of ceramic materials of a wide range of composition, sizes, and morphology. Production of ceramic powders with controlled characteristics require control over atomization variables (temperature, pressure), such as pH, concentration, and calcination temperature. The chemical properties and thermal characteristics of the solution and the precursor must be known, because they affect the particle morphology through all stages of the process. Water-based solutions are preferred over organic because of the low cost and environmental benefits. Furthermore, high solubility salts must be used for achieving optimal mixing of the ions, and it is advisable to use dilute precursor solutions to reduce the agglomerate mean size.

Spray drying is a practical and economical method to achieve ceramic products with superior properties. Developing the characteristics of the precursors and the process technology is possible to obtain advanced materials with improved chemical homogeinity, controllable size morphology and distribution at low crystallization temperatures.

REFERENCES

[1] R. Vehring, Pharmaceutical Particle Engineering via Spray Drying, *Pharm. Res.* 25 (2007) 999-1022.

[2] J. Elversson, A. Millqvist-Fureby, Particle Size and Density in Spray Drying Effects of Carbohydrate Properties, *J. Pharm. Sci.* 94 (2005) 2049-2060.

[3] R. Sun, Y. Lu, K. Chen, Preparation and characterization of hollow hydroxyapatite microspheres by spray drying method, *Mater. Sci. Eng. C* 29 (2009) 1088–1092.

[4] H. Schiffter, J. Condliffe and S. Vonhoff, Spray-freeze-drying of nanosuspensions: the manufacture of insulin particles for needle-free ballistic powder delivery, *J. R. Soc. Interface* 7 (2010) S483–S500.

[5] L. C. Chow and L. Sun, Properties of Nanostructured Hydroxyapatite Prepared by a Spray Drying Technique, *J. Res. Natl. Inst. Stand. Technol.* 109 (2004) 543-551.

[6] K. Keogh, C. Murray, J. Kelly, B. O'Kennedy , Effect of the particle size of spray-dried milk powder on some properties of chocolate, *Lait* 84 (2004) 375–384.

[7] P. M. Kelly, Innovation in milk powder technology, *Int. J. Dairy Technol.* 59 (2006) 70-75.

[8] Sudhagar Mani, S. Jaya, H. Das, Sticky Issues on Spray Drying of Fruit Juices, In ASAE Meeting Presentation, Paper No: MBSK 02-201(2002) 1–18.

[9] G. R. Rodríguez-Hernández, R. González-García, A. Grajales-Lagunes, M. A. Ruiz-Cabrera, M. Abud-Archila, Spray-Drying of Cactus Pear Juice (Opuntia streptacantha): Effect on the Physicochemical Properties of Powder and Reconstituted Product, *Drying Technol.* 23 (2005) 955–973.

[10] A.L. Costa, C. Galassi, E. Roncari, Direct synthesis of PMN samples by spray-drying, *J. Eur. Ceram. Soc.* 22 (2002) 2093–2100.

[11] F. Bezzi, A.L. Costa, D. Piazza, A. Ruffini, S. Albonetti, C. Galassi, PZT prepared by spray drying: From powder synthesis to electromechanical properties, *J. Eur. Ceram. Soc.* 25 (2005) 3323–3334.

[12] R. López, F. González, M.P. Cruz, M.E. Villafuerte-Castrejon, Piezoelectric and ferroelectric properties of $K_{0.5}Na_{0.5}NbO_3$ ceramics synthesized by spray drying method, *Mater. Res. Bull.* 46 (2011) 70–74.

[13] R. López, F. González, M.E. Villafuerte-Castrejón, Structural and electrical characterization of $(K_{0.48}Na_{0.52})_{0.96}Li_{0.04}Nb_{0.85}Ta_{0.15}O_3$ synthesized by spray drying, *J. Eur. Ceram. Soc.* 30 (2010) 1549–1553.

[14] Van Driessche, R. Mouton and S. Hoste, Rapid formation of the $Bi_{2-x}Pb_xSr_2Ca_2Cu_3O_y$ high T_C-Phase, using Spray-Dried Nitrate Precursor Powders, *Mater. Res. Bull.* 31 (1996) 979-992.

[15] Tsetsekou, E. Georgiopoulos, C. Andreouli, Development of superconducting $YBa_2Cu_3O_{7-x}$ powders by spray drying, *Supercond. Sci. Technol.* 15 (2002) 1610–1616.

[16] F. Zhang, Y. Xie, W. Lu, X. Wang, S. Xu, X. Lei, Preparation of microspherical α-zirconium phosphate catalysts for conversion of fatty acid methyl esters to monoethanolamides, *J. Colloid Interface Sci.* 349 (2010) 571–577.

[17] B. Rivas-Murias, J.F. Fagnard, Ph. Vanderbemden, M. Traianidis, C. Henrist,R. Cloots, B. Vertruyen, Spray drying: An alternative synthesis method for polycationic oxide compounds, *J. Phys. Chem. Solids* (2010), doi:10.1016/j.jpcs.2010.12.001.

[18] T. S. Lyubenova, F. Matteucci, A.L. Costa, M. Dondi, M. Ocaña, J. Carda, Synthesis of Cr-doped $CaTiSiO_5$ ceramic pigments by spray drying, *Mater. Res. Bull.* 44 (2009) 918–924.

[19] R. López, E.A. Aguilar, J. Zárate-Medina, J. Muñoz-Saldaña, D. Lozano-Mandujano, Nanoindentation of melt-extracted amorphous YAG and YAG:Eu, Nd micrometric fibers synthesized by the citrate precursor method, *J. Eur. Ceram. Soc.* 30 (2010) 73–79.

[20] R. López, J. Zárate, E. A. Aguilar, J. Muñoz-Saldaña, Preparation of neodymium-doped yttrium aluminum garnet powders and fibers, *J. Rare Earth* 26 (2008) 670-673.

[21] K. Prabhakaran, M.O. Beigh, J. Lakra, N.M. Gokhale, S.C. Sharma, Characteristics of 8 mol% yttria stabilized zirconia powder prepared by spray drying process, *J. Mater. Process. Technol.* 189 (2007) 178–181.

[22] V. Sharma, K. M. Eberhardt, R. Sharma, J. B. Adams, P. A. Crozier, A spray drying system for synthesis of rare-earth doped cerium oxide nanoparticles, *Chem. Phys. Lett.* 495 (2010) 280–286.

[23] J.M. Kim, N. Kumagai, Y. Kadoma, H. Yashiro, Synthesis and electrochemical properties of lithium non-stoichiometric $Li_{1+x}(Ni_{1/3}Co_{1/3}Mn_{1/3})O_{2+\delta}$ prepared by a spray drying method, *J. Power Sources* 174 (2007) 473–479.

[24] S. Y. Reyes, J. Serrato, S. Sugita, Low-Temperature Formation of Alpha Alumina Powders Via Metal Organic Synthesis, *Adv. In Tech. Of Mat. And Mat.* 8 (2006) 55-62.

[25] T. A. Ring, Fundamentals of Ceramic Powder Processing and Synthesis, Academic Press, 1996, pp 307-334.

[26] S. Mujumdar, Handbook of Industrial Drying, Taylor and Francis, Third Edition (2006) 215-254.

[27] J. Straatsma, G. Van Houwelingen, A.E. Steenbergen, P. De Jong, Spray drying of food products: 1. Simulation model, *J. Food Eng.* 42 (1999) 67-72.

[28] J. Straatsma, G. Van Houwelingen, A.E. Steenbergen, P. De Jong, Spray drying of food products: 2. Prediction of insolubility index, *J. Food Eng.* 42 (1999) 73-77.

[29] M. Goula, K. G. Adamopoulos, Spray drying of tomato pulp in dehumidified air: II. The effect on powder properties, *J. Food Eng.* 66 (2005) 35–42.

[30] D. Velić, M. Bilić, S. Tomas, M. Planinić, Simulation, calculation and possibilities of energy saving in spray drying process, *Appl. Therm. Eng.* 23 (2003) 2119–2131.

[31] F. Cavani, F. Trifiro, A. Vaccari, Hydrotalcite-type anionic clays: Preparation, properties and applications, *Catal. Today* 11 (1991) 173-301.

[32] J.A. Rivera, G. Fetter, Y. Jiménez, M.M. Xochipa, P. Bosch, Nickel distribution in (Ni,Mg)/Al-layered double hydroxides, *Appl. Catal. A-Gen.* 316 (2007) 207–211.

[33] T.W. Kim, M. Sahimi, T.T. Tsotsis, The preparation and characterization of hydrotalcite micromembranes, *Chem. Eng. Sci.* 64 (2009) 1585 – 1590.

[34] U. Costantino, V. Ambrogi, M. Nocchetti , L. Perioli, Hydrotalcite-like compounds: Versatile layered hosts of molecular anions with biological activity, *Micropor. Mesopor. Mater.* 107 (2008) 149–160.

[35] T. Wu, D. Sun, Y. Li, H. Zhang, F. Lu, Thiocyanate removal from aqueous solution by a synthetic hydrotalcite sol, *J. Colloid Interf. Sci.* 355 (2011) 198-203.

[36] M.J. Climent, A. Corma, S. Iborra, and A. Velty, Activated hydrotalcites as catalysts for the synthesis of chalcones of pharmaceutical interest, *J. Catal.* 221 (2004) 474–482.

[37] L. Perioli, T. Posati, M. Nocchetti, F. Bellezza, U. Costantino, A. Cipiciani, Intercalation and release of antiinflammatory drug diclofenac into nanosized ZnAl hydrotalcite-like compound, *Appl. Clay Sci.* (2010), doi:10.1016/j.clay.2010.06.028.

[38] E. Scavetta, M. Berrettoni, R. Seeber, D. Tonelli, [Ni/Al-Cl]-based hydrotalcite electrodes as amperometric sensors: preparation and electrochemical study, *Electrochim. Acta* 46 (2001) 2681-2692.

[39] M. Hesiquio-Garduño, B. Zeifert, J. Salmones, H. Reza, Synthesis and characterization of Co-Hydrotalcite-Like compounds, *J. of Metastable Nanocryst. Mater.* 20-21 (2004) 257-262.

[40] J. Zhang, N. Zhao, W. Wei, Y. Sun, Partial oxidation of methane over Ni/Mg/Al/La mixed oxides prepared from layered double hydrotalcites, *Int. J. Hydrogen Energ.* 35 (2010) 11776-11786.

[41] M. Mokhtar, A. Inayat, J. Ofili, W. Schwieger, Thermal decomposition, gas phase hydration and liquid phase reconstruction in the system Mg/Al hydrotalcite/mixed oxide: A comparative study, *Appl. Clay Sci.* 50 (2010) 176–181.

[42] M.J. Climent, A. Corma, S. Iborra, K. Epping, A. Velty, Increasing the basicity and catalytic activity of hydrotalcites by different synthesis procedures, *J. Catal.* 225 (2004) 316–326.

[43] B. Montanari, A. Vaccari, M. Ganzzano, P. Käbner, H. Papp, J. Pasel, R. Dziembaj, W. Makowski, T. Lojewski, Characterization and activity of novel copper-containing

catalysts for selective catalytic reduction of NO with NH_3, *Appl. Catal. B-Environ.* 13(1997) 205-217.

[44] D. Rosales Suárez, B.H. Zeifert, M. Hesiquio Garduño, J. Salmones Blázquez, A. Romero Serrano, Cu hydrotalcite-like compounds: Morphological, structural and microstructural properties, *J. Alloy Compd.* 434-435 (2007) 783-787.

[45] L. Chmielarz, P. Kuśtrowski, A. Rafalska-Łasocha, D. Majda, R. Dziembaj, Catalytic activity of Co-Mg-Al, Cu-Mg-Al and Cu-Co-Mg-Al mixed oxides derived from hydrotalcites in SCR of NO with ammonia, *Appl. Catal. B-Environ.* 35 (2002) 195–210.

[46] A.E. Palomares, A. Uzcátegui, A. Corma, NO_x storage/reduction catalysts based in cobalt/copper hydrotalcites, *Catal. Today* 137 (2008) 261–266.

[47] Aristizábal, N. Barrabés, S. Contreras, M. Kolafa, D. Tichit, F. Medina, J. Sueiras, Pt/CuZnAl mixed oxides for the catalytic reduction of nitrates in water: Study of the incidence of the Cu/Zn atomic ratio, *Phys. Procedia* 8 (2010) 44–48.

[48] A.E. Palomares, J.G. Prato, F. Rey, A. Corma, Using the "memory effect" of hydrotalcites for improving the catalytic reduction of nitrates in water, *J. Catal.* 221 (2004) 62–66.

[49] Alejandre, F. Medina, X. Rodriguez, P. Salagre, Y. Cesteros, J.E. Sueiras, Cu/Ni/Al layered double hydroxides as precursors of catalysts for the wet air oxidation of phenol aqueous solutions, *Appl. Catal. B-Environ.* 30 (2001) 195–207.

[50] Dubey, S. Kannan, S. Velu, K. Suzuki, Catalytic hydroxylation of phenol over CuM(II)M(III) ternary hydrotalcites, where M(II) = Ni or Co and M(III) = Al, Cr or Fe, *Appl. Catal. A-Gen.* 238 (2003) 319–326.

[51] Taguchi, F. Schüth, Ordered mesoporous materials in catalysis, *Micropor. Mesopor. Mater.* 77 (2005) 1–45.

[52] V.V. Guliants, M.A. Carreón, Y.S. Lin, Ordered mesoporous and macroporous inorganic films and membranes, *J. Membrane Sci.* 235 (2004) 53–72.

[53] E. Géraud, V. Prévot, F. Leroux, Synthesis and characterization of macroporous MgAl LDH using polystyrene spheres as template, *J. Phys. Chem. Solids* 67 (2006) 903–908.

[54] M. Halma, K.A. Dias de Freitas Castro, V. Prévot, C. Forano, F. Wypych, S. Nakagaki, Immobilization of anionic iron(III) porphyrins into ordered macroporous layered double hydroxides and investigation of catalytic activity in oxidation reactions, *J. Mol. Catal. A-Chem.* 310 (2009) 42–50.

[55] E. Géraud, V. Prévot, J. Ghanbaja, F. Leroux, Macroscopically Ordered Hydrotalcite-Type Materials Using Self-Assembled Colloidal Crystal Template, *Chem. Mater.* 18 (2006) 238-240.

[56] Y. Fu, Z. Jin, Z. Liu, W. Li, Preparation of ordered porous SnO_2 films by dip-drawing method with PS colloid crystal templates, *J. Eur. Ceram. Soc.* 27 (2007) 2223-2228.

[57] Z. Liu, Z. Jin, W. Li, J. Qiu, J. Zhao, X. Liu, Synthesis of PS colloidal crystal templates and ordered ZnO porous thin films by dip-drawing method, *Appl. Surf. Sci.* 252 (2006) 5002-5009.

[58] Y. Jia, C. Duran, Y. Hotta, K. Sato, K. Watari, Macroporous ZrO_2 ceramics prepared from colloidally stable nanoparticles building blocks and organic templates, *J. Coll. Interface Sci.* 291 (2005) 292-295.

[59] A.B.D. Nandiyanto, K. Okuyama, Progress in developing spray-drying methods for the production of controlled morphology particles: From the nanometer to submicrometer size ranges, *Adv. Powder Technol.* (2010), doi:10.1016/j.apt.2010.09.011.

[60] Y. Saito, H. Takao, T. Tani, T. Nonoyama, K. Takatori, T. Homma, T. Nagaya, M. Nakamura, Lead-free piezoceramics, *Nature* 432 (2004) 84–87.
[61] Chowdhury, J. Bould, Y. Zhang, C. James, S.J. Milne, Nano-powders of $Na_{0.5}K_{0.5}NbO_3$ made by a sol–gel method, *J. Nanopart. Res.* 12 (2009) 209–215.
[62] Y. Shiratori, A. Magrez, C. Pithan, Particle size effect on the crystal structure symmetry of $K_{0.5}Na_{0.5}NbO_3$. *J. Eur. Ceram. Soc.* 25 (2005) 2075–2079.
[63] Y.P. Fu, S. Tsao, C.T. Hu, Preparation of $Y_3Al_5O_{12}$:Cr powders by microwave-induced combustion process and their luminescent properties, *J. Alloys Compd.* 395 (2005) 227–230.
[64] Y.P. Fu, S.B. Wen, C.S. Hsu, Preparation and characterization of $Y_3Al_5O_{12}$:Ce and Y_2O_3:Eu phosphors powders by combustion process, *J. Alloys Compd.* 458 (2008) 318–322.
[65] D. Jia, Y. Wang, X. Guo, K. Li, Y.K. Zou, W. Jia, Synthesis and characterization of YAG:Ce^{3+} LED nanophosphors, *J. Electrochem. Soc.*154 (2007) J1–J4.
[66] Z. Na, W. Dajian, L. Lan, M. Yanshuang, Z. Xiaosong, M. Nan, YAG:Ce phosphors for WLED via nano-pseudoboehmite sol–gel route, *J. Rare Earths* 24 (2006) 294–297.
[67] Y. Pan, M. Wu, Q. Su, Comparative investigation on synthesis and photoluminiscence of YAG:Ce phosphor, *Mater. Sci. Eng. B* 106 (2004) 251–256.
[68] Ikesue, I. Furusato, K. Kamata, Fabrication of polycrystalline, transparent YAG ceramics by a solid-state reaction method, *J. Am. Ceram. Soc.* 78 (1995) 225–228.
[69] J. Zárate, R. López, E.A. Aguilar, Synthesis of yttrium aluminum garnet by modifying the citrate precursor method, *Adv. Tech. Mat. Mat. Proc.* 7 (2005) 53–56.
[70] Y. Liu, L. Gao, Low-temperature synthesis of nanocrystalline yttrium aluminum garnet powder using triethanolamine, *J. Am. Ceram. Soc.* 86 (2003) 1651–1653.
[71] Q. Lu, W. Dong, H. Wang, X. Wang, A novel way to synthesize yttrium aluminum garnet from metal–inorganic precursors, *J. Am. Ceram. Soc.* 85 (2002) 490–492.
[72] K.R. Han, H.J. Koo, C.S. Lim, A simple way to synthesize yttrium aluminum garnet by dissolving yttria powder in alumina sol, *J. Am. Ceram. Soc.* 82 (1999)1598–1600.
[73] S.A. Hassanzadeh-Tabrizi, E. Taheri-Nassaj, H. Sarpoolaky, Synthesis of an alumina-YAG nanopowder via sol–gel method. *J. Alloys Compd.* 456 (2008) 282–285.
[74] M.K. Cinibulk, Synthesis of yttrium aluminum garnet from a mixed-metal citrate precursor, *J. Am. Ceram. Soc.* 83 (2000) 1276–1278.
[75] H.G. Jung, Y.H. Cheong, I.D. Han, S.J. Kim, S.G. Kang, Lowtemperature fabrication of polycrystalline yttrium aluminum garnet powder via a mechanochemical solid reaction of nanocrystalline yttria with transition alumina, *Solid State Phenom.* 135 (2003) 7–10.
[76] X. Li, H. Liu, J. Wang, H. Cui, X. Zhang, F. Han, Preparation of YAG:Nd nano-sized powder by co-precipitation method, *Mater. Sci. Eng. A* 379 (2004) 347–350.
[77] M. Shojaie-Bahaabad, E. Taheri-Nassaj, R. Naghizadeh, Effect of yttria on crystallization and microstructure of an alumina-YAG fiber prepared by aqueous sol–gel process, *Ceram. Int.* 35 (2009) 391–396.
[78] Towata, H.J. Hwang, M. Yasuoka, M. Sando, K. Niihara, Preparation of polycrystalline YAG/alumina composite fibers and YAG fiber by sol–gel method, *Compos. Part A-Appl. S.* 32 (2001)1127–1131.
[79] S. J. Lukasiewicz, Spray-Drying Ceramic Powders, *J. Am. Ceram. Soc.* 72 (1989) 617-624.

[80] S. Kannan, V. Rives, H. Knözinger, High-temperature transformations of Cu-rich hydrotalcites, *J. Solid State Chem.* 177 (2004) 319–331.
[81] L. Chmielarz, P. Kustrowski, A. Rafalska-Łasocha, R. Dziembaj, Influence of Cu, Co and Ni cations incorporated in brucite-type layers on thermal behaviour of hydrotalcites and reducibility of the derived mixed oxide systems, *Thermochim. Acta* 395 (2003) 225–236.
[82] S. Kannan, Decomposition of nitrous oxide over the catalysts derived from hydrotalcite-like compounds, *Appl. Clay Sci.* 13 (1998) 347–362.
[83] H. Du, Z. Li, F. Tang, S. Qu, Z. Pei, W. Zhou, Preparation and piezoelectric properties of $(K_{0.5}Na_{0.5})NbO3$ lead-free piezoelectric ceramics with pressure-less sintering, *Mater. Sci. Eng. B* 131 (2006) 83–87.
[84] H. Du, F. Tang, F. Luo, D. Zhu, S. Qu, Z. Pei, W. Zhou, Influence of sintering temperature on piezoelectric properties of $(K_{0.5}Na_{0.5})NbO_3$–$LiNbO_3$ lead-free piezoelectric ceramics, *Mater. Res. Bull.* 42 (2007)1594–1601.

In: Sprays: Types, Technology and Modeling
Editor: Maria C. Vella, pp. 209-227
ISBN 978-1-61324-345-9

Chapter 5

CONTROL OF ATMOSPHERIC PLASMA SPRAY PROCESS: HOW TO CORRELATE COATING PROPERTIES WITH PROCESS PARAMETERS?

Abdoul-Fatah Kanta **[*]** ***and Chao Zhang***
Service de Science des Matériaux, Faculté Polytechnique,
Université de Mons, Mons, Belgium

ABSTRACT

The microstructure and in-service properties of plasma-sprayed coatings are derived from an amalgamation of intrinsic and extrinsic spray parameters. These parameters are interrelated, which follow mostly non-linear relationships. The interactions among the spray parameters make the optimization and control of this process quite complex. Understanding relationships between coating properties and process parameters is mandatory to optimize the spray process and ultimate product quality. Process control consists of defining unique combination of parameter sets and maintaining them as a constant during the entire spray process. This unique combination must take the in-service coating properties into consideration. Artificial intelligence is a suitable approach to predict operating parameters of atmospheric plasma spray to attain required coating characteristics.

Keywords: Atmospheric plasma spray; process control; system expert; artificial neural networks, fuzzy logic controller

---

* Corresponding author: 56 rue de l'Epargne, 7000 Mons, Belgium. Tel. +32 65 37 44 47, fax. +32 65 37 44 16, e-mail: abdoul.kanta@umons.ac.be.

1. INTRODUCTION

Thermal spray is a technique which permits to deposit coatings onto components to confer or enhance their functional or mechanical properties, such as wear and corrosion resistances, thermal and electrical insulation, biocompatibility and catalytic ability [1]. The coating may be metal alloys, ceramic carbides or oxides, thermoplastic polymer or any other associated composites. Atmospheric plasma spray (APS) is one of the most commonly used thermal spray processes [2]. The case of this specific process is considered in this study.

Atmospheric plasma spray is a complex high-temperature process, which has many spray parameters could influence the ultimate coatings. Spray parameters can be divided into feedstock material parameters, feedstock injection parameters, kinematics parameters, substrate/coating parameters, environmental parameters and energy parameters [3]. The quality of the atmospheric plasma-sprayed coatings [4] directly deriving from their microstructure are indirectly linked to the spray parameters. Take energy parameters as an example, they directly determine the thermodynamic characteristics of plasma jet. And, the coating properties depend on the in-flight particle characteristics (velocity, temperature and size) at impact which are influenced by the thermodynamic characteristics of plasma jet as well as the others spraying parameters.

The spray parameters are interrelated, *via* complex – non-linear – relationships. Understanding and controlling these correlations is essential for a robust quality control of the process. Indeed, a better robustness and higher performances are recurring themes for whatever manufactured systems. Parametric drifts and fluctuations occur during atmospheric plasma spray. These drifts and fluctuations originate mainly from the electrode wear and intrinsic plasma jet instabilities: the plasma net power varies in this case modifying significantly its thermodynamic properties and therefore modifying the momentum and heat transfers to the particles. It is possible to control the in-flight particle characteristics by adjusting continuously the operating parameters, in particular the energetic parameters. Due to the large amplitudes of these drifts and fluctuations, the strategy to adopt will depend on the required corrections to apply to the particle characteristics.

Developing a reliable controller requires: (i) implementation of sensors to accurately diagnose in-flight particle characteristics; (ii) development of a robust command to insure the stability of the control system, (Fuzzy logic permits to define parametric correction rules and the command can be based on these algorithms); (iii) linking of the robust command to a predictive model (Artificial neural networks, among many artificial intelligence protocols such as genetic algorithms, is proved to be able to predict in-flight particle and coating characteristics); (iv) validation of the corrections with an extensive database used as a reference.

This paper first summarizes the main spray parameters involved in atmospheric plasma spray as well as the drifts and the fluctuations associated with this process. Then, the principle techniques of diagnostic and control are reviewed.

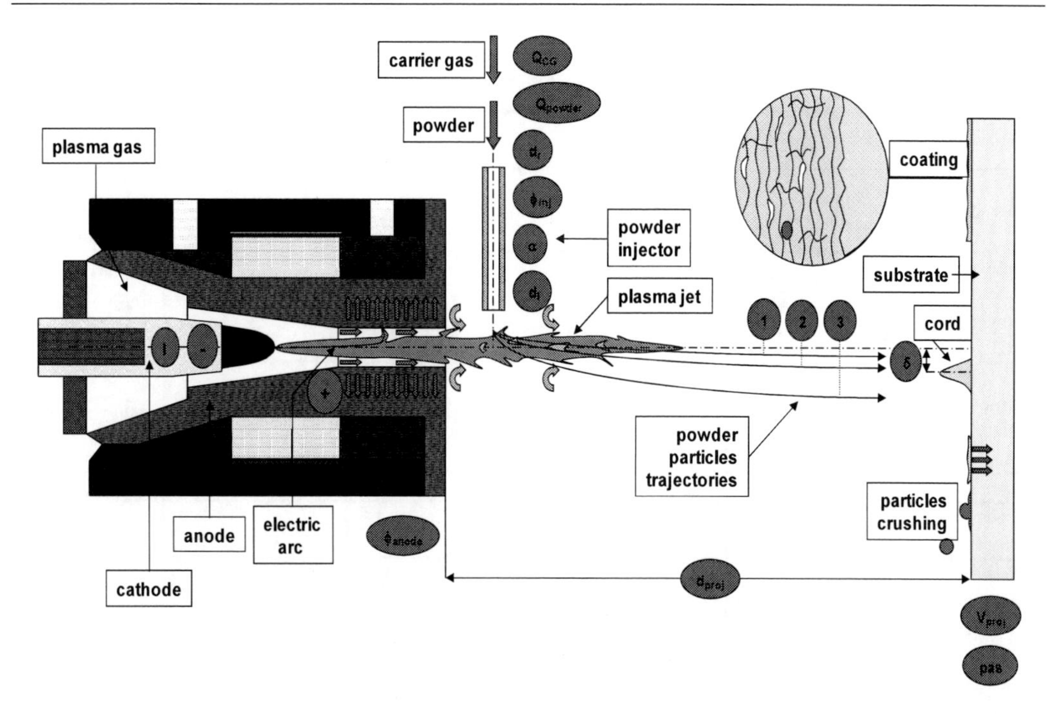

Figure 1. Scheme of atmospheric plasma spray [6].

2. PRINCIPLE OF ATMOSPHERIC PLASMA SPRAY

There are a number of variations of thermal spray processes. Among them, the most common are twin-wire arc spraying (TWAS), atmospheric plasma spraying (APS), low pressure plasma spraying (LPPS), flame spraying (FS), high velocity oxygen-fuel spraying (HVOF), detonation gun (D-Gun). The main characteristics of these processes are summarized in Table 1 [5]. The atmospheric plasma spraying is the case reviewed in this study. Figure 1 illustrates the basic principle of the APS process [6].

The APS uses a plasma torch as an heat source to melt and accelerate feedstock powders [1, 7, 8]. Tubular injector allows injection of the material to be sprayed in the form of powder in the plasma jet. The powder is carried by a feed gas. An electrical arc with high intensity is generated between two electrodes (the electrodes are cooled by circulating water) to ionize the gas mixture (argon, hydrogen, helium or nitrogen) mainly following electron avalanche mechanism. The created flow is electrically neutral and consists of a mixture of electrons, ions and also of excited molecules and atoms since ionization is not totally complete. APS coating is formed by a stream of total/partial molten droplets impacting on a substrate. After impact, the individual droplet spreads and solidifies to a thin lamella. The stacking of lamellae constitutes the coating. A plasma-sprayed coating is generally of lamellar structure. Pores are always present in the obtained coatings. A fraction of voids from several percent to 20% can be formed in the deposit [9]. Some of the voids result from insufficient filling and incomplete wetting of molten liquid to previously formed rough deposit surfaces. The micro-cracks can be formed easily in the splat of ceramic materials because of quenching stress that occurs in the splats. Such micro-cracks are also a kind of void that appears in the coating and constitutes a fraction of porosity.

Table 1. Characteristics of the different variations of thermal spray [10]

Characteristics	Flame-powder	Flame-wire	Electric arc	Atmospheric plasma	Supersonic flame	Detonation gun
heat source	flame oxy-acetylene	flame oxygen-acetylene	electric arc	plasma torch	flame	flame oxygen-acetylene
approximate maximum temperature (°C)	3000	3000	6000	12000	3000	3000
average particle velocity ($m.s^{-1}$)	40	150	250	200	700	950
material form	powder	wire or cord	wire	powder	powder	powder
deposition rates ($kg.h^{-1}$)	1-3	1 to 10	5-15	1-4	3-5	3-5
deposition efficiency (%)	50	70	80	70	70	70
average adhesive strength (MPa)	20-40	20-40	40	30-70	50-80	50-80
porosity (%)	2-10	2-10	1 to 10	1 to 10	0.5 to 2	0.5 to 2
example of deposited materials	• alloys • oxides • Polymers	• alloys • oxide • carbide in metal matrix	• alloys	• alloys • oxides • carbides	• alloys • oxides • carbide in metal matrix	• carbides • alloys in metal matrix

3. Process Parameters

Process parameters of thermal spray include part or all of the steps of surface treatment such as the surface preparation, the definition of the torch/piece, the realization of the deposit and the final machining [4]. Figure 2 summarizes the main process parameters in APS process [11]. These parameters are interdependent. Some of them can be controlled by the operator (extrinsic parameters) such as arc current intensity, plasma gas flow rates and gas compositions. Others are not directly accessible by the operator (intrinsic parameters), which include the voltage at the extremity of the electrodes, the effectiveness of the cooling, the wear of electrodes, etc.

The APS process is characterized by random phenomena and sometimes it can be unstable. The causes of these instabilities are numerous: arc electric instability in the nozzle depending in particular on anode wear and plasma forming gases, instabilities in the powder distributions, drifts from current, drifts from plasma gas flow rates due to potential leakage, random deviation of the particle trajectories due to an accumulation of powder on the injector tip, etc.

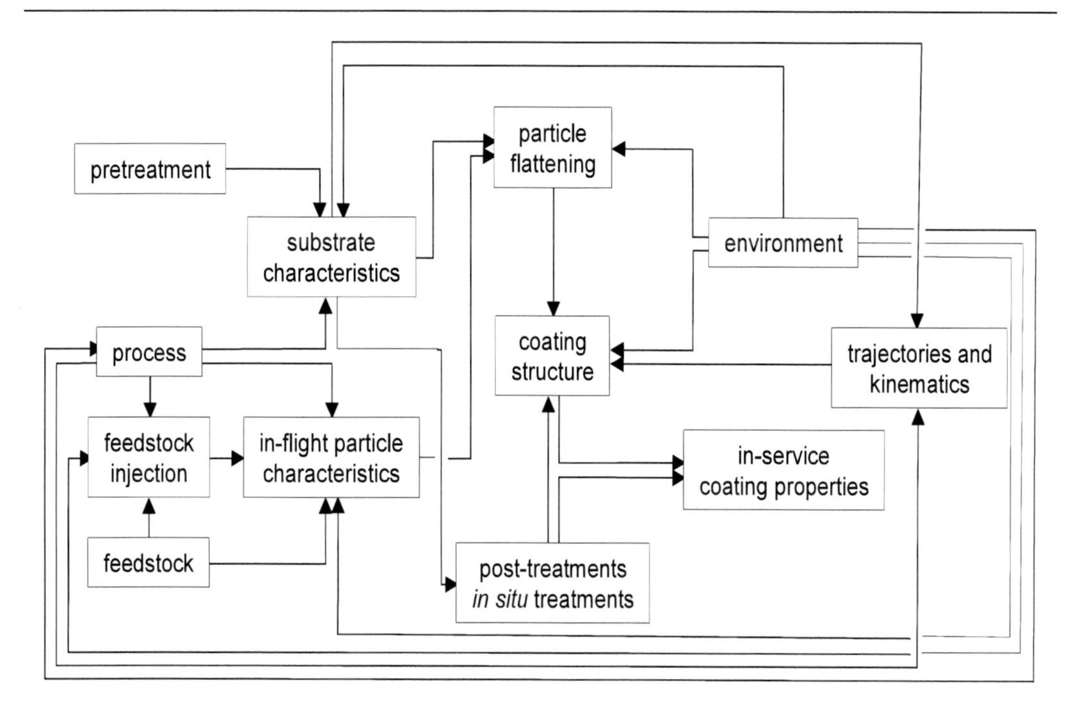

Figure 2. Interaction between the different parameters involved in APS process [11].

3.1. Feedstock Material Parameters

Feedstock materials are mainly characterized by their chemical composition, their shape (the mode of elaboration) and their size (particle size distribution). The powders are elaborated following different methods. These methods are chosen to get the specific powder morphology for the application based on the physico-chemical characteristics of materials and economic criteria. The most commonly used methods for the ceramic powders are agglomeration-drying and fusion-grinding.

The feedstock material characteristics are of great importance in determining energetic and injection parameters, which control the efficiency of heat transfer and deposition efficiency. Melting and vaporization temperatures have crucial importance because the spray material must melt without decomposition (the difference between the melting temperature and vaporization temperature must be greater than or equal to 200° C to reasonably envisage to elaborate layers of this material by APS process). Thus the heat source must provide sufficient heat enthalpy to the material so that it can achieve melting (thermal transfer) and produce a gas jet capable to accelerate the particles toward the substrate (kinetics transfers).

3.2. Powder Injection Parameters

Feedstock injection parameters permit to optimize the feedstock injection into the plasma jet core. The powder injection parameters control the trajectory of particles in the plasma jet and therefore transfers, particularly influence the heat transfer from plasma to in-flight particles. The extrinsic parameters mainly concern the distance, angle and internal diameter of the injector, carrier gas nature, carrier gas flow rate and feedstock feed rate. These parameters allow adjusting the momentum and trajectory of the particles when they penetrate the plasma jet.

The injector is generally perpendicular to the plasma jet, with a reduced internal diameter (1.2-2 mm) to limit the dispersion of the powder. The position of the injector output (i.e. the distance from the nozzle output to the axis of the torch) is also important since it influences the trajectory of the powder and the heat-treatment of the powder by plasma.

The powder feed rate has an influence on the efficiency of spray. The powders with a diameter less than 5 μm also present difficulties to flow properly. The problems of small particle injection led to consider the use of specific distributors [12].

The choice of an internal diameter of injector coupled with the carrier gas allows adjustment of the initial velocity of particle ejections. The momentum communicated to the particles affects their trajectory in the plasma jet and thus their in-flight characteristics [13]. They must be adjusted to enable an optimal stay of particles in the plasma and get the best efficiency of spray.

The distance between the powder injector and axis of the torch allows adjusting the injection characteristics (diameter of the injector, carrier gas flow, etc.), the width of the cone particle and their penetration in the plasma jet. A long distance of injection is accompanied by a large dispersion of particles in the jet of plasma. The distance of particle injection also depends on powder size. The smaller the powder size, the higher injection velocity should be employed to maintain their momentum to an appropriate value which permits their introduction in the plasma jet. It is commonly accepted that injection parameters should be adjusted to obtain a trajectory forming an angle of 3.5° to axis of the torch. This trajectory is considered as ideal, which permits to obtain maximum velocity and surface temperature of the particles [14].

3.3. Kinematics Parameters

Kinematics parameters (e.g. gun transverse velocity and scanning step, spray distance, spray angle) govern the heat flux into the part to be covered *via* the latent heat transmitted to the solidification of lamellae and *via* the heat flux transmitted by the warm recombined plasma gases diluted with the surrounding atmosphere gases [15].

Among these parameters, the spray distance is the parameter which is easily adjustable to vary the state of particle at impact onto substrate. Indeed, heat transfer between environment and aerodynamic braking of particles modifies the characteristics of particles at impact. The control of the spray distance permits to modify the velocity and temperature of particles at impact and therefore the structure of the coating.

Spray angle has also an influence on the profile of the deposited splats [16, 17] and consequently on coating structure, particularly in terms of porosity. It should also be noted that deposition efficiency decreased significantly as the spray angle becomes less than a critical value.

Relative velocity between torch and substrate influences the radial geometry of formed splats as well as on the characteristics of the heat flow transmitted to the substrate by latent heat of the particles solidification and their cooling during the spray [18].

Scanning step has identical influences on the quality of the deposit material and on the heat flux transmitted to the substrate. Also, it has an influence on the homogeneous distribution of the deposit material in macroscopic scale [19].

3.4. Parameters Relative to the Coating-Substrate Interaction

Coating-substrate interaction parameters mainly concern the average temperature of the substrate on which particles are deposited, the physiochemical properties of the substrate surface, the surface state at nano and micrometric scales as well as thermal and mechanical characteristics of substrate, or composite of substrate and deposit, or deposit when a layer with significant thickness already been formed.

At the macroscopic scale, the control of the temperature of the deposition and the piece to be coated allows to control the level of residual stresses in the coating [20]. Indeed, during deposition, the particles heat is transmitted to substrate during their solidification and cooling [21]. During cooling, the difference of thermal expansion coefficient between the substrate (often metallic nature) and the coating (often ceramic nature) leads to the development of so-called thermal residual stresses. These constraints increase with the coating thickness. The consecutively solidification and rapid cooling by oriented conductive transfer may lead to the delamination of the deposit from the part when the level locally exceeds the level of adherence. Studies have shown that particles spread better, i.e. with a largest crash rates on a hot substrate [22, 23, 24]. The preheating of the substrate surface causes a slight superficial oxidation (therefore a change in surface physico-chemistry as well as in its nanoscale topology) and desorption of molecules such as water or any other adsorbed contaminants [25, 26, 27]. Thus, critical temperatures of substrate are defined below which half of the collected lamellae present with a significant splash rate [28, 29]. These splash (flattening splashing), which originally derived from the premature local solidification in contact with the substrate disrupting the viscous flow, lead to the formation of pores at the interface [30]. The nature of the substrate influences the force of liquid. These forces affect the wetting of the substrate by liquid and thus the quality of the spreading splats. They also influence the adherence quality of substrate/deposit [31, 32].

3.5. Environmental Parameters

The spray environment is also one of parameters which influence the characteristics of the coatings elaborated by APS process. This parameter is commonly adjusted using forced cooling by compressed dry air and in some cases using vaporization of cryogenic liquids [33].

Due to the high viscosity of the plasma jet coupled with its high velocity of ejection, the atmosphere (cold) surrounding the jet is driven by pumping in the jet [33]. When atmosphere is air, as in the case of APS, the mixture of cold air with plasma leads to decrease in velocity and temperature of the jet due to the consumed enthalpy in order to dissociate a large number of molecules, e.g. oxygen at 3500 K, a small fraction of nitrogen at 7500 K and a high turbulence at the jet periphery. A first consequence, in the treatment of in-flight particles, is lower heat and momentum transfers [34]. A second consequence could be the development of chemical reactions of oxidation and/or nitriding as a function of the nature of the introduced particles. Some of these reactions may be continued when particles leave the plasma jet (generally defined by gas temperature below 8000 K) prior to reach the substrate [33, 35]. In some cases, it is also possible to observe preferential vaporization of metal alloys or elements of ceramics as a function of elements which constitute plasma jet, oxygen in particular. For example, particles of zirconia partially stabilized with yttria can impact the substrate in a sub-stoechimetric state with vaporization of the oxygen ions. The transformation of tungsten carbide (WC) to carbide (W_2C) with decarburization is another significant example of this phenomenon.

Although the pressure and atmosphere is not adjustable in APS process, it is easily controllable in the chamber of low pressure plasma spray (LPPS) system. It permits to prevent or greatly limit the oxidation of metal powders and substrates when the atmosphere is based on neutral gas. There are other benefits when using a controlled atmosphere in low pressure, which includes a lower cooling of plasma jet, an increase of plasma jet length and a prolongation of the isotherms and velocities.

At pressure below atmospheric pressure, the expansion of the plasma jet is much greater than at atmospheric pressure. This permits to obtain more important momentum transfers from the plasma jet to the particles. In some cases, it can improve the spreading of particles and consequently density and adhesion of the coating to the substrate. However, the low density of jet energy reduces the heat enthalpy transferred to the in-flight particles. Meanwhile, convective exchanges between the substrate and the surrounding atmosphere are also more limited and this induced a higher heating of the substrate. In contrast, there also exists high pressure plasma spray system. Over-pressure of surrounding atmosphere conducts to lower expansion of plasma. One observed thus the increase of energy density inside the jet, which should improve the heating of in-flight particles. The surface temperature of particle may in some cases approach to vaporization temperature of the material whereas the particles velocity is low. [36]. These particular conditions permit, in some specific cases, to obtain coatings which are dense [37]. However, the gains obtained are not always significant and there are more difficulties of implementation. This is why these working conditions are less used in practice.

3.6. Energetic Parameters

Energetic parameters (i.e., arc current, nature and flow rates of plasma forming gases) govern thermodynamic properties (enthalpy) and transport coefficients (thermal conductivity, viscosity, etc.) of the plasma jet. They should be optimized to transfer appropriate momentum and heat to feedstock particles [1, 38]. These parameters permit to control the temperature and the velocity of the in-flight particles by adjustment of the plasma jet characteristics. These adjustable parameters, for fixed electrodes and injection mode of plasma forming gases (longitudinal or radial injections are most commonly used), are arc current, the mass flow rates and nature of plasma forming gases. Arc current must be chosen judiciously. Low current leads to inadequate melting of particles and high current may result in vaporization of particles [39].

Heat and momentum transfers from the plasma jet to the particles depend on the duration of interaction and the nature of the plasma forming gases. Good transfers are required to obtain the largest number of particles completely melted when their impact onto substrate. In most cases nowadays, primary plasma gas is argon because it forms stable viscous plasma (μ_{10000K} = 2.7 kg.m^{-1}.s) and easily to be ionized. It enables better stabilization of electric arc in comparison with nitrogen for example and brings momentum to the jet. However, the argon provides low thermal conductivity (k_{10000K} = 0.6 W.m^{-1}.K). Hydrogen is often used as a secondary plasma gas mixed with argon [40]. Hydrogen mixed to argon allows to increase the enthalpy and thermal conductivity of plasma and thus to improve thermal transfer between plasma and particles [41]. However, in some cases, excessive heat can result in a decrease of thermal transfer due to the formation of peripheral vapor cloud [42]. The helium is also often used. The addition of helium increases the plasma jet velocity owing to helium was light and easily accelerated. Meanwhile, the high viscosity of helium (μ_{15000K} = 4.3 kg.m^{-1}.s) will increase the plasma viscosity and enhance the acceleration effect of powder. As a result, the particle velocity increased [43]. Because of its high thermal conductivity (k_{10000K} = 2.4 W.m^{-1}.K) compared with argon, it could increase plasma heat enthalpy of argon plasma [44]. As a result, the particle temperature may maintain stable due to helium's dual effect of plasma's enthalpy increase and particle's dwelling time reduction.

4. Arc Root Fluctuations and Instabilities

The APS torch transfers the energy of electric arc, which is created between the cathode and the anode, to plasma forming gases. Plasma is sensitive to various instabilities which differ by several magnitudes. The conditions of powder injection (position and diameter of the injector, angle to the axis of the torch, carrier gas flow rate) have influences on in-flight particles when the powder nature and grain size are fixed. The conditions of powder injection should be optimised in order to give the powder a momentum at the point of injection adapted to the plasma and thus ensure a good penetration in the plasma jet. The composition and their proportion of plasma forming gases have a direct effect on the regime of operating torch [45]. Jet instability manifests by deviation and dispersion of the particles' trajectories due to the momentum variation of the plasma jet.

Electric arc is composed of a main arc column, fixed at the extremity of the cathode which is the source of electrons, and a column of connection which ends by arc root on the anode surface. The arc root moves permanently on the anode surface in back and forth movement with an extension phase of downstream from the anode, interrupted by a breakdown of arc upstream or downstream of the previous arc root. The fluctuations of arc are translated by fluctuating voltage at the extremities of the electrodes.

The voltage between the electrodes is the sum of the cathodic fall, anodic fall and arc voltage. It is not possible to decouple these three components by measuring the voltage. Often the sum of the cathodic and anodic falls is assumed to be constant that only variation of arc voltage is responsible for the overall voltage variation at the electrode extremities.

Studies [46, 47] revealed the effect of the fluctuations of the arc position on particle treatments: increase in the length of the arc causes an acceleration and better gas heating. Wear, due to the stagnation of the arc root on the anode, gradually widening the anodic wall and after a few dozen hours of operating, arc root fluctuation increases with anode wear making wear exponential increase. The melting of the wall where arc root clings creates asperities which constituted preferential attachment points for the arc root.

The arc fluctuation has an effect on the particle treatments. This effect translates into spatial and temporal variation of particles characteristics (trajectory, velocity, temperature, melting state, etc.), which may be harmful to the homogeneity and reproducibility of the deposits realized by APS process. In addition, the difference in time scales between arc root fluctuations (from 100 to 500 μs) and the development of erosion (dozens of hours) makes measure complexes.

5. Electrode Erosions

For use in industrial environments, the anode life is approximately 30-60 h [48]. The duration depends on the operating conditions (arc current intensity, plasma gas flow rate and nature, etc.), the number of the torch startups (in order of 300 to 400 for 30 to 50 h) and how these startups are realized.

Anode, collector of electrons, is subjected to high thermal flows. The temperature of these thermal flows can be higher than those incurred by the cathode which is cooled by the emission of electrons [49, 50].

Studies on cathode (which provide electrons to the arc) demonstrate [51] that:

- the shape and the size of cathode have low influence on the characteristics of the arc but strongly affect the erosion of the cathode;
- they have little influence on the temperature of cathodic spot which is mainly controlled by the current density and the nature of the material. However, large diameter of cathode leads to a higher temperature gradient and a more limited cathodic spot;
- The temperature in the arc is often higher than that at the edge of this spot, therefore the erosion is most important;

When the cathode is new, its extremity is smooth and sharp. It follows therefore high velocity gradient with a maximum relatively elevated axis. As time elapses, the cathode extremity becomes rounded due to the material melting. One notes the attenuation of maximum velocities and distribution of more homogeneous velocities.

Cathode erosion is mainly determined by the characteristics of the arc (i.e. the current) and the cathode material [52, 53]. High hydrogen proportion in the $Ar+H_2$ plasma forming gases induces higher temperature at the extremity of cathode. With arc current higher than 500 A, the electronic emission is predominant mechanism of the cooling in the centre of cathodic spot and changes in cathode geometry (widest diameter for example) have only a minimal effect on the temperature of the cathode extremity.

So far, there are no methods to measure directly the erosion of the electrodes. To estimate the effectiveness of the torch and correct the effects of the electrode erosions, one uses techniques based on online control which mainly capture in-flight particle characteristics.

Erosion of the electrodes induced the decrease of plasma enthalpy. Therefore the temperature of in-flight particles decreased. This can change the final properties of the deposit or lead to its destruction by incorporation of the electrode particles.

The *modus operandi* [54] of a plasma torch can be characterized by its signal voltage. The shape and the fluctuation amplitudes of tensions permit to estimate the movement and therefore the electric arc dynamic state following operating conditions.

6. On-Line Process Control

Coating quality can be assessed by its structural characteristics (porosity, percentage of oxides, thickness, residual stresses, etc.), physicochemical and thermomechanical properties (bond strength, hardness, module of Young, thermal conductivity, etc.) and by its in-service properties (corrosion resistance, wear resistance, etc.). All of these characteristics and properties are conditioned by particle characteristics at impact and the physicochemical properties of substrate. Relationship between operating parameter and the coating properties allows defining optimal operating window. Controllers permit to maintain the operating parameters in these ranges.

6.1. Measurement Apparatus

To control the process, several one-line techniques of diagnostic capable to operate in the thermal spray conditions have been developed [55]. These techniques allow to characterize the state of in-flight particle (velocity, surface temperatures and diameter), or to monitor the deposit or substrate temperatures or to measure the coating thickness after each pass. These techniques are based essentially on the technology of CCD cameras and/or pyrometry. There are many on-line monitoring systems:

- DPV2000 (TECNAR[1]) [56] permits to obtain particles velocity (measurements of flight time), their temperature (by bichromatic pyrometry) and to evaluate their diameter (from the thermal emission of particles). It also follows the evolution of the particles jet position through a CCD camera;
- Accuraspray (TECNAR) [57] allows the control of temperature and velocity of particles respectively by two colors pyrometry (T > 900°C) and the measurement of the flight time;
- Plumespector (TECNAR) [58] allows to control of the jet of hot particles with a CCD camera;
- IPP - In-Flight Particle Pyrometer [59]. The measured temperature is an average value of all particles that pass through a large volume (5 mm of diameter and 50 mm length). It is a global measure;
- Spray-Watch (OSEIR[2]) [60]. It is composed of black-and-white rapid shutter CCD camera and a two colours pyrometer which respectively provide particles velocity using the lengths of the traces left by in-flight particles during a known time of exposure at high temperature (T > 1300°C);

[1] Tecnar, 1321 Hocquart Street, St-Bruno, Québec J3V 6B5, Canada.

[2] Oseir Ltd, Hermiankatu 6A, FIN-33720 Tempere, Finland.

- Ignatiev et al. [61] developed a CCD camera which is capable to measure size, velocity and temperature of in-flight hot particles (between 1200 and 3500°C);
- STRATONICS[3] [62] proposed a two colors pyrometer using a CCD camera and which allows to measure temperature of particles between 1000 and 2700 K;
- SDC[4] (Spray Deposit Control) [63]. It consists of a CCD camera provides distribution of heat flow associated with in-flight warm particles. The velocity of the particles can be determined by traces left over a known time of exposure. A pyrometer measures the temperature of the substrate and the coating under construction. Finally the measurement of the displacement of a witness sample (deflection of the beam) permits to deduce thermal stresses and tempering stresses to which it is subjected during the steps of preheating, deposition and cooling.

Utilization of these devices in practice depends on continuous or sequentially monitoring of measured characteristics and their comparison with reference values previously established during the optimization of the operating conditions as to develop coatings with desired characteristics and properties. Admissible variation ranges are defined during the optimization of these operating conditions. When parametric drift is detected and if it is in the range of admissible variation, (i) operator adjusts the process parameters based on his experience or based on formalized rules, (ii) no adjustment is made. In the case of process parameters adjustment, fluctuations in deposit characteristics and properties which vary as a function of adjustment pertinence are observed, even if they remain confined to admissible variations. In the second case, variations in deposit characteristics and properties which are amplified during time are recorded and actions are decided therefore that measured values differ from the admissible window, the deposit process is stopped. Causes is searched and corrective actions are initiated (e.g. change the electrodes of plasma torch, etc.).

6.2. New Process Control Concept

Recently artificial intelligence algorithms are proposed to predict and adjust some operating parameters taking into account drifts and fluctuations which can be diagnosed by in-flight particle characteristics. Due the complexity of the synergism of operating parameter, the study is confined mainly on in-flight particle diagnosis (surface temperature and velocity at fixed spray distances) and their adjustment to their reference values by acting on the power parameters, Figure 3.

During control process, the priority is to adjust the current intensity initially, then the hydrogen rate and finally the total plasma flow rate. According to the conditions (in-flight particle characteristics variations), only one of these three power parameters or the combination of the two parameters could be adjusted.

---

[3] Stratonics Inc., 23151 Verdugo Dr. #114 Laguna Hills, CA 92653, USA.

[4] SPCTS-UMR CNRS 6638, Université de Limoges, 123 Avenue Albert Thomas, 87060 Limoges cedex, France.

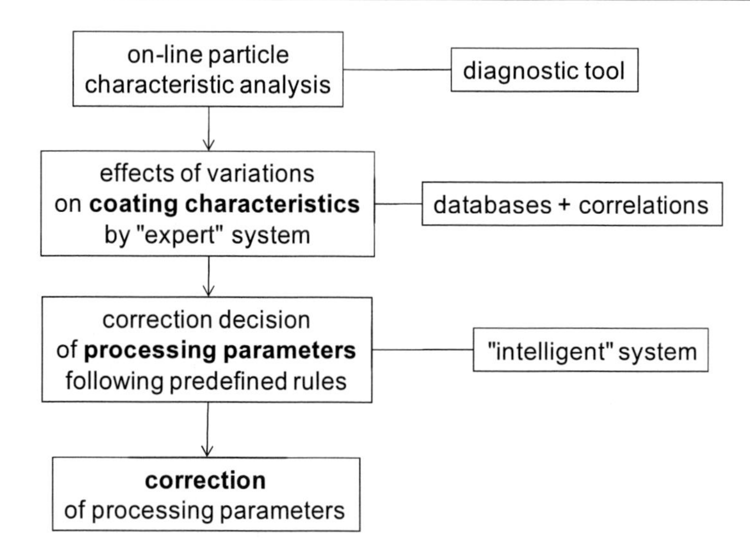

Figure 3. Process control requirement.

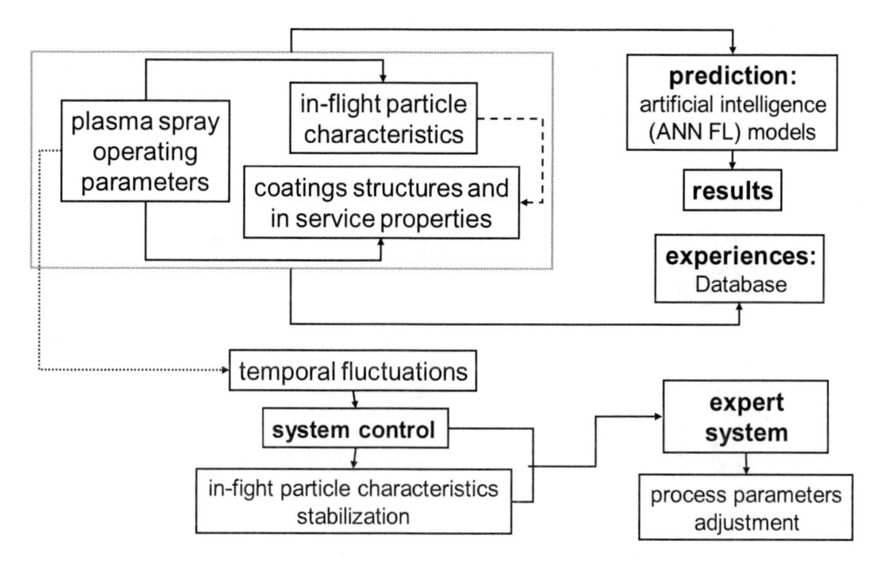

Figure 4. Control flow chart.

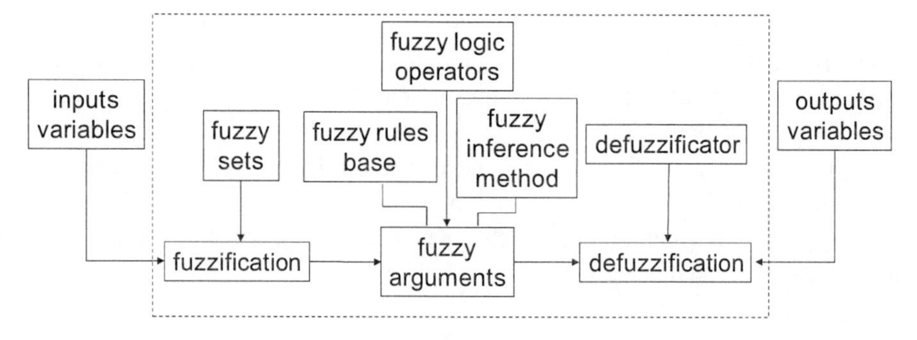

Figure 5. Architecture of the fuzzy logic controller.

The methodology and key process steps should be validated by experimental measurements. Figure 4 summarizes the control basis. The system is based on the cooperation of two models based on artificial intelligence (AI): artificial neural networks (ANN) to predict the desired parameters (what adjustments should be made to move the parameters towards the target values?) and fuzzy logic (FL) to define the strategy for controlling the power parameters (which parameters act and with what amplitudes?).

Artificial neural networks proved to be a pertinent tool to predict parameters and coating structural attributes from the knowledge of processing parameters [64]. This approach has been adopted to estimate the adjustments. Fuzzy logic allows to model the imperfections by taking into account the intermediate states between "all and nothing" [65]. Fuzzy logic controller (FLC) requires some numerical parameters in order to operate according to what is considered as significant error and as significant rate-of-change-of-error. Nevertheless, exact values of these numbers are usually not critical unless very responsive performance is required in which case empirical tuning would determine them. Any sensor data that provides some indication of system actions and reactions is sufficient.

The purpose of control is to vary the behaviour of a system by changing an input or inputs of this system according to a rule or a set of rules that model how the system operates [66].

To develop the fuzzy logic methodology, diverse steps are essential as shown in Figure 5. Since the FLC processes use defined rules governing the target control system, it can be modified and tweaked easily to improve or drastically alter the system performances. A fuzzy expert system is an expert system that uses a collection of fuzzy membership functions and rules. Fuzzy knowledge-based controller is used to compute the mapping from the input values to the output values, and typically consists of three sub-processes, namely fuzzification, inference and defuzzification [67].

7. OUTLINE

This paper gives a review on the progress of the process control of atmospheric plasma spray. Firstly, the process parameters involved in APS process are reviewed. The structure and in-service properties of plasma-sprayed coatings depend on thermodynamic properties and transport coefficients of the plasma jet which are controlled by the process parameters. The process is also characterized by several parametric drifts and fluctuations at different characteristic times, especially, due to the electrode erosion and intrinsic plasma jet instabilities. Secondly, the efforts to establish relationships between operating parameters and the coating properties are introduced. Methodologies based on artificial intelligence appear as the more robust ones to define and control the complex correlations between in-flight particle characteristics and the power parameters. The developed expert system allows full process control on the basis of pre-defined rules. The rules control the "real time" adjustment of the power parameters so that constant values for the in-flight particle characteristics can be maintained. The required correction, which depends on measured magnitudes of drifts and fluctuations, can be implemented.

REFERENCES

[1] B. Normand, V. Fervel, C. Coddet, V. Nikitine, Tribological properties of plasma sprayed alumina-titania coatings: role and control of the microstructure, *Surface and Coatings Technology*, 123 (2000) 278-287.

[2] P. Fauchais, A. Vardelle. Heat, mass and momentum transfer in coating formation by plasma spraying, *International Journal of Thermal Sciences*, 39 (2000) 852-870.

[3] I.A. Fisher, Variables Influencing the Characteristics of Plasma-Sprayed Coatings, *Inter. Met. Rev.* 17 (1972) 117-129.

[4] P. Fauchais, J.F. Coudert, A. Vardelle, M. Vardelle, A. Grimaud, P. Roumilhac, State of the art for the under-standing of the physical phenomena involved in plasma spraying at atmospheric pressure, Thermal Spray: Advances in Coatings Technology, September 14-17, 1987, Orlando, USA, Ed. D.L. Houck, ASM International, Materials Park, OH. USA, 1988, pp. 11-19

[5] A. Proner, Revêtement par projection thermique, Techniques de l'Ingénieur, traité Matériaux métalliques, M 1 645, pp. 6.

[6] A.-F. Kanta, G. Montavon, C.C. Berndt, M.-P. Planche, C. Coddet, Intelligent system for prediction and control: Application in plasma spray process, *Expert Systems with Applications*, 38 (2011) 260-271.

[7] S. Audisio, M. Caillet, A. Galerie, H. Mazille, Revêtements et traitements de surface, Presses polytechniques et universitaires Romandes, Lausanne, Suisse, 1998.

[8] A. Proner, Revêtements par projection thermique, Techniques de l'ingénieur, MD, M 1645, 1999, pp. 1-20

[9] C.-J. Li, A. Ohmori and R. McPherson, The relationship between microstructure and Young's modulus of thermally sprayed ceramic coatings, *J. Mater. Sci.* 32 (1997) 997–1004.

[10] A. Proner, Revêtement par projection thermique, Techniques de l'Ingénieur, traité Matériaux métalliques, M 1 645, pp. 6.

[11] A.-F. Kanta, G. Montavon, C. Coddet, Predicting spray processing parameters from required coating structural attributes by artificial intelligence, *Advanceed Engineering Materials* 8 (2006) 628-635.

[12] M. Dvorak, F. Dietrich, A New Powder Feeder for the Transport of Ultrafine Powders, Thermal Spray 2001: New Surfaces for a New Millennium, May 28-30, 2001, Singapore, Ed. C. C. Berndt, K. A. Khor, E. F. Lugscheider, ASM International, Materials Park, OH., USA, 2001, pp. 539-541

[13] T. Zhang, D.T Gawne, B. Liu, Computer modelling of the influence of process parameters on the heating and acceleration of particles during plasma spraying, *Surface and Coatings Technology* 132(2000)233-243.

[14] M. Vardelle, Etude de la structure des dépôts d'alumine obtenus par projection plasma en fonction des températures et des vitesses des particules au moment de leur impact sur la cible, Thèse de troisième cycle, Université de Limoges, France, 1980.

[15] G. Montavon, S. Sampath, C.C. Berndt, H. Herman, C. Coddet, "Effect Of the spray angle on splat morphology during thermal spraying, *Surface and Coatings Technology* 91 (1997) 107-115.

[16] J. Ilavsky, A. Allen, G. G. Long, S. Krueger, Influence of spray angle on the pore and

crack microstructure of plasma- sprayed deposits, *Journal of the American Ceramic Society,* 80 (1977) 733-742.

[17] M. F Smith, R.A Neister, R.C. Dykhuizen, An investigation of the effects of droplet impact angle in thermal spray deposition, Thermal Spray Industrial Applications, June 20-24, 1994, Boston, USA, Ed. C. C Berndt, S. Sampath, ASM International, Materials Park, OH., USA, 1994

[18] M. P. Planche, B. Normand, E. Suzon, C. Coddet, The relation between in-flight particles characteristics and coatings properties under plasma spraying conditions, In Thermal Spray 2001: New Surfaces for a New Millennium, May 28-30, 2001, Singapore, Ed. C. C. Berndt, K. A. Khor, E. F. lugscheider, ASM International, Materials Park, OH, USA, 2001, pp. 771-777.

[19] M. Friis, C. Persson, J. Wigren, Influence of particle in-flight characteristics on the microstructure of atmospheric plasma sprayed yttria stabilized ZrO2, *Surface and Coatings Technology*, 141 (2001) 115-127.

[20] Y. Zhu, H. Liao, C. Coddet, Transient thermal analysis and coating formation simulation of thermal spray process by finite difference method, *Surface and Coatings Technology* 200 (2006) 4665-4673.

[21] L. Pawloski, The science and engineering of thermal spray coatings, John Wiley, 1995.

[22] A. Hasui, S. Kitahara, T. Fukushima, On relation between properties of coating and spraying angle in plasma jet spraying, *Transaction of National Research Institute for Metals* 12 (1970) 9-20

[23] M. Mellali, Influence de la rugosité et de la température de surface du substrat sur l'adhérence et les contraintes résiduelles au sein de dépôts d'alumine projetés par plasma, Thèse de doctorat, Université de Limoges, France, 1994.

[24] L. Bianchi, Projection par plasma d'arc et plasma inductif de dépôts céramiques : mécanismes de formation de la première couche et relation avec les propriétés mécaniques des dépôts, Thèse de doctorat, Université de Limoges, 1995.

[25] P. Fauchais, M. Fukumoto, A. Vardelle, M. Vardelle, Knowledge concerning splat formation, *Journal of Thermal Spray Technology* 13 (2003) 337-360.

[26] S. Sampath, X. Jiang, Splat formation and microstructure development during plasma spraying: deposition temperature effects, *Materials Science and Engineering A-Structural Materials* 304-306 (2001) 144-150

[27] X. Jiang, Y. Wan, H. Herman, S. Sampath, Role of condensates on substrate surface on fragmentation of impinging molten droplets during thermal spray, *Thin Solid Films* 385 (2001) 132-141

[28] L. Bianchi, A. Grimaud, F. Blein, P. Lucchese, P. Fauchais, Comparison of plasma sprayed alumina coatings by RF and DC plasma spraying, *Journal of Thermal Spray Technology* 4 (1995) 59-66.

[29] N. Sakakibara, H. Tsukuda, A. Notomi, The splat morphology of plasma sprayed particles and the relation to coating properties, Thermal Spray: Surface Engineering via Applied Research, May 8-11, 2000, Montréal, Canada, Ed. C. C. Berndt, ASM International, Material Park, OH, USA, 2000, pp. 753-758.

[30] M. Fukumoto, Y. Huang, Flattening mechanism in thermal sprayed nickel particle impinging on flat substrate surface, *Journal of Thermal Spray Technology* 8 (1999). 427-432

[31] V. V. Sobolov, J. M. Guilemany, Flattening of droplets of splats in thermal spraying: a

review of recent work-Part 2, *Journal of Thermal Spray Technology*. 8 (1999) 301-314.

[32] I. Hofinger, K. Raab, J. Möller, M.Bobeth, Effect of substrate roughness on the adherence of NiCrAlY thermal spray coatings, *Journal of Thermal Spray Technology* 11 (2002) 387-392.

[33] E. Pfender, J. R. Fincke, R. Spores, Entrainment of cold gas into thermal plasma jets, Plasma Chemistry and. *Plasma Processing* 11 (1991) 529-543.

[34] J. R. Fincke, W. D. Swank, The effect of plasma jet fluctuations on particle time histories, Thermal Spray Coatings: Properties, Processes and Applications, May 4-10, 1991, Pittsburgh, PA, Ed. T. F. Bernecki, ASM International, Materials Park, OH. USA, 1992, pp. 193-198.

[35] V. Guipont, R. Mollins, M. Jeandin, G. Barbezat, Plasma-Sprayed Ti-6Al-4V Coatings in a reactive nitrogen atmosphere up to 250 kPa, International Thermal Spray Conference, March 4-6, 2002, Essen, Allemagne, Ed. E. Lugscheider, C.C. Berndt, , DVS-German Welding Society, Düsseldorf, Allemagne, 2002, pp. 247-252.

[36] S. Sodeoka, M. Suzuki, T. Inoue, Effect of chamber pressure and spray distance on the plasma sprayed alumina deposition, Thermal Spray 2003: Advancing the Science and Applying the Technology, May 5-8, 2003 Orlando, USA, Ed. B. R. Marple, C. Moreau, ASM International, Materials Park, OH., USA, Vol. 1, 2003, pp. 597-601.

[37] S. Sodeoka, M. Suzuki, T. Inoue, Control of plasma sprayed particles temperature and velocity by chamber pressure, Thermal Spray 2001: New Surfaces for a New Millennium, May 28-30, 2001 Singapore, Ed. C. C. Berndt, K. A. Khor, E. F. Lugscheider, ASM International, Materials Park, OH, 2001, pp. 737-741.

[38] R. Bolot, Modélisation des écoulements de plasmas d'arc soufflé : Application à la projection de matériaux pulvérulents, Ph.D. Thesis, Université de Technologie de Belfort-Montbéliard, France, 1999.

[39] H. Chen, C. H. Choi, S. W. Lee, C. X. Ding, Deposition efficiency and microhardness of plasma sprayed zirconia coatings using different powders as feedstocks, Thermal Spray 2004: Advances in Technology and Application, May 10-12, 2004, Osaka, Japon, ASM International, Materials Park, OH. USA, 2004, pp. 19-20.

[40] M. Tului, F. Ruffni, F. Arezzo, S. Lasisz, Z. Znamirowski, L. Pawlowski, Some properties of atmospheric air and inert gas high-pressure plasma sprayed ZrB2 coatings, *Surface and Coatings Technology* 151-152 (2002) 483-489.

[41] B. Pateyron, G. Delluc, M. F. Elchinger, P. Fauchais, Study of the behavior of the heat conductivity and transport properties of a simple reacting system: H2 - Ar and H2 - Ar - Air. Dilution effect in spraying process at atmospheric pressure, *J. High Temp. Chem. Proc.* 1 (1992) 325-332.

[42] M. Vardelle, A. Vardelle, P. Fauchais, Diagnostics for particulate vaporization and interactions with surfaces, *Pure and Applied Chemistry* 64 (1992) 637-644.

[43] C. Zhang, C.-J. Li, H. Liao, M.-P. Planche, C.-X. Li, C. Coddet, Effect of in-flight particle velocity on the performance of plasma-sprayed YSZ electrolyte coating for solid oxide fuel cells, *Surface and Coatings Technology* 202 (2008) 2654-2660.

[44] S. Janisson, E. Meillot, A. Vardelle, J. F. Coudert, B. Pateyron, P. Fauchais, Plasma Spraying Using Ar-He-H2 Gas Mixtures, Thermal Spray: Meeting the Challenges of the 21st Century, May 25-29, 1998, Nice, France, Ed. C. Coddet, ASM Int., Materials Park, Oh. USA, 1998, pp.803-808.

[45] B. Pateyron, M. F. Elchinger, G. Delluc, P. Fauchais, Thermodynamic and transport

properties of Ar-H2 and Ar-He plasma gases used for spraying at atmospheric pressure. I : Properties of the mixtures, *Plasma Chemistry and Plasma Processing* 12 (1992). 421-449.

[46] B. Dussoubs, Modélisation tri-dimensionnelle du procédé de projection plasma : influence des conditions d'injection de la poudre et des paramètres de projection sur le traitement et la répartition des particules dans l'écoulement, Thèse de doctorat de l'Université de Limoges, France, 1998.

[47] J. F. Bisson, B. Gauthier, C. Moreau, Effect of Plasma Fluctuations on In-flight Particle Parameters, Thermal Spray 2001: New Surfaces for a New Millennium, May 28-30, 2001, Singapour, Ed. C.C. Berndt, K.A. Khor, E.F. Lugscheider, ASM International, Materials Park, OH. USA, 2001, pp. 715-721.

[48] D. Rigot, B. Pateyron, J. F. Coudert, P. Fauchais, J. Wigren, Evolutions et dérive des signaux émis par une torche à plasma à courant continu (type PTF4), 6èmes Journées d'études sur les fluctuations des arcs, 17-18 mars 2003, Clermont-Ferrand, France

[49] X. Zhou, J. Heberlein, Characterization of the arc cathode attachment by emission spectroscopy and comparison to theoretical predictions, *Plasma Chemistry and Plasma Processing* 16 (1996) 229S-244S.

[50] H. P. Li, E. Pfender, X. Chen, Application of Steenbeck's Minimum Principle for Three-dimensional Modelling of DC Arc Plasma Torches, *Journal of Physics D-Applied Physics* 36 (2003) 1084-1096

[51] J. Heberlein, Electrode phenomena in DC Arcs and their influence on plasma torch design progress, Plasma Processing of Materials, 2003, Ed. P. Fauchais, Begell House, pp. 147-164.

[52] B. Rethfeld, J. Wendelstorf, T. Klein, G. Simon, A self-consistent model for the cathode fall region of a electric arc, *Journal of Physics D-Applied Physics* 29 (1996) 121-128.

[53] K. C. Hsu, E. Pfender, Analysis of the cathode region of a free-burning high intensity argon arc, *Journal of Applied Physics* 54 (1983) 3818-3824.

[54] P. Fauchais, G. Montavon, M. Vardelle, J. Cedelle, Developments in direct current plasma spraying, *Surface and Coatings Technology* 201 (2006) 1908-1921.

[55] P. Fauchais, M. Vardelle, How to improve the reliability and reproducibility of plasma sprayed coatings, Thermal Spray 2003: Advancing the Science and Applying the Technology, May 5-8, 2003, Orlando, USA, Ed. B. R. Marple, C. Moreau, ASM International, Materials Park, OH, USA, Vol. 2, 2003, pp. 1165-1173.

[56] C. Moreau, P. Gougeon, M. Lamontagne, V. Lacasee, G. Vaudreuil, P. Cielo, On-line control of the plasma spraying process by monitoring the temperature, velocity and trajectory of the in-flight particles, Thermal Spray Industrial Applications, June 20-24, 1994, Boston, USA, Ed. C. C. Berndt, S. Sampath, ASM International, Materials Park, OH, USA, 1994, pp. 431-437.

[57] J. F. Bisson, M. Lamontagne, C. Moreau, L. Pouliot, J. Blain, F. Nadeau, Ensemble in-flight particle diagnostics under thermal spray conditions, Thermal Spray 2001: New Surfaces for a New Millennium, May 28-30, 2001, Singapore, Ed. C. C. Berndt, K. A. Khor, E. F. Lugscheider, ASM International, Materials Park, OH, USA, 2001, pp. 705-714.

[58] Tecnar Automation, 1321 Hocquart Street, St-Bruno, QC, Canada J3V 6B5

[59] W.D. Swank, J.R. Fincke, D.C. Haggard, A particle temperature sensor for monitoring and control of the plasma spray process, Advances in Thermal Spray Science and Technology, September 11-15, 1995, Houston, USA, Ed. C.C. Berndt, S. Sampath, ASM International Materials Park-OH, USA, 1995, pp. 111-116.

[60] J. Vuttulainen, E. Hämäläinen, R. Hernberg, P. Vuoristo, T. Müntyla, Novel method for in-flight particle temperature and velocity measurements in plasma spraying using a single CCD camera, *Journal of Thermal Spray Technology* 10 (2001) 94-104.

[61] M. Ignatiev, V. Senchenko, V. Dozhdikov, I. Smurov, Ph. Bertrand, Digital diagnostic system based on advanced CCD image sensor for thermal spraying monitoring, International Thermal Spray Conference, March 4-6, 2002, Essen, Germany, Ed. E. Lugscheider, C. C. Berndt, DVS Deutscher Verband für Schweißen, 2002, pp. 1001-1006.

[62] J. E. Craig, R. A. Parker, D. Y. Lee, Thermal Spray: Surface Engineering via Applied Research, May 8-11, 2000, Montréal, Canada, Ed. C. C. Berndt, ASM International, Materials Park, OH, USA, 2000, pp. 51-66.

[63] T. Renault, M. Vardelle, P. Fauchais, H. Hoffmann, F. Braillard, On-line monitoring (SDC) through coating surface temperature of residual stresses in APS WC-Co17wt% coatings on Hastelloy X, Thermal Spray 2001: New Surfaces for a New Millennium, May 28-30, 2001 Singapore, Ed. C.C. Berndt, K. A. Khor, E. F. Lugscheider, ASM International, Materials Park, OH, USA, 2001, pp. 743-750.

[64] A.-F. KANTA, Développement d'un système expert basé sur l'intelligence artificielle pour la prédiction et le contrôle du procédé de projection plasma, Thèse de doctorat, l'Université de Technologie de Belfort-Montbéliard, N° 75, 2007

[65] L. A. Zadeh, Outline of a new approach to the analysis of complex systems and decision processes, *IEEE Transactions on Systems, Man, and Cybernetics* 3 (1973) 28-44.

[66] P. Nopporn, P. Suttichai, S. Yosanai, Maximum power point tracking using adaptive fuzzy logic control for grid-connected photovoltaic system Renew. *Energy* 30 (2005) 1771-1778.

[67] S.O.T. Ogaji, L. Marinai, S. Sampath, R. Singh, S.D. Prober, Gas-turbine fault diagnostics: a fuzzy-logic approach, *Applied Energy* 82 (2005) 81-89.

In: Sprays: Types, Technology and Modeling
Editor: Maria C. Vella, pp. 229-255
ISBN 978-1-61324-345-9

Chapter 6

LIQUID FLOW STRUCTURE IN PRESSURE SWIRL SPRAYS: STUDY OF DROPLET COLLISION PHENOMENA

J. L. Santolaya*[*], L. A. Aísa, E. Calvo, L. M. Cerecedo, J. A. García, and J. I. García
Departament of Materials and Fluids Science and Technology
University of Zaragoza, Zaragoza, Spain

ABSTRACT

In this work, a detailed characterization of the liquid flow structure in sprays generated by pressure swirl nozzles was performed through experimental techniques. The overall droplet size distributions across the spray were obtained and the effects of the droplet collision phenomena on the spray development were studied. A Phase Doppler particle analyzer was used to obtain simultaneous measurements of size and droplet velocity and a data post-processing, applying the generalized integral method, was used to evaluate number concentrations and liquid volume fluxes for different droplet size classes. The parameters of radial mean spread and spatial dispersion of the sprays were also calculated.

A number of liquid injection conditions, which included two atomization regimes, were investigated. High collision rates were estimated in the spray densest zones and some collision outcomes, as the droplet coalescence and the separation with satellite droplet formation were detected. A liquid flow rate transfer between size classes, due to the coalescing collisions, was obtained as the sprays developed. This process, which favored the large size classes, caused the progressive growth of the droplet mean diameter along the spray. The separation was found a more important droplet collision outcome as droplet velocities increased.

Keywords: Pressure swirl nozzle, droplet size classes, liquid flow structure, droplet collisions, coalescence

[*] CPS – Edif. Torres Quevedo, C/ María de Luna 3, Tel: +34 976761881, e-mail: jlsanto@unizar.es.

INTRODUCTION

The performance of many industrial processes involves the formation of two-phase flows. A multitude of applications such as spray-drying, scrubbing towers for dust and gas, fuel injection in combustion systems, spray painting, powder metallurgy, chemical processing, industrial washers or spray coating require the generation and development of liquid droplets in a surrounding gas flow. Specific physical phenomena may considerably affect the liquid flow structure of the spray. Particularly, the interaction of the atomized liquid with the gas flow field redistributes spatially the spray droplets due to the differences in droplet inertia, momentum and drag, and the droplet collision phenomena cause a droplet velocity exchange and can intimately affect the droplet size distribution.

Among the wide variety of devices used to achieve the disintegration of a liquid volume, pressure swirl nozzles are satisfactory for producing well-atomized sprays through the break-up of thin conical sheets. The operation of these injectors is based on the high angular momentum that acquires the liquid inside the atomizer. A distributor, generally having spirally shaped grooves, generates a rotational flow which gets accelerated as the nozzle diameter decrease towards the outlet. Depending on the operation data, the transport properties and the geometry a rotationally symmetric gas core is formed in the area of the nozzle axis. The characteristics of this complex internal flow were studied experimentally and theoretically in many works [1-3].

The liquid, forced to follow a helical path, emerges from the nozzle in the form of a hollow conical sheet that soon becomes unstable and disintegrates because of different mechanisms. The breakup phenomenon has been extensively studied, particularly in plane configurations. For conical sheets [4], it was observed that a collapsed tulip-shape sheet changed to a diverging wavy sheet as the liquid injection pressure was increased.

The effect of fluid properties on the sheet characteristics was also examined through experimental methods [5, 6]. It was demonstrated that an increase on the viscosity or on the surface tension of the liquid injected inhibited the growth of surface waves as breakup mechanism and generated more developed sheets.

In the literature one finds a significant number of investigations [7-9] to experimentally determine, by non-intrusive optical measurement techniques, the droplet size distribution resulting of the sheet breakup process. Among the optical techniques, small angle light-scatter detection and phase Doppler anemometry (PDA) were mainly used. Systems based on the light scattering can be appropriate for a relatively rapid characterization of droplet sizes. For high-resolution and time averaged measurements of spray droplet size and velocity, the phase Doppler particle analyzer is a very powerful tool.

By means of PDA, the study of the spray near field was performed in various works [10 – 12]. Measurements showed that, in the absence of any significant external air flow field, the large droplets tend to maintain the high velocity of the liquid sheet whereas small droplets couple to the local induced air flow. The droplet number concentration increased almost continuously from the edge toward the core region of the spray because of the preferential accumulation of small droplets which are transported by the continuous phase. The entrained air flow rate [13], which has a decisive influence on the heat and mass transfer between phases, was also obtained.

On the other hand, axial volume fluxes presented initially local maxima at the edge of the spray. In this region, droplet collision was considered as a very frequent event, due mainly to the relative velocities between droplets. Typically, the droplet interaction process is separated into two parts: first, calculating the collision rate between droplets and second, analyzing the droplet collision outcome.

In order to obtain the collision rate of particles quite a number of theoretical studies, modelling and simulations were already carried out [14, 15]. By using the Stokes number, which is defined as the ratio of particle response time to the relevant scale of turbulence, two limiting cases for the occurrence of particle collisions can be identified: case 1, for $St \rightarrow 0$, where particles completely follow the flow turbulence and case 2, for $St \rightarrow \infty$, where the particle motion is completely uncorrelated with the fluid. In practical two-phase flows, droplets partially respond to turbulence and the velocities of colliding droplets will be correlated to a certain degree.

Regarding the collision outcome, previous works [16, 17] characterized the possible outcomes of droplet impacts, under atmospheric conditions and for higher background pressures. Experiments enabled to observe that most interactions between two liquid droplets resulted in the formation of a deforming post-collision liquid mass that evolved in different form according three main non-dimensional parameters: the droplet Weber number, We, which is the ratio of the inertia force to the surface tension force, the droplet size ratio, Δ, which is the ratio of the smaller droplet to that of the larger one, and the impact parameter, B, which defines the geometrical orientation of the interacting droplets. B=0 and 1, respectively, designate head-on and grazing collisions.

The final collision outcome was classified into five different regimes, which were named: slow coalescence, bouncing, coalescence after substantial deformation, reflexive or near head-on separation and stretching or off-centre separation. Collision outcomes on the B-We plane are qualitatively described through the conceptual sketch of Figure 1.

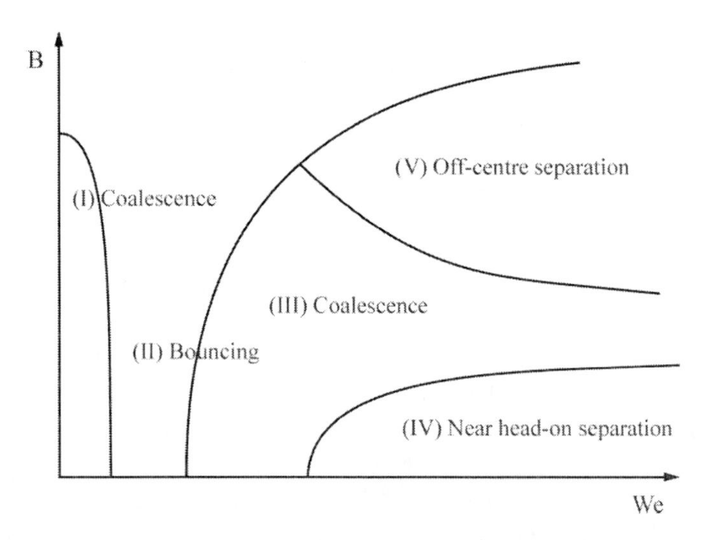

Figure 1. Collision regimes of hydrocarbon droplets.

The material properties of both the droplet and ambient fluid were additional relevant parameters [18]. Hence, the value of the critical We and B, which define the transitions between regions varied with droplet fluid and background gas conditions. The droplet size ratio can be also taking into account to obtain three-dimensional maps of droplet collision outcomes [19].

Stable coalescence between two colliding droplets is achieved in two cases: if the droplets are travelling slowly enough and the air between them has time to exit before the droplets touch and if the droplets have high velocities and the gas film can actually be absorbed into the liquid. Separation is promoted for high values of We when the collision is either near head-on or near grazing. At high impact parameters the droplets tend to stretch apart, while for near head-on collisions the droplets can oscillate and undergo a reflexive separation. This process is characterized by the formation of a connecting ligament which can either contract to form a single droplet or further break up into more satellite droplets [20].

The droplet head-on collision sequence that results in separation was divided into three periods [17] shown schematically in Figure 2. In period 1, two droplets first impinge head on and form an outwardly spreading disk; in period 2, the disk contracts under surface tension to recover the droplet shape and in period 3, a stretched liquid cylinder is formed with its two ends moving outward. The critical transition from coalescence to separation is achieved if the initial kinetic energy is just large enough to overcome the viscous dissipation after coalescence and break the inter-connecting ligament.

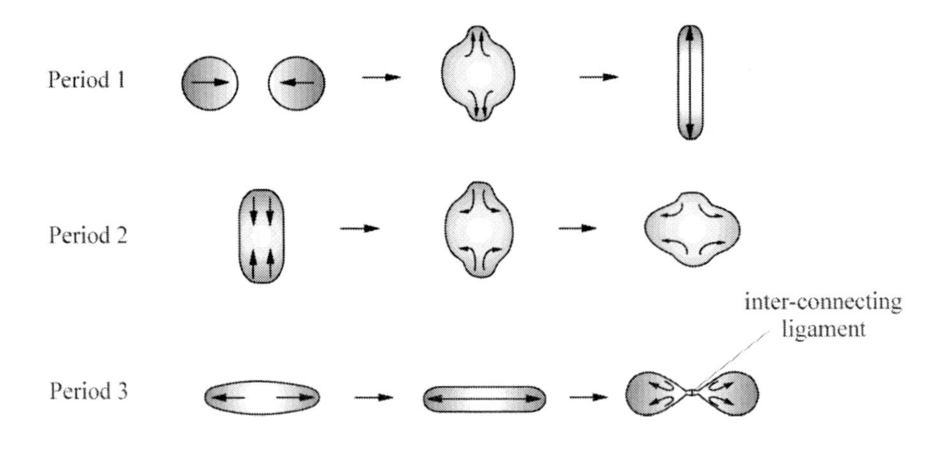

Figure 2. Schematic of the droplet separation collision sequence.

Stable coalescence (regimes I and III), and fragmentation with subsequent formation of satellite droplets (regimes IV and V) can significantly modify the droplet size distribution as spray develops and, therefore, affects critically spray mixing and subsequent processes. Previous spray investigations analyzed the effects of collision phenomena on the droplet size distribution by means of predictive models [12], [21] and experimental methods [22]. The relevance of the liquid transfer process between droplet size classes as spray develops was confirmed. Here, an exhaustive valuation of collision rates between droplet size classes was performed. In addition, the incidence of different droplet collision outcomes on the spray evolution was examined.

EXPERIMENTAL FACILITY

Experiments were performed at a specially designed facility for spray generation under controlled conditions. The test liquid was oil of industrial origin, used previously in lubrication and refrigeration applications. The oil properties measured at the injection temperature are listed in Table 1.

Table 1. Oil properties at T=95°C

Density	$\rho_l = 847$ Kg/m3
Dynamic viscosity	$\mu_l = 0.0166$ Kg/ms
Surface tension	$\sigma_l = 0.032$ N/m
Vapor pressure	$P_v = 98.13$ pa
Refractive index	m = 1.483 + 0.00072i

The liquid was driven from the tank to the nozzle through a pumping line (Figure 3). It was filtered to eliminate the undesirable solid particles and it was heated by an electrical resistance in order to decrease its high viscosity and to improve atomization performance. The injection conditions were controlled by a set of sensors. Temperature was fixed at 95° C and injection pressure was varied from 4 to 20 bar. Sprays developed inside a transparent chamber 340 mm square, in which a surrounding airflow (co-flow), with a mean velocity of 0.15 m/s, was generated by an exhaust system.

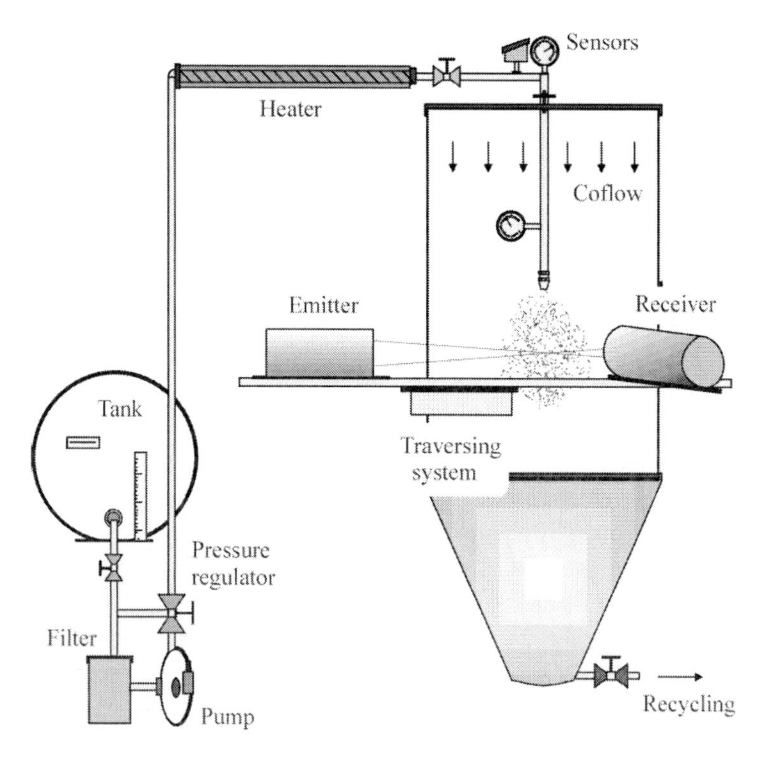

Figure 3. Experimental setup.

An auxiliary equipment was used to generate a very fine aerosol of water-glycerin mixture (D_{10}=3.4 μm), which was added to the co-flow allowing the continuous phase measurements by the PDPA technique. After atomization, the injected liquid was recovered for recycling.

A commercial pressure swirl nozzle of low flow rate and exit cone angle of 80° was used in this study. In Figure 4 we can observe the nozzle internal geometry. The pressurized liquid enters through three small grooves into the convergent swirl chamber where it is accelerated to the nozzle exit orifice.

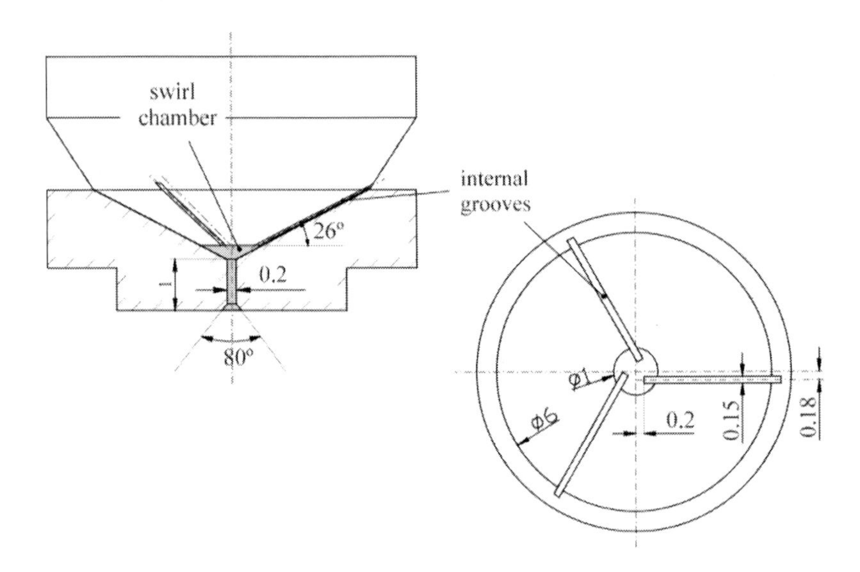

Figure 4. Nozzle internal geometry.

In order to visualize the emerging liquid and to examine the mechanism of sheet disruption, a high-speed photography system composed by a low noise CCD and a stroboscopic light source of 0.5 μs pulse time were used.

The spray measurements were carried out by means of a TSI-Aerometrics PDPA system which allowed us to simultaneously measure the drop size and two components of velocity. The receiving optics was placed to collect scattered light at 70° from the forward direction (Figure 5). At this angle the light scattering is dominated by first order of refraction and that due to reflection is a minimum. Thus, errors associated with trajectory ambiguities due to the gaussian beam effect were reduced [23]. Other geometrical settings were also tested but a not lineal relationship between phase shift and droplet diameter was detected. The beam separation was 40 mm and a transmissions lens with 500 mm of focal length was used. The focal length of the receiving optics was 300 mm. Both emitter and receiver were mounted on a traversing system controlled by a computer.

The region where the laser beams intersect is the sample volume or illumination volume. For laser beams with a Gaussian intensity profile, their diameter is typically defined as the location where the beam intensity falls to $1/e^2$ of the maximum intensity. The detection volume o probe volume is the region of space in which valid measurements can be obtained and their size depends on both, the operating parameters and the droplet diameter to be measured.

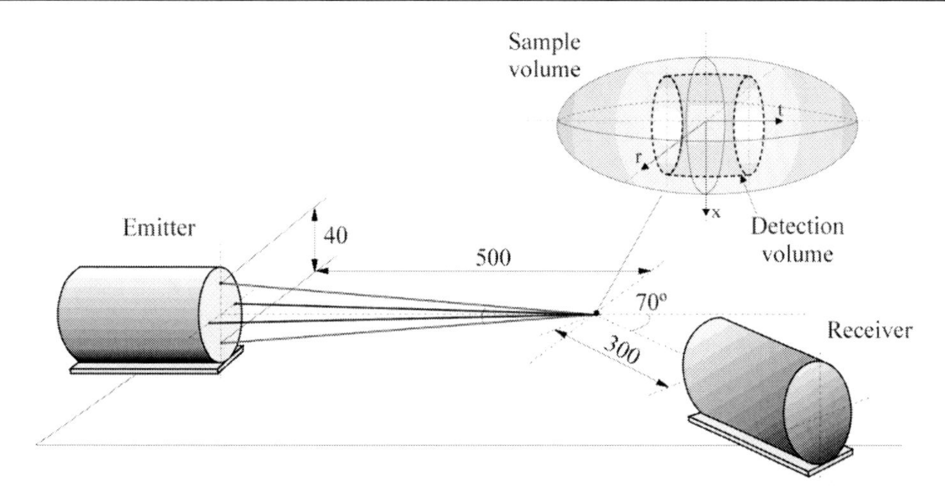

Figure 5. Optical configuration of the PDPA system.

Further details of the PDA system can be consulted in a number of works [23, 24].

STUDY AND ANALYSIS METHOD

Measurements on the sprays were performed at four axial stations located at 9, 18, 36 and 72 mm from the atomizer exit. The droplet diameter and both, the axial and radial velocities, were measured every 2 mm in radial direction at each axial station using the PDPA system. A typical validation percentage along a generic radius in the spray field was 85-95% for velocities and 65-85% for drop sizes. It should be noted that the PDPA rejects signals from non-spherical droplets and those due to multiple droplets in the probe volume.

The mean velocities were determined by averaging overall droplet size distribution at the respective locations. The Sauter mean diameter, SMD or D_{32}, was taken as characteristic parameter of the local droplet size distribution. D_{32} is inversely proportional to the interfacial area per unit volume, consequently is a decisive parameter in the heat and mass transfer process between droplets and gas. In addition both, D_{43} and D_{V50} were used to characterize the global droplet size distributions across the spray.

The droplet number concentration and the liquid volume flux were determined by the method based on the transit time of the droplet at the probe volume [25, 26]. This method, which avoids errors due to the droplet trajectory through the probe volume, can be formulated as a particular case of the General Integral Method, [27]. Nevertheless, the accuracy of droplet volume flux measurements depends on a perfect signal detection and size evaluation of all droplets crossing the probe volume and on the determination of the probe volume size itself.

The number concentration of each droplet size class was obtained by the following expression:

$$C_{Ni} = \frac{1}{\Delta t} \frac{\sum_{\forall kj} tt_{kj}}{Vol_i} \qquad (1)$$

where Δt is the measurement time at each location, Vol_i is the PDPA detection volume for each droplet size class and tt_{kj} is the transit time though the detection volume for droplets corresponding to each velocity direction class and each velocity module class.

The number concentration for all droplets was calculated as:

$$C_N = \sum_{\forall i} C_{Ni} \tag{2}$$

And the liquid volume fluxes, for each droplet size class and for all droplets, were obtained respectively as:

$$\mathbf{f}_{Vi} = \frac{1}{\Delta t} \frac{\sum_{\forall kj} tt_{kj} \cdot \mathbf{V}_{Dj}}{Vol_i} \cdot \frac{\pi D_i^3}{6} \tag{3}$$

$$\mathbf{f}_V = \sum_{\forall i} \mathbf{f}_{Vi} \tag{4}$$

where $\mathbf{V}_{Dj}$ is the velocity vector for each droplet velocity module class and D_i is the droplet diameter.

Droplet volume fluxes were evaluated in axial direction. The total liquid flow rates at each measurement axial station, for each droplet size class and for all droplets, were calculated respectively as:

$$F_{Ti} = \int_{S_T} f_{Vxi} \cdot ds \tag{5}$$

$$F_T = \int_{S_T} f_{Vx} \cdot ds \tag{6}$$

where S_T is the spray cross transversal area. The spray boundary at each axial station was located from PDPA measurements as the radial position where the data rate was less than 10 droplets per minute.

The percentage of total liquid flow rate not measured was calculated as:

$$def = 100 \cdot \left[1 - \frac{F_T}{Q}\right] \tag{7}$$

where Q is the liquid volumetric flow rate injected through the atomizer.

Finally, the percentage of cumulative liquid flow rate by droplet size classes was obtained as:

$$\%F_{ac} = 100 \cdot \sum_{i=1}^{n} \frac{F_{Ti}}{F_T} \tag{8}$$

The analysis of the sprays was carried out under the hypothesis of axis-symmetry, plotting variables evolution profiles on the dimensionless radial coordinate r*. This coordinate was obtained as:

$$r^* = \frac{r}{R_{50}} \tag{9}$$

where R_{50} is the radial position that contains 50% of the total liquid volumetric flow rate measured at each axial station.

In a similar way, the radii R_{10} and R_{90}, which are the radial positions including, respectively, 10 and 90% of the collected liquid flow rate at each axial station, were defined. Hence, the spray spatial dispersion parameter was calculated as follows:

$$\Delta R = R_{90} - R_{10} \tag{10}$$

The analysis of dispersed phase included the detailed study of three droplet size classes that were named small, medium and large. The small class integrated droplet diameters from 5 to 10 μm; the medium class, diameters from 20 to 30 μm, and the large class, diameters from 50 to 60 μm.

The characterization of the continuous phase and subsequent determination of the air velocity fields was performed analyzing the signals of the smallest droplets (smaller than 5 μm) which supposedly follow the instantaneous changes in gas velocity.

The entrained air flow rate was calculated by integrating the local axial air velocities across the spray, such as shows the following expression:

$$F_{Ta} = \int_{S_T} V_{ax} \cdot ds \tag{11}$$

Air mass flow rate balances were checked in several cylindrical control volumes in order to validate the continuous phase measurements. Average errors of 8% were obtained between incoming and outgoing air flow rates to the control volumes.

According to the classic kinetic theory of gases, droplet collision phenomena were studied assuming droplets are not deflected due to their interaction with the surrounding gas flow. Thus, collision events of one droplet with another one occur inside a collision cylinder, which is defined by the cross-section area $\pi/4\left(D_1 + D_2\right)^2$ and the length $|\mathbf{V}_{D1} - \mathbf{V}_{D2}|\,\Delta t$, where D_1 and D_2 are the diameters of the colliding droplets and $|\mathbf{V}_{D1} - \mathbf{V}_{D2}|$ is their relative velocity. Hence, the collision rates between two different droplet size classes can be obtained as:

$$Z_{ii'} = \sum_{\forall j,j'} \frac{\pi}{4}(D_i + D_{i'})^2 \mid \mathbf{V}_{Dj} - \mathbf{V}_{Dj'} \mid C_{Nij} C_{Ni'j'} \tag{12}$$

where C_{Nij} is the droplet number concentration corresponding to each size class and each droplet velocity module class.

The collision frequency between two droplet size classes can be calculated as:

$$f_{Cii'} = \sum_{\forall j,j'} \frac{\pi}{4}(D_i + D_{i'})^2 \mid \mathbf{V}_{Dj} - \mathbf{V}_{Dj'} \mid C_{Ni'j'} \tag{13}$$

And the collision frequency between one droplet size class and all size classes is:

$$f_{Ci} = \sum_{\forall i'} f_{Cii'} \tag{14}$$

Additionally, the axial mean free paths between subsequent droplet collisions were calculated according to the following expression:

$$L_{Cxi} = \overline{V}_{Dxi}\, \tau_{Ci} \tag{15}$$

where $\overline{V}_{Dxi}$ is the axial mean velocity for each droplet size class and τ_{Ci} is the time between successive droplet collisions, which was calculated as:

$$\tau_{Ci} = \frac{1}{f_{Ci}} \tag{16}$$

Finally, it was also possible to evaluate the influence of collision phenomena about the continuous phase flow development taking into account the ratio of droplet response time, τ_D, to the time between collisions, τ_C.

The droplet response time, τ_D, can be calculated as follows:

$$\tau_D = \frac{\rho_l D^2}{18\, \mu_a} \tag{17}$$

where ρ_l is the liquid density and μ_a is the gas dynamic viscosity.

SPRAY FORMATION AND ATOMIZATION QUALITY

A liquid hollow conical sheet was generated at the atomizer exit only if the liquid tangential velocity was high enough. This happened when the injection pressure was higher than 10 bar. Figure 6 shows the sequence of the liquid sheet instantaneous disintegration for injection pressures from 10 to 20 bar.

Two main sheet disrupting mechanisms can be observed: perforations and amplification of surface waves. Thus, two different atomization regimes (I and II) were established. At 10 and 12 bar, the tulip-shaped sheet breaks-up by the growth of perforations mainly. As the injection pressure raises 14 bar, the diverging conical-shaped sheet breaks-up via a wave growth process. In addition, an intensification of the wave motion on the sheet surface and a reduction of the breakup length are produced as the pressure increases.

In order to study the initial spray structure and atomization quality in each regime, three injection pressures were selected: 12, 16 and 20 bar. The liquid volumetric flow rate injected through the atomizer was measured at each case. Other parameters as the Reynolds number at the exit orifice, Re_o, and the nozzle discharge coefficient, C_D, were also calculated [28]. Values are summarized in the Table 2.

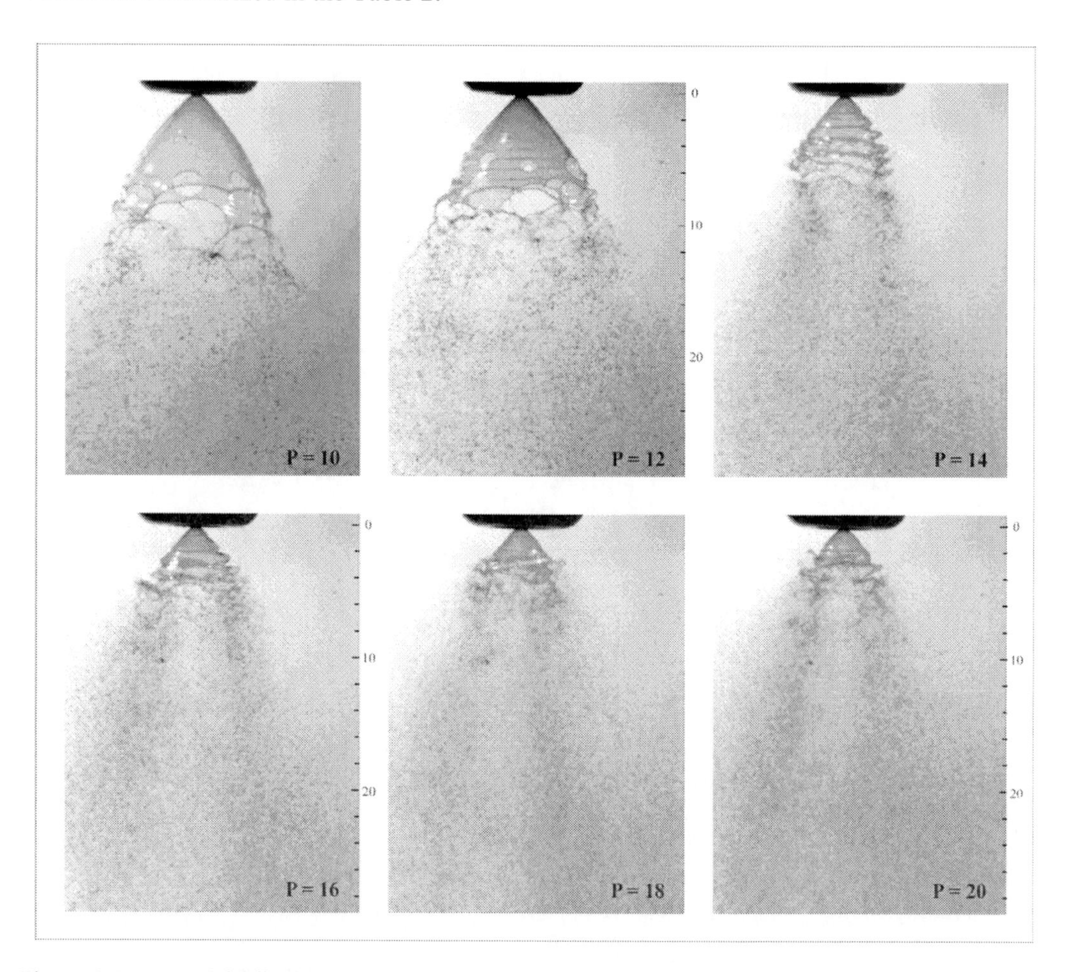

Figure 6. Images of the liquid sheet instantaneous disintegration at different injection pressures.

Table 2. Injected liquid flow rate, Reynolds number at the exit and nozzle discharge coefficient

	P=12	P=16	P=20
Q (cm^3/s)	0.662	0.777	0.854
$Re_o = \rho_l Q d_o / A_o \mu_l$	214	252	277
$C_D = Q / A_o \sqrt{2P/\rho_l}$	0.39	0.40	0.39

According to the regular performance of pressure atomizers, Q and Re_o increased as the pressure did. Seeing that the discharge coefficient acquired almost constant values at the tested pressure range, it was considered that the nozzle was working in its stable zone.

The interaction of the atomized liquid with the gas flow field was analyzed for the three injection pressures. Figure 7 represents the continuous phase velocity maps at the first axial measurement station located at 9 mm from the exit. The air velocity fields show that surrounding air flows into the spray. The interaction between the incoming air and the dispersed phase gives rise to the smallest droplets transport towards the spray core. If in regime I, a vortex zone behind the leading edge of the conical sheet is detected, in regime II, a high velocity air jet is formed at the central zone of the spray.

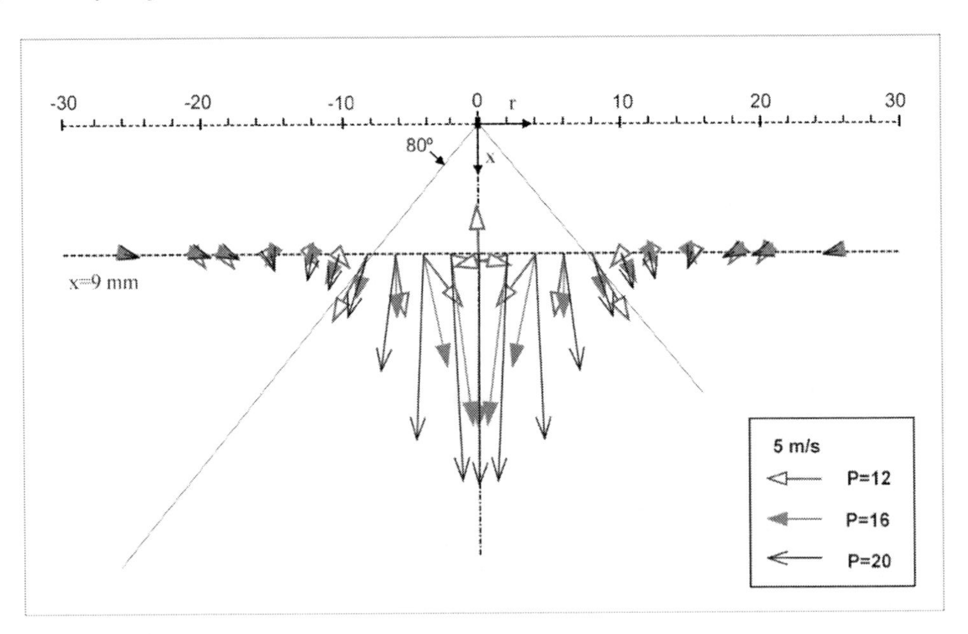

Figure 7. Continuous phase velocity maps. Axial station: 9 mm.

The entrained air flow rates, obtained from Eq. (11), are summarized in Table 3. In all cases, the entrained air flow rates increase as the injection pressure does. A bigger entrainment is noted for the regime II sprays, which present larger radial dispersion and higher axial air velocities than that of regime I.

The R_{50} and ΔR parameters, which represent the spray radial mean spread and the spray zone where the liquid flow rate is mainly concentrate, are also shown in Table 3. It can be noted that the spray of regime I spreads-out quite less than those of regime II. As a consequence, the cross-transversal area of the spray is clearly more reduced.

Table 3. Experimental data of the sprays at the axial station of 9 mm

	P=12	P=16	P=20
F_{Ta} (l/s)	0.35	1.52	2.28
R_{50} (mm)	6.21	9.13	8.40
ΔR (mm)	2.70	6.84	11.09

In Figure 8, the profiles of mean and rms droplet velocity are plotted using the r* coordinate. In regime I, where a very long liquid sheet was generated, the hollow cone structure of the spray is clearly perceptible on the axial mean velocity profile, with a peak at r*=1 (Figure 8a).

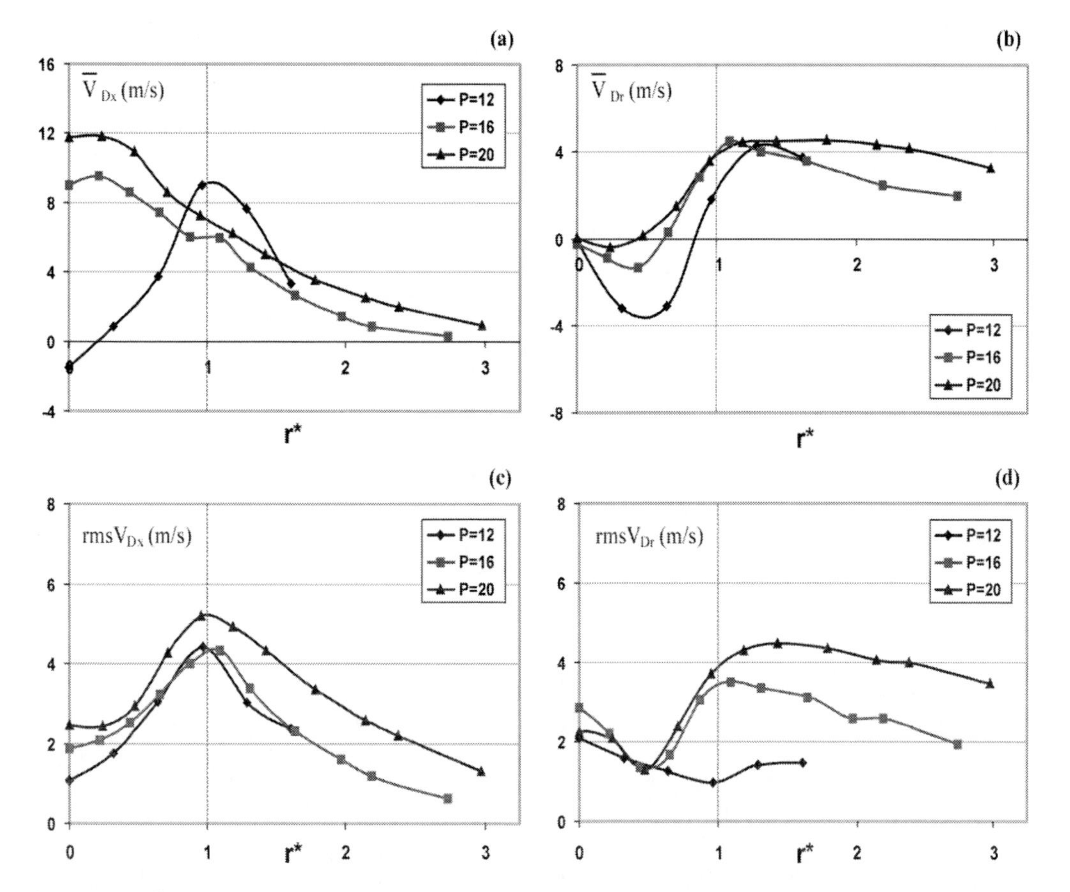

Figure 8. Profiles of mean and rms droplet velocity. Axial station: 9 mm.

a) axial mean velocities; b) radial mean velocities; c) axial rms velocities; d) radial rms velocities.

The axial mean velocity decays quickly and becomes negative at the axis due to the presence of a vortex zone. In regime II, the highest axial velocities correspond to small droplets which are transported by the central high velocity air jet. Nevertheless, a plateau around r*=1 due to the initial structure of the injected liquid is noted at P=16 bar.

Radial mean velocities (Figure 8b) have negative values at r*<1 as a result of the smallest droplets motion toward the spray axis. The spray generated at P=12 bar shows the widest negative zone with the highest velocity modules. This is due to the proximity of the sheet disintegration zone where the central air jet is initiated.

The axial rms velocity profiles (Figure 8c) show the maximum around r*=1. These high droplet velocity fluctuations, located at the highest volume flux zone, were also observed by other researchers [29] and could be initially attributed to the high droplet size-velocity correlation resulting from the drag forces. In the P=12 bar case, the axial velocity is clearly dominant, consequently the radial rms droplet velocity (Figure 8d) reduces notably.

In Figure 9, the axial volume fluxes, for all droplets and for three different droplet size classes, named small, medium and large, respectively, are plotted. Axial volume flux profiles for all droplets (Figure 9a), show the typical configuration observed for hollow cone sprays. The liquid flow rates are initially concentrated in a small region of annular cross section shape (densest spray region) with the highest volume fluxes around R_{50}. As the pressure increases, the maximum axial volume flux decreases due to progressive spatial distribution of the liquid droplets.

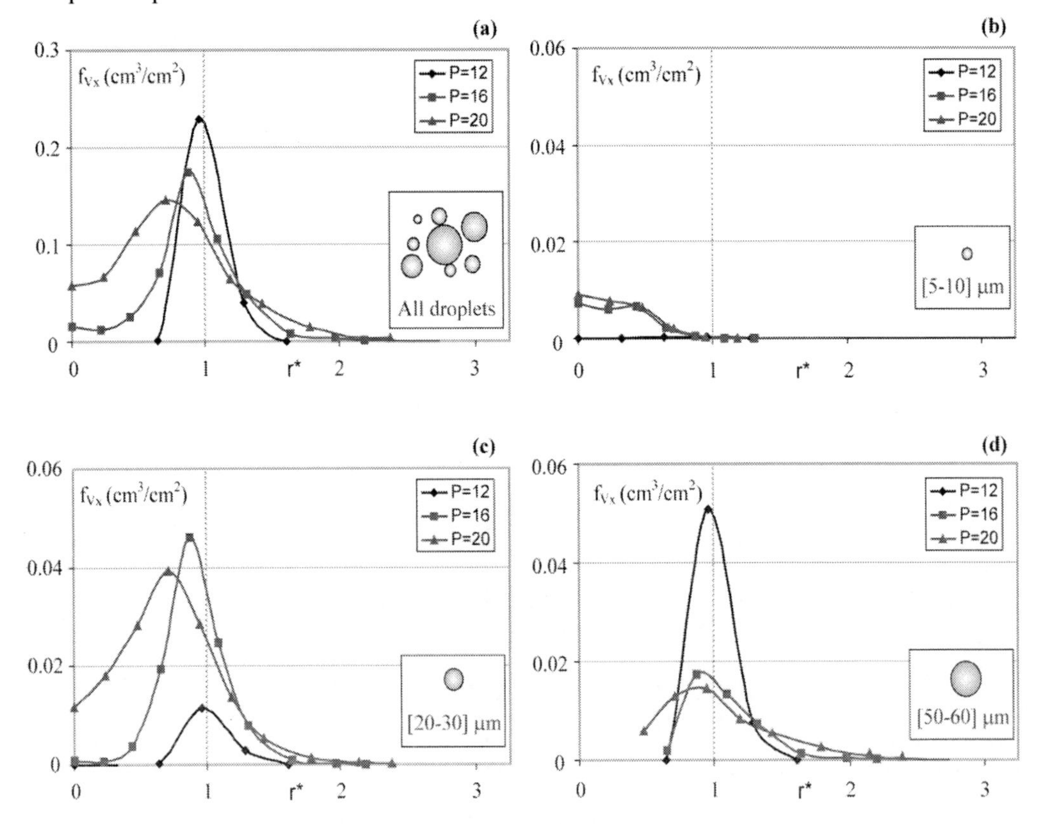

Figure 9. Droplet axial volume flux profiles. Axial station: 9 mm.

a) All droplets; b) small size class; c) medium size class; d) large size class.

The small size class, concentrated at the spray center, only presents significant values of f_{Vx} for the regime II (Figure 9b). The hollow cone configuration of the sprays can also be perceived from the medium and large classes f_{Vx} profiles (Figure 9c-9d). Since the large size class shows very much higher f_{Vx} in regime I than in regime II, a worse-quality spray is noted for the regime I.

In order to compare the atomization quality for the sprays generated at each injection pressure, D_{32} and D_{V50} parameters were considered. First, the profiles of the Sauter mean diameter are represented in Figure 10. A progressive radial increase of D_{32} is measured since the smaller droplets are collected at the spray core transported by the continuous phase, whereas the larger droplets move to the spray edge, relatively unperturbed, according to a process of droplet inertial classification. If the characteristic local position of r*=1 is examined, considerably smaller droplet diameters are measured for the regime II sprays. Values can be consulted in Table 4.

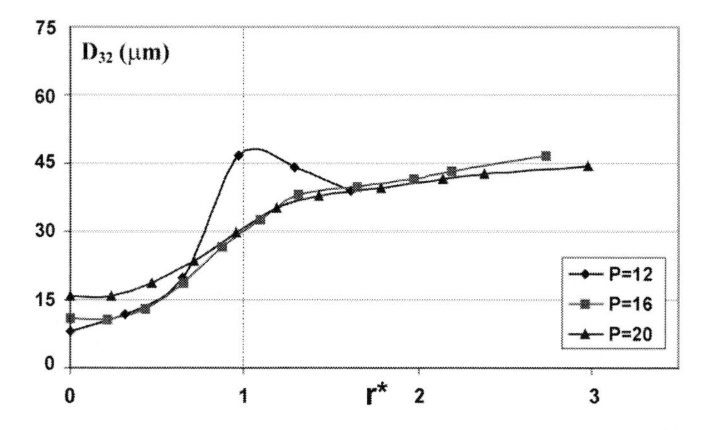

Figure 10. D_{32} profiles. Axial station: 9 mm.

The percentages of cumulative liquid flow rate by droplet size classes could be obtained from Eq. 8. The resulting cumulative distributions, plotted in Figure 11, show that a clearly finer spray is obtained for the atomization regime II.

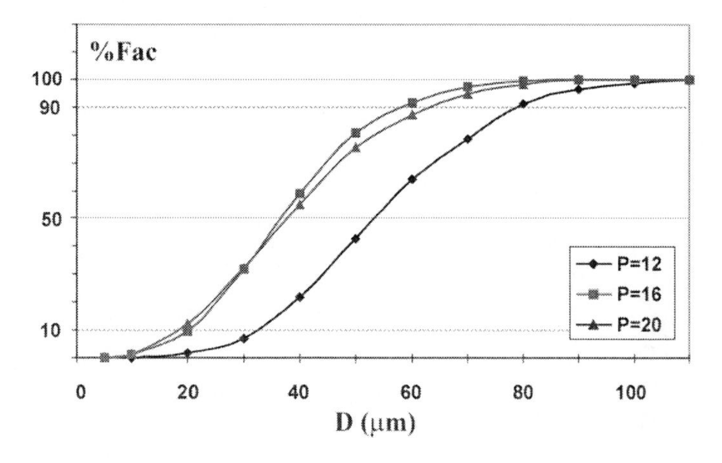

Figure 11. Cumulative liquid flow rate distributions. Axial station: 9 mm.

Table 4. Atomization quality of the sprays. Axial station: 9 mm

	P=12	P=16	P=20
D_{32} (μm) at r*=1	52.1	36.5	35.7
D_{V50} (μm)	53.2	36.8	37.5
$S = (D_{V90} - D_{V10}) / D_{V50}$	0.86	1.04	1.18

Thus, the generation and amplification of surface waves on the liquid sheet is a more efficient sheet breakup mechanism. The statistical parameters of the cumulative distributions are listed in Table 4. D_{32} and D_{V50} reduce notably when pressure increases from 12 to 16 bar. Span rises due mainly to the D_{V50} reduction. Within regime II, very small variations of D_{V50} and S are obtained as the pressure goes up. These results not agree with the common idea that a pressure increase leads to finer droplet distributions. However, it is according to theoretical predictions of the sheet disintegration process [30] that found a rather constant value of the droplet mean diameter for high injection pressure values. When pressure increases, two phenomena operate simultaneously with opposite effects. If the increasing liquid velocity enhances the destabilizing effects of the aerodynamic forces and reduces the population of large droplets, the disintegration occurs sooner and thereby with a thicker liquid sheet.

ANALYSIS OF THE SPRAY STRUCTURE

The structures of the much finer sprays generated by the atomization regime II were analyzed in detail. The main characteristics of the sprays obtained at 16 and 20 bar, can be compared in Figure 12.

The images of the sheet instantaneous disintegration, the mean velocity fields of both phases (droplets and air) and the profiles of droplet axial volume flux, f_{Vx}, at each axial measurement station are represented for the two cases. It is also shown the axial evolution of the R_{50} and ΔR parameters.

As exposed previously, the surrounding air flows into the spray near field causing the formation of a high velocity jet at the central zone. The interaction between the incoming air and the disperse phase gives rise to the transport of the smallest droplets towards the spray core and the consequent coupling of the velocities of both phases at this region. High relative velocities between phases can be initially detected around the densest spray region and towards the spray periphery. Nevertheless, droplets quickly decelerate downstream by the strong momentum exchange with the slower moving air and droplet velocities approach those of the continuous phase in the spray far field.

The hollow cone structure of the sprays is remarked by the axial volume flux profile at the nearest axial location to the sheet break-up point. As sprays develop, the transport of small droplets into the spray core by the incoming air and the droplet inertial classification cause the spatial redistribution of the liquid flow rate and, consequently, more uniform profiles of the axial volume flux. A progressive increase of R_{50} and ΔR (dashed region) was also obtained in each case. The values corresponding to each axial station can be reviewed in Table 5.

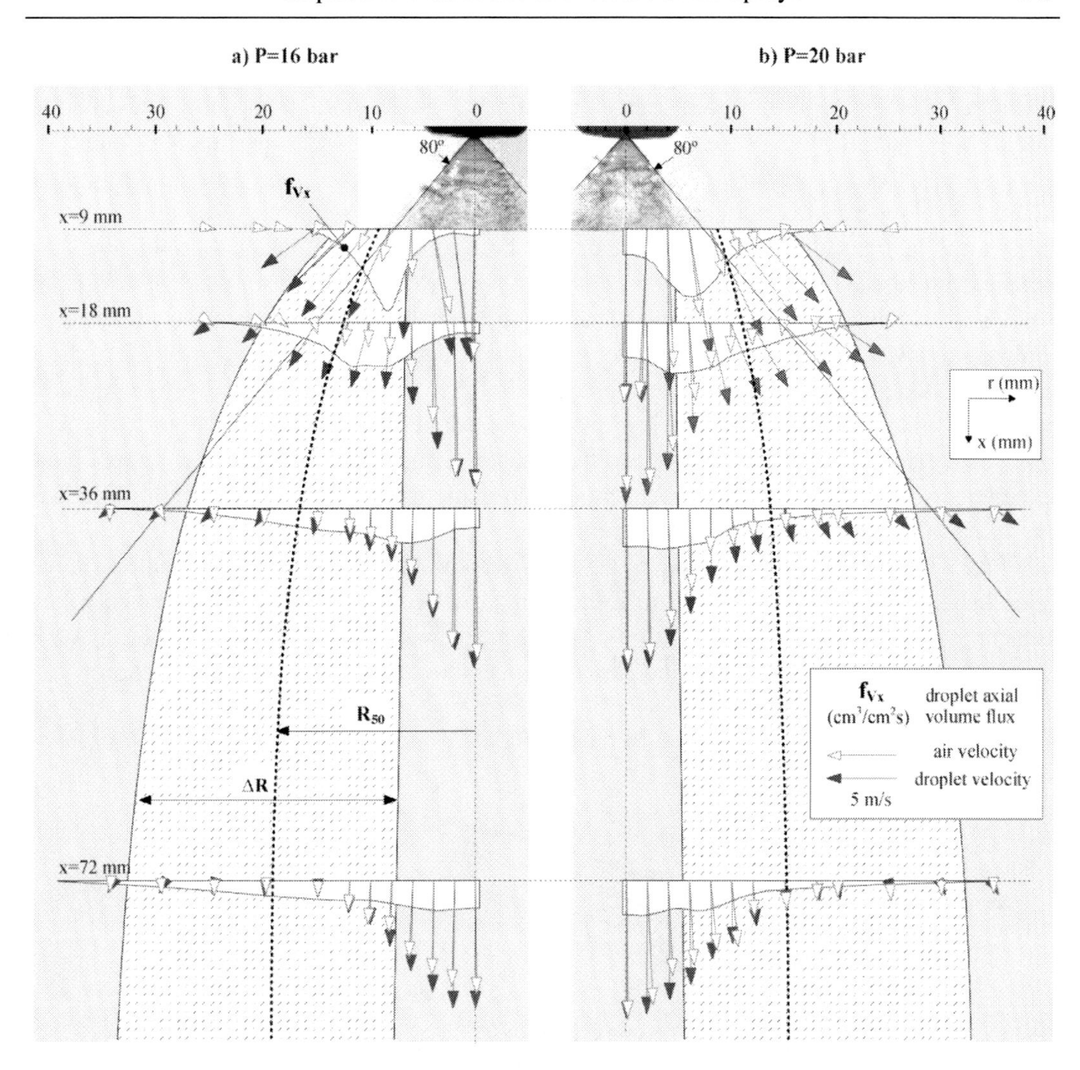

Figure 12. Structure of the sprays. a) P=16 bar, b) P=20 bar.

Table 5. Experimental data of the sprays at each axial measurement station

	P=16				P=20			
x (mm)	R_{50} (mm)	ΔR (mm)	F_T (cm^3/s)	% def	R_{50} (mm)	ΔR (mm)	F_T (cm^3/s)	% def
9	9.13	6.84	0.51	34.4	8.4	11.09	0.55	35.9
18	12.11	12.39	0.62	20.3	11	15.56	0.65	23.7
36	16.31	20.88	0.63	17.8	13.46	22.81	0.68	20.6
72	18.49	24.70	0.61	22	15.04	27.16	0.68	20.9

We observe that R_{50} reduces and ΔR increases as pressure goes up from 16 to 20 bar, since the sheet disintegration occurs nearer the nozzle exit. The total liquid flow rates and the liquid flow rate deficits were calculated at each axial station from Eqs. (6) and (7), respectively. Results are presented in Table 5. It can be noted that the injected liquid volumetric flow rate was not completely captured by the measurements. With the exception of x=9 mm, deficits about 20% were obtained. These results are similar to those obtained by other researchers [31].

Usual explanations for no droplets validation by PDPA technique are the presence of multiple droplets in the probe volume or non-spherical droplets produced in the first sheet break-up. Here, non-spherical droplets generated by the collision phenomena are believed to be a very likely cause of data reject. As discussed below, droplet collisions will be a particularly important phenomenon in the densest zone of the spray and as previous experiments showed, [17]; [32], the temporal formation of a mass of deforming liquid is produced in the most droplet collision events.

In Figure 13, the representation of the D_{32} profiles using the dimensionless radial coordinate, allow us to detect the significant axial growth of the droplet mean diameter, which is a consequence of the droplet collision-coalescence phenomena that occur along the spray. It should be also noted that the axial increase of D_{32} is more important for the 16 bar case.

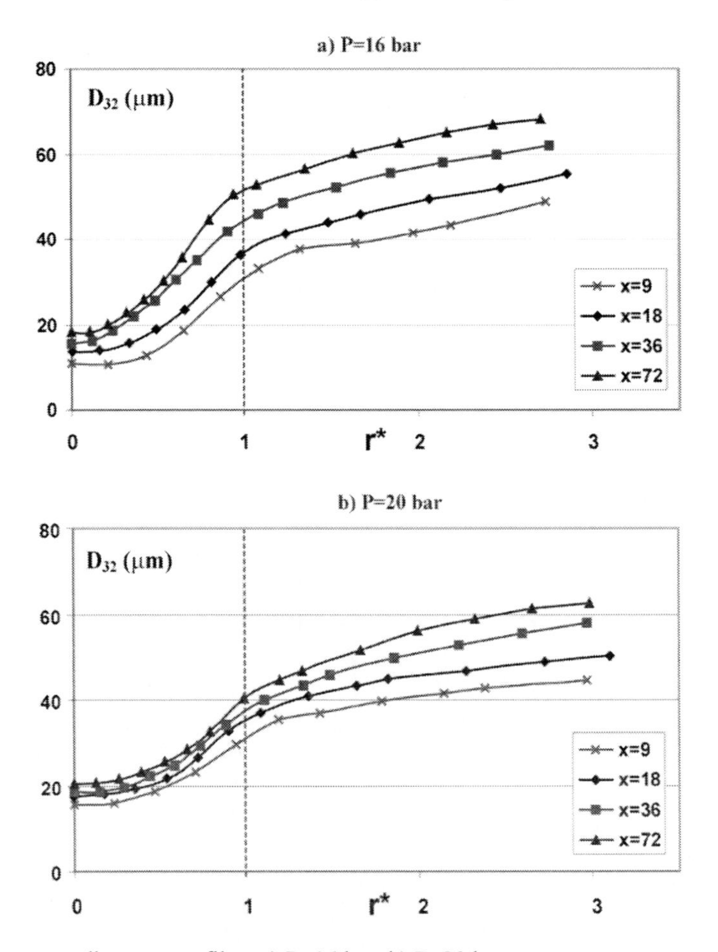

Figure 13. Sauter mean diameter profiles. a) P=16 bar, b) P=20 bar.

For analyzing in detail the liquid flow structure three different droplet size classes: small, medium and large, were considered. The axial volume flux profiles, denoted as f_{Vxs} for the small, f_{Vxm} for the medium and f_{Vxl} for the large size class, are plotted in Figure 14 at each axial measurement station. The radian mean spread of each size class, designated R_{50s}, R_{50m} and R_{50l}, respectively, was also calculated and their axial evolution was compared with the radial mean spread of the spray.

The f_{Vxm} and f_{Vxl} profiles, in Figure 14a and 14b, show the initial hollow cone structure of the liquid flow rate for both sprays.

In the 16 bar case, the following behaviour is observed as spray develops: the peak of the fv_{xm} profile moves progressively to the spray centre, while the fv_{xl} peak moves to the periphery. This trend is also observed for R_{50m} and R_{50l} that evolve, respectively, to the core and to the edge of the spray. The axial evolution of the f_{Vxm} profile must be analyzed taking into account the variation of the total liquid flow rate of this size class, (see Figure 16a). A progressive transfer of liquid flow rate from the medium droplet size class to other size classes was produced along the spray, which specifically affected their spatial distribution.

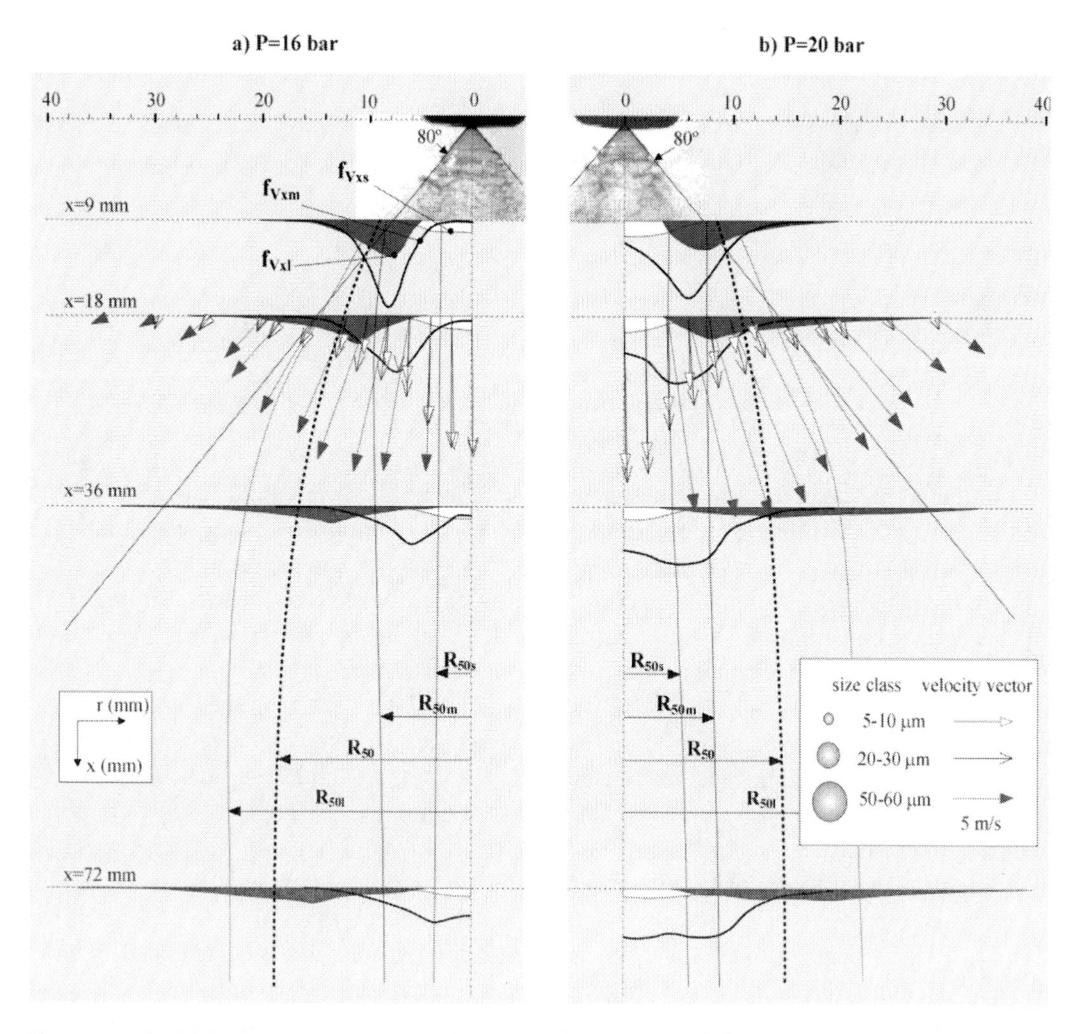

Figure 14. Liquid flow structure. Study of three different droplet size classes: small, medium and large. a) P=16 bar, b) P=20 bar.

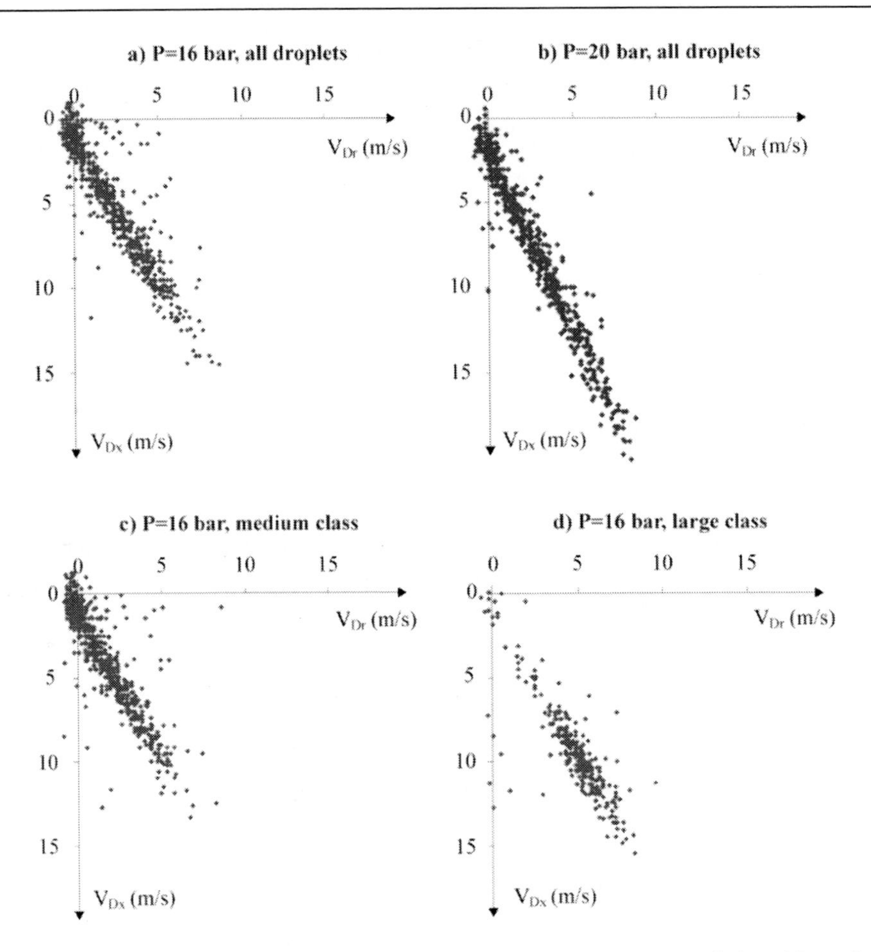

Figure 15. Vertexes of the droplet velocity vectors. Axial station: x=18 mm. Radial position: r*=1.

In the 20 bar case, R_{50m} remain almost constant along the spray and high values of f_{Vxm} are detected at downstream sections from the nozzle exit. If we observe the $\%F_T$ variation of the medium size class (see Figure 16b) a less rigorous reduction can be also noted.

On the other hand, the small size class, with low axial fluxes, is concentrated at the centre of the sprays. Nevertheless, a light increase of R_{50s} should be noted for the 20 bar case, suggesting the presence of small droplets in outer radial positions.

If the mean velocity maps of the three droplet size classes are examined in Figure 14, a behavior clearly dominated by the aerodynamic effects can be noted in the velocity vectors of the densest spray region. The axial station of 18 mm was selected for analyzing the correlation between droplet velocity components and for the subsequent study of droplet collision phenomena.

The vertexes of the velocity vectors for a number of droplets moving at the local position of R_{50} were represented in Figure 15a and 15b, for the injection pressures of 16 and 20 bar, respectively. Strong correlation between velocity components and elevated dispersion of velocity modules can be observed in two cases. According to the classic kinetic theory of gases the collision rate between droplets can be related to their on relative velocity. Therefore, high droplet collision rates are expected at the densest spray region.

It is noted that the relative velocity between droplets increases substantially at P=20 bar, so higher inertia collisions, which could affect the collision outcome, will be produced in this case. An important dispersion of velocity modules can be also observed within the droplets of the medium and large size classes (Figure 15c and 15d). Thus, high droplet collision rates can be also expected within each droplet size class.

ANALYSIS OF DROPLET COLLISION PHENOMENA

In order to elucidate the effects of the droplet collisions on the spray characteristics, the axial evolution of the overall volume droplet size distribution was analyzed.

By means of Eq. (5), the total liquid flow rates for ten droplet size classes were calculated at each axial measurement station. The resulting overall droplet size distributions of the sprays generated at 16 and 20 bar, are represented in Figure 16a and 16b, respectively.

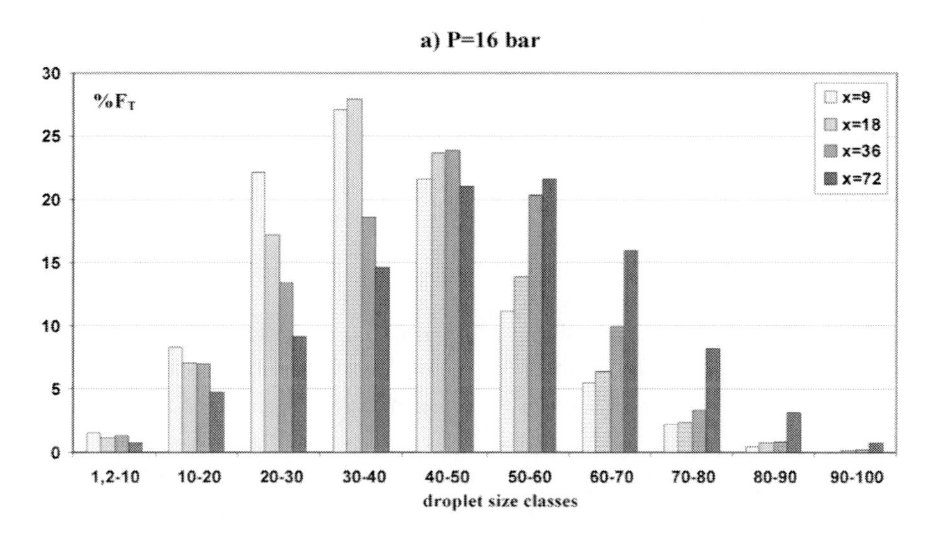

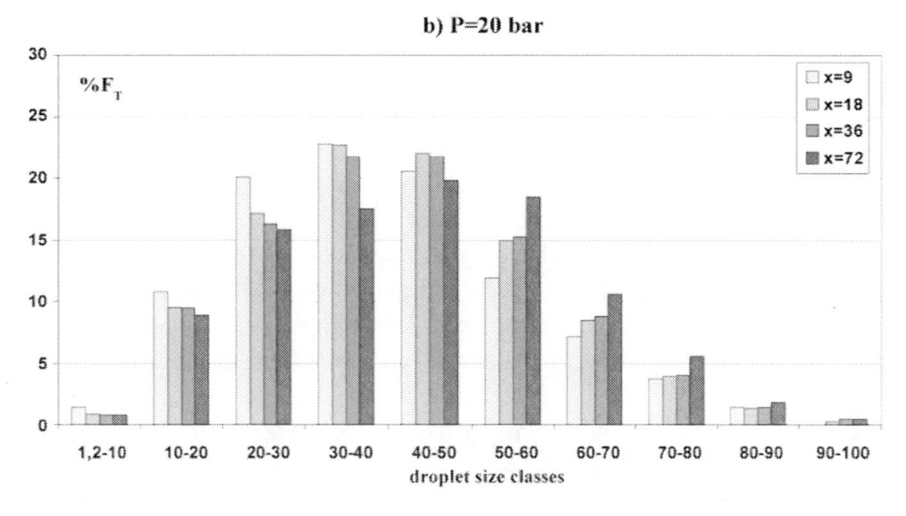

Figure 16. Axial evolution of the overall droplet size distribution across the spray.

The examination of the total liquid flow rate distributions by droplet size classes allow us to discover the progressive liquid flow rate transfer between size classes as spray develops. Particularly, regarding the $\%F_T$ axial evolution in 16 bar case, a drastic reduction of 13% is produced for droplet diameters from 20 to 30 μm, while a considerable increase of 10.5% is measured for the droplets of the large size class.

Measuring values of $\% F_T$ at each axial station can be seen in Table 6. The corresponding mean diameters and typical deviations of the overall droplet size distributions are also summarized in Table 6.

Table 6. Data of the overall droplet size distributions at different axial stations

	P=16				P=20			
x (mm)	D_{43} (μm)	rms_F (μm)	$\%F_T$ (20-30)	$\%F_T$ (50-60)	D_{43} (μm)	rms_F (μm)	$\%F_T$ (20-30)	$\%F_T$ (50-60)
9	37.8	14.9	22.1	11.1	39.3	16.9	20.1	11.8
18	39.9	15	17.1	13.8	41.4	16.6	17.2	14.9
36	42.7	16.3	13.3	20.3	41.7	17.1	16.2	15.2
72	49.6	17.9	9.1	21.6	43.4	18	15.8	18.5

In two cases, statistical moments show that spray evolution leads to coarser and wider volume droplet size distributions, due to the droplet collision-coalescence phenomena. Nevertheless, if the growth of D_{43} is examined along the two sprays, a substantial increase of around 12 μm is obtained at P=16 bar while a more moderated growth of approximately 4 μm is detected at P=20 bar.

For understanding the causes of this different behavior in the overall droplet size distributions, first a study of droplet collision frequencies and mean free paths between subsequent realizations of droplet-droplet collisions was performed. Following, the possible collision outcome was analyzed by examination the collision Weber number.

Collision frequencies of the medium and large droplet size classes were determined from Eq. (14) and mean free paths in the axial direction of these size classes, were obtained from Eq. (15). The resulting profiles at the axial station of x=18 mm, are plotted in Figure 17a and 17b, for the injection pressures of 16 and 20 bar, respectively.

In both cases, the highest collision frequencies of the two droplet size classes were located at radial positions where high volume fluxes and high relative velocities between droplets were measured. At P=16 bar, maxima are obtained at R_{50}. Hence, the lowest values for the time between subsequent collisions and for the axial mean free paths will be also detected at this spray zone. Towards the core and the edge of the spray, L_{Cx} increase considerably.

Special attention should be put in maxima and minima of the profiles plotted in Figure16, which are listed in Table 7.

Table 7. Collision data for the medium and large droplet size classes at the axial location of 18 mm

	P=16				P=20			
Size class	f_{Cmax} (1/s)	τ_{Cmin} (s)·10^{-3}	L_{Cxmin} (mm)	τ_D /τ_{Cmin}	f_{Cmax} (1/s)	τ_{Cmin} (s)·10^{-3}	L_{Cxmin} (mm)	τ_D /τ_{Cmin}
Medium	802	1.24	3.8	1.3	1020	0.98	5.9	1.7
Large	2160	0.46	3.1	17.8	2173	0.46	7.1	17.8

L_{Cxmin} are bigger in the 20 bar case, since the droplets present higher velocities. Therefore, the number of collision events per unity of length reduces at P=20 bar and give us explanation for the more moderated growth of D_{43} by collision-coalescence phenomena, which was detected in this case.

On the other hand, the droplet response times were compared with minima of the times between successive collisions. As shown in Table 7, τ_D values are bigger than τ_{Cmin} around R_{50} position. Thereby, droplets are not able to completely respond to the fluid flow between successive collisions and the flow properties as mean velocity and turbulence can be considerably modulated by the dispersed phase at this region of the sprays.

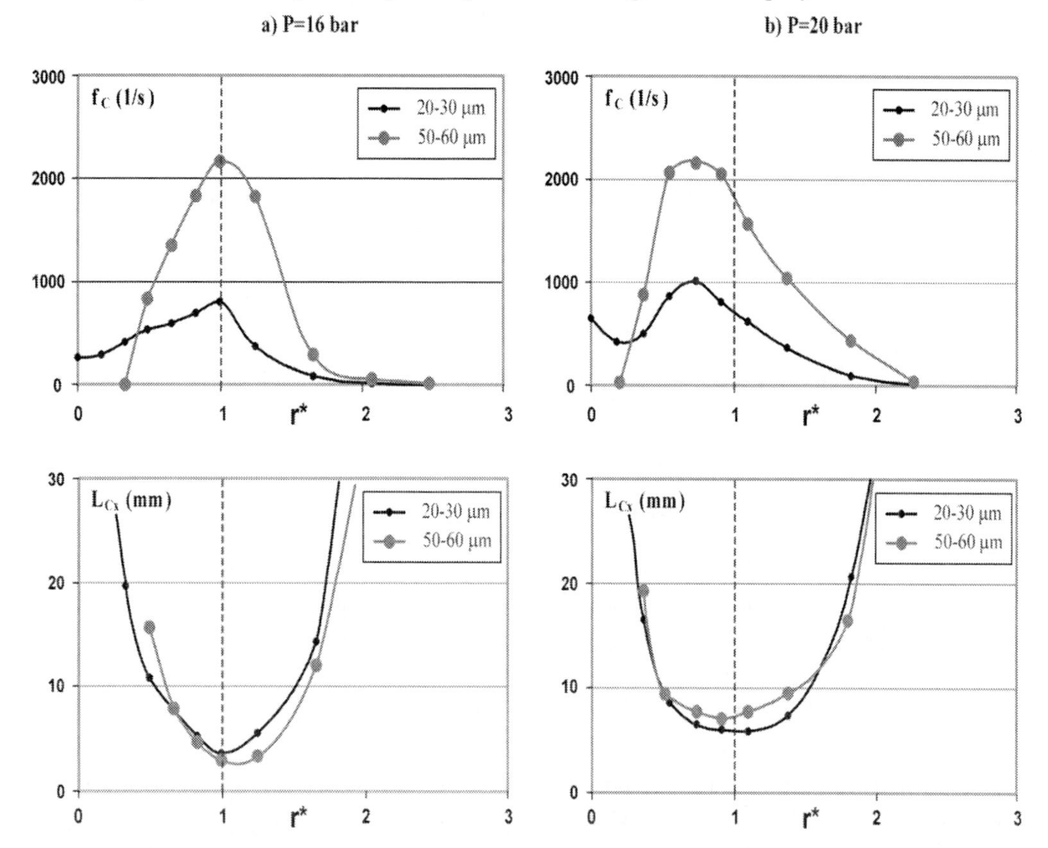

Figure 17. Profiles of droplet collision frequencies and axial mean free paths.

Medium and large droplet size classes. Axial location, x=18 mm. a) P=16 bar, b) P=20 bar.

Finally, the droplet collision outcomes at the particular location of R_{50} were investigated. The number size distributions were plotted in Figure 18 and the mean collision Weber numbers, We_C, were calculated for the two injection pressures. Results for collisions between droplets of the medium size class, between droplets of the large size class and collisions between droplets of the medium and large classes can be compared in Table 8.

It should be noted that as pressure increases, higher We_C are obtained and the local droplet number distribution tends to acquire a bimodal shape.

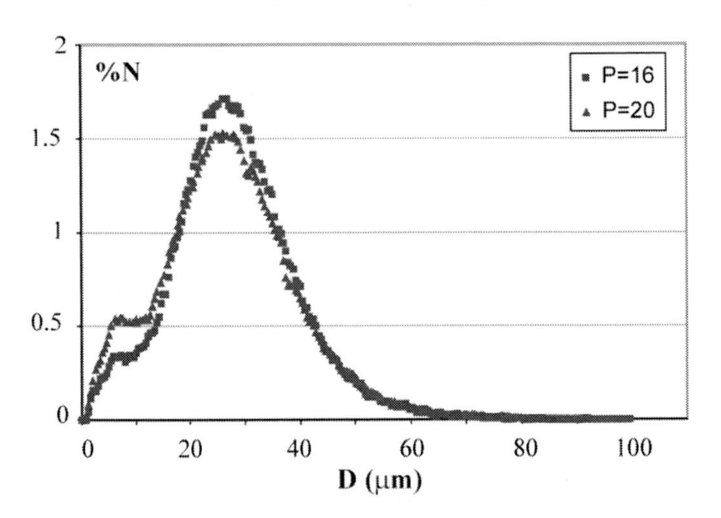

Figure 18. Number size distributions. Axial station: 18 mm, radial location: $r=R_{50}$.

Table 8. Mean collision Weber and mean droplet Weber numbers. Local position: x=18 mm, $r=R_{50}$

	P=16			P=20		
Weber number	Medium	Large	Medium/ Large	Medium	Large	Medium/ Large
$We_C = \rho_l V_{rel} D/\sigma_l$	14.9	70.2	60.6	21.6	158	123.2
$We_D = \rho_a V_{rel}^2 D/\sigma_l$	0.008	0.11	-	0.025	0.51	-

The presence of small droplets could be due to the secondary atomization process, as it is appointed in other investigations [38]. However, the low Weber numbers of the droplets generated in this experiment (see Table 8) do not suggests this possibility. Rüger et al, (2000) assumed that this is a result of the Gaussian beam effect in the PDA measurements, but this problem was practically removed in the related work due to the selected optical configuration of the PDA.

In this work, the significant increase of small droplets for higher pressures is believed to be the result of the increase of satellite droplets generation by collision phenomena. The rate of small droplets produced after an impact is higher as the collision Weber number increases and this occurs when the relative velocities of the colliding droplets is higher.

A concern of the majority of recent studies in the field of droplet collisions is the identification of the critical values of We and B, which determines the boundaries between the different type of collision outcomes. The application of these critical values in hydrocarbon droplet collisions for the prediction of the size distribution evolution will be object of investigation in future works.

CONCLUSION

Hollow cone pressure swirl sprays developing in a slow moving air were investigated using experimental techniques. The spray near field was analyzed at three different cases, which included two distinct sheet atomization regimes: perforations and surface wave instabilities.

All sprays showed the typical hollow cone structure of the injected liquid. The smallest droplets, transported by the entrained air, concentrated around the central zone. Wide drop size distributions were obtained toward the spray periphery and a continuous increase of the droplet mean diameter was measured in the radial direction. The parameter of spray radial mean spread was located near the position of maximum axial volume flux (spray densest region).

When perforated-sheet changes to wavy-sheet disintegration, a notably finer spray with a higher radial dispersion was obtained. The entrained air flow rate showed a significant increase with higher axial velocities at the central zone. The air entrainment process notably affected to the droplet mean velocity profiles.

High dispersion of velocity modules and strong directional correlation were detected for the droplets of the spray densest region, even within each droplet size class. The relative velocities between droplets promoted high collision rates, which decisively affected the axial evolution of the droplet size distribution. As a consequence of the coalescing collisions, coarser and wider distributions were obtained along the spray.

The pressure increase generated a higher number of droplet collisions with separation outcome and a reduction of the number of collision events per unit of length. Hence, a more moderated growth of the droplet mean diameter along the spray was obtained.

REFERENCES

[1] Yule, A.J., Chinn, J.J. The internal flow and exit conditions of pressure swirl atomizers. *Atomization Sprays* 2000, vol 10, pp 121-146.

[2] Nonnenmacher, S., Piesche, M. Design of hollow cone pressure swirl nozzles to atomize Newtonian fluids. *Chem. Eng. Sc.* 2000, vol 55, pp 4339-4348.

[3] Halder, M.R., Dash, S.K., Som, S.K. Initiation of air core in a simplex nozzle and the effects of operating and geometrical parameters on its shape and size. *Exp. Th. Fluid Sc.* 2002, vol 26, pp 871-878.

[4] Lefebvre A. H. Atomization and sprays. Hemisphere Publishing Corporation. New York, 1989.

[5] Chung, I.P., Presser, C. Fluid property effects on sheet disintegration of a simplex pressure-swirl atomizer. *AIAA J. Prop. Power* 2001, vol 17, n° 1, pp 212-216.

[6] Inamura, T., Tamura, H., Sakamoto, H. Characteristics of liquid film and spray injected from swirl coaxial injector. *AIAA J. Prop. Power* 2003, vol 19, n° 4, pp 632-639.

[7] Ballester, J.M., Dopazo, C. Drop size measurements in heavy oil sprays for pressure-swirl nozzles. *Atomization Sprays* 1996, vol 6, pp 377-408.

[8] Bates, C.J. The effect of aperture setting on pressure atomizer phase Doppler anemometry measurements. ICLASS-1994, Rouen, France. *Begell House*, pp 429-436.

[9] Sommerfeld, M. Analysis of isothermal and evaporating turbulent sprays by phase-Doppler anemometry and numerical calculations. *Int. J. Heat Fluid Flow* 1998, vol 19, pp 173-186.

[10] Santolaya, J.L., Aisa, L., Calvo, E., García, I., Cerecedo, L.M. Experimental study of near-field flow structure in hollow cone swirl sprays. *AIAA J. Prop. Power* 2007, vol 23, pp 382-389.

[11] Feikema, D.A., Eskridge, R., Hutt, J.J. Structure of a nonevaporating swirl injector spray. *Atomization Sprays* 1997, vol 7, pp 77-95.

[12] Rüger, M., Hohmann, S., Sommerfeld, M., Kohnen, G. Euler/Lagrange calculations of turbulent sprays: the effect of droplet collisions and coalescence. *Atomization Sprays* 2000, vol 10, pp 47-81.

[13] Schelling, J., Reh, L. Influence of atomizer design and coaxial gas velocity on gas entrainment into sprays. *Chem. Eng. Proc.* 1999, vol 38, pp 383-393.

[14] Gourdel, C., Simonin, O., Brunier, E. Modelling and simulation of gas-solid turbulent flows with a binary mixture of particles. *Int. Conf. Multiphase Flows*, ICMF-1998, Lyon, France.

[15] Sommerfeld, M. Validation of a stochastic Lagrangian modelling approach for inter-particle collisions in homogeneous isotropic turbulence. *Int. J. Multiphase Flow* 2001, vol 27, pp1829-1858.

[16] Jiang, Y.J., Umemura, A., Law, C.K. An experimental investigation on the collision behaviour of hydrocarbon droplets. *J. Fluid Mech.* 1991, vol 234, pp 171-190.

[17] Qian, J., Law, C.K. Regimes of coalescence and separation in droplet collision. *J. Fluid Mech.* 1997, vol 331, pp 59-80.

[18] Orme, M. Experiments on droplet collisions, bounce, coalescence and disruption. *Progess Energy Combustion Sc.* 1997, vol 23, pp 65-79.

[19] Post, S.L., Abraham, J. Modelling the outcome of drop-drop collisions in diesel sprays. *Int. J. Multiphase Flow* 2002, vol 28, pp 997-1019.

[20] Munnannur, A., Reitz, R. D. A new predictive model for fragmenting and non-fragmenting binary droplet collisions. *Int. J. Multiphase Flow* 2007, vol 33, pp 873-890.

[21] Kollár, L.E., Farzaneh, M. Modeling the evolution of droplet size distribution in two-phase flows. *Int. J. Multiphase Flow* 2007, vol 33, pp 1255-1270.

[22] Santolaya, J.L., Aisa, L., Calvo, E., García, I., García, J.A. Analysis by droplet size classes of the liquid flow structure in a pressure swirl hollow cone spray. *Chem. Eng. Proc.* 2010, vol 49, pp125-131.

[23] Bachalo, W.D. Spray diagnostics for the twenty-first century. *Atomization Sprays* 2000, vol 10, pp 439-474.

[24] Widmann, J.F., Presser, C., Leigh, S.D. Improving phase Doppler volume flux measurements in low data rate applications. *Meas Sc. Tech.* 2001, vol 12, pp 1180-1190.

[25] Qiu, H.H., Sommerfeld, M. A reliable method for determining the measurement volume size and particle mass fluxes using the phase-Doppler anemometry. *Exp. Fluids* 1992, vol 13, pp 393-404.

[26] Zhu, J.Y., Rudoff, R.C., Bachalo, E.J., Bachalo, W.D. Number density and mass flux measurements using the phase doppler particle analyzer en reacting and non-reacting swirling flows. 31st Aerospace Science Meeting 1993, Reno, USA.

[27] Aísa, L.A., García, J.A., Cerecedo, L.M., García, J.I., Calvo, E. Particle concentration and local mass flux measurements in two-phase flows with PDA. Application to a study on the dispersion of spherical particles in a turbulent air jet. *Int. J. Multiphase Flow* 2002, vol 28, pp 301-324.

[28] Dupoy, D., Florès, B., Lisiecki, D., Dumouchel, C. Behaviour of swirl atomizers of small dimensions. ICLASS-1994, Rouen, France. Begell House, pp 374-381.

[29] Lai, W.H., Wang, M.R., Huang, D.Y. Turbulence modulation in a simplex spray. ICLASS-1997, Seoul, Korea. Begell House, pp 409-416.

[30] Cousin, J., Yoon, S.J., Dumouchel, C. Coupling of classical linear theory and maximum entropy formalism for prediction of drop size distribution in sprays: application to pressure-swirl atomizers. *Atomization Sprays* 1996, vol 6, pp 601-622.

[31] Domnick, J. Some comments concerning the state-of-the-art of phase Doppler anemometry applied to liquid sprays. ILASS-1997, Florence, Italy. *Enel,* pp 366-374.

[32] Willis, K.D., Orme, M.E. Experiments on the dynamics o droplet collisions in a vacuum. *Exp. Fluids* 2000, vol 29, pp 347-358.

[33] Dumouchel, C., Sindayihebura, D. Drop size distribution characteristics of sprays produced by swirl atomizers of small dimensions. ILASS-1998, Manchester, England. UMIST, pp 206-211.

In: Sprays: Types, Technology and Modeling
Editor: Maria C. Vella, pp. 257-286
ISBN 978-1-61324-345-9

Chapter 7

MODELING ASPECTS OF THE INJECTION OF UREA-SPRAY FOR NO_X ABATEMENT FOR HEAVY DUTY DIESEL ENGINES

Andreas Lundström** and Henrik Ström
Chemical and Biological Engineering,
Chalmers University of Technology,
Gothenburg, Sweden

ABSTRACT

Eulerian-Lagrangian simulations of an evaporating water-urea spray have been performed. Velocity, temperature, composition and size distribution are simulated. The chapter contains an extensive discussion of what forces and mass/heat transfer models that should be included.

Evaporation of water is included and evaporation and reaction of urea is modeled using a new model for the urea vapor pressure, previously developed by the authors [17].

The main objective of this chapter is to present a guide how to model a water-urea spray including mass and heat transfer. It is shown how the modeling of turbulent dispersion and the choices of injector position affects the model predictions in terms of spray uniformity and residence time.

---

* Corresponding author: andreas.lundstrom@chalmers.se, +46 (0)31-772 3094.

NOMENCLATURE

Variables – Latin Letters

Variable	Units	Description
A	m^2	area
B_M		mass transfer number
B_T		heat transfer number
C_D		drag coefficient
c_p	J/kg·K	specific heat
D	m	pipe diameter
D_{AB}	m^2/s	diffusion coefficient of vapor in the bulk
$\overline{d}$	m	Rosin-Rammler diameter
d	m	diameter
F	N	force
F_M		film thickness correction factor
F_T		film thickness correction factor
g	m/s^2	gravitational acceleration
h	W/K	heat transfer coefficient
Δh_{vap}	J/kg	heat of vaporization
k	W/m·K	thermal conductivity
k	m^2/s^2	turbulent kinetic energy
k_c	m/s	mass transfer coefficient
m	kg	mass
n		Rosin-Rammler exponent
P	Pa	total pressure
p	Pa	partial pressure
t	s	time
T	K	temperature
u	m/s	velocity
X		mass fraction
Y		mole fraction

Variables – Greek Letters

Variable	Units	Description
μ	$kg·m/s^2$	molecular viscosity
ν	m^2/s	kinematic viscosity
ρ	kg/m^3	density
σ	N/m	surface tension between droplet and gas
ξ		uniformly distributed random number

Superscripts and Subscripts

Subscript	Meaning
o	saturation
film	film
g	gas
i	coordinate direction (i = x, y, z)
p	particle (droplet)
r	relative
s	surface
urea	urea
vap	vapor
x	subscript for identification of forces acting on droplets D = drag, B = buoyancy, L = lift, T = thermophoresis, H = history, AM = added mass, G = gravity, P = pressure gradient
0	initial

Dimensionless Numbers

Abbreviation	Name	Definition
Nu	Nusselt number	$\frac{hd_p}{k_g}$
Pr	Prandtl number	$\frac{c_{p,g}\mu_g}{k_g}$
Re	Reynolds number	$\frac{\rho_g D u_g}{\mu_g}$
Re_p	Droplet Reynolds number	$\frac{\rho_g d_p U_c}{\mu_g}$
Sc	Schmidt number	$\frac{\mu_g}{\rho_g D_{AB}}$
Sh	Sherwood number	$\frac{k_c d_p}{D_{AB}}$
We	Weber number	$\frac{\rho_g d_p U_c^2}{\sigma}$

1. Introduction

Today there exists an ever increasing demand on the automotive industry to reduce the levels of harmful emissions from diesel engines. To meet these new and harder regulations new strategies for the exhaust gas aftertreatment systems are being implemented, adding to the ever growing complexity and design of such systems. One of the major pollutants in diesel exhaust is nitrogen oxides (NO_X). NO_X has adverse health effects on humans and contributes to smog and acid rain.

One suitable technique to reduce NO_X emissions for mobile applications is selective catalytic reduction (SCR) with urea. In this application urea is injected upstream of the SCR catalyst. When the urea is heated by the warm exhaust gases it decomposes into ammonia (NH_3) and isocyanic acid (HNCO). NH_3 is the reducing agent needed to transform the NO_X to N_2 on the downstream SCR catalyst.

Due to the complexity of urea-SCR systems computational fluid dynamics (CFD) will most probably be a valuable tool in the design process. The Eulerian-Lagrangian framework is in this context probably the most viable approach to model the injection of urea-spray due to the relative low mass loadings exhibited during urea dosage. In essence this framework captures the most important interactions between the fluid flow and the urea particles i.e. mass, heat and momentum in a straightforward method.

In this work a review of the Eulerian-Lagrangian approach for modeling of a urea-spray for mobile exhaust gas aftertreatment applications is presented. This includes aspects of mass, heat and momentum transfer to and from the urea particles as well as effects of turbulence on the dispersion of the urea. Different models for the decomposition rate of urea are compared and results from simulations evaluating effects of different injection position and turbulent dispersion are discussed.

1.1. Decomposition of Urea

Commonly for a urea-SCR system an aqueous 32.5 wt% urea and water solution (UWS) is injected into the hot exhaust gases upstream of the catalyst. In the ideal process water evaporates leaving solid urea which then melts and undergoes thermal decomposition into gaseous ammonia and isocyanic acid [1, 2]:

$$NH_2 - CO - NH_2(s) \rightarrow NH_3(g) + \ HNCO\ (g), \Delta H = 185.5\ kJ/mol \qquad (1)$$

Trough hydrolysis on the downstream SCR catalyst, or slowly in the gas phase at high temperatures, the formed isocyanic acid may be converted to additional ammonia [1]:

$$HNCO\ (g) + H_2O(g) \rightarrow NH_3(g) + CO_2(g) \qquad (2)$$

This reaction is limited mainly by external and internal mass transfer [3] and is much faster than the SCR reactions. This means that every mole of decomposed urea ultimately results in two moles of ammonia taking part in the SCR reactions.

There exists some discern to what state the urea particles are in after the water has evaporated. When all of the water in the UWS droplets has evaporated more or less completely there exist several suggestions in the literature about the state of the residue urea. These suggestions range from solid [1, 3, 4] or molten [3-5] urea to a very concentrated UWS solution [6]. Urea has a melting point of approximately 406 K [7]. After the melting point is reached there exist several suggestions to at which temperature the actual decomposition starts e.g. 406 K [8], 416 K [9], 425 K [2, 10], 433 K [11, 12] and even at as low temperature as 353 K [13].

In previous studies urea decomposition has been modeled (in order of increasing complexity) simply by introduction of a conversion efficiency factor tuned against experimental data [3], as controlled by turbulent mixing [4], treating the decomposition as a heat transferred limited process with a fixed decomposition temperature at 425 K [14], using an Arrhenius expression [8, 15] or as an evaporation process with a saturation pressure determined from experimental data [16].

Both of the models proposed by Birkhold et. al [15, 16] were tuned against experimental data but produce particle temperatures during urea decomposition far above any experimental values reported in the literature. The model in [14] employed a fixed decomposition temperature in the range of reported values of urea decomposition and thus avoids the problems with unrealistically high decomposition temperatures.

In this work a vapor pressure for urea recently proposed by Lundström et al. [17] is used to model the evaporation rate of urea. The model by Lundström et al. uses the extrapolated vapor pressure above solid urea to obtain the corresponding liquid vapor pressure. In [17] the model is compared to recently published data by Wang et al. [18] and exhibit remarkable agreement.

1.2. By-Product Formation

Depending on temperature and local urea and isocyanic acid concentrations the efficiency of urea decomposition varies. This is due to the formation of high molecular by-products that form during urea decomposition. Several studies report the formation of biuret, ammeline, ammelide and cyanurric acid (CYA) during urea decomposition [10, 12, 19-21]. By product formation is reported to start at temperatures above 406 K i.e. the melting point of urea [10]. Formation of biuret is believed to proceed via the reaction of gas phase isocyanic acid and molten urea:

$$NH_2 - CO - NH_2 + HNCO\,(m) + H_2O(g) \rightarrow NH_2 - CO - NH - NH_2(m) \qquad (3)$$

Biuret has been suggested to work as the precursor in the formation of CYA which is stable to high temperatures (above 550 K) and the deposition product found in greatest abundance in unfavorably tuned urea-SCR systems [21]. Thus, accurate prediction of local HNCO and urea concentrations are of great importance during the design process.

1.3. Wall Effects

Deposits are usually found on regions heavily wetted by the UWS spray i.e. on the exhaust pipe walls, the catalyst or on mixing devices [21]. Deposition is generally attributed to the spray being poorly adjusted leading to a mixture of urea and water hitting the walls [1, 22]. This accumulation of urea in the system and the consequent enhanced possibility for by-product formation further complicates the modeling of the urea-SCR system. The only previous work addressing this issue is presented in [16]. Today most commercial CFD codes include the possibility to model film formation (traditionally developed for in-cylinder calculations). The interaction between the spray and the walls is determined depending on the temperatures of the impinging droplet and the wall and the extent of droplet deformation. The end result ranges from film formation, rebounding and splashing. On top of this framework the complexity of the urea chemistry has to be added including the evaporation of water which changes the rheology of the film, solidification and subsequent melting of urea and the formation of by-products. To further increase the complexity also local film boiling during the different evolution phases of the film might have to be modeled. Thus, given the above presented difficulties and the ambiguous results such endeavors might bring, no attempt to model wall wetting was considered in the current work.

2. Modeling

2.1. System Description

Since the scope of this work is to give a more general overview of the Eulerian-Lagrangian frame work applied to the urea-SCR system a simple geometry consisting of a straight pipe with two different possible spray injection points was considered. Thus effects on evaporation rates and uniformity due to the different models will be easier to quantify and evaluate.

The model system consisted of a straight pipe with a diameter of 120 mm and a length of 1 meter from the exit to the injection point. Two different injection points were investigated, on where the injector is placed at the top of the pipe at a 45° angle to the wall and one where the spray is situated in the centerline directed in the streamwise direction. The injection was modeled as a hollow cone pulse-injector with a mean half cone angle of 10°. Injected droplets were taken from a Rosin-Ramler diameter distribution:

$$d_p = \bar{d}(-\ln(1-\xi))^{1/n} \tag{1}$$

A mean diameter of 100 μm and n=1.21 were used throughout the simulations. The total time for a droplet to evaporate depends heavily on the diameter of the droplets. Of course, these parameters depend on the type and specifications of the injector employed. An air assisted injector would typically produce a finer spray than a single phase mechanical injector. Which type of injector that is used in the real application is commonly manufacture specific and depends on many aspects concerning the final product, i.e. the vehicle. Typical aspects would include the size and packing of the aftertreatment system. However, it is for the purpose of this work not important that the above presented spray exactly matches a single vendor's specifications but should merely form a realistic base for comparison. The flow rate of the injector was 2.5 g/s of UWS and the pulse-width was 10 ms. Droplets were injected with a velocity of 17.75 m/s and 10,000 parcels were used.

2.2. Eulerian-Lagrangian Spray Modeling

The most common approach to modeling an evaporating spray is within the Eulerian-Lagrangian framework, where the continuous phase is solved for in an Eulerian (stationary) reference frame whereas the droplets are tracked in a Lagrangian (moving) reference frame. In this section, we will introduce a Eulerian-Lagrangian framework for simulations of urea-SCR systems, with special focus on the description of the motion of individual droplets.

2.2.1. The Exhaust Gas Flow Field

Complete information about a fluid flow field can be obtained from the Navier-Stokes and continuity equations with the appropriate boundary conditions. However, the computational cost of solving for a highly turbulent flow in an application of industrial scale is typically overwhelming [23]. The gas flow in an exhaust pipe in an aftertreatment system is turbulent, with bulk Reynolds numbers on the order of 100,000 [14]. It is therefore not realistic to resolve the flow directly, and it is typically assumed that the mean gas flow field can be obtained from the Reynolds-averaged Navier-Stokes (RANS) equations with a suitable turbulence model instead.

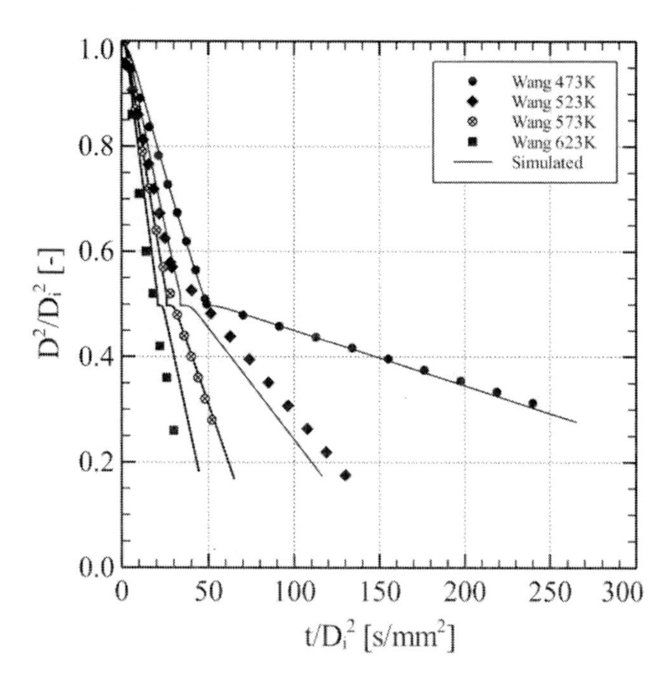

Figure 1. Simulated evaporation rates for UWS droplets compared to data by Wang et al. [18].

The incompressible RANS equations are:

$$\frac{\partial U_i}{\partial x_i} = 0 \tag{2}$$

$$\rho\left(\frac{\partial U_i}{\partial t} + U_j \frac{\partial U_i}{\partial x_j}\right) = -\frac{\partial P}{\partial x_i} + \frac{\partial}{\partial x_j}\left[(\mu + \mu_t)\frac{\partial U_i}{\partial x_j}\right] - \frac{2\rho}{3}\frac{\partial k}{\partial x_i} \tag{3}$$

The influence of the unresolved turbulence on the mean flow field is accounted for via the introduction of a turbulent viscosity, μ_t. In the present work, the turbulent viscosity is calculated from the turbulent kinetic energy, k, and the turbulent dissipation rate, ε. Hence, we use a k-ε turbulence model variant, and solve two additional modeled transport equations (one for k and one for ε) [24].

In the current work, the RNG k-ε turbulence model is chosen, since it responds better to the higher strain rates associated with spray-generated turbulence [25]. This turbulence model has been used extensively in the literature for simulations of turbulent exhaust gas flows [14, 26, 27].

2.2.2. The Droplet Equation of Motion

In a urea-SCR system, depending on the operating conditions, many thousands of droplets may be injected each second. The motion of a droplet in a fluid flow field can be determined from Newton's second law of motion, which states that the acceleration of the droplet is given by the net force on the droplet (per unit mass). In addition, external forces due to gravitation, temperature gradients and non-continuum effects (i.e. Brownian motion) may affect the droplet trajectory. The full equations of motion for the droplet become:

$$\frac{dx_{p,i}}{dt} = u_{p,i} \tag{4}$$

$$m_p \frac{du_{p,i}}{dt} = \oint_S \left[-p\delta_{ij} + \mu\left(\frac{\partial u_i}{\partial x_j} + \frac{\partial u_j}{\partial x_i}\right)\right] n_j dS + F_G + F_T + F_{BM} \tag{5}$$

Here, the different terms on the right hand side of Eq. (5) correspond to a number of different forces on the droplets. The first term is an integral over the droplet surface of the pressure and viscous forces, and is commonly referred to as the aerodynamic force. The other three terms represent the gravitational acceleration, the thermophoretic force and the Brownian motion, respectively. These are not a result of the fluid flow field around the droplet, and must therefore be added explicitly to the force balance.

2.2.3. The Aerodynamic Force

Due to the large number of droplets in a typical urea-SCR system, it is not feasible to resolve the actual flow field around each individual droplet. In fact, it is not even possible in the current framework, since in the current RANS approach, only the mean gas flow field is resolved. Therefore, the aerodynamic force cannot be calculated directly, but must be approximated from the averaged gas flow field interpolated to the point where the droplet is currently located.

In the general case, the aerodynamic force can be decomposed into a number of contributions of different physical interpretation. In the limit of zero particle Reynolds number and for particles smaller than the Kolmogorov scale, this decomposition has been derived analytically by Maxey and Riley [28]. However, in the urea-spray application, the particle Reynolds numbers do not remain small and the droplets may well be larger than the Kolmogorov scale [14]. In the particle Reynolds number range in question ($Re_p < 300$), semi-empirical models for the different contributions to the total aerodynamic force are typically used instead.

The sum of the aerodynamic force and the external force due to gravity may be decomposed as:

$$\oint_S \left[-p\delta_{ij} + \mu\left(\frac{\partial u_i}{\partial x_j} + \frac{\partial u_j}{\partial x_i}\right)\right] n_j dS + F_G = F_B + F_{AM} + F_P + F_D + F_L + F_H \tag{6}$$

We shall now go through these forces and discuss their relative importance in the description of droplet motion in urea-SCR systems.

The first term on the right hand side is the so-called buoyancy force, F_B. It is the resultant force to the gravitational force on the droplet (the droplet weight) and the buoyant force (the weight of the displaced gas), and may be calculated from Eq. (10). In the urea-SCR system, this force is responsible for making the heavier droplets "fall" in the direction of the gravitational acceleration. It might therefore affect the deposition of droplets onto walls and the distribution of droplets over the cross-section of the exhaust pipe. In addition, gravity is known to cause a "crossing trajectory" effect in turbulent flows, by which droplets are moved from one turbulent eddy to another via the action of gravity, hence reducing the turbulent dispersion [29]. It has been established that this force is necessary to include to obtain an accurate description of droplet motion in a urea-spray [14].

The second term is the added (or virtual) mass force, F_{AM}. This force represent part of the unsteady contribution to the drag force on the droplet, and is important only during acceleration if the continuous phase is of similar or higher density to that of the droplets. As this is not the case for a urea-spray (the droplet density is approximately 2000 times larger than the gas phase density), the virtual mass force is insignificant in urea-SCR applications [14].

The third term, F_P, stems from the pressure gradient of the undisturbed fluid and is commonly denoted the pressure gradient force. In the limit of droplet diameters smaller than the spatial variations of the (undisturbed) gas flow, this force will be identical to the pressure force on a fluid sphere in the position of the droplet [28]. Hence, it can be shown that the pressure gradient force is insignificant when the droplet density is much larger than the exhaust gas density. It is therefore not necessary to include in simulations of a urea-spray.

The fourth term is the drag force, F_D, which is the most important force on the droplets in the spray [14]. It is given by Eq. (9) together with a sub-model for the droplet drag coefficient (C_D). In general, the drag coefficient for a spherical non-evaporating particle is a function of the particle Reynolds number [30]. In the urea-SCR system, where the droplets may distort and evaporation takes place from the droplet surface, the drag coefficient must be adjusted to also take these effects into account [14]. This issue is further discussed in *Section 2.3.1.*

The fifth term, F_L, is the lift force on the droplet. The lift force is by definition the force in a plane perpendicular to the drag force. It is due to a velocity gradient (shear) in the continuous phase over the droplet and/or to droplet rotation. Ström et al. [14] showed, using an expression from Kurose and Komori [31], that the total momentum transfer to the droplet from the gas because of the shear (Saffman) lift force is below 0.1% in a generic urea-SCR system. The shear lift force therefore may be neglected. Since the rotational (Magnus) lift force is often less by an order of magnitude to that of shear [32], also this lift force may be neglected [14].

The last term is the so-called history force, F_H. This force accounts for history effects in the boundary layer around the droplet, and represents together with the added mass force the unsteady contribution to the droplet drag. Since droplets will be continuously accelerated and decelerated in the turbulent exhaust gas, it is apparent that this force could become important. The history effects on droplet motion may be assessed by an *a posteriori* analysis based on the expression for the history force from Odar and Hamilton [33], which is valid also at the droplet Reynolds numbers attained in the urea-SCR system. Such an analysis has shown that the total momentum transfer to the droplets from history effects is less than 1% of that transferred by the steady drag force in a typical urea-SCR system [14]. It is therefore concluded that history effects need not be included.

Consequently, the aerodynamic force on the droplets in a urea-SCR spray may be approximated by:

$$\oint_S \left[-p\delta_{ij} + \mu\left(\frac{\partial u_i}{\partial x_j} + \frac{\partial u_j}{\partial x_i}\right)\right] n_j dS + F_G \approx F_B + F_D \tag{7}$$

2.2.4. Other Forces

The remaining two forces to discuss are now the thermophoretic force (F_T) and the Brownian motion (F_{BM}). The former of these two is the force experienced by a particle in a temperature gradient. The total momentum transfer due to thermophoretic effects was estimated by Ström et al. [14] by logging the thermophoretic force (evaluated from the expression of Talbot et al. [34]) under generic urea-spray simulations. It was concluded that the momentum transfer due to the thermophoretic force is negligible when compared to the steady drag force [14].

Brownian motion is a phenomenon that may be observed when a particle trajectory deviates from that predicted by continuum theory in a seemingly random fashion, causing it to follow a meandering path. It is a molecular phenomenon that becomes important when the particle size is comparable to the mean free path of the gas. In a typical exhaust gas aftertreatment system, Brownian motion starts to become important when the droplets become smaller than 1 - 2 μm [14]. If the remaining lifetime of the droplet is negligible when it becomes this small, Brownian motion may be neglected. If that is not the case, however, it must be included.

If Brownian motion is to be accounted for in simulations of droplet motion in urea-SCR systems, two changes are necessary in the particle equation of motion. Since Brownian motion does not naturally appear from the aerodynamic force, it must be modelled via an *ad hoc* fictitious force, F_{BM} [35]. In addition, rarefaction effects on the drag coefficient must be accounted for. This can be done using the so-called Cunningham correction factor [36].

In conclusion, no forces other than the aerodynamic force are typically important in a urea-SCR system. However, depending on the droplet size distribution and lifetimes in the system, the Brownian motion may become important. Throughout the rest of this work, it will be assumed that the Brownian motion is not important and can be neglected.

2.2.5. Final Equation of Motion for the Droplet

As the preceding discussion has shown, a great number of the different forces acting on the droplets in the urea-spray can be neglected in comparison to the largest ones. This significantly facilitates the modeling of the urea-spray and decreases the computational load associated with solving for the droplet trajectories. The final equation of motion for the droplet becomes:

$$m_p \frac{du_{p,i}}{dt} = F_D + F_B \tag{8}$$

where the drag force is given by Eq. (9) and the buoyancy force by Eq. (10).

2.3. Sub-Models to the Droplet Equation of Motion

What remains to be established now is how the droplet drag coefficient, C_D in Eq. (9), shall be evaluated, and how the turbulent dispersion of droplets shall be accounted for. The latter issue is not related to the current description of the droplet motion, but to the fact that in the RANS-approach to the turbulent flow field the turbulent fluctuations are not resolved. Since the turbulent motion of the gas phase will significantly influence the trajectories of the droplets [14], these turbulent fluctuations must be reintroduced (i.e. modeled) in some way.

2.3.1. The Droplet Drag Coefficient

Urea-spray droplets are not rigid, they will both deform and experience internal circulation during their lifetime in the urea-SCR system. Since they are evaporating and decomposing, the boundary condition for fluid flow at the droplet surface will be affected by the mass transfer. In addition, the turbulent fluctuations of the continuous phase may affect the drag coefficient for droplets which are large compared to the Kolmogorov scales. All of these effects may influence the droplet drag coefficient and how it varies with time and position in a given system.

Ström et al. [14] assessed the influence of each of the mentioned phenomena on the drag coefficient for a typical urea-spray droplet. It was concluded that the effects of internal circulation and the continuous phase turbulence could be neglected [14]. The effects of deformation and evaporation should however be accounted for, and this can be accomplished by using the drag coefficient correlation of Haywood et al. [37] (Eq. (12)). This is also the correlation used in the present work.

With this correlation, the droplet drag coefficient is first evaluated using Eq. (11) (from Clift et al. [30]), which gives the drag coefficient for a rigid, non-evaporating sphere at the current droplet Reynolds number. This coefficient is then adjusted for evaporation (via the heat transfer number, B_T) and for deformation (via the droplet Reynolds and Weber numbers) as prescribed by Eq. (12). Note that the droplet Reynolds number, Re_p, shall be evaluated at the free stream density and the film viscosity using the 1/3 rule (explained in *Section 2.4*) to achieve the highest possible level of accuracy [37, 38].

When all water has evaporated and only pure urea is left, the droplet will have been accelerated almost to the local gas velocity. Since the viscosity of the urea melt is also very high, it is reasonable to assume that the droplet does not distort any more at this point. Therefore, the factor that accounts for droplet distortion is dropped from Eq. (12) after all the water is gone.

It should be mentioned at this point that it is assumed in the present work that the average droplet spacing is much larger than the droplet radius, so that the spray can be regarded as dilute. If this assumption is violated, it becomes necessary to also adjust the drag coefficient for the presence of other droplets. For more information, the reader is referred to [39].

2.3.2. Turbulent Dispersion

The droplets from the urea-spray will soon be accelerated by the exhaust gas. As an example, a 5 μm droplet injected into a 75 m/s gas flow has obtained the velocity of the gas already after approximately 0.2 ms [14]. Therefore, during most of the droplet lifetime in the urea-SCR system, the turbulent fluctuations of the gas phase will be significant to the relative velocity between the droplet and the gas, and the droplet motion will be influenced by these fluctuations to a very high degree. This will affect not only the droplet trajectories but also the heat and mass transfer situation [14]. In other words, turbulent dispersion of the droplets from the urea-spray is a key phenomenon in the urea-SCR system, and it must be taken into account.

We shall here focus only on the modeling of turbulent dispersion for Lagrangian particle tracking with RANS-based turbulence models, since this is the main viable alternative for simulations of real industrial systems today. For discussions on more elaborate treatments of the turbulent exhaust gas flow and the implications for the resolution of turbulent dispersion phenomena, see e.g. [40-43].

A convenient and straightforward way to accommodate turbulent dispersion of droplets is to use a so-called Discrete Random Walk (DRW) model [40]. In essence, a turbulent eddy is sampled from a Gaussian distribution based on the local turbulence quantities (i.e. k and ε). This "synthetic" turbulent eddy is characterized by a velocity fluctuation (given by $\sqrt{2k/3}$) and a lifetime (given by e.g. $0.3\,k/\varepsilon$). The fluid velocity experienced by the droplet is then taken as the sum of the mean fluid velocity as resolved by the RANS-simulation and the current eddy velocity. The droplet interacts with the eddy for the shortest of the eddy lifetime and the time it would take for the droplet to cross through the eddy. After this time has elapsed, the droplet enters a new eddy, whose characteristics are sampled at the current droplet location.

Unfortunately, DRW models are known to have serious flaws when used together with RANS-based turbulence models. Since in a typical k-ε variant (such as the one used in the present work) the turbulence is assumed to be isotropic, there is no discernable difference between the turbulent characteristics in the different coordinate directions. This causes problems in the near-wall regions, where the wall-normal velocity fluctuations are typically overpredicted [41]. As a result, the extent of the droplet-wall interactions is overestimated. On the other hand, when no model for turbulent dispersion is used, the transport of droplets towards the walls by turbulent fluctuations in the gas phase that actually is present remains unresolved. In conclusion, if droplet-wall interactions are expected to be of significance in a simulation of a urea-SCR system, a more elaborate turbulence model or a DRW model that has been tuned to experimental data or DNS simulations [41] will be necessary.

In addition, the decomposition efficiencies predicted when turbulent dispersion is accounted for is typically higher than if turbulent dispersion is erroneously neglected. This phenomenon is related to the enhanced heat and mass transfer rates predicted in the heat and mass balances for the droplets when there is a relative velocity between the droplet and the gas throughout the droplet's lifetime [14].

In the present work, results are presented from simulations with and without a DRW model, in order to establish the explicit influence this type of model has on the overall simulation results for a urea-spray.

2.4. UWS Evaporation

In the simulations two-way coupling was employed between the dispersed and continuous phase. It was assumed that the evaporation process could be divided into two separate regimes were water first evaporates and when all water is gone urea starts to evaporate. This was justified by the much lower vapor pressure of urea than that of water. Experiments performed by Lundström et al. [2] using a cordierite monolith impregnated with UWS exhibit a large separation between the peaks of water evaporation and urea thermolysis, further adding validity to this assumption. No attempts to model the melting of urea was made in the current procedure, as may be verified, a quick estimation revealed that the time for melting was only about 5% of the total time for evaporation for the droplets and thus this was ignored in the present work.

Tabel 1. Drag and buoyancy forces used with correlations for the drag coefficient

Momentum	Comments	
$\mathbf{F}_D = C_D \frac{\pi d_p^2}{4} \frac{\rho_g |\mathbf{U}_c| \mathbf{U}_c}{2}$	drag force	(9)
$\mathbf{F}_B = m_p \frac{\mathbf{g}(\rho_p - \rho_g)}{\rho_p}$	buoyancy force	(10)
$C_{D,sphere} = \frac{24}{Re_p}\left(1 + 0.2 Re_p^{0.63}\right)$	drag coefficient for rigid sphere	(11)
$C_D = C_{D,sphere}\left(1 + B_T\right)^{-0.2}\left(1 + 0.06 Re_p^{-0.12} We^{1.4}\right)$	drag coefficient compensated for evaporation	(12)

Tabel 2. Antoine parameters with pressure in Pa and Temperature in K

Parameter	Value
A	$1.1663 \cdot 10^{1}$
B	$4.0854 \cdot 10^{3}$
C	$3.3953 \cdot 10^{-3}$

Table 3. Equations for modeling of water and urea evaporation

Mass and Heat	Comments	
$Sh_0 = 2.0 + 0.552 Re_p^{0.5} Sc^{1/3}$	Sherwood correlation from [48]	(13)
$Nu_0 = 2.0 + 0.552 Re_p^{0.5} Pr^{1/3}$	Nusselt correlation from [48]	(14)
$B_M = \frac{X_{vap,s} - X_{vap,g}}{1 - X_{vap,s}}$	mass transfer number	(15)
$B_T = (1 + B_M)^{\phi} - 1$	heat transfer number	(16)
$F_T = (1 + B_T)^{0.7} \ln(1 + B_T) / B_T$	heat transfer enhancement factor	(17)
$F_M = (1 + B_M)^{0.7} \ln(1 + B_M) / B_M$	mass transfer enhancement factor	(18)
$Sh_s = 2.0 + (Sh_0 - 2.0) / F_M$	adjusted Sherwood number	(19)
$Nu_s = 2.0 + (Nu_0 - 2.0) / F_T$	adjusted Nusselt number	(20)
$\phi = \frac{C_{p,vap,ref}}{C_{p,f,ref}} \frac{Sh_s}{Nu_s} \frac{Pr_{f,ref}}{Sc_{f,ref}}$		(21)
$\frac{dT_p}{dt} = \frac{-\dot{m}_p}{m_p C_{p,i}} \left(\frac{C_{p,vap,ref}(T_g - T_p)}{B_T} - \Delta h_{vap,i} \right)$	droplet heat balance, index i = water or urea evaporation	(22)
$\frac{dm_p}{dt} = -\pi d_p \rho_{f,ref} D_{AB,ref} Sh_s \ln(1 + B_M)$	droplet mass balance	(23)
$\frac{dX_{p,urea}}{dt} = -\frac{\dot{m}_p}{m_p} X_{p,urea}$	urea mass fraction during water evap.	(24)

The modeling approach for evaporation of the UWS droplets was the same as that presented in [14] except for the urea evaporation phase. In the current procedure the boiling rate equation used to model urea thermolysis in [14] was replaced and modeled as an evaporation process using the vapor pressure recently presented by Lundström et al. [17].

$$\log_{10} p^{\circ}_{urea}(T) = A - \frac{B}{C + T} \quad (2)$$

Table 4. Equations for evaluation of film properties including the 1/3 and the employed mixing rule

Mass and Heat	Comments	
$T_{film} = T_p + \frac{1}{3}(T_g - T_p)$	1/3 rule for temperature	(25)
$X_{film} = X_{vap,s} + \frac{1}{3}(X_{vap,g} - X_{vap,s})$	1/3 rule for mass fraction	(26)
$\Phi_{film} = \frac{Y_{film}\Phi_{vap}}{Y_{film} + (1-Y_{film})\left(\frac{M_g}{M_{vap}}\right)^{1/2}} + \frac{(1-Y_{film})\Phi_g}{Y_{film}\left(\frac{M_{vap}}{M_g}\right)^{1/2} + (1-Y_{film})}$	Herning and Zipper mixture model, ϕ=property	(27)

Table 5. Material data for the gas phase. All properties are functions of temperature and the value given is for a temperature of 673 K

Material property	Value at T=673 K	Reference
Density (kg/m^3)	0.52	[49]
Viscosity (Pa s)	3.23 x 10-5	[49]
Conductivity (W/m K)	0.052	[49]
Specific heat (J/kg K)	1069	[49]

Table 6. Material data for the dispersed phase

Material property	Urea/Water mixture	Urea
Density (kg/m^3)	[50]	13330.0
Conductivity (W/m K)	*	*
Specific heat (J/kg K)	*	
Vapor Pressure (Pa)	[50]	
Heat of vaporization (J/kg)	•	[51, 52]
Heat of decomposition (J/mol)	-	[20]

An asterisk (*) implies that the property for pure water was used. A bullet (•) means that the values were calculated using the data for pure water but at the saturation temperature and pressure of the urea/water mixture. The heat of vaporization for urea was calculated as the difference between the fusion and sublimation enthalpy.

The Antoine parameters used are given in Table 2.

For completeness, all of the equations involved in solving the heat and mass balances for the UWS droplets are given in Table 3. The procedure is the same as proposed in [44] which accounts for effects of variable thermophysical properties, non-unit Lewis number and Stefan flow. Effects of internal mixing on mass and heat transfer are neglected based on the results presented in [15]. Thus the process of finding the mass and heat transfer coefficients becomes an iterative procedure between Eqns (13)-(21). When the final mass and heat transfer coefficients are obtained the mass and heat balances for the droplet may be evaluated and if water still remains the mass fraction of urea updated. In the correlations Re_p was evaluated at the free stream density and the film viscosity. This approach has been shown to improve predictions [45].

The material properties of the film were evaluated using the 1/3 rule [45] and the Herning and Zipper model [46] was used as the mixing rule since this model has been proven to give good results in the evaluation of physical properties of gas films of droplets in turbulent evaporating sprays [47]. The 1/3 rule for the evaluation of film properties and the mixing rule are given in detail in Table 3. The different material properties for the gas and dispersed phases used in the simulations are presented in Table 5-6.

Keeping in accordance with the results presented in [17] the best agreement compared to experimental data is obtained when viewing the urea decomposition into NH_3 and HNCO as occurring outside of the droplet film i.e. gas phase urea decomposing. Thus, in the current modeling the heat of evaporation for urea was taken from the droplet whereas the heat of decomposition was evaluated outside of the droplet i.e. in the gas bulk. In Figure 1 the predicted rate of evaporation is compared to the experimental data by Wang et al. [18] and shows good agreement. As previously stated, one concern with previous models was the prediction of high decomposition temperatures (i.e. close to the bulk temperature), as can be seen in Figure 2 the new model predicts an equilibrium temperature ("decomposition"), i.e. the temperature reached when the endothermic evaporation is balanced by the heat transfer to the droplet, much closer to the temperatures reported in the literature. Finally in Figure 3 a complete simulation carried out for a 70 μm droplet in a quiescent atmosphere at 673 K and 101325 Pa is presented.

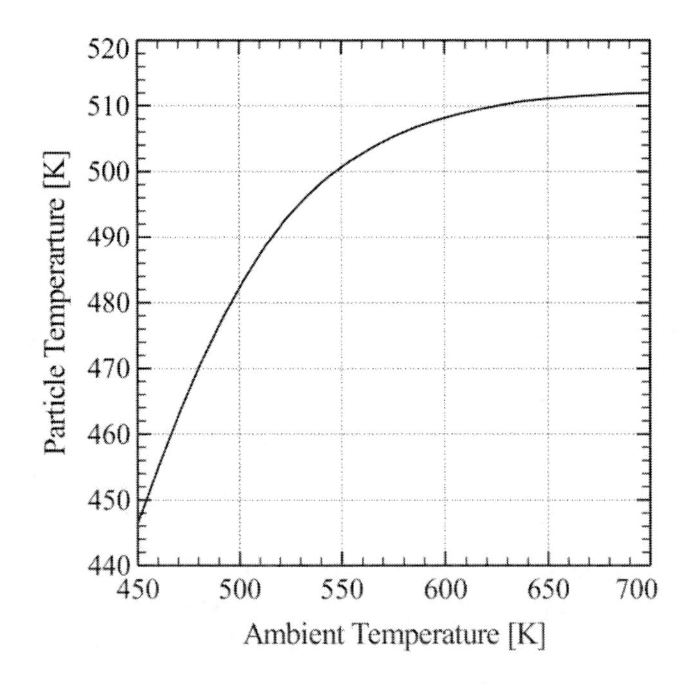

Figure 2. Particle temperature during the equilibrium part of urea evaporation as a function of ambient temperature.

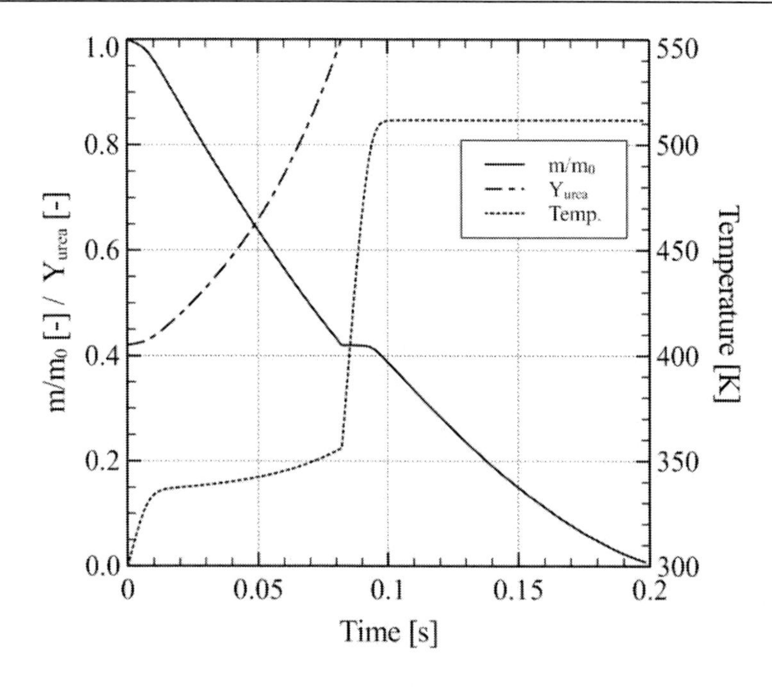

Figure 3. Simulation of a 70 μm UWS droplet in a quiescent atmosphere at 673 K and 101325 Pa.

Table 7. Detailed summary of simulation parameters for the examined cases

Case	Linear gas velocity	Gas temperature	Injection point	DRW
1	75 m/s	673 K	Top	No
2	75 m/s	673 K	Top	Yes
3	75 m/s	673 K	Center	No
4	75 m/s	673 K	Center	Yes

3. Results and Discussion

3.1. Simulation Conditions

In the current work a total of four different cases were investigated. As previously mentioned in *Section 2.1* two different injection points were considered. The reason for choosing two different injection points, at different wall normal angles, was to investigate the possible trade-off between particle residence time and spray uniformity at the outlet i.e. NH_3 uniformity. In order to investigate the effect of turbulent dispersion on the uniformity the same two injection points were also simulated with the DRW model. A detailed summary of the simulation conditions is presented in Table 7.

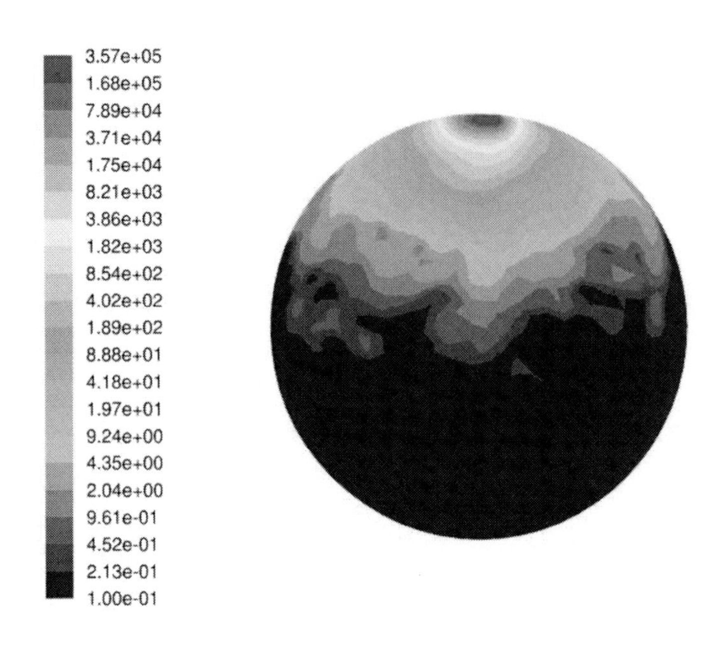

Figure 4. Distribution of all particles injected into the system after reaching the outlet for case 1.

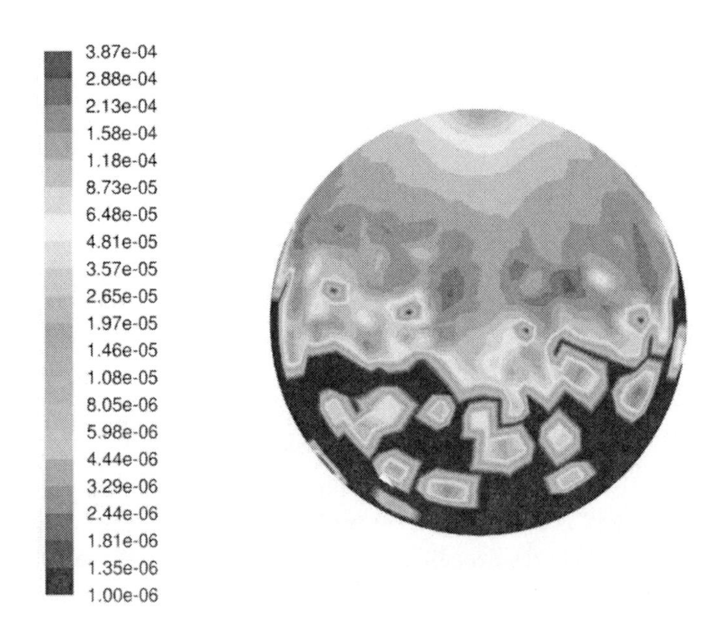

Figure 5. Distribution of number average droplet diameter at the outlet for all injected droplets reaching the outlet for case 1.

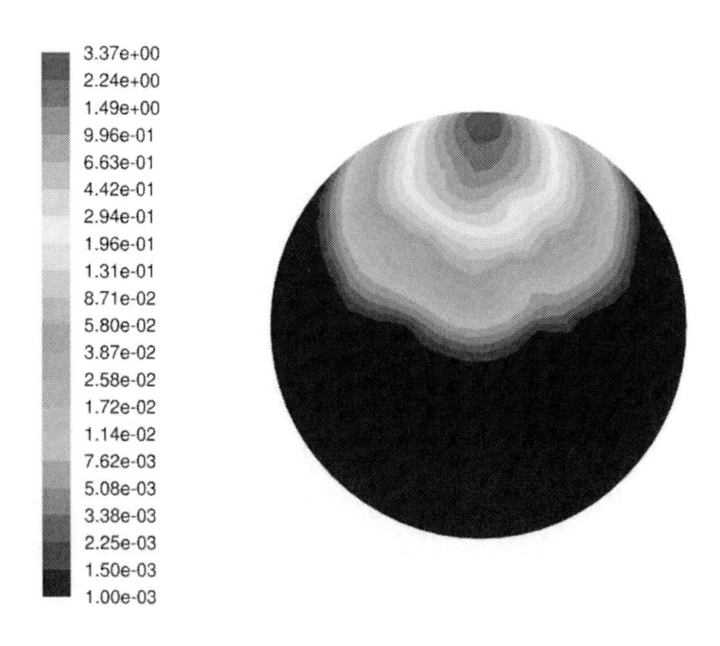

Figure 6. Distribution of all NH_3 after reaching the outlet for case 1.

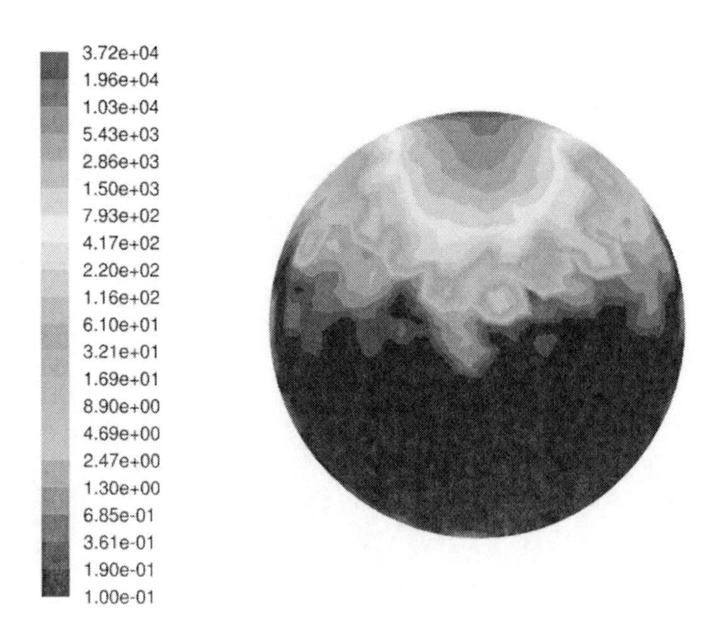

Figure 7. Distribution of all particles injected in to the system after reaching the outlet for case 2.

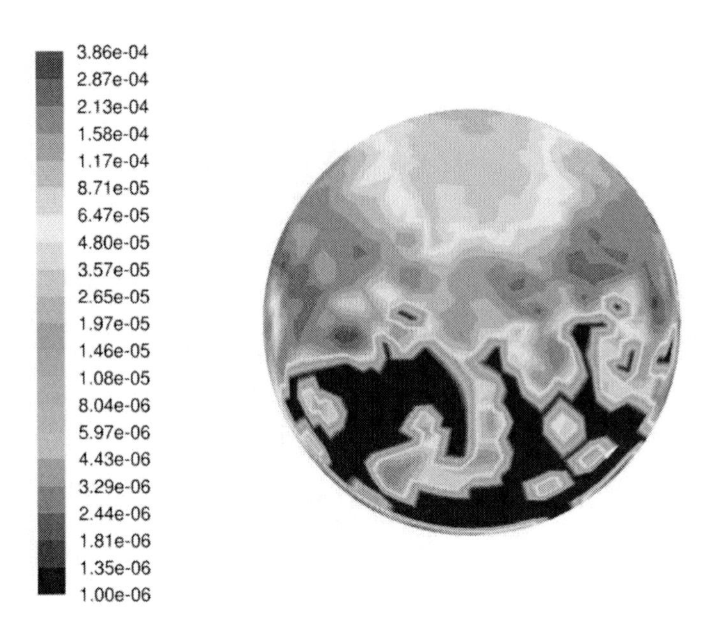

Figure 8. Distribution of number average droplet diameter at the outlet for all injected droplets reaching the outlet for case 2.

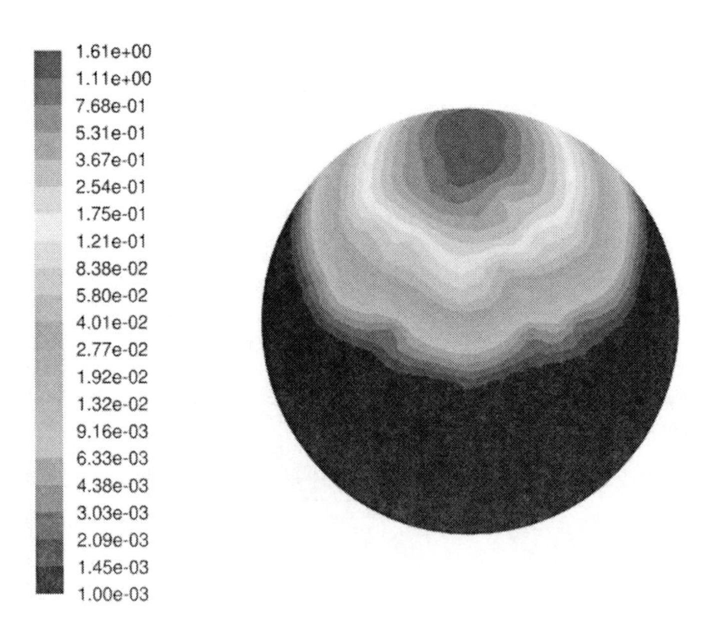

Figure 9. Distribution of all NH_3 after reaching the outlet for case 2.

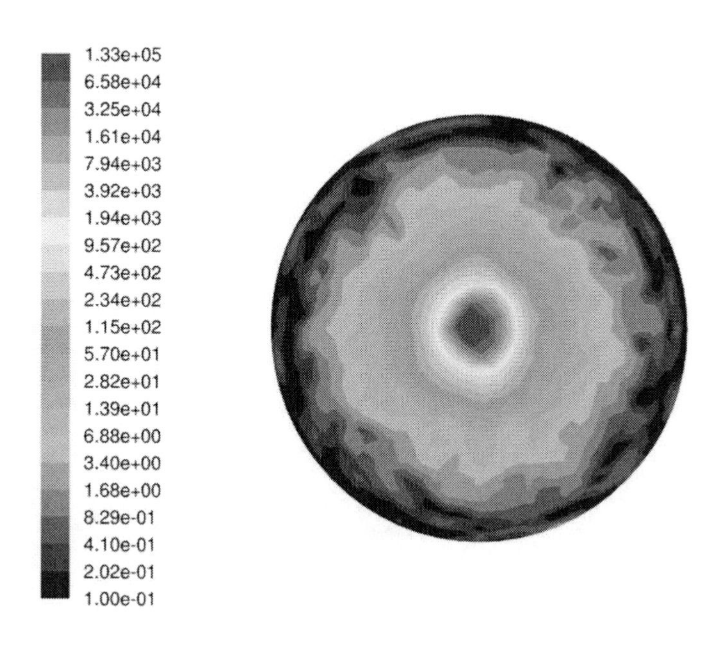

Figure 10. Distribution of all particles injected into the system after reaching the outlet for case 3.

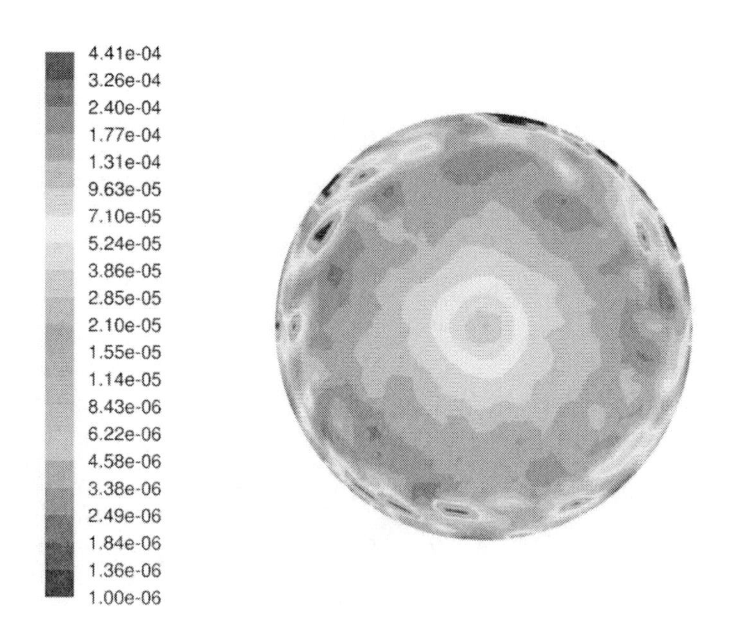

Figure 11. Distribution of number average droplet diameter at the outlet for all injected droplets reaching the outlet for case 3.

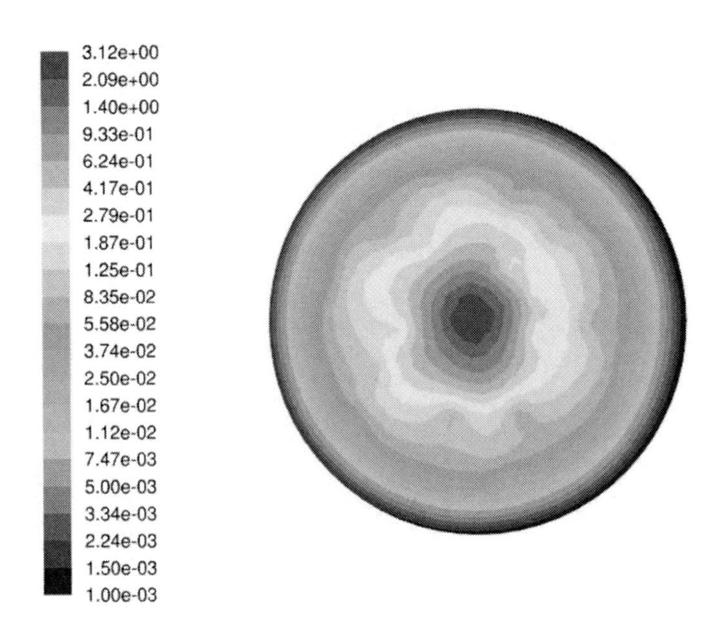

Figure 12. Distribution of all NH_3 after reaching the outlet for case 3.

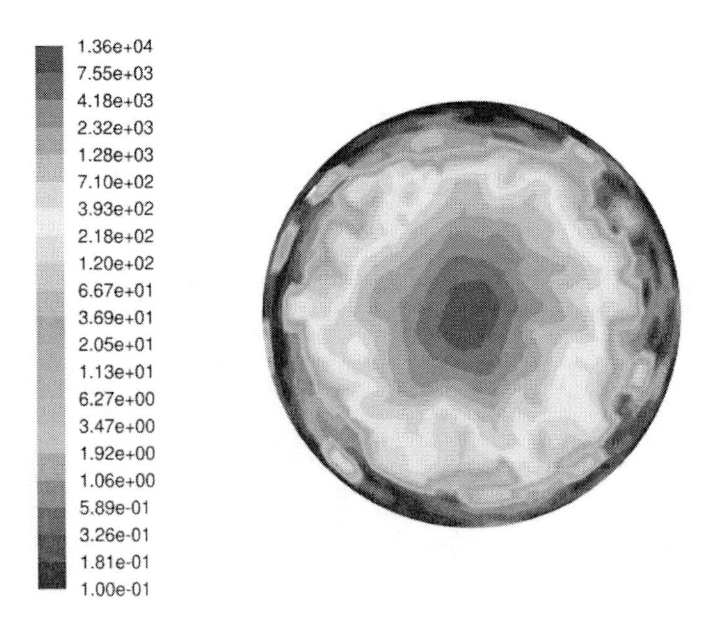

Figure 13. Distribution of all particles injected into the system after reaching the outlet for case 4.

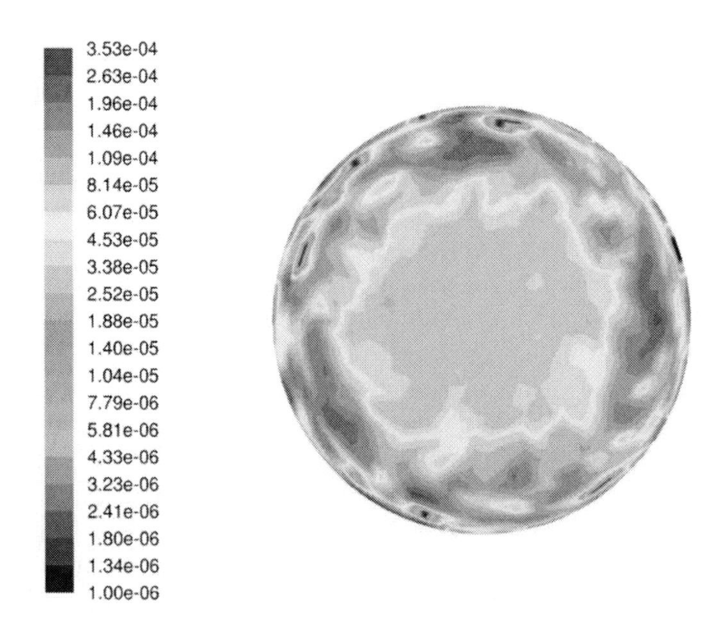

Figure 14. Distribution of number average droplet diameter at the outlet for all injected droplets reaching the outlet for case 4.

3.2. Spray Uniformity Results

To evaluate the different effects of the model choices, i.e. injection position and turbulent dispersion, on the decomposition efficiency and uniformity the total accumulated number of particles, the number average particle diameter and total accumulated NH_3 mass at the outlet were used.

When comparing the distribution of the number of particles that reached the outlet for the cases were no turbulent dispersion was modeled, Figures 4 and 10, it becomes evident that the effect of injection position greatly affects the uniformity at the outlet. In the case where the injection is positioned at the wall i.e. *case 1* most of the droplets appear in the top part of the outlet as opposed to *case 3* where the majority of the particles exit at the center of the pipe. The most striking difference between these two cases is however the absence of droplets in the lower part of the pipe in Figure 4, whereas in Figure 10 droplets are present over the entire cross-section. It can therefore be concluded that when the spray is positioned at the wall, the droplets do not penetrate very far into the gas stream at the current gas flow conditions.

Since turbulent dispersion was neglected in *case 1* and *case 3*, and it is exactly this mechanism which is responsible for the main transport of droplets in the planes perpendicular to the streamwise gas flow direction, we compare the previous findings to the results from Figure 7 and Figure 13. In the cases when turbulent dispersion is accounted for, the situation is similar, although a visible change to the better in terms of uniformity may be discerned. It should be noted that the effects of turbulent dispersion as shown here are overestimations, as they are based on predictions from a k-ε model, which is known to overpredict the near-wall normal turbulent fluctuations.

The Figures 4, 7, 10 and 13 show only the droplets that exit the pipe at the outlet, and do not take into account the droplet history inside the pipe or the droplets that have been totally evaporated and decomposed before they reach the outlet. It is therefore of great interest to study the accumulated NH_3 that reached the outlet under the duration of the simulation. These plots are provided as Figures 6, 9, 12 and 15. As can be seen from these contour plots, they correlate highly with the accumulated number of droplets on the outlet previously discussed. This is not a surprising result, since the highest number of droplets on the outlet corresponds to the droplet diameters belonging to the smallest range. It is thus mainly the smallest droplets in the spray that have time to release all water and start decomposing into ammonia, and hence the NH_3 is observed in the regions of the small droplets.

Finally, the number-averaged droplet diameter of the droplets that reach the outlet has been provided as Figures 5, 8, 11 and 14. From these figures, it may indeed be concluded that the areas of a large number of droplets and high NH_3 concentration correspond to the regions of smaller droplet diameters. It can also be seen from Figure 5 that although there in fact is a smaller number of large droplets spread over the entire pipe cross-section, these droplets do not contribute significantly neither to the number of droplets exiting the pipe (cf. Figure 4) nor the NH_3 released (cf. Figure 6).

Another conclusion that may be drawn from Figures 5, 8, 11 and 14, is that the turbulent dispersion acts so as to smooth out gradients in droplet diameter. These gradients are otherwise caused by the different inertia of the different droplet sizes. For example, a large gradient in droplet diameter is clearly visible in Figure 5, when the DRW model is not used. In contrast, the smaller droplets have spread over a significantly larger area when DRW was used (Figure 8).

3.3. Decomposition Efficiencies

The decomposition efficiencies, divided into an evaporation efficiency for water and a decomposition efficiency of urea, are reported in Table 8. There are two main effects governing the decomposition efficiency in the system: turbulent dispersion and residence time of the evaporating droplets.

Table 8. Summary of total amount evaporated water and urea

Case	% Evaporated water	% Evaporated Urea
1	46.7	13.7
2	48.0	15.8
3	42.3	13.0
4	43.7	14.5

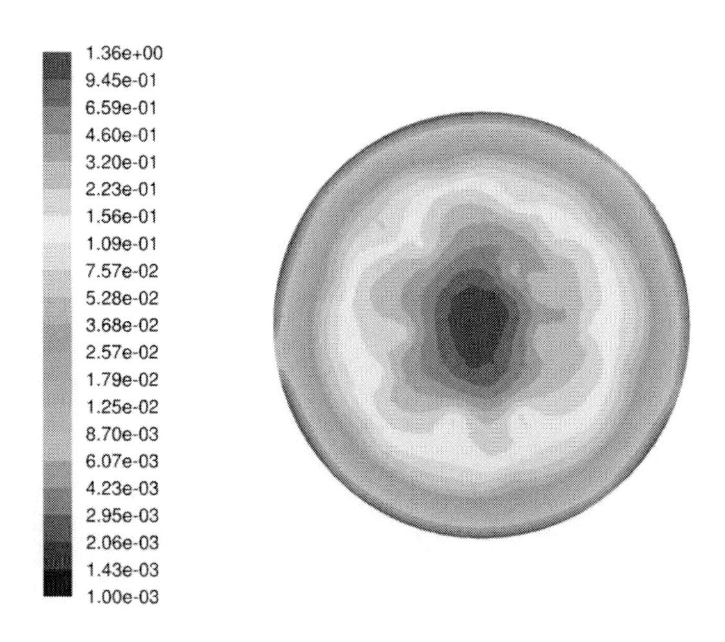

Figure 15. Distribution of all NH_3 after reaching the outlet for case 4.

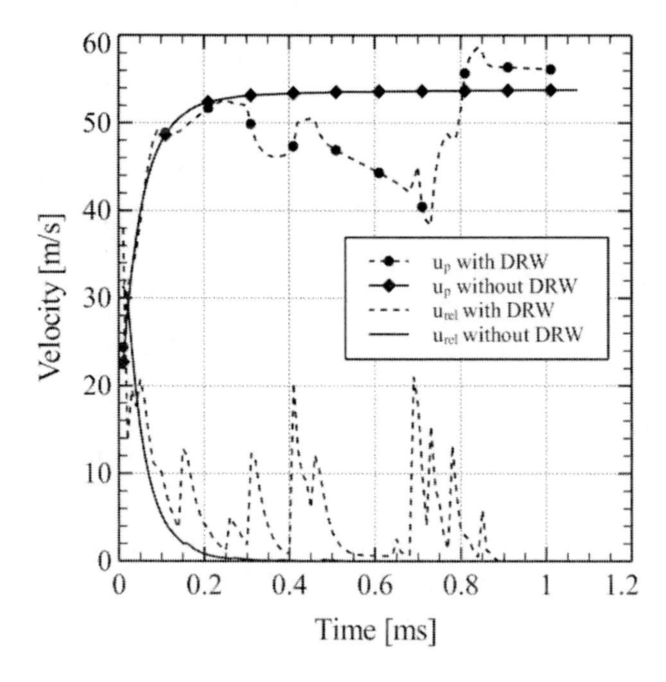

Figure 16. Droplet velocity and relative velocity between droplet and gas for a 5 μm droplet, as simulated with and without the DRW model for turbulent dispersion.

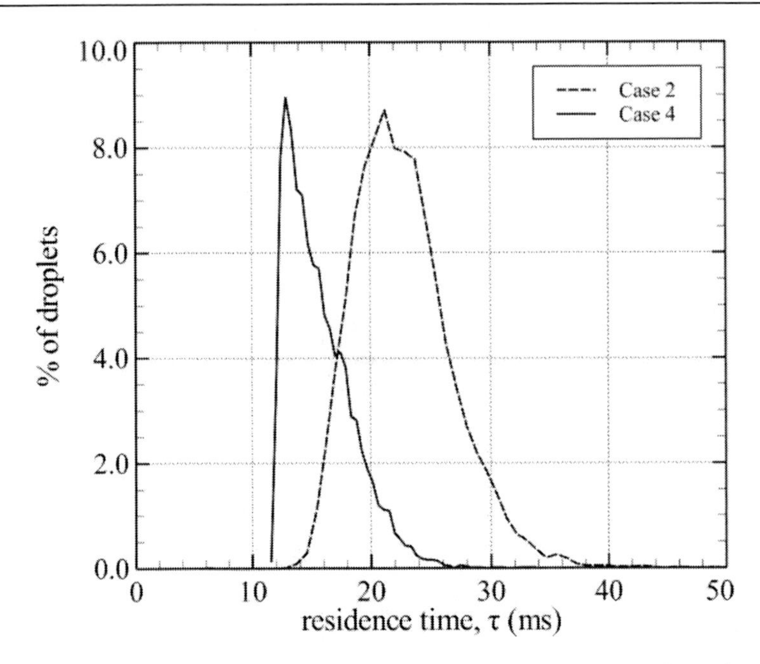

Figure 17. Particle residence time distributions for the two cases were turbulent dispersion was modeled i.e. case 2 and 4.

By comparing *cases 1* and *3* to *cases 2* and *4*, the effect of turbulent dispersion on the decomposition efficiency is found. It is observed that the decomposition efficiency increases when turbulent dispersion is accounted for. This is due to the enhanced heat and mass transfer to and from the droplets as continuously accelerated and decelerated in the gas flow. The difference in experienced relative velocities with and without a DRW model is shown in Figure 16.

The difference in decomposition efficiency between the two spray positions can be found from a comparison between *cases 1* and *2* to *cases 3* and *4*. The decomposition efficiency is always lower when the spray is located at the center of the pipe. The explanation is found from an investigation of the droplet residence times, as shown in Figure 17. When the spray is situated at the wall, the residence time distribution is rather broad and with a higher average residence time. For the spray in the center, the residence time distribution has a sharp peak and a tail towards longer residence times, but the average is significantly shorter than for the wall-located spray. The differences in decomposition efficiencies thus highlight the sensitivity of the decomposition efficiency to the droplet residence time in the pipe.

4. Conclusions

The current work outlines a suitable modeling framework for simulations of urea-sprays in urea-SCR systems for typical automotive applications. A recently published model for urea evaporation and decomposition, exhibiting a good agreement with previously published experimental data, has been implemented into a CFD framework. The effects of evaporation on the droplet heat and mass transfer rates are accounted for, and the droplet equation of motion takes droplet distortion and evaporation into account.

The simulation results have shown that:

- The location of the spray injector has significant effects on the uniformity of NH_3 passed on to the catalyst and the decomposition efficiency of the droplets in the spray. If the spray is located in the near-wall region, the droplets will have a longer residence time before the catalyst, and the decomposition efficiency will be higher. On the other hand, the distribution of the droplets over the cross-section of the pipe can be very poor (depending on the exhaust gas flow conditions). The situation is reversed for a spray located at the center of the pipe.
- Turbulent dispersion increases the heat and mass transfer to the droplets, and thus also the decomposition efficiency. It also has a profound effect on the uniformity of the NH_3 distribution before the catalyst.
- The sensitivity of the decomposition efficiency to the droplet size has been discussed. It is shown that in order to obtain a high degree of decomposition for these relatively short residence times, a spray producing a large number of small droplets is necessary.

REFERENCES

[1] Koebel, M., M. Elsener, and M. Kleemann, Urea-SCR: a promising technique to reduce NOx emissions from automotive diesel engines. *Catalysis Today,* 2000. 59(3-4): p. 335-345.

[2] Lundström, A., B. Andersson, and L. Olsson, Urea thermolysis studied under flow reactor conditions using DSC and FT-IR. *Chemical Engineering Journal*, 2009. 150(2-3): p. 544-550.

[3] Chi, J.N. and H.F.M. DaCosta, Modeling and Control of a Urea-SCR Aftertreatment System. *SAE Technical Paper* (2005), 2005. 2005-01-0966.

[4] Chen, M. and S. Williams, Modelling and Optimization of SCR-Exhaust Aftertreatment Systems. . *SAE Technical Paper* (2005), 2005. 2005-01-0969.

[5] Chigier, N.A., The Physics of Atomization, in 5th International Conference on Liquid Atomization and Spray Systems (ICLASS-91). 1991, *National Institute of Standards and Technology.*

[6] Koebel, M., M. Elsener, and G. Madia, Recent advances in the development of urea-SCR for automotive applications, in SAE International Fall Fuels and Lubricants Meeting and Exhibition. 2001, SAE International, Warrendale, Pennsylvania, USA: San Antonio, Texas, USA.

[7] Linde (editor), D.R., CRC Handbook of Chemistry and Physics. Internet Version 2007, 87th Edition ed. 2007, ed. D.R. Linde. 2007, Boca Rato, FL, 2007: Taylor and Francis.

[8] Oh, T.Y., et al. Design optimization of the mixing chamber in SCR system for marine diesel engine. in Proceedings of the 6th International Symposium on Diagnostics and Modeling of Combustion in Internal Combustion Engines. 2004: The Japan Society of Mechanical Engineers.

[9] Sun, W.H., et al. Small Scale Test Results From New Selective Catalytic NOx Reduction Process Using Urea. in MEGA Symposium. 2001. Chicago, IL.

[10] Schaber, P.M., et al., Thermal decomposition (pyrolysis) of urea in an open reaction vessel. *Thermochimica Acta*, 2004. 424(1-2): p. 131-142.

[11] Calabrese, J.L., et al., The Influence of Injector Operating Conditions on the Performance of a Urea-Water Selective Catalytic Reduction (SCR) system. *SAE Technical Paper* (2000), 2005. 2000-01-2814.

[12] Stradella, L. and M. Argentero, A study of the thermal decomposition of urea, of related compounds and thiourea using DSC and TG-EGA. *Thermochimica Acta,* 1993. 219: p. 315-323.

[13] Sluder, C.S., et al., Low-Temperature Urea Decomposition and SCR Performance, in SAE World Congress. 2005, SAE International, Warrendale, Pennsylvania, USA: Detroit, Michigan, USA.

[14] Ström, H., A. Lundström, and B. Andersson, Choice of urea-spray models in CFD simulations of urea-SCR systems. *Chemical Engineering Journal*, 2009. 150(1): p. 69-82.

[15] Birkhold, F., et al., Modeling and simulation of the injection of urea-water-solution for automotive SCR DeNOx-systems. *Applied Catalysis B: Environmental*, 2007. 70(1-4): p. 119-127.

[16] Birkhold, F., et al., Analysis of the Injection of Urea-Water-Solution for Automotive SCR DeNOx-Systems: Modeling of Two-Phase Flow and Spray/Wall Interaction. *SAE Technical Paper* (2006) 2006. 2006-01-0643.

[17] Lundström, A., Urea Decomposition for Urea-SCR Applications, Ph.D Thesis 2010, Chalmers University of Technology: Gothenburg.

[18] Wang, T.J., et al., Experimental investigation on evaporation of urea-water-solution droplet for SCR applications. *AIChE Journal*, 2009. 55(12): p. 3267-3276.

[19] Fang, H.L. and H.F.M. DaCosta, Urea thermolysis and NOx reduction with and without SCR catalysts. *Applied Catalysis B: Environmental*, 2003. 46(1): p. 17-34.

[20] Koebel, M. and E.O. Strutz, Thermal and Hydrolytic Decomposition of Urea for Automotive Selective Catalytic Reduction Systems: Thermochemical and Practical Aspects. *Industrial and Engineering Chemistry Research*, 2003. 42(10): p. 2093-2100.

[21] Dong, H., S. Shuai, and J. Wang, Effect of Urea Thermal Decomposition on Diesel NOx-SCR Aftertreatment Systems, in SAE International Powertrains, Fuels and Lubricants Congress. 2008, SAE International, Warrendale, Pennsylvania, USA: Shanghai, China.

[22] Ball, J.C., A toxicological evaluation of potential thermal degradation products of urea. *SAE Technical Paper* (2001), 2001. 2001-01-3621.

[23] Pope, S.B., Turbulent Flows. 2000: Cambridge University Press.

[24] Wilcox, D.C., Turbulence Modeling for CFD. 1998: DCW Industries.

[25] Stiesch, G., Modelling Engine Spray and Combustion Processes. 2003, Berlin: Springer-Verlag.

[26] Ekström, F. and B. Andersson, Pressure drop of monolithic catalytic converters - experiments and modeling. *SAE Technical Paper* (2002), 2002. 2002-01-1010.

[27] Reitz, R.D. and C.J. Rutland, Development and testing of diesel engine CFD models. *Progress in Energy and Combustion Science*, 1995. 21(2): p. 173-196.

[28] Maxey, M.R. and J.J. Riley, Equation of motion for a small rigid sphere in a nonuniform flow. *Physics of Fluids*, 1983. 26(4): p. 883-889.

[29] Csanady, G.T., Turbulent Diffusion of Heavy Particles in the Atmosphere. *Journal of the Atmospheric Sciences,* 1963. 20(3): p. 201-208.
[30] Clift, R. and M.E. Weber, Bubbles, drops and particles. 1978, New York: Academic Press.
[31] Kurose, R. and S. Komori, Drag and lift forces on a rotating sphere in a linear shear flow. *Journal of Fluid Mechanics*, 1999. 384(-1): p. 183-206.
[32] Fan, L.-S. and C. Zhu, Principles of Gas-solid Flows. 1998, Cambridge: Cambridge University Press.
[33] Odar, F. and W.S. Hamilton, Forces on a sphere accelerating in a viscous fluid. *Journal of Fluid Mechanics*, 1964. 18(02): p. 302-314.
[34] Talbot, L., et al., Thermophoresis of particles in a heated boundary layer. *Journal of Fluid Mechanics*, 1980. 101(04): p. 737-758.
[35] Li, A. and G. Ahmadi, Dispersion and Deposition of Spherical Particles from Point Sources in a Turbulent Channel Flow. *Aerosol Science and Technology*, 1992. 16(4): p. 209 - 226.
[36] Davies, G.W., Definitive equations for the fluid resistance of spheres. *Proceedings of the Physical Society,* 1945. 57(4): p. 259-270.
[37] Haywood, R.J., M. Renksizbulut, and G.D. Raithby, Numerical Solution of Deforming Evaporating Droplets at Intermediate Reynolds Numbers. Numerical Heat Transfer, Part A: Applications: *An International Journal of Computation and Methodology,* 1994. 26(3): p. 253 - 272.
[38] Yuen, M.C. and L.W. Chen, On Drag of Evaporating Liquid Droplets. *Combustion Science and Technology*, 1976. 14(4): p. 147 - 154.
[39] Virepinte, J.F., et al., A Rectilinear Droplet Stream in Combustion: Droplet and Gas Phase Properties. *Combustion Science and Technology*, 2000. 150(1): p. 143 - 159.
[40] Crowe, C., M. Sommerfeld, and Y. Tsuji, Multihpase flows with droplets and particles. 1998, Florida: CRC Press.
[41] Dehbi, A., A CFD model for particle dispersion in turbulent boundary layer flows. *Nuclear Engineering and Design*, 2008. 238(3): p. 707-715.
[42] Tian, L. and G. Ahmadi, Particle deposition in turbulent duct flows--comparisons of different model predictions. *Journal of Aerosol Science*, 2007. 38(4): p. 377-397.
[43] Wang, Q. and K.D. Squires, Large eddy simulation of particle-laden turbulent channel flow. *Physics of Fluids*, 1996. 8(5): p. 1207-1223.
[44] Abramzon, B. and W.A. Sirignano, Droplet vaporization model for spray combustion calculations. *International Journal of Heat and Mass Transfer*, 1989. 32(9): p. 1605-1618.
[45] Kolaitis, D.I. and M.A. Founti, A comparative study of numerical models for Eulerian-Lagrangian simulations of turbulent evaporating sprays. *International Journal of Heat and Fluid Flow*, 2006. 27(3): p. 424-435.
[46] Reid, R.C., J.M. Prausnutz, and T.K. Sherwood, The Properties of Gases and Liquids. Third ed. ed. 1977: McGraw-Hill Book Company.
[47] Sparrow, E.M. and J.L. Gregg. The variable fluid-property problem in free convection. in Trans. *ASME* 80. 1958.
[48] Frössling, N., On the Evaporation of Falling Drops. Gerl. Beitr. *Zur Geophysik*, 1938. 52: p. 170-216.
[49] Hellsten, G., Tabeller och Diagram. 1st ed. ed. 1992: Almqvist och Wiksell Förlag AB.

[50] Perman, E.P. and T. Lovett, Vapour pressure and heat of dilution of aqueous solutions. *Transactions of the Farady Society*, 1926. 22: p. 1-19.
[51] Della Gatta, G. and D. Ferro, Enthalpies of fusion and solid-to-solid transition, entropies of fusion for urea and twelve alkylureas. *Thermochimica Acta*, 1987. 122(1): p. 143-152.
[52] Ferro, D., et al., Vapour pressures and sublimation enthalpies of urea and some of its derivatives. *The Journal of Chemical Thermodynamics*, 1987. 19(9): p. 915-923.

In: Sprays: Types, Technology and Modeling
Editor: Maria C. Vella, pp. 287-310
ISBN 978-1-61324-345-9

Chapter 8

PROCESSING AND PARTICLE CHARACTERIZATION OF NANOPOWDERS BY SPRAY PYROLYSIS ROUTE

Takashi Ogihara*
Graduate School of Material Science and Engineering,
University of Fukui, Fukui, Japan

ABSTRACT

Oxide and metal nanopowders were synthesized by spray pyrolysis using various types of atomizers and heating sources. In the spray pyrolysis process, the aerosols generate from the aqueous solution that contains dissolved precursors by using an atomizer such as an ultrasonic transducer and two-fluid nozzle. The advantages of spray pyrolysis are that the controls of particle morphology, size, size distribution, composition of resulting powders are possible. The diameter of aerosol is dependent on the frequency of the piezoelectric transducer or the size of two-fluid nozzle. Therefore, the particle size is controlled by adjusting them. In the pyrolysis, the aerosols were directly dried and heated to form spherical oxide nanoparticles at high temperatures. Homogenous, spherical oxide and alloy powders with narrow particle size distribution were successfully prepared by the spray pyrolysis. The particle microstructure of as-prepared powders was influenced by solution concentration, residence time in the furnace, pyrolysis temperature of the starting precursor. However, the stoichiometric composition of multi-component oxide or alloy was maintained regardless of particle microstructure. The particle size depended on the concentration of starting solution and flow rate of carrier air, but the particle size distribution was independent. We have tried various types of oxide and metal powders by the spray pyrolysis. In this paper, synthesis and characterization of oxide powders derived from spray pyrolysis were described. Furthermore, the mass production system was developed by flame spray pyrolysis using gas burners. The various types of cathode materials for an lithium ion battery were continuously produced. The excellent rechargeable performance of them was demonstrated by electrochemical measurement.

---

* 9-1 Bunkyo 3, Fukui-shi, Fukui, 910-8507, Japan. E-mail ogihara@matse.u-fukui.ac.jp.

1. Introduction

Spray pyrolysis is a versatile process regarding the synthesis of ceramic and metal fine powders [1-5]. An ultrasonic atomizer [6] or two-fluid nozzle [7] has been used as a mist generator. The mist was directly dried and pyrolyzed to form oxide or metal powders at a temperature less than 1000°C. The advantages of spray pyrolysis are that the control of particle size, particle size distribution and morphology are possible. Furthermore, the fine powders with homogeneous composition can be easily obtained because the component of starting solution is kept in the mist derived from an ultrasonic atomizer or two-fluid nozzle. Each metal ion was homogeneously blended in each mist, which played a role as the microreactor. The production time was very short (less than 1 min). The other solution processes such as sol-gel, hydrothermal, precipitation, hydrolysis was order of hour for the synthesis of oxide powders. In addition, the process such as the separation, the drying and the firing step must be done after the chemical reaction in the solution. The oxide powders are continuously obtained without these steps in the spray pyrolysis. So far, it has been reported that this process is effective in the multi-component oxide powders such as $BaTiO_3$ [8, 9] and alloy powders such as Ag-Pd [10, 11]. Recently, layered types of lithium transition metal oxides such as $LiCoO_2$ [12], $LiNiO_2$ [13, 14], $LiNi_{0.5}Mn_{1.5}O_4$ [15] $LiNi_{1/3}Mn_{1/3}Co_{1/3}O_2$ [16, 17] and spinel types of lithium transition metal oxides such as $LiMn_2O_4$ [18 - 21], which are used as the cathode materials for Li ion batteries, also have been synthesized by spray pyrolysis. It was confirmed that these cathode materials derived from spray pyrolysis showed excellent rechargeable performances. This revealed that the particle characteristics such as uniform particle morphology, narrow size distribution and homogeneous chemical composition led to higher rechargeable capacity, higher efficiency, long life cycle and higher thermal stability.

Recently, a large spray pyrolysis system was also developed for mass production of spherical fine oxide and metal powders with homogeneous composition. In spray pyrolysis, the electric furnace has been often used to pyrolyze the mist. The difference of the pyrolysis temperature between inside and outside the electric furnace increases with increasing dimensions of the electrical furnace, leading to a fluctuation in the chemical and physical properties of the resulting powders. This problem was solved by the flame spray pyrolysis in which the mist was decomposed in the flame using the gas burner [22 - 25]. The uniform temperature profile in the furnace is given regardless of furnace scale. In this paper, the physical and chemical properties of oxide and metal powders synthesized by various types of spray pyrolysis were described. Furthermore, the possibility of spray pyrolysis as a mass production system was also described.

2. Spray Pyrolysis

2.1. Ultrasonic Spray Pyrolysis

Figure 1 shows the schematic diagram of the ultrasonic spray pyrolysis apparatus used with this work. This apparatus consisted of a mist box with ultrasonic transducer, electrical furnaces, quartz tube and cyclone.

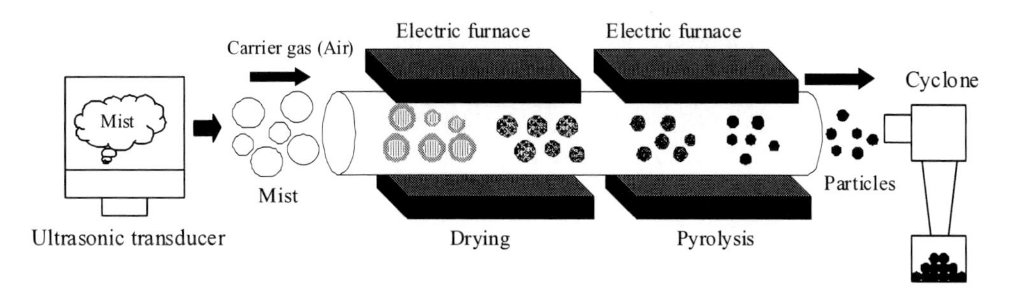

Figure 1. Schematic diagram of ultrasonic spray pyrolysis.

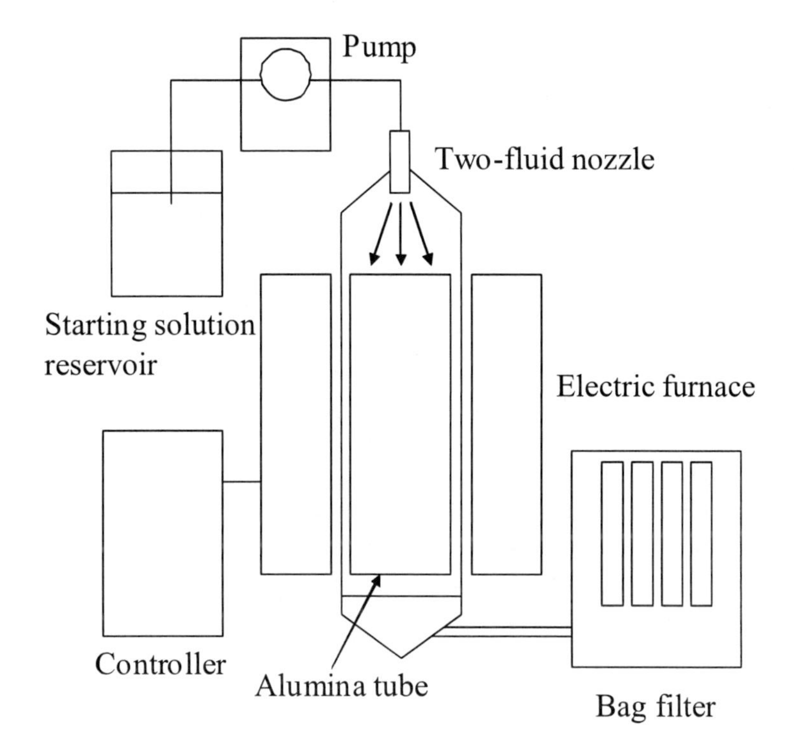

Figure 2. Schematic diagram of the two-fluid type spray pyrolysis.

A mist of the starting solution was generated by the piezoelectric transducer (1.6 MHz) at 25 °C and was introduced into two electrical furnaces through a quartz tube by carrier air. The flow rate was from 2 to 8 dm^3/min. The temperature of the electrical furnace to dry the mist was set at 400 °C and that of the furnace to pyrolyze the mist was changed from 500 to 900 °C. The as-prepared powders were continuously collected using a cyclone with an aspirator. The droplet-sized mist (*D*), generated using an ultrasonic transducer [26]. were very small and can be determined by equation (1).

$$D = 0.34\left(\frac{8\pi\gamma}{\rho f^2}\right)^{1/3} \quad (1)$$

where $\tilde{n}$ is the density of water as a solvent, $\tilde{a}$ is the surface tension of water, f is the frequency of the piezoelectric transducer. The value of ρ and $\tilde{a}$ was 0.997×10^3 kg/m^3 and 7.96×10^{-2} N/m at 25 °C, respectively. In this case, $f = 1.6$ MHz and the droplet size was calculated to be about 3 μm at 25 °C. Assuming that droplets of an inorganic salt solution are pyrolyzed completely to form dense oxide, the particle size of oxide (D_P) can be calculated by equation (2)

$$D_P = \left(\frac{M_O C}{\rho_O}\right)^{1/3} D \qquad (2)$$

where M_O is the molecular weight of oxide, $\tilde{n}_O$ (kg/m^3) is the density of oxide formed and C (mol/dm^3) is the concentration of the starting solution.

2.2. Two-Fluid Type Spray Pyrolysis

Figure 2 shows the schematic diagram of the two-fluid type spray pyrolysis apparatus. It consisted of a two-fluid nozzle, electric furnace and bag filter. The two-fluid nozzle has been often used for a spray dryer. Generally, the size of the mist is dependent on the diameter of the nozzle. The starting solution was introduced into the two-fluid nozzle using a roller pump and then the mist was continuously generated from the two-fluid nozzle under high pressure. The two-fluid nozzle is able to generate the mist of the solution with high viscosity. The viscosity of the starting solution increased with increasing the concentration of the starting solution. The mist generation by ultrasonic transducer is limited when the viscosity of solution became high. The mist of the aqueous solution with high viscosity was generated in the two-fluid nozzle with air carrier gas. The mist generated from two-fluid nozzle was continuously introduced to the electric furnace and then pyrolyzed at the temperature ranged from 500 °C to 900 °C. As-prepared powders were collected using the bag filter.

2.3. Plasma Assisted Spray Pyrolysis

Figure 3 shows the schematic diagram of spray pyrolysis apparatus with arc plasma as a heating zone [27]. The apparatus consisted of a mist box, drying furnace, plasma reactor and powder collector. The starting solution was misted at 2.4 MHz by an ultrasonic transducer in a mist box and dried in the drying furnace at 200 °C. After drying, the mist was introduced into the plasma reactor by air carrier gas. The flow rate ranged from 1 to 10 L/min. The six electrodes were evenly set (60°) in the plasma reactor. The mist was continuously pyrolyzed by the arc plasma under an air atmosphere. The pyrolysis temperature ranged from 7000 to 10000 °C. The oxide nanoparticles were collected using a collision-type filter in powder collector.

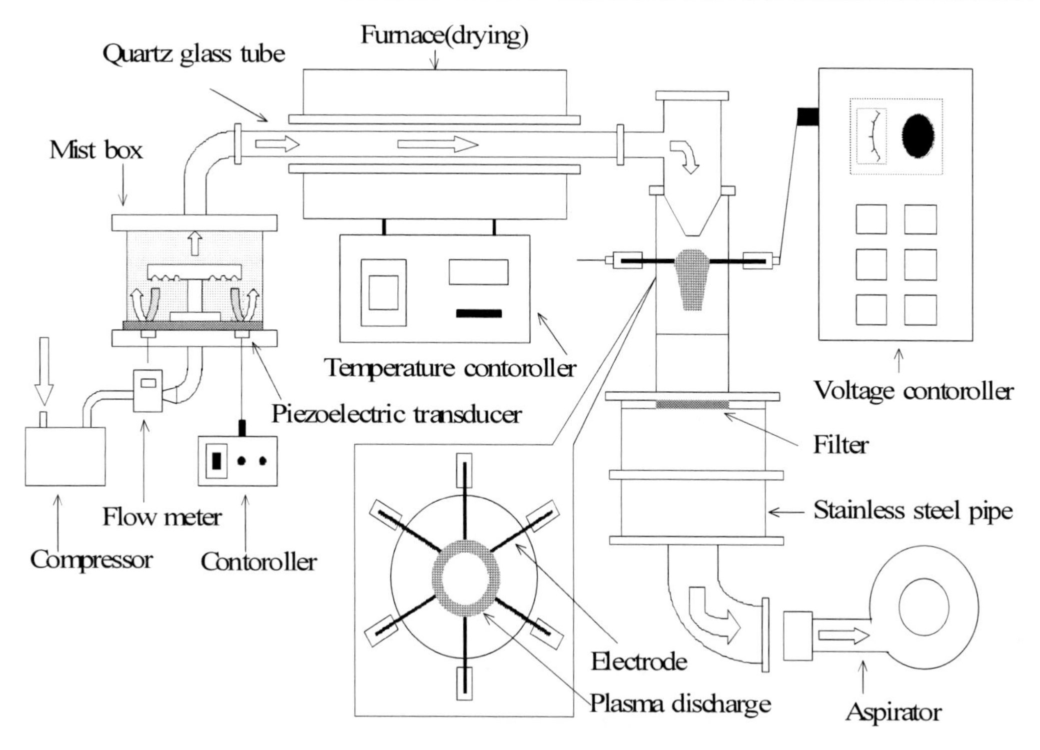

Figure 3. Schematic diagram of plasma-assisted spray pyrolysis.

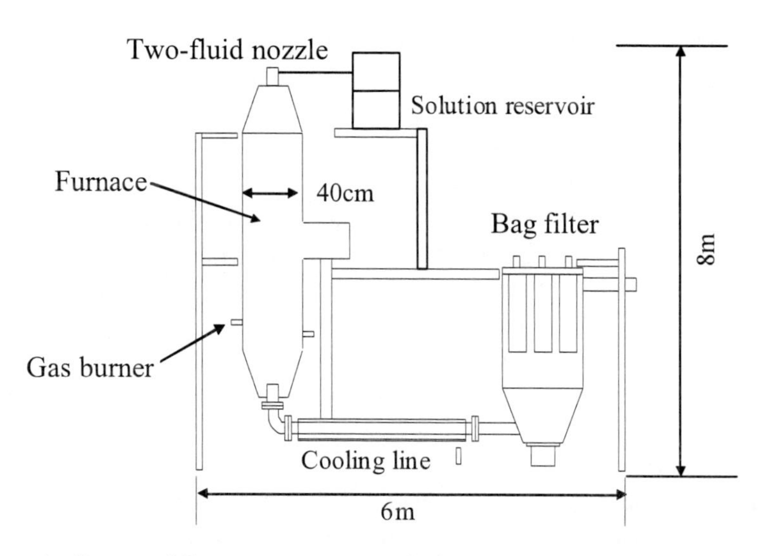

Figure 4. Schematic diagram of flame types spray pyrolysis.

2.4. Flame Type Spray Pyrolysis

It is difficult for spray pyrolysis to homogeneously pyrolyze a large quantity of aerosol in a short time in the electrical furnace that the scale-up was done. In the spray pyrolysis, the difference of pyrolysis temperature between inside and outside of the electrical furnace increases with increasing the dimension of electrical furnace, leading to the fluctuation of the properties for the resulting powders. We have developed the flame type spray pyrolysis apparatus by using a gas burner to produce homogeneous multi-component ceramic powders for a large-scale furnace.

Figure 4 shows the schematic diagram of flame types spray pyrolysis apparatus. This apparatus (0.4 mϕ × 6 m × 8 m) consisted of a two-fluid nozzle atomizer (a), combustion furnace with gas burners using city gas (b) and a powder collection box using bag filter (c). The starting solution prepared was atomized to generate the aerosol by using a two-fluid nozzle with a nozzle size of 20 μm. The aerosol was continuously sprayed from the upper part of the combustion furnace to the downward at carrier gas flow rate of 10 dm^3/h and then pyrolyzed when the aerosol passed through the flame set at 700°C. The decomposition temperature of aerosol was 500 °C. These apparatus have the powder production potential of 2 kg/hr. It is possible to produce lithium manganate cathode materials of 1 ton/month with this apparatus.

3. Preparation and Characterization of Oxide and Metal Powders by Ultrasonic Spray Pyrolysis

3.1. Metal Powders for LTCC

The advantage of low-temperature, co-fired ceramics (LTCC), which enables sintering at less than 1000 °C, is the possibility to use the conductive materials with low resistivity. Metals such as Ag, Pd, Ag-Pd alloy, and Ni have been used as conductive materials because their sintering temperatures are close to that of conventionally-used substrates. So far, several types of Ag powder have been used as electrode paste for LTCC substrates. The particle characteristics of Ag powders affect their shrinkage behavior during the sintering, and when coarse Ag powders with a broad size distribution and aggregation are used, they frequently show bending and cracking on the substrate. For these coarse Ag powders, surface treatment is carried out using various additives. Therefore, it is important to understand the sintering behavior of Ag particles of submicron size and narrow size distribution in order to solve such problems. In this work, various types of silver powder were prepared by chemical reduction and spray pyrolysis and then their sintering behavior was investigated in detail. The electrical properties of Ag paste were also determined.

Ag-Pd alloy fine particles were synthesized by ultrasonic spray pyrolysis using aqueous nitrate solutions of Ag and Pd as starting materials. $AgNO_3$ and $Pd(NO_3)_2$ were weighed and then dissolved in double distilled water to prepare aqueous solutions of 0.1 to 2.0 mol/dm^3. $Pd(NO_3)_2$ was added within the range of 5 to 30 wt%. Mist of aqueous nitrate solution was generated by an ultrasonic vibrator (1.6 MHz) at 25 °C and introduced into an electric furnace at a flow rate of 15 dm^3/min while air was used as a carrier gas and pyrolyzed at 800 °C. As-prepared powders were continuously collected by use of a cyclone separator. Figure 5 shows SEM of typical Ag-Pd alloy fine powders. As-prepared powders had spherical morphology with narrow size distribution and non-agglomeration. SEM showed that the microstructure of as-prepared particles was also dense. Table 1 shows the chemical composition of as-prepared powders. Their chemical composition was in good agreement with that of the starting solution. Pd was homogeneously doped in Ag particles. XRD showed that as-prepared powders were well-crystallized and had a single phase. The crystal phase of as-prepared Ag-Pd powders has an alloy composition and a solid solution is obtained. Other oxide phases such as PdO and Ag_2O are not observed. The change of the particle size and particle size distribution for the concentration of the starting solution was shown in Figure 6. Figure 7 shows the relation between the particle size and concentration of starting materials. The particle size increased with increasing the concentration. The particle size was in a good agreement with the calculated value. The particle size distribution (σ_g) decreased with increasing the concentration. Slurry was prepared by mixing as-prepared powders, ethyl cellulose, and 2.2.4-trimethyl-1.3-pentandiol-monoisobuthylate. Paste was obtained from the slurry by use of an applicator and dried for 2 hours at 100 °C and sintered at 800 - 1000 °C. Figure 7 shows the relation between the shrinkage and sintering temperature.

Figure 5. SEM of Ag-Pd (95:5) powders.

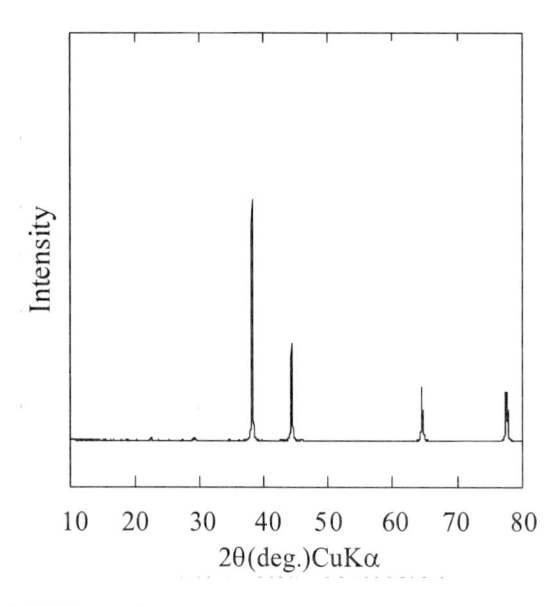

Figure 6. XRD of Ag-Pd (95:5) powders.

Table 1. Chemical composition of Ag-Pd powders

	Concentration (mol%)	
	Ag	Pd
Theoretical	95.00	5.00
Ag-Pd particles	95.16	4.84

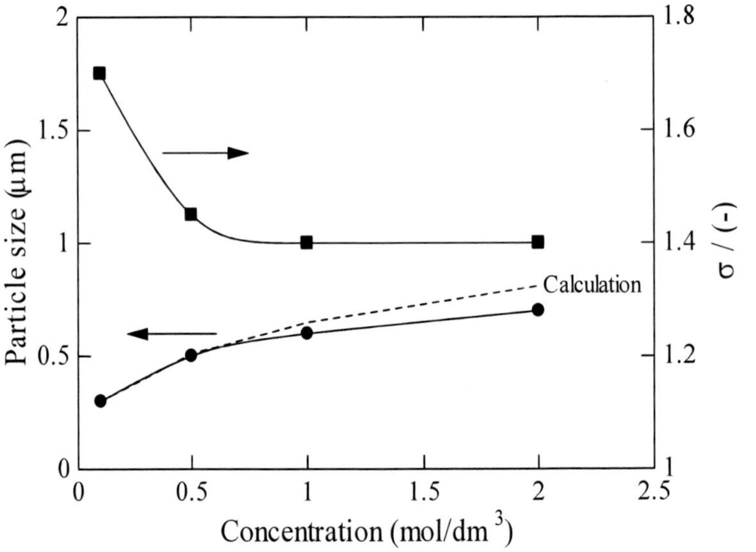

Figure 7. Change of particle size and σ_g as a function of concentration.

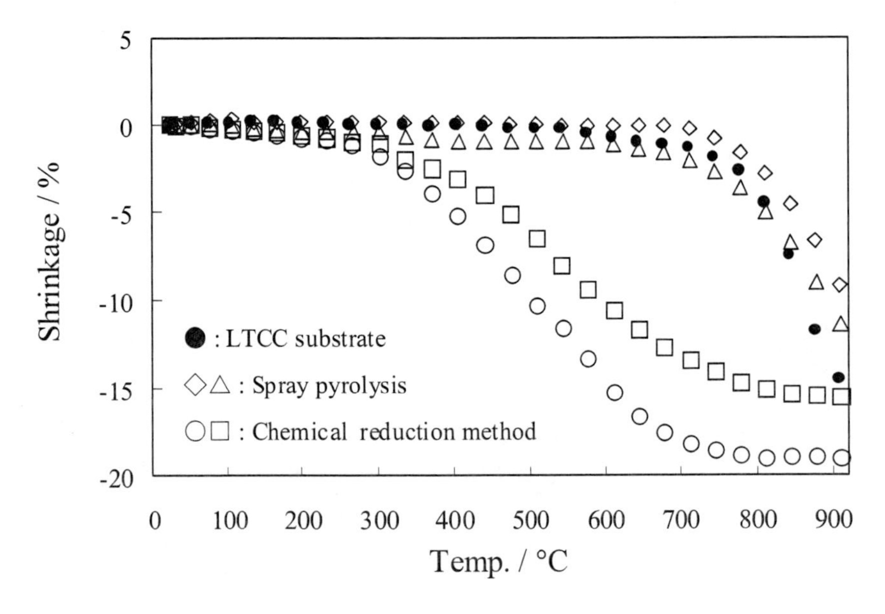

Figure8. Relation between the shrinkage and sintering temperature.

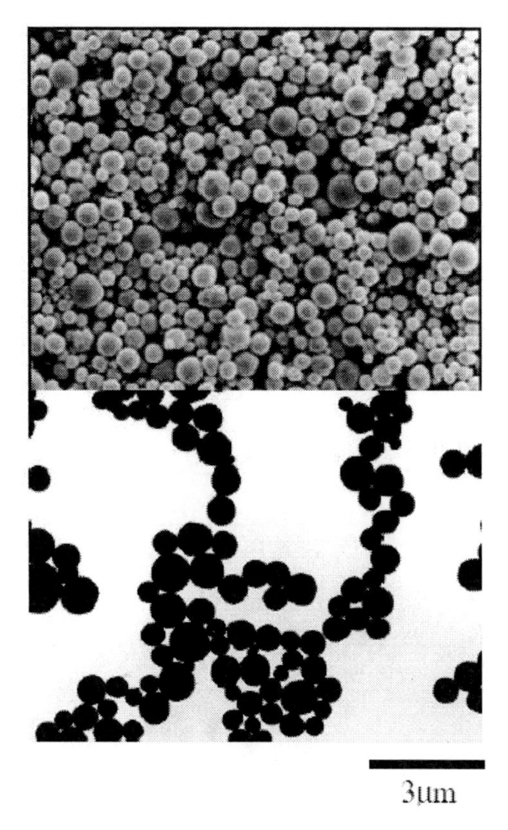

Figure 9. SEM and TEM photographs of $BaTiO_3$ particles.

3.2. Oxide Powders

3.2.1. $BaTiO_3$ Powders for Dielectric Ceramics

It is well known that $BaTiO_3$ powders are widely used as the electronic ceramic materials because of its excellent dielectric properties. Electrical properties of $BaTiO_3$ ceramics strongly depend on microstructures as well as chemical compositions. The preparation of $BaTiO_3$ powders by spray pyrolysis using two-fluid nozzle led to the formation of large particles with hollow microstructure. This resulted in low sinterability. To improve the sinterability, it is necessary to prepare fine particles with a dense microstructure and narrow size distribution. On the other hand, $BaTiO_3$ fine particles with a dense microstructure are offered by ultrasonic spray pyrolysis. Metal organic compound was used as a starting material. Titanium tetra isopropoxide ($Ti(iso\text{-}OC_3H_7)_4$, denoted as TTIP) was used. Generally, TTIP was unstable in the aqueous solution and precipitated as titanium hydroxide. In this work, TTIP was dissolved in the aqueous solution with lactic acid. Ti complex was formed by lactic acid in the aqueous solution and then stabilized without the precipitation of hydroxide. Barium nitrate or barium acetate was also used as a barium source. Barium source was mixed with Ti complex solution to prepare the starting solution. The concentration of starting solution ranged from 0.1 to 0.5 mol/dm^3.

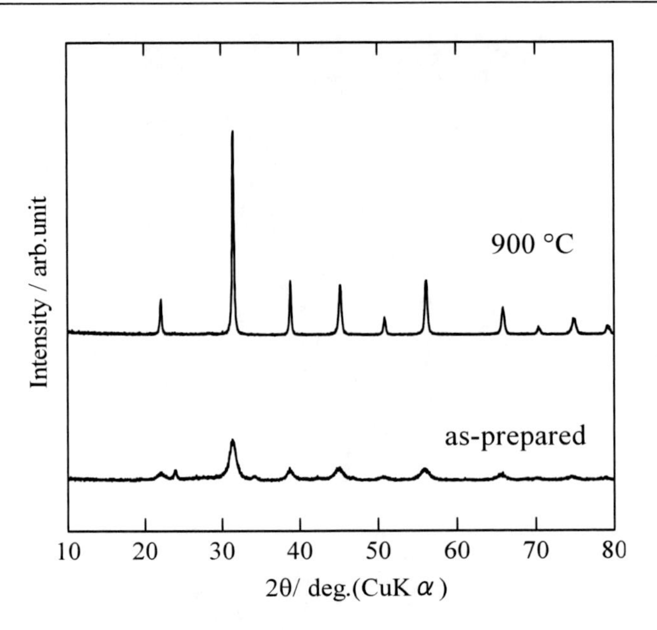

Figure 10. XRD of as-prepared and calcined powders.

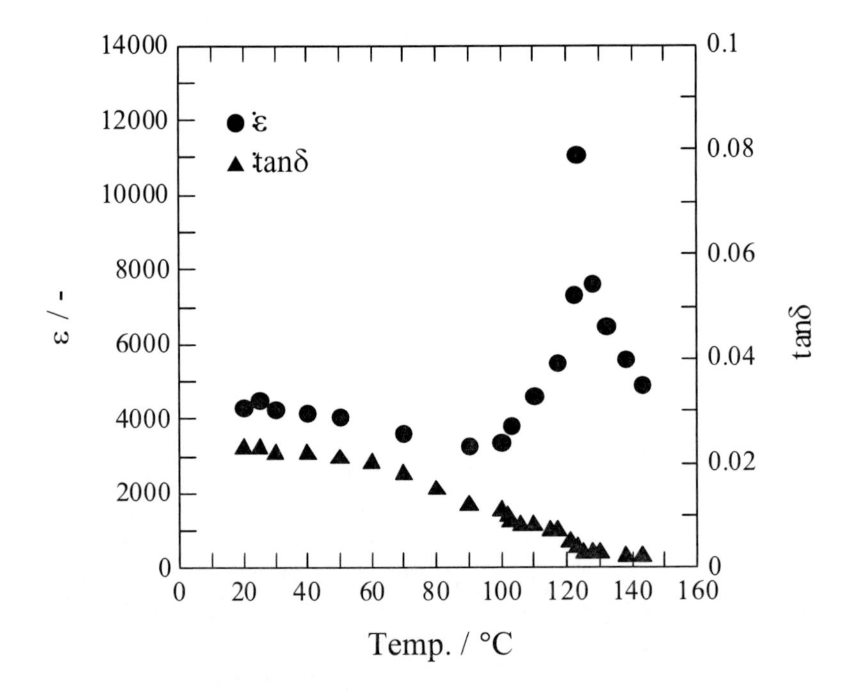

Figure 11. Change of dielectric constant and tanδ for the temperature.

The mist of the starting solution was generated by the ultrasonic transducer with 2.4 MHz and introduced into an alumina tube in the electrical furnace with air carrier gas (7 dm^3/min). The mist was drying at 400 °C and then pyrolyzed at 700 °C.

Figure 12. SEM of $LiAl_{0.05}Mn_{1.95}O_4$ powders.

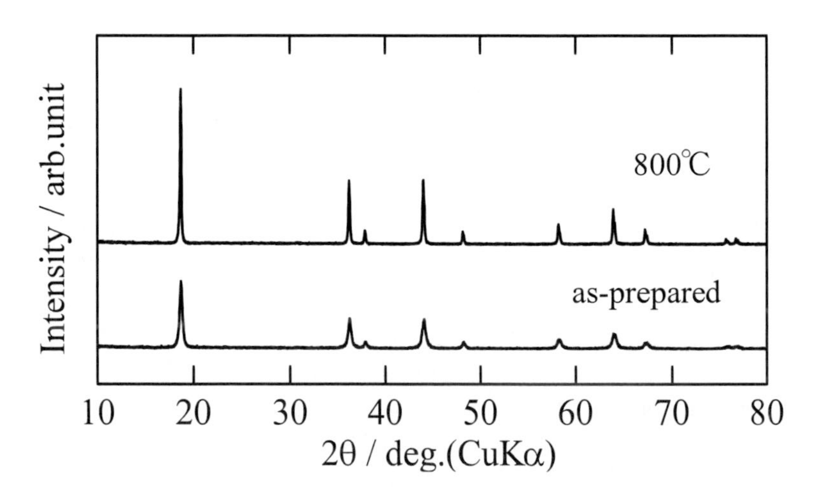

Figure 13. XRD of as-prepared powders and $LiAl_{0.05}Mn_{1.95}O_4$ powders calcined at 800 °C.

Table 2. Chemical composition of $LiAl_{0.05}Mn_{1.95}O_4$ powders

	Atomic concentration. (mol%) Li Mn Al
Theoretical	1 1.95 0.05
As-prepared	1 1.95 0.05

Figure 9 shows SEM and TEM of $BaTiO_3$ particles obtained by spray pyrolysis using aqueous solution of barium nitrate and Ti complex. As-prepared particles had spherical morphology and non-agglomeration. It was confirmed that as-prepared particles also had a dense microstructure. The molar ratio of Ba and Ti determined by ICP analysis was 0.999:1.001. It was confirmed that the spray pyrolysis technique led to form the double oxide with homogeneous composition. Figure 10 shows XRD of as-prepared and calcined powders. As-prepared powders had tetragonal $BaTiO_3$ phase with low crystallinity. After the calcination at 900 °C, $BaTiO_3$ powders with high crystallinity were given. $BaTiO_3$ powders were sintered at 1200 °C for 2 h after the calcination at 900 °C. The relative density of the green body was 58 %. The relative density of the sintered body was also 98 %. Figure 10 shows the change of the dielectric constant and tan δ for the temperature. The dielectric constant and tanδ of the sintered body obtained from $BaTiO_3$ nanopowders was about 4900 and 0.022 at 25 °C, respectively. That of the $BaTiO_3$ sintered body was also 11000 and 0.002, respectively at 120 °C.

3.2.2. $LiMn_2O_4$ Powders for Lithium Ion Battery

$LiMn_2O_4$ is regarded as a promising cathode material for large lithium ion batteries due to their advantages such as low cost, abundance, non-toxicity and being thermally stable. The rechargeable capacity of $LiMn_2O_4$ deteriorates because of the unstableness of the crystal structure in the charged state and the Jahn-Teller effect of high spin Mn^{3+}. To improve this problem, numerous attempts to keep the cycle life by adding the foreign metals to Mn site have been done. The foreign metals such as Mg, Al and Cr have been used as the doping agent. The doping of aluminum ion was most effective to keep the life cycle. Aluminum doped $LiMn_2O_4$ powders were synthesized by spray pyrolysis using metal nitrate. Lithium nitrate, manganese nitrate and aluminum nitrate were used as starting materials. They were dissolved in an appropriate atomic molar ratio to prepare the starting solution. The mist of starting solution was generated with an ultrasonic vibrator (2.4 MHz) and introduced into quartz tube in the electrical furnace with the air carrier (6 dm^3/min). The mist was drying at 400 °C and then pyrolyzed at 900 °C. As-prepared powders were continuously collected using the cyclone. After the synthesis of $LiAl_{0.05}Mn_{1.95}O_4$ powders, the calcination was carried out at 800 °C to eliminate the water and increase the crystallinity. Figure 12 shows SEM of $LiAl_{0.05}Mn_{1.95}O_4$ powders synthesized by spray pyrolysis. $LiAl_{0.05}Mn_{1.95}O_4$ powders had spherical morphology with porous microstructures which consisted of the primary particles ranging from 50 nm and 100 nm. Figure 13 shows XRD of as-prepared powders and $LiAl_{0.05}Mn_{1.95}O_4$ powders calcined at 800 °C As-prepared powders were crystallized to spinel phase (Fd3m). After the calcination, the crystallinity of $LiAl_{0.05}Mn_{1.95}O_4$ powders significantly increased. The other oxide and carbonate phases such as Al_2O_3, Mn_2O_3, Li_2CO_3 were not observed.

The cathode was prepared using 80 wt% $LiAl_{0.05}Mn_{1.95}O_4$, 10 wt% acetylene black and 10 wt% fluorine resin. Metal lithium sheet was used as an anode. The polypropylene sheet (Celgard 2400 Heist) was used as a separator. 1 mol/dm^3 $LiPF_6$ in ethylene carbonate / 1,2-dimethoxyethane (EC : DEC = 1 : 1) was used as the electrolyte. 2032 type coin cell was built up in the glove box under an argon atmosphere. The rechargeable capacity of $LiAl_{0.05}Mn_{1.95}O_4$ cathode was measured with a battery tester between 3.5 V and 4.3 V. The current density ranged from 0.03 (0.1 C) to 6 (10 C) mA/cm^2.

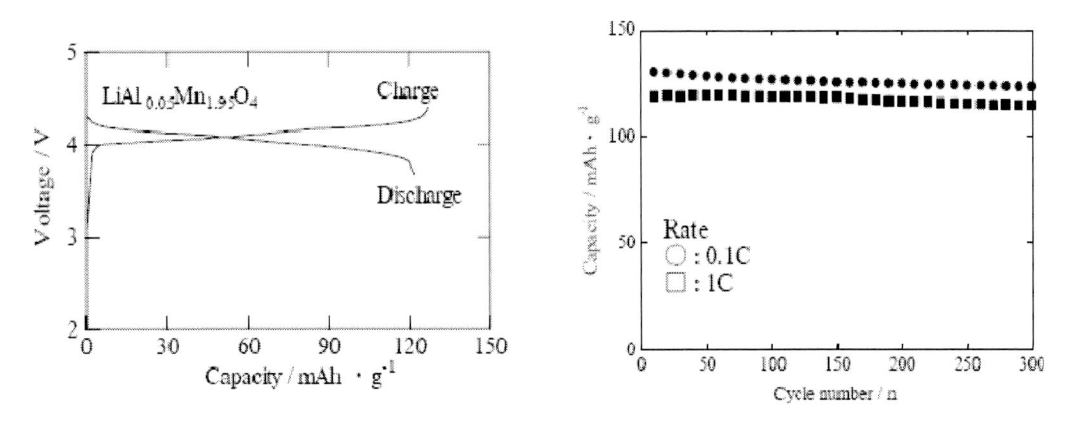

Figure 14. Rechargeable curve and cycle performance of $LiAl_{0.05}Mn_{1.95}O_4$ cathode.

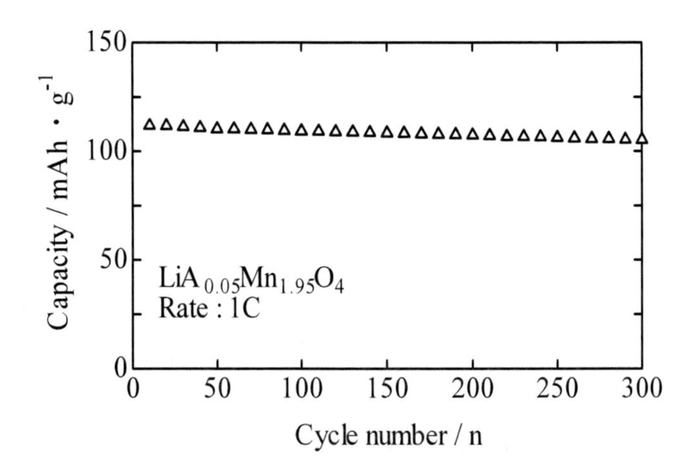

Figure 15. Cycle performance of $LiAl_{0.05}Mn_{1.95}O_4$ cathode at 50 °C.

Figure 14 shows the rechargeable curve and cycle performance of $LiAl_{0.05}Mn_{1.95}O_4$ cathode. Aluminum doping led to the disappearance of the typical voltage jump at 4 V and then the S curve with average operating voltage of 3.6 was observed. The charge and discharge capacity of $LiAl_{0.05}Mn_{1.95}O_4$ cathode was 126 and 121 mAh/g, respectively. The rechargeable efficiency of $LiAl_{0.05}Mn_{1.95}O_4$ cathode was 96 %. The discharge capacity was reduced to 95 % of initial discharge capacity after about 300 cycle number at 1 C. It was clear that the excellent cycle stability was shown by the addition of aluminum ion. The addition of aluminum ion resulted in the increase of more than 3.5 of Mn valence.

Figure 15 shows the relation between the cycle number and discharge capacity of $LiAl_{0.05}Mn_{1.95}O_4$ cathode at 50 °C. The rechargeable test carried out up to 300 cycles at 1 C. The discharge capacity reduced to 112 mAh/g at 50 °C. The discharge capacity was reduced to 94% of the initial discharge capacity after about 300 cycle number at 1 C. The cycle stability of $LiAl_{0.05}Mn_{1.95}O_4$ cathode at 50 °C was maintained as well as the room temperature.

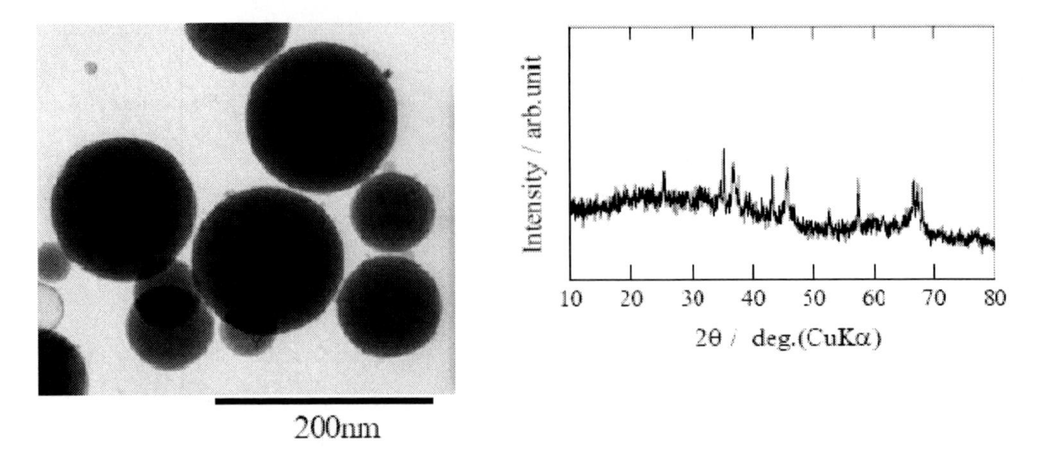

Figure 16. TEM and XRD of alumina nanoparticles.

4. Preparation of Oxide Nanopowders by Plasma-Assisted Spray Pyrolysis

Alumina nanopowders were prepared by plasma-assisted spray pyrolysis using an aqueous solution of aluminum nitrate. So far, amorphous spherical alumina powders with submicrometer size were generally given by spray pyrolysis using nitrate. To prepare crystalline alumina nanopowders directly, the arc plasma was used as the heat source of pyrolysis. The pyrolysis temperature was about 5000 °C. Figure 16 shows TEM photograph and XRD of alumina nanoparticles obtained by plasma-assisted spray pyrolysis. TEM revealed that the as-prepared particles had spherical morphology with dense microstructure and non-aggregation. The average size and σ_g of as-prepared particles was about 100 nm and 1.3, respectively. XRD shows that as-prepared particles were crystallized. The crystal phase of α-alumina was observed by the pyrolysis using the arc plasma.

5. Preparation and Characterization of $LiFePO_4$ Cathode Powders by Two-Fluid Type Spray Pyrolysis

Cathode powders for lithium ion battery were produced by two-fluid type spray pyrolysis. Olivine-type $LiFePO_4$ cathode powders were successfully produced by this method. Olivine-type $LiFePO_4$ exhibits a relatively high theoretical capacity of 170 mAh/g and a stable cycle performance at high temperatures. However, the low electrical conductivity of $LiFePO_4$ prevented its application as a cathode material for the lithium ion battery. Therefore, the conductive materials such as carbon and metals were added to $LiFePO_4$ in order to enhance its electrical conductivity. It has been known that spray pyrolysis lead to directly form $LiFePO_4$/C at one step [28 - 32]. $LiFePO_4$/C powders were produced by using a large two-fluid type spray pyrolysis apparatus and then characterized the particles of the prepared $LiFePO_4$/C powders.

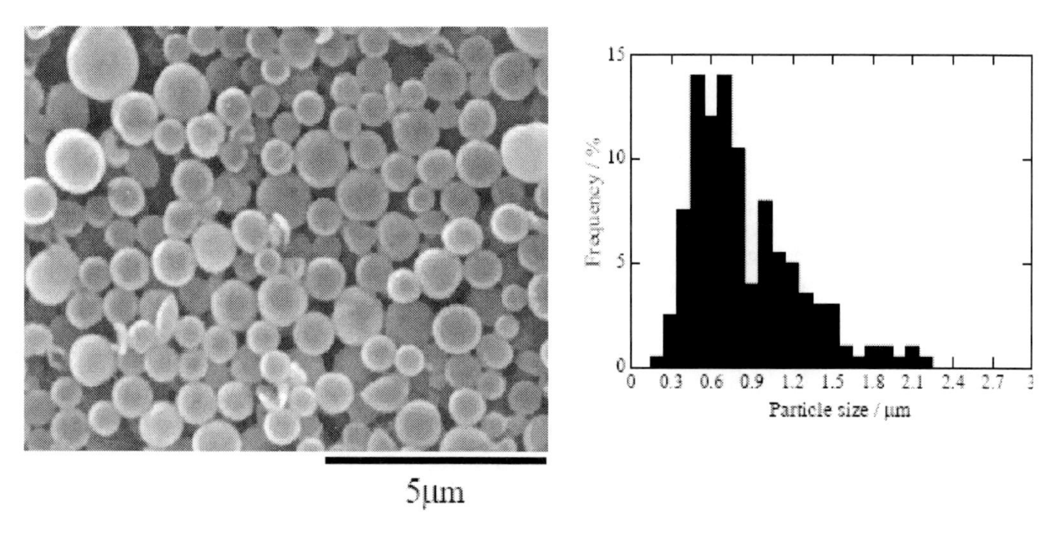

Figure 17. SEM photograph and particle size distribution of as-prepared powders.

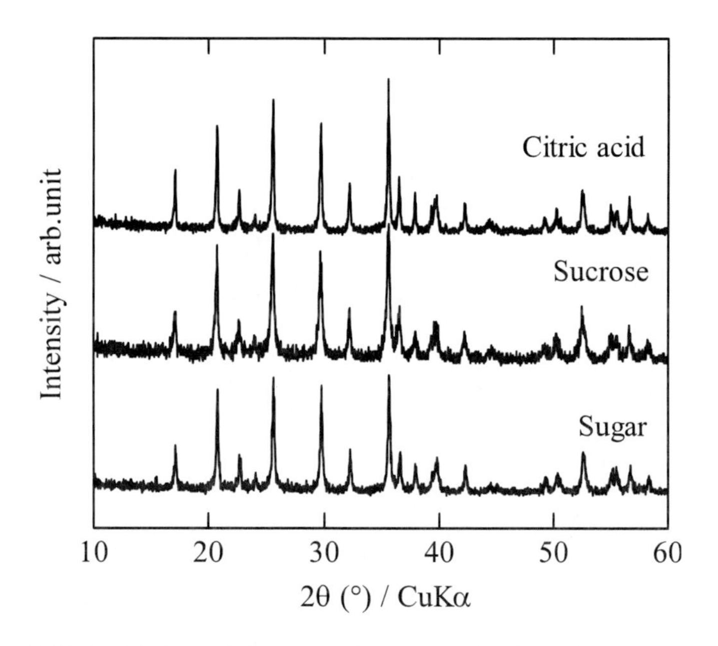

Figure 18. XRD of $LiFePO_4/C$ obtained from organic compound indicated.

Lithium nitrate, iron nitrate and phosphoric acid were used as starting materials. They were weighted out to attain a molar ratio of Li:Fe:P = 1:1:1 and then dissolved in water to prepare 1 mol/dm^3 of aqueous solutions. Various types of organic compounds such as sucrose, fructose, white sugar, and citric acid were used as carbon sources. The contents of the organic compound was 10 wt%. The starting solution was introduced into the two-fluid nozzle using a roller pump. The mist of the aqueous solution was generated in the two-fluid nozzle with air carrier gas and then pyrolyzed at 500 °C. As-prepared $LiFePO_4/C$ powders were produced at 100 g/hr and collected in the bag filter.

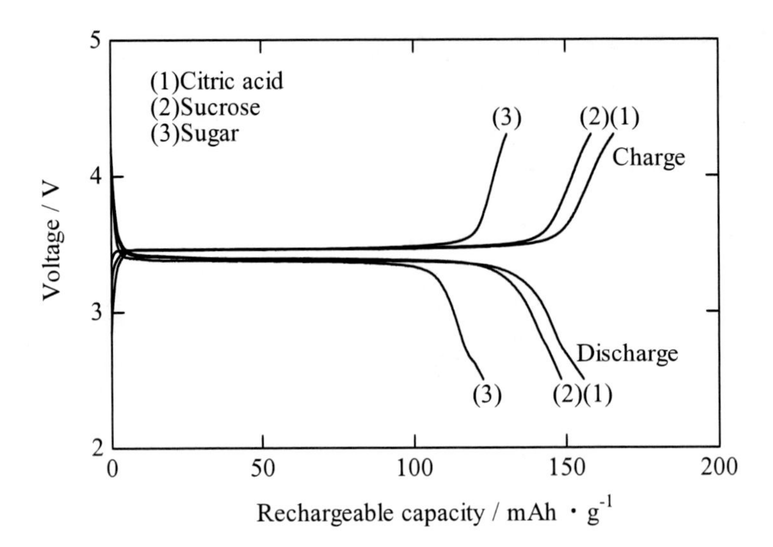

Figure 19. Rechargeable curves of $LiFePO_4$ cathodes.

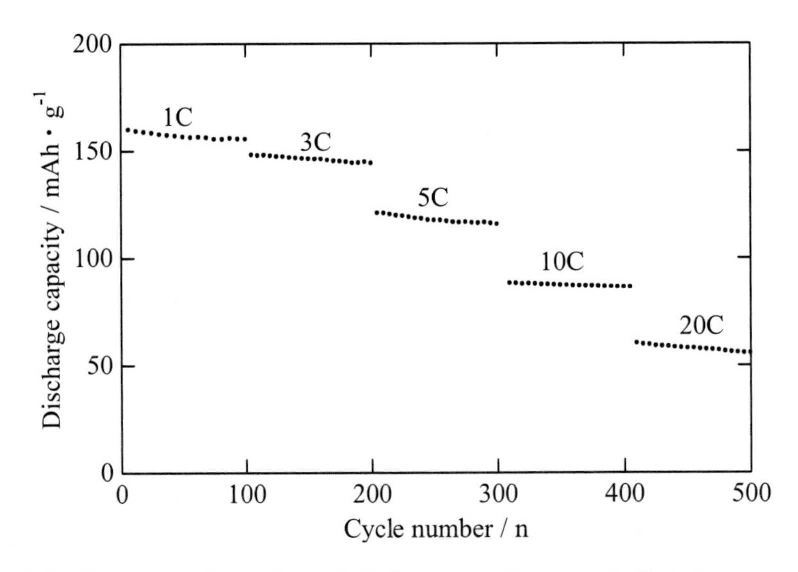

Figure 20. Relation between cycle number and discharge capacity at rate indicated.

Figure 17 shows typical SEM photograph and particle size distribution of as-prepared powders produced at 100 g/hr by spray pyrolysis of an aqueous solution of citric acid. The yield of the as-prepared powders was 95 % regardless of the types of carbon sources. As-prepared particles had a spherical morphology with a smooth surface and non-aggregation regardless of the types of carbon sources used. Figure 17 also shows these hollow particles. This resulted in the drastic decomposition of organic acid in the step of pyrolysis. The average particle size of as-prepared powders obtained from citric acid was about 1.2 μm. The particle size of $LiFePO_4/C$ powders ranged from 0.2 μm to 3 μm. $LiFePO_4/C$ powders had a broad size distribution because of the broad size distribution of the mist generated by the two-fluid nozzle. The specific surface area of the $LiFePO_4/C$ powders was about 10 m^2/g regardless of the types of carbon sources used; this suggested that the particle microstructure of $LiFePO_4/C$ powders was porous. The particle densities of the $LiFePO_4/C$ powders obtained from sucrose and citric acid were 3.5 kg/m^3 and 3.2 kg/m^3, respectively. It is considered that the hollow or porous microstructure led to a reduced particle density of $LiFePO_4/C$ powders.

Figure 18 shows XRD patterns of $LiFePO_4/C$ powders prepared by spray pyrolysis of an aqueous solution with the indicated organic compound. $LiFePO_4/C$ powders were crystallized to the olivine phase by the calcination under argon (95 %)/hydrogen (5 %) atmosphere. XRD revealed that the diffraction patterns of $LiFePO_4/C$ powders were in a good agreement with the olivine structure (space group: Pnma) and that other phases were not observed. The chemical composition of $LiFePO_4/C$ powders was in a good agreement with that chemical composition of the starting solution determined by inductively coupled plasma analysis. Therefore, no evidence for the diffraction peaks of carbon was found in the diffraction pattern; this indicates that the carbon contained in the organic compounds is amorphous and that the presence of carbon does not influence the formation of $LiFePO_4$.

The electrochemical properties of $LiFePO_4/C$ cathode for lithium-ion batteries were examined. Figure 19 shows the rechargeable curves of $LiFePO_4/C$ cathodes at 1 C. A long plateau was observed at approximately 3.5 V in each rechargeable curve. The discharge capacity of $LiFePO_4/C$ cathode obtained from citric acid was approximately 155 mAh/g. That of $LiFePO_4/C$ cathode obtained from sucrose was approximately 148 mAh/g. That of $LiFePO_4/C$ cathode obtained from sugar was approximately 122 mAh/g. The rechargeable efficiency was 93% for the $LiFePO_4/C$ cathode obtained from citric acid and greater than 90% for the $LiFePO_4/C$ cathodes obtained from other organic compounds. The rechargeable capacity of the $LiFePO_4/C$ cathode obtained from citric acid was higher than that obtained from sucrose. TG analysis showed that the concentrations of residual organic species in as-prepared powders obtained from citric acid and sucrose were approximately 20 wt% and 18 wt%, respectively. The carbon contents of $LiFePO_4/C$ powders obtained from citric acid and sucrose were 14 wt% and 10 wt%, respectively, after calcination at 700 °C for 2 h. It is considered that this difference in carbon contents is related to the difference in rechargeable capacities.

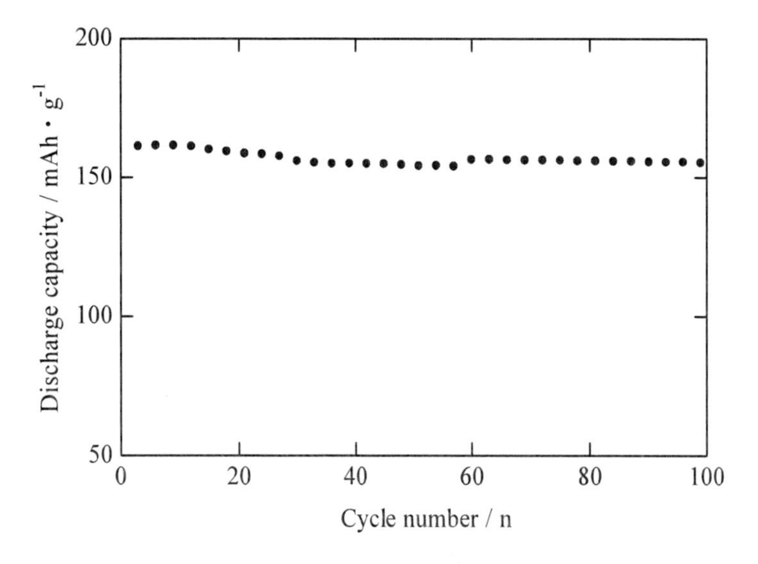

Figure 21. Relation between cycle number and discharge capacity at 50 °C.

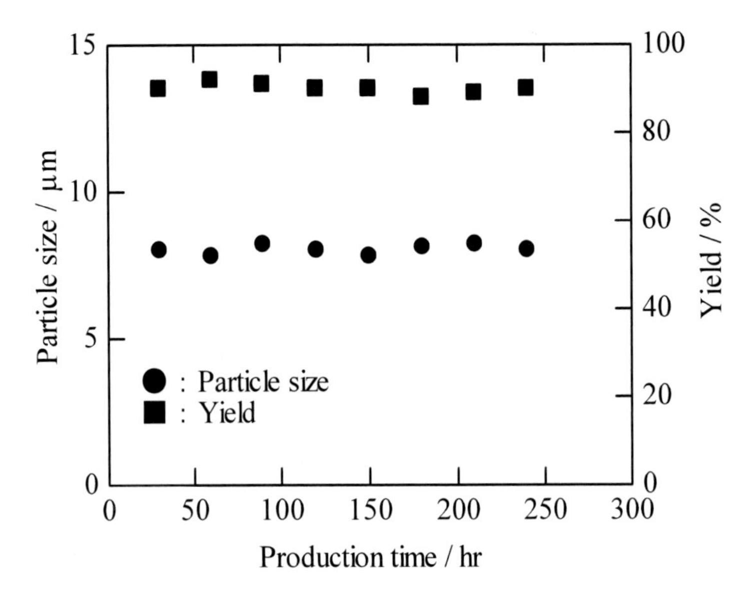

Figure 22. Change of particle size and yield for production time.

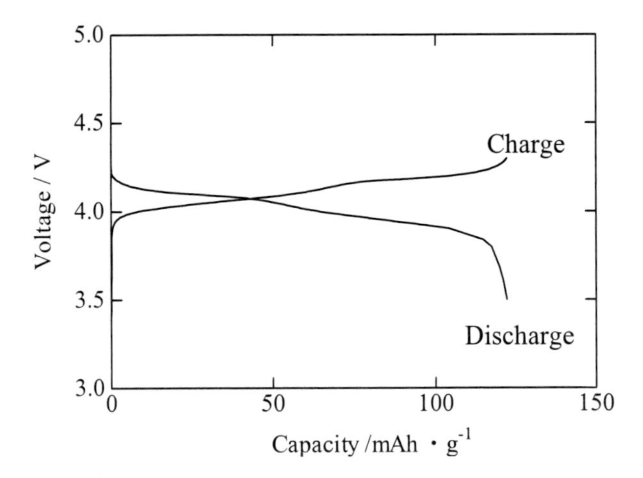

Figure 23. Rechargeable curves of $LiAl_{0.05}Mn_{1.95}O_4$ cathode at 1 C.

Figure 20 shows the relation between the cycle number and the discharge capacity of the $LiFePO_4$/C cathode obtained from citric acid at the indicated discharge rate. A rechargeable test was carried out for up to 100 cycles at each discharge rate. The discharge capacity decreased with the increasing discharge rate. The discharge capacity was 60 mAh/g at 20 C. The discharge capacity of the $LiFePO_4$/C cathode after 100 cycles at 1 C was 96 % of the initial discharge capacity. The same tendency of cycle stability was also observed in the cycle data from 3 C to 20 C. The retention of the discharge capacity was greater than 90 %. It was found that the $LiFePO_4$/C cathode exhibited excellent cycle stability at each discharge rate. Figure 21 shows the relation between the cycle number and the discharge capacity of the $LiFePO_4$/C cathode obtained from citric acid at 50 °C. The rechargeable test of the coin cell was performed at 1 C for up to 100 cycles while being heated on a plate maintained at temperatures from 48 °C to 52 °C. The $LiFePO_4$/C cathode obtained from citric acid exhibited a discharge capacity of 150 mAh/g and a stable cycle life. This suggests that the electrical conductivity of the $LiFePO_4/C$ cathode might be enhanced at 50 °C. The discharge capacity of the $LiFePO_4/C$ cathode was 96 % of its initial discharge capacity after 100 cycles at 50 °C.

6. Mass Production and Characterization of Cathode Powders by Flame Type Spray Pyrolysis

The cathode powders were continuously produced using flame type spray pyrolysis. In order to confirm the reliability of production, the mass production of $LiAl_{0.05}Mn_{1.95}O_4$ powders was done at a rate of 2 kg/h for 250 hrs. Figure 22 shows the particle size and yield of $LiAl_{0.05}Mn_{1.95}O_4$ powders as a function of a production time. $LiAl_{0.05}Mn_{1.95}O_4$ powders were taken out from the bag filter every 30 hrs and then average particle size was measured by SEM photographs. The fluctuation of the particle size and yield was in the range of 3 %. This suggested that the apparatus exhibited good production stability for mass production. The average particle size of $LiAl_{0.05}Mn_{1.95}O_4$ powders was about 8 μm and had broad size distribution because of the use of a larger sized two-fluid nozzle with 20 μm in diameter. The geometrical standard deviation of average particle size ranged from 1.2 to 1.5. These powders had relatively narrow size distribution. SSA of them measured by BET method ranged from 1 to 10 m^2/g. XRD patterns revealed that the diffraction peaks of $LiA_{l0.05}Mn_{1.95}O_4$ were in a good agreement with spinel structure (space group Fd3m). The diffraction peaks of impurities (e.g. Al_2O_3, Mn_2O_3, Li_2CO_3) were not identified. Aluminum was uniformly doped and nickel was substituted in the Mn site. The chemical composition of $LiAl_{0.05}Mn_{1.95}O_4$ powders were determined by Atomic adsorption spectrum (AAS) analysis. AAS showed that the molar ratio of Li : Al: Mn was 0.99 : 0.051 : 1.95. The chemical composition was in a good agreement with the starting solution composition. It was confirmed that this process was effective for the rapid preparation of multi-component oxide powders such as metal doped lithium manganate. The demonstration of rapid preparation offers the improvement of the quantity of production and the decrease of production cost.

Figure 23 shows the charge and discharge curves of $LiAl_{0.05}Mn_{1.95}O_4$ cathode at 1 C. The discharge capacity of $LiAl_{0.05}Mn_{1.95}O_4$ cathode was 125 mAh/g. The doping of aluminum ion led to the disappearance of the typical voltage jump at 4 V in the rechargeable curves and then the S curve with average operating voltage of 3.6 V was observed. On the reason why these cathode materials have a high capacity, it was considered that the particle characteristics such as uniform spherical morphology and fine size lead to the rapid diffusion of Li^+ ion at particles/electrolyte interface. Figure 24 shows the relation between cycle number and discharge capacity of cathode at 1 C. The rechargeable test was carried out up to 500 times. It was clear that the excellent cycle stability was shown. The discharge capacity was reduced to 80 % of initial discharge capacity after about 500 cycle numbers at 1 C. It was considered that the rechargeable reaction uniformly occurred at the level of particle because of narrow size distribution. The discharge capacity of $LiAl_{0.05}Mn_{1.95}O_4$ cathode decreased with increasing charging rate.

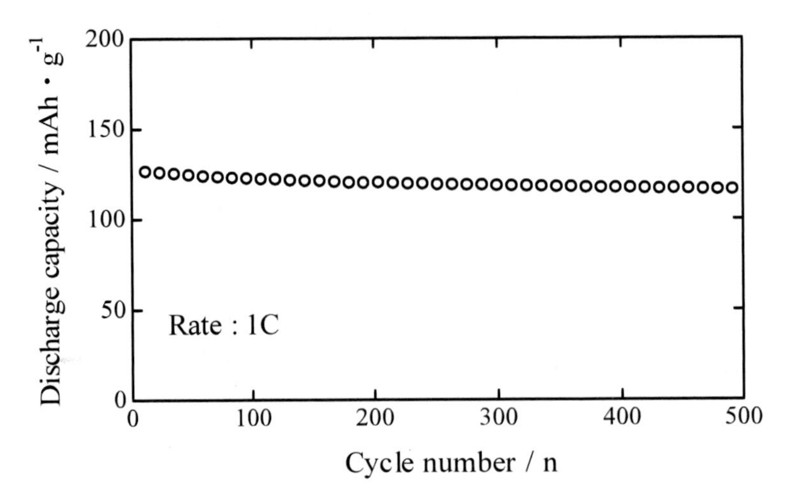

Figure 24. Relation between cycle number and discharge capacity.

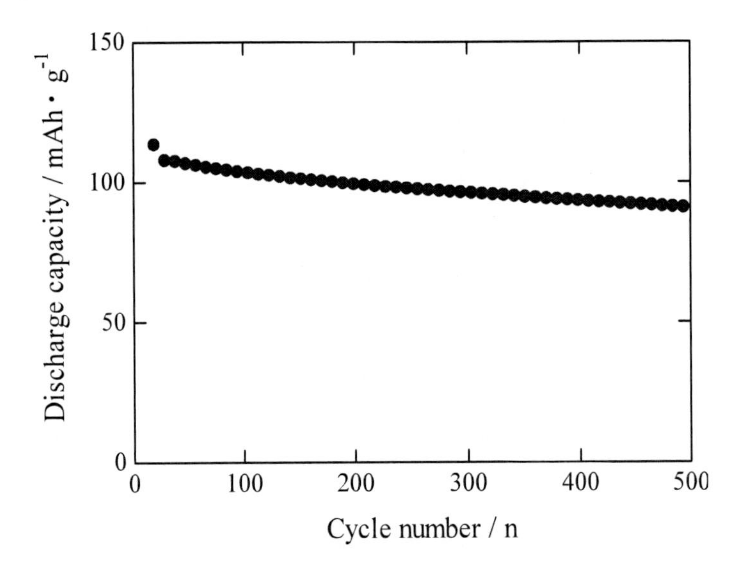

Figure 25. Relation between cycle number and discharge capacity at 50 °C.

The discharge capacity reduced to 80 mAh/g at 10 C. The discharge capacity was maintained over 90 % of initial discharge capacity at 10 C. This suggested that the addition of Al led to its effectiveness to maintain the stability of rechargeable cycle at high rate. Figure 25 shows the relation between cycle number and discharge capacity of the cathode at 50 °C. The rechargeable test carried out up to 500 cycles at a rate of 1C. The cycle stability of $LiAl_{0.05}Mn_{1.95}O_4$ cathodes at 50 °C decreased with increasing cycle number, but 80 % of initial discharge capacity was still maintained after 500 cycles.

7. SUMMARY

Spray pyrolysis technique offered various types of oxide and metal powders with spherical morphology and homogeneous composition. The particle size and size distribution were significantly influenced by the mist generation process, concentration and pyrolysis temperature. The pyrolysis via arc plasma was effective for the formation of crystalline nanoparticles. The oxide powders derived from spray pyrolysis led to the improved sinterability and various electrical properties such as dielectric and electrochemistry. It was demonstrated that the spray pyrolysis using flame was a useful process for mass production of oxide powders.

REFERENCES

[1] G.L. Messing, S.C. Zhang, G.V. Javanthi, Ceramic Powder Synthesis by Spray Pyrolysis, *J. Am. Ceram. Soc.*, 76, 2707-2726 (1993)

[2] H. Ishizawa, O. Sakurai, N. Mizutani, M. Kato. Homogeneous Y_2O_3 Stabilized ZrO_2 Powders by Spray Pyrolysis Method, *Am. Ceram. Soc. Bull.*, 65, 1399-1402 (1986)

[3] B. Dubois, D. Ruffier, P. Odier. Preparation of Fine, Spherical Yttria-Stabilized Zirconia by The Spray-Pyrolysis Method, *J. Am. Ceram. Soc.,* 72, 713-715 (1989)

[4] T. C.Pluym, S. W.Lyons, Q.H. Powell, A.S. Gurav, T.T. Kodas, Palladium Metal and Palladium Oxide Particle Production by Spray Pyrolysis, *Mater. Res. Bull.*, 28, 369-376 (1993)

[5] S. Stopic, J. Nedeljkovic, Z. Rakocevic, D. Uskokovic, Influence of Additives on The Properties of Spherical Nickel Particles Prepared by Ultrasonic Spray Pyrolysis, *J. Mater. Res.,* 7, 3059-3065 (1999)

[6] H. Ishizawa, O. Sakurai, N. Mizutani, M. Kato, Preparation and Formation Mechanism of TiO_2 Fine Particles by Spray Pyrolysis of Metal Alkoxide, Yogyo-Kyokai-shi, 93, 382-386 (1985)

[7] D. M.Roy, R.R. Neurgaonkar, T.P. O'holleran, R. Roy, Preparation of Fine Oxide Powders by Evaporative Decomposition of Solution, *Am. Ceram. Soc. Bull.*, 56, 1023 - 1024 (1977)

[8] S. Wada, A. Kubota, T. Suzuki, T. Noma, Preparation of Barium Titanate Fine Particles from Chelated Titanium and Barium Nitrates by The Mist Decomposition Method in Air, *J. Mater. Sci. Lett.*, 13, 193-194 (1994)

[9] T. Ogihara, H. Aikiyo, N. Ogata , N.Mizutani, Synthesis and Sintering of Barium Titanate Fine Powders by Ultrasonic Spray Pyrolysis, *Adv. Powder Technol.*, 10, 37-50 (1999)

[10] N. Iida, K. Nakayama, I. W. Lenggoro, K. Okuyama, Oxide Behavior of Spray Pyrolyzed Ag-Pd Alloy Particles, *J. Soc. Powder Technol., Japan*, 38, 542-547 (2001)

[11] N. Aoyagi, T. Ookawa, R. Ueyama, N. Ogata , T. Ogihara, Preparation of Ag-Pd Particles by Ultrasonic Spray Pyrolysis and Application to Electrode for LTCC, *Key Eng. Mater.*, 248, 187-190 (2003)

[1 2] T. Ogihara, Y. Saito, T. Yanagwa, N. Ogata, K. Yoshida, M. Takashima, S. Yonezawa, Y. Mizuno, N. Nagata, K. Ogawa, Preparation of Spherical $LiCoO_2$

Powders by The Ultrasonic Spray Decomposition and Its Application to Cathode-Active Material in Lithium Secondary Battery, *J. Ceram. Soc. Jpn.,* 101, 1159-1163 (1993)

[13] T. Ogihara, N. Ogata, S. Yonezawa, M. Takashima, N.Mizutani. Synthesis of Porous $LiNiO_2$ Powders by Ultrasonic Spray Pyrolysis and Their Cathode Property, Denki Kagaku, 66, 1202-1205 (1998)

[14] T. Ogihara, N. Ogata, N. Mizutani, Synthesis of $LiNi_XM_{1-X}O_2$, (M=Co,Mn) Precursor by Ultrasonic Spray Pyrolysis and Electrochemical Property, *Key Eng. Mater.*, 157-158, 303-308 (1999)

[15] S.H. Park, Y.K. Sun, Synthesis and Electrochemical Properties of 5V spinel $LiNi_{0.5}Mn_{1.5}O_4$ Cathode Materials Prepared by Ultrasonic Spray Pyrolysis Method, *Electrochim. Acta* 50, 431-434 (2004)

[16] S.H. Park, C.S. Yoon, S.G. Kang, H.S. Kim, S.I. Moon, Y.K. Sun, Synthesis and Structural Characterization of Layered $Li[Ni_{1/3}Co_{1/3}Mn_{1/3}]O_2$ Cathode Materials by Ultrasonic Spray Pyrolysis Method, *Electrochim. Acta*, 49, 557–563 (2004)

[17] I. Mukoyama, T. Ogihara, N. Ogata, K. Nakane, Synthesis and Characterization of a Metal (Fe, Al or Mg)-doped $Li[Ni_{1/3}Mn_{1/3}Co_{1/3}]O_2$ Particle by Ultrasonic Spray Pyrolysis, *Key Eng. Mater.*, 350, 203-206 (2007)

[18] H. Aikiyo, K. Nakane, N. Ogata, T.Ogihara, Effect of the Doping of Magnesium on Electrochemical Property of $LiMn_2O_4$, *J. Ceram. Soc. Jpn.*, 109, 197-200 (2001)

[19] I. Mukoyama, T. Kodera, N. Ogata, T. Ogihara, Synthesis and Lithium Battery Properties of $LiM(M=Fe,Al,Mg)_XMn_{2-X}O_4$ Powders by Spray Pyrolysis, *Key Eng. Mater.,* 301, 167-170 (2006)

[20] K. Myojin, T. Ogihara, N. Ogata, N. Aoyagi, H. Aikiyo, T. Ookawa, S. Omura, M. Yanagimoto, M. Uede, T. Oohara, Synthesis of Non-stoichiometric Lithium Manganate Fine Powders by Internal Combustion Type Spray Pyrolysis Using Gas Burner, *Adv. Powder Technol.*, 15, 397-403 (2004)

[21] K. Matsuda, I. Taniguchi, Relationship between the Electrochemical and Particle Properties of $LiMn_2O_4$ Prepared by Ultrasonic Spray Pyrolysis, *J. Power Sources*, 132, 156–160 (2004)

[22] I. Mukoyama, K. Myoujin, N. Ogata, T. Ogihara, Large-Scale Synthesis and Electrochemical Properties of $LiAl_XMn_{2-X}O_4$ Powders by Internal Combustion Type Spray Pyrolysis Apparatus Using Gas Burner, *Key Eng. Mater.*, 320, 251-254 (2006)

[23] I. Mukoyama, K. Myoujin, T. Nakamura, H. Ozawa, T. Ogihara, Lithium Battery properties of $LiNi_{0.5}Mn_{1.5}O_4$ Powders synthesized by Internal Combustion Type Spray Apparatus Using Gas Burner, *Key Eng. Mater.*, 350, 191-194 (2007)

[24] T. Kodera, I. Mukoyama, K. Myoujin, T. Ogihara, M. Uede, Electrochemical Properties of $LiNi_{1/3}Co_{1/3}Mn_{1/3}O_2$ Powders Prepared by Internal Combustion Type Spray Pyrolysis, *Key Eng. Mater.*, 388, 89-92 (2009)

[25] M. Yamada, B. Dongying, T. Kodera, K. Myouijn, T. Ogihara, Mass Production of Cathode Materials for Lithium Ion Battery by Flame Type Spray Pyrolysis, *J. Ceram. Soc. Jpn.*, 17, 1017-1020 (2009)

[26] R.J. Lang, Ultrasonic Atomization of Liquids, J. Acoust. Soc. Am., 34, 6-8 (1962)

[27] T. Kodera, H. Horikawa, T. Nakamura, K. Myoujin, T. Ogihara, Synthesis and Characterization of Oxide Nanoparticles by Plasma Assisted Spray Pyrolysis, *Trans. Mater. Res. Soc.*, 32, 151-153 (2007)

[28] M.R. Yang, T.H. Teng, S.H. Wu, $LiFePO_4$/Carbon Cathode Materials Prepared by Ultrasonic Spray Pyrolysis, *J. Power Sources*, 159, 307–311 (2006)

[29] M. Konarova, I. Taniguchi, Preparation of $LiFePO_4$/C Composite Powders by Ultrasonic Spray Pyrolysis Followed by Heat Treatment and Their Electrochemical Properties, *Mater. Res. Bull.*, 43, 3305–3317 (2008)

[30] T. Kodera, K. Egawa, K. Myoujin, T. Ogihara, Synthesis and Electrochemical Property of $LiFePO_4$/Carbon Composite Powder Spray Pyrolysis, *Trans. Mater. Res. Soc.*, 33, 989-992 (2008)

[31] K. Egawa, I. Mukoyama, T. Kodera, K. Myoujin, T. Ogihara, Electrochemical Properties of Carbon-Doped $LiFePO_4$ Cathode Materials Prepared by Spray Pyrolysis, *Key Eng. Mater.,* 388, 81-84 (2009)

[32] S. Akao, M. Yamada, T. Kodera, T. Ogihara, Mass Production of $LiFePO_4$/C Powders by Large Type Spray Pyrolysis Apparatus and Its Application to Cathode for Lithium Ion Battery, *Intl. J. Chem. Eng.*, 2010, ID 175914 (2010)

In: Sprays: Types, Technology and Modeling
Editor: Maria C. Vella, pp. 311-324
ISBN 978-1-61324-345-9

Chapter 9

THICKNESS EVOLUTION IN SPRAY PYROLYTICALLY DEPOSITED FLUORINE DOPED TIN DIOXIDE FILMS

Chitra Agashe*
Solcoat Consultants, Rambaag Colony,
Kothrud, Pune, India

ABSTRACT

Fluorine doped tin dioxide (SnO2:F) films were grown using spray pyrolysis technique to obtain a high quality transparent conductor (TCO). During optimization of each process parameter it was observed that the film thickness and growth rate are crucial in assigning the structural properties, which thereby govern the electro-optical properties of the films. Presented efforts are to understand the structural evolution through the competitive roles of film thickness and growth rate when governed through the time of deposition and the precursor concentration, respectively. For this, the films were grown in three ways. Set A contains films of different thickness in a range ~50-1000 nm, grown using nearly constant growth rate ~100 nm/min. Set B consists of films of thickness ~500 nm, grown using different growth rates in a range ~20-200 nm/min. Set C consists of films of varied thickness up to ~1000 nm, grown using different growth rates in a range ~20-200 nm/min and with a fixed time of deposition. Competitive effects of both the film thickness and growth rate in set C films were isolated by comparing their results with those for set A and set B films. Film properties were investigated using x-ray diffraction technique, scanning electron microscopy and Hall effect measurements. The structural evolution helped to achieve a technologically important textured growth in these films viz. along [200] orientation. The growth rate induced effects accelerated this growth.

PACS No.: 81.15.Rs, 71.20.Nr, 85.60.-q,

* Solcoat Consultants, Flat 6, Ronak Apartment, Rambaag Colony, Kothrud, PUNE -411038, INDIA. telephone: +919890827797, e-mail : chitra_epf2003@yahoo.com, cma_ipv@yahoo.co.in.

Keywords: tin dioxide films, resistivity, growth mechanism, structural properties, optoelectronic devices

1. INTRODUCTION

Transparent conducting oxides (TCOs) form a unique group of materials having a very interesting combination of high visible transparency and high electrical conductivity. This makes them widely applicable in a variety of optoelectronic devices like solar cells, thin film transistors, light emitting diodes, gas sensors, etc. [1-4]. Mainly studied materials are thin films of cation/anion doped binary oxides like tin dioxide, indium oxide, zinc oxide, together with ternary oxides [5-6]. The commercialization demands growing these films by a cost effective and simple technique, such as spray pyrolysis. This is a fast deposition process and easily applicable for large-area deposition at atmospheric conditions. Hence, for spray deposited polycrystalline TCO films, understanding the interdependent roles of process parameters is of vital importance.

The electrical conductivity of TCO films can be improved through that of carrier concentration and / or that of carrier mobility. A high carrier concentration is obtained through non-stoichiometry and extrinsic doping. Improving the carrier concentration beyond certain extent also leads to an undesired increase in the near-infrared absorbance [6]. On the contrary, efforts to improve the carrier mobility are of significance, since this approach has no adverse effects on the optical properties [6]. The carrier mobility can be improved by investigating the structural properties in detail, such as orientation and grain size. For this one needs to concentrate on the film growth. Hence, film thickness and growth rate stand as the basic parameters of this study.

Present work is an attempt to understand the competitive effects of film thickness and growth rate in spray deposited fluorine doped tin dioxide (SnO_2:F) films. Our previous work on un-doped SnO_2 films has shown that the concentration of precursor solution (Sn incorporation) was the key parameter in defining the growth [7]. It was also observed that doping by fluorine [8] and antimony [9] through a wide range could not change the preferred growth except the extent of prominence. Here we extend these efforts to study the growth in further details.

2. EXPERIMENTAL PROCEDURE

The SnO_2:F films were deposited using spray pyrolysis technique. The precursor consisted of tin tetrachloride ($SnCl_4.5H_2O$) dissolved in a mixture of methanol and double distilled water (volume ratio 9:1). The precursor concentration varied from 0.015M to 0.35M. Ammonium fluoride (NH_4F) was added to the precursor solution to maintain the F/Sn atomic ratio as 30 at.%. This dopant ratio does not change the growth rate [8]. The actual incorporation of fluorine into the film was expected to be low due to the high volatility of fluorine compounds produced during the deposition [10]. Corning 7059 glass substrates were used. The deposition temperature and solution flow rate were 450(+/-5) °C and 6 ml/min, respectively, for the range of precursor concentration. Other process parameters were kept constant as given earlier [11]. Set A films of thickness ~50-1000 nm were grown by spraying solution of a precursor concentration 0.12M in different amounts and using a growth rate ~100 nm/min. Set B consists of films of thickness ~500 nm, grown using different growth rates in a range ~20-200 nm/min. The film thickness was controlled through the time of deposition. The set C films of varied thickness up to ~1000 nm were obtained by spraying precursor solutions of different concentrations in a fixed amount and hence, consequent to different growth rates in a range ~20-200 nm/min.

Film thickness was measured using a Talystep detector (conventional roughness detector with a stylus - Taylor Hobson Model). Structural properties were investigated using a low angle x-ray diffractometer with grazing angle 1.0° to confirm full penetration. CuK_α (λ=1.542 Å) radiation was used. The films showed prominent grain growth along [110], [200], [101], [211], accompanied by a less prominent growth along [220], [310], [301] and [321]. This nomenclature is used as per the JCPDS standards [12]. The preferred growth along the orientation [hkl] was expressed in terms of a parameter `relative prominence (RP[hkl])', defined as,

$$RP[hkl] = (I[hkl]) / (\sum_{hkl} (I[hkl])) (\%) \quad (1)$$

where, I[hkl] is the relative intensity of a prominent [hkl]. Morphology of different films was studied on a JEOL JSM-6360A scanning electron microscope. The electrical conductivity and Hall voltage at room temperature were determined using the Van der Pauw geometry [13]. From these, the carrier concentration and carrier mobility were calculated.

3. Results and Discussion

In the following we discuss 1) orientational, 2) morphological and 3) electrical properties.

3.1.1. Orientational Properties of Set A Films

Set A films were grown using precursor concentration 0.12M and growth rate ~100 nm/min. Earlier we observed that the growth rate is linearly dependent on the precursor concentration up to 0.17 M, beyond which there is an onset of saturation [7]. This implies that for 0.12M concentration, the growth is governed by only one species viz. Sn containing species which reacts with adsorbed H_2O molecules and forms the film [7]. So, all these set A films will have a similar starting layer and will contain only the thickness related effects.

X-ray diffractograms of typical films from set A are shown in Figure 1a-c. The films were mainly oriented along [110], [200], [101] or [211], accompanied by a quite less prominent presence of [220], [310], [301] and [321]. For thinner to thicker films, the orientation with the highest relative intensity viz. [110] shifted to [200] and then to [101]. Korotcenkov et al. reported a similar thickness dependent presence of [110] and [200] for spray deposited SnO_2 films with thickness ~160 nm [14]. Figure 2a shows the RP[110], RP[200], RP[101] and RP[211] for these films. Re-orientational effects seem to occur around film thickness 250 nm. The preferred growth changed from [110] to [200]. RP[110] continually decreased up to film thickness ~750 nm, whereas RP[200] increased and saturated. The RP[101] showed a gradual increase beyond ~250 nm thickness. RP[211] also increased but marginally. This thickness evolution shows that with a growth rate ~100 nm/min, one can grow films of thickness ~250-900 nm, preferentially oriented along [200].

In the following we discuss why one may observe these orientational changes. We consider the aspect of thin film growth alone viz. the thickness related effects.

The spray pyrolytic SnO_2 deposition from hydrous $SnCl_4$ is an endothermic reaction. For deposition temperatures ~425 °C, the reactants' interaction with the substrate surface results in a heterogeneous type of nucleation in very early stages of growth [15-16]. As per the JCPDS data for rutile SnO_2, [110] is the strongest orientation [14, 17]. Our results suggest a nucleation controlled random growth for very thin films. When the thickness is increased only by increasing the amount of solution sprayed as in this set A, the arrangement of grains in succeeding layers gets gradually more defined by the grains in the preceding layers.

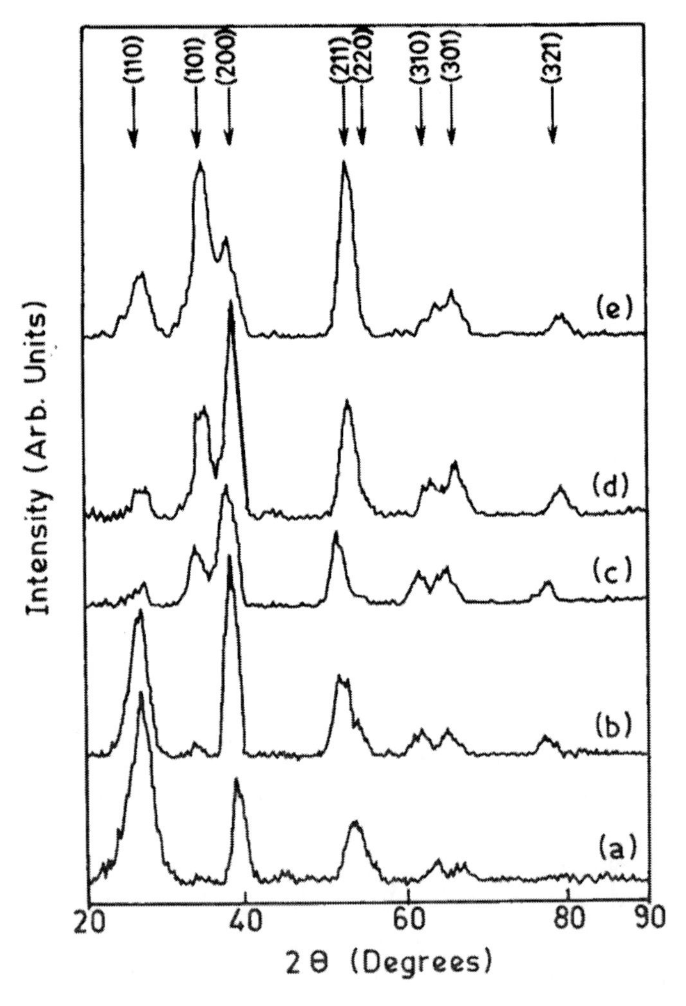

Figure 1. X-ray diffractograms of typical films grown up to different thickness (a-c : set A) or using different growth rates (c-e : set B), a) ~75 nm and 100 nm/min, b) ~265 nm and 100 nm/min, c) ~480 nm and 100 nm/min, d) ~520 nm and 150 nm/min, and e) ~510 nm and 190 nm/min.

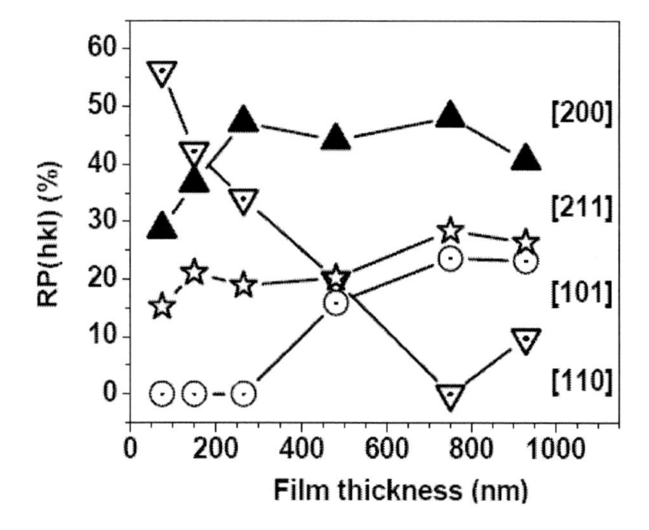

a.

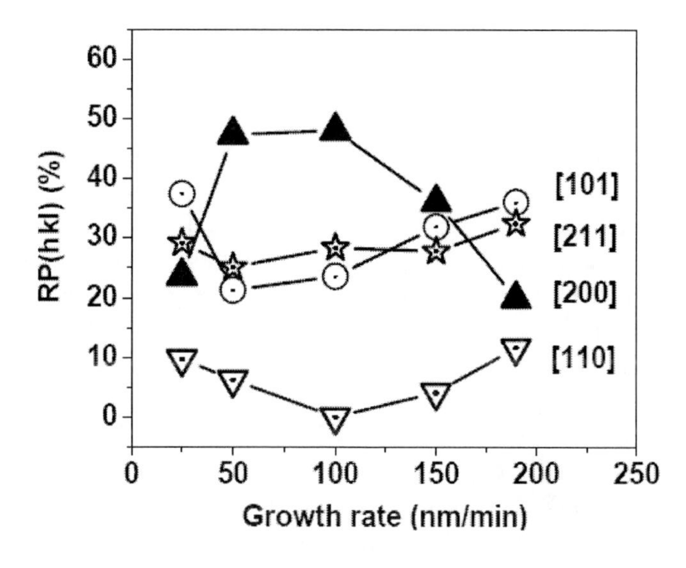

b.

Figure 2. Relative prominence RP[110], RP[200], RP[101] and RP[211] for (a) 75-930 nm-thick films grown at a rate of ~100 nm/min (set A) and (b) ~500 nm-thick films deposited at rates ~20-190 nm/min (set B).

The constraints developed by the preceding layers unlike those on an un-deposited glass substrate, define the growth to occur in a particular way, resulting perhaps in a preferred growth. In such a case, the orientation with a minimum interfacial energy is favored. For tetragonal rutile SnO_2, [200] orientation has a low atomic density and a minimum interfacial energy [18-20]. Hence, the increased prominence of [200] is anticipated. For thickness beyond a particular one, the re-orientational effects may occur, which probably lead to a different packing of the grains. The non-equilibrium conditions of thin film deposition will decide which orientation shall prevail. One can observe such a changing growth pattern only if films are grown `enough thick', where the thickness limits are decided by the growth rate.

Further, among many technological applications of SnO_2 based films, they are used as a `transparent and conducting' coating, with a typically required thickness between ~400 nm and ~800 nm. Our results from set A show that in this range of thickness, when the films are grown with an intermediate precursor concentration like 0.12 M, they are preferentially oriented along [200]. As discussed by Belanger et al., in presence of oxygen vacancies in SnO_2, the orientations [200] and [400] do not contain Sn^{2+} trap states [21]. This is unlike other orientations in SnO_2 and is a consequence of the multi-valency of Sn. This suggests that [200] oriented films will also have better electrical properties, which is of prime interest for any TCO application. Hence, understanding the [200] preferential growth in SnO_2 films is important.

Considering the commercial applicability of transparent conducting SnO_2:F films in a variety of optoelectronic devices, it was important to also find the range of growth rate such that one can grow high quality [200] oriented SnO_2 films. Our studies on the role of growth rate in thickness evolution are discussed in the following.

3.1.2. Orientational Properties of Set B Films

The growth rate of spray pyrolytically deposited SnO_2 based films can be controlled in 4 ways – 1) type of precursor, 2) concentration of precursor, 3) constituents of the precursor solution and 4) dopant incorporation. A variety of inorganic and organic tin compounds have been tried as precursor source in spray pyrolytic deposition [22-24]. Different decomposition products of the reactants and also their impinging flux define the growth process. The chemistry of the reactive species in the vicinity of the growing surface as well as their concentration (per unit area) determines the way in which the films will grow. Among the inorganic compounds, tin tetrachloride (hydrated) seems to be the best precursor [16, 22] and was used here.

The set B films of thickness ~500 nm (520 nm, 515 nm, 480 nm, 520 nm and 510 nm) were deposited using growth rates ~25, 50, 100, 150 and 190 nm/min. This was achieved by using precursor concentrations, 0.03 M, 0.06 M, 0.12 M, 0.17 M and 0.3 M, respectively. Hence, each film was grown with a different amount of Sn in the impinging flux. The incorporation of oxygen in all cases is solely through the ambient surrounding the heated substrate. The water in the precursor is not the source of oxygen but is used only to increase the solubility of $SnCl_4.5H_2O$ in methanol. Both methanol and water will evaporate when the droplets gradually reach the hot substrate. The use of methanol leads to a chemically reducing atmosphere. The pyrolysis of precursor needs to take place instantaneously as the decomposition of reactant species and formation of the product [25]. The increasing precursor concentration will increasingly reduce the oxygen partial pressure and will favor the Sn rich non-stoichiometry. This will govern the inclusion of oxygen vacancy as well as that of the excess Sn at interstitial sites. When beyond a limit, the deviation from stoichiometry `may' affect the construction (growth) of the films.

The effect of growth rate on the orientational properties is typically seen from Figure 1c-e. Figure 2b reveals the growth rate effect on the thickness evolution through the respective variations in RP[110], RP[200], RP[101] and RP[211]. RP[110] remained the lowermost over the entire range of growth rate. The RP[200] is maximum over the growth rate range ~50-150 nm/min. For higher growth rates [101] became prominent whereas [211] possessed an intermediate prominence. This implies that the films with the desired preferred growth along [200] can be obtained using growth rate ~50-150 nm/min.

We observed that the growth rate is linearly dependent on the precursor concentration up to 0.17 M, beyond which there is an onset of saturation [7]. This type of growth is known as the Rideal-Eley mechanism [26]. In the spray pyrolysis deposition (at the atmospheric conditions), it is difficult to control the incorporation of oxygen into the film. Hence, the stoichiometry also (in addition to the growth rate) is governed by the Sn containing species. When the Sn concentration is too high, the surface concentration of reactants is saturated which limits the rate of formation of SnO_2 layers. Hence, the growth rate deviates from its linear dependence and gradually saturates for higher precursor concentrations. This also indicates that the orderly formation of SnO_2:F films may get gradually disturbed as we go to the growth rates beyond ~150 nm/min, using higher precursor concentrations.

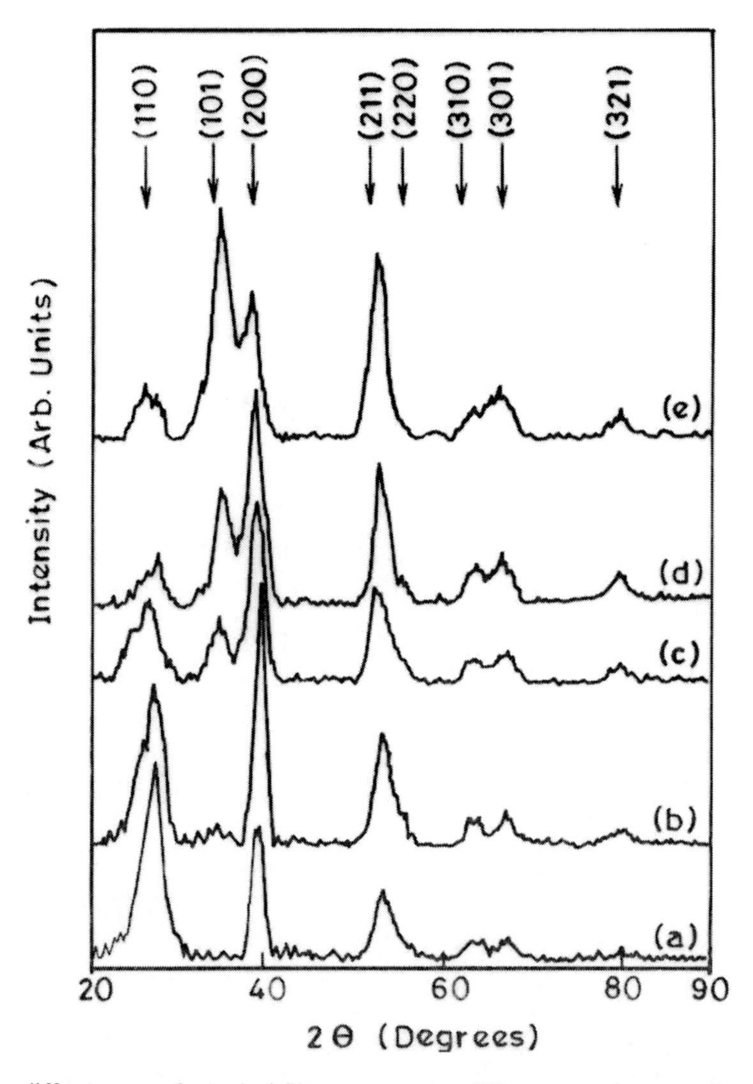

Figure 3. X-ray diffractograms for typical films grown using different growth rates and up to different thicknesses: (a) 25 nm/min, 120 nm, (b) 50 nm/min, 240 nm, (c) 150 nm/min, 765 nm, (d) 170 nm/min, 840 nm and (e) 190 nm/min, 930 nm (set C).

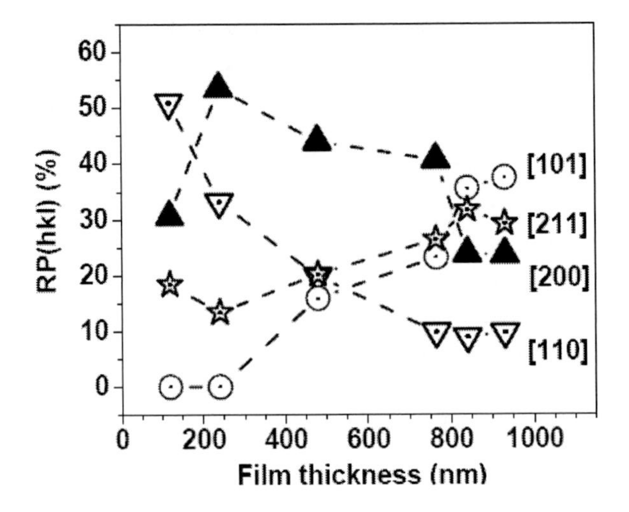

a.

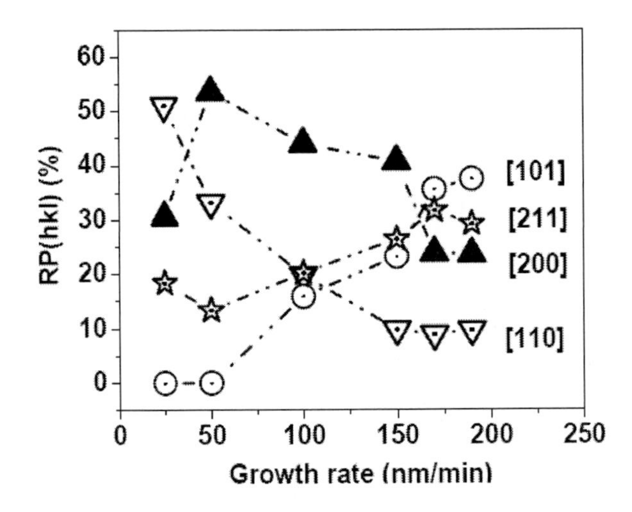

b.

Figure 4. Relative prominence RP[110], RP[200], RP[101] and RP[211] with (a) film thickness and (b) growth rate for films from set C.

Such a growth rate dependent growth pattern suggests that the mass transfer through the impinging flux of precursor species (cation to anion ratio) builds succeeding layers of crystallites depending on the nucleation sites provided by the crystallites from the preceding layers. This supports that the growth of spray deposited SnO_2:F films is a `nucleation and growth' process. The growth along [200] realized on the basis of minimum interfacial energy considerations, seems to be disturbed at higher growth rates consequent to the crowding of Sn containing reactive species impinging through the flux. This may favor the growth of crystallites `also' along some other orientations, as observed here beyond the growth rate ~150 nm/min. Kaneko et al. stated that a preferred growth along [200] seems to arise from the O-Sn-O species forming at the decomposition states of the source compound [27]. If the films are grown using different Sn flux through the impinging species, the formation of O-Sn-O species can get modified, resulting in the favor of other orientations like [101] and [211]. Fantini et al. also discussed this issue for SnO_2 films grown by chemical vapor deposition

[28]. This explains why and how the growth rate modifies the orientational properties of SnO_2:F films. Figure 2a has revealed that the [101] and [211] growth is a result of thickness evolution also. An increase in the growth rate through the precursor concentration, maintained the basic growth pattern, but modified the change in prominence of different orientations.

3.1.3. Orientational Properties of Set C Films

The set C films were deposited by spraying a fixed amount of precursor solution of different concentrations and using a fixed time of deposition. Hence, each of these films will have a different thickness consequent to different growth rate. The films for this set were grown up to the thickness ~120, 240, 510, 760, 840 and 930 nm, using growth rate of 25, 50, 100, 150, 170 and 190 nm/min, respectively. Their starting layers of grains as also the consecutive layers can be different. These films will contain the thickness related effects as well as the growth rate related effects. Hence, comparing these films with set A films of similar thickness will give the difference due to the growth rate and a comparison with set B films will show the difference due to the film thickness.

a) sample - 6-20-2.

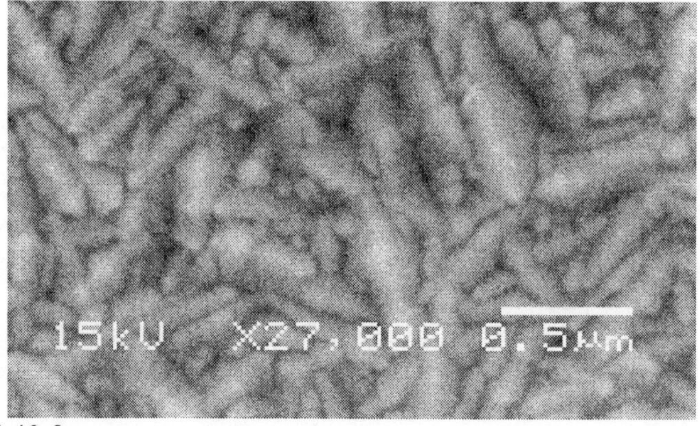

b) sample - 6-40-3.

Figure 5. Scanning electron morphological features of SnO_2 : F films with a) thickness ~480 nm and growth rate ~100 nm/min, b) thickness ~930 nm and growth rate ~100 nm/min.

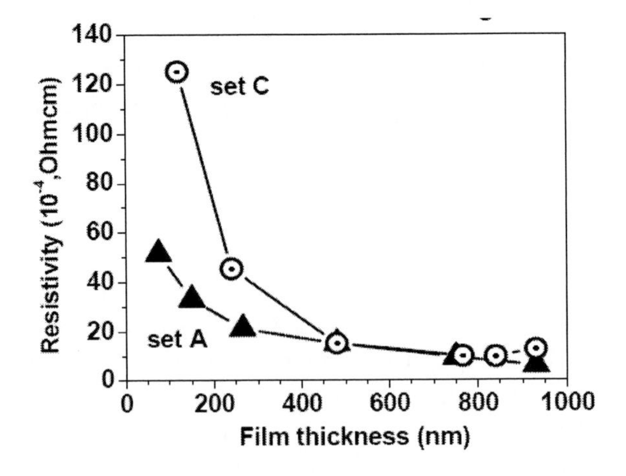

a.

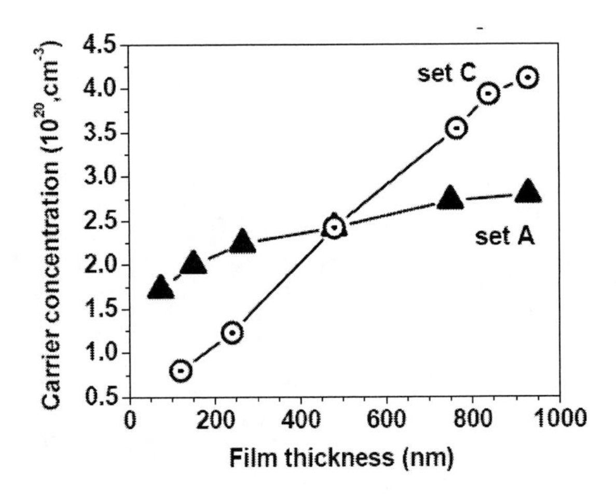

b.

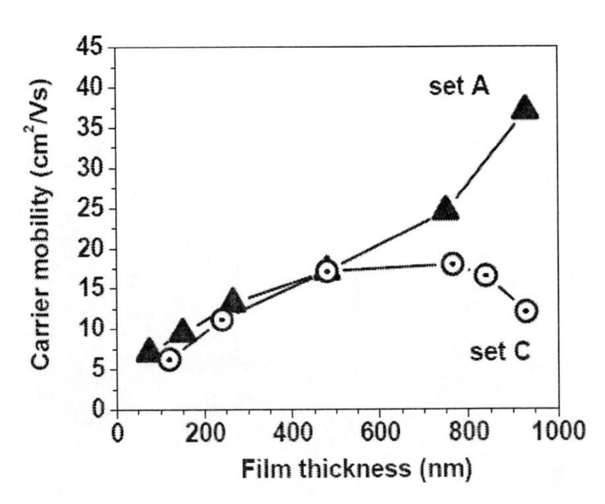

c.

Figure 6. Resistivity (a), carrier concentration (b) and carrier mobility (c) for films of different thickness prepared at same growth rate ~100 nm/min (set A) and for films of different thickness prepared at different growth rates ~20-190 nm/min (set C).

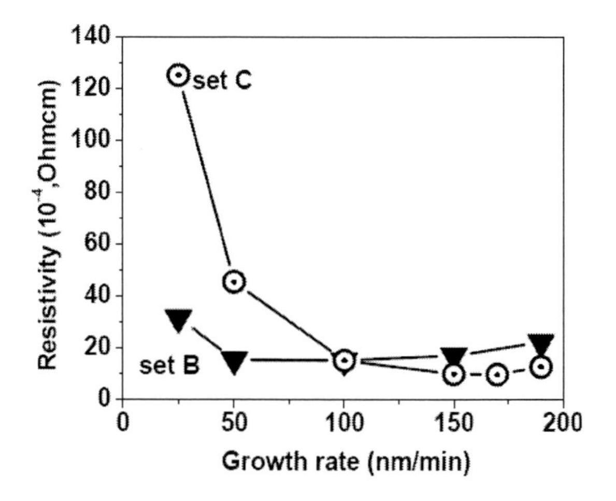

a.

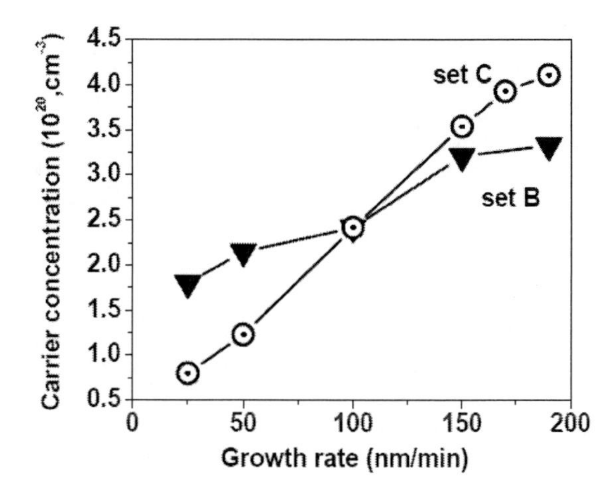

b.

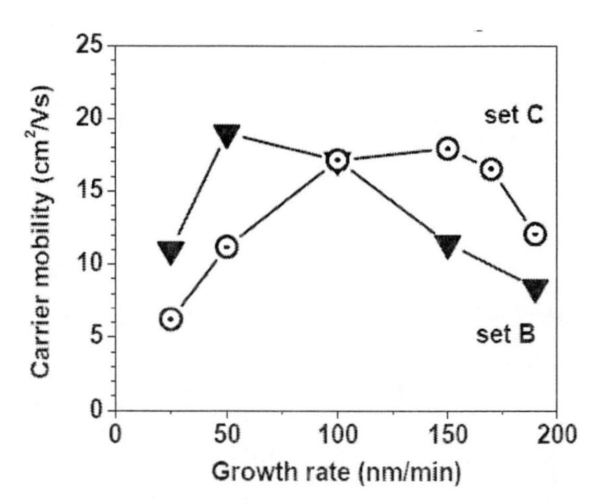

c.

Figure 7. Resistivity (a), carrier concentration (b) and carrier mobility (c) for ~500 nm thick films prepared at different growth rates (set B) and for ~100-950 nm thick films prepared at different growth rates (set C).

X-ray diffractograms for set C films are depicted in Figure 3. These films also show a change in the high intensity orientation from [110] to [200] and then to [101]. The intensity of [211] was always in a medium range. The variation in RP[110], RP[200], RP[101] and RP[211] with the film thickness as well as with the growth rate is shown in Figure 4a and Figure 4b, respectively. The RP[200] was maximum for films of thickness ~250-750 nm grown using growth rates 50-150 nm/min. This is accompanied by the increased prominence of [101] and [211] and the gradual decrease in RP[110]. One can directly compare Figure 4a with Figure 2a and Figure 4b with Figure 2b, to see the difference due to growth rate and that due to film thickness, respectively. These studies showed us that the SnO_2 based films of thickness ~250-750 nm, could be grown preferentially along [200] using the growth rate ~50-150 nm/min.

Several reports on the structural / orientational properties of SnO_2 based films reveal that depending on the deposition technique and the chemical nature of the precursor, the films show diverse patterns of growth. The effect of growth rate has also been studied but mainly with reference to the deposition temperature. The collective efforts to separate the effect of growth rate (when controlled by the concentration of precursor solution) and the effect of film thickness are performed here. To find a wider acceptable range of growth rate is a technologically important outcome of present efforts.

3.2. Morphological Features

Typical scanning electron morphological (surface) features of a film with ~480 nm thickness grown using the growth rate ~100 nm/min are depicted in Figure 5a. Figure 5b reveals how the grains grow laterally as one grows films of thickness ~930 nm with growth rate ~ 100 nm/min. Former film is the common one for all sets whereas later film corresponds to the high mobility case from set A. The orientational properties discussed in previous section revealed that as one grows the thicker films, the [200] orientation wins the growth race. These [200] grains grow from substrate to surface as also laterally. This is seen as elongated regions in Figure 5b. A `secondary' prominence of other orientations may also contribute to such a growth pattern. An enhanced grain size and prominent oriented growth provide an organized material and is desired for improved electronic transport. This is an important aspect while applying SnO_2:F films as transparent electrode in different opto-electronic devices.

3.3. Electrical Properties of Set A, Set B and Set C Films

Figures 6a and 7a show the dependence of electrical resistivity on the film thickness and the growth rate for sets A and C, and sets B and C, respectively. The set C films possess a cumulative effect of film thickness and growth rate and hence, their resistivity shows a steeper reduction than that for set A films and for set B films.

Figures 6b and 7b reveal the dependence of carrier concentration on the film thickness and the growth rate for sets A and C, and sets B and C, respectively. Since, the set C films of increasing thickness were made using increasing growth rate (controlled through the precursor concentration), the Sn incorporation in these films increased and as a consequence the carrier concentration increased as 0.8-4.1 x $10^{20}cm^{-3}$. On the other hand, the carrier concentration in set A and set B films varied differently. Set A films were grown using a fixed precursor concentration, anticipating a fixed inclusion of tin, oxygen and fluorine. This effectively resulted in a high carrier concentration. There is an increase in carrier concentration for set B films also, less prominent but similar to that for set C films, which again owes to the way these films were prepared viz. with increased growth rate achieved through the increased precursor concentration. The small rise in carrier concentration of set A films of increasing thickness as also the difference in the carrier concentration for set B and set C films suggests that film thickness also has a role in assigning the carrier concentration.

Figures 6c and 7c reveal a dependence of carrier mobility on the film thickness and growth rate for sets A and C, and sets B and C, respectively. The mobility may be governed by – 1) orientational changes, 2) grain size variation and 3) scattering due to carriers. The orientational changes affect the carrier mobility through the grain boundary related electrostatic and structural aspects. The grain boundary scattering is negligible in these degenerate semiconductors. The structural aspect like the mismatching of differently oriented grains' surfaces creates a mechanical obstacle to the carrier transport. This would reduce the carrier mobility.

The set C films of increasing thickness were prepared using increasing growth rates and this led to an increase in grain size up to ~ 250 nm. The orientational changes in set C films were a bit intense. The change in mobility in Figure 6c, suggests that the possible improvement due to grain size is suppressed by the orientational changes and the scattering due to carriers. This leads to a reduction in carrier mobility for set C thicker films. For set A films the increase in film thickness also led to an increase in grain size which will favor an improvement in carrier mobility. This will be assisted by the orientational change to [200]. For set A films, a contribution of carrier scattering which will reduce the mobility seems to be less significant. The set C and set B films were prepared using comparable growth rates. The carrier mobility in set B films is more sensitive to growth rate induced changes in the structural properties. The set B films were of same thickness ~500 nm and hence, possess no major advantage of an increase in grain size. Here the drop due to orientational changes and scattering due to carriers is evident. The set C films showed a drop in mobility for thicker films grown using higher growth rates.

Conclusions

Comparative effects of film thickness and growth rate in spray pyrolytically deposited SnO_2:F thin films were studied for thickness up to ~930 nm and growth rate up to ~190 nm/min. It was observed that the growth rate as controlled through the concentration of precursor solution, plays a prominent role in determining the range of film thickness where the thickness evolution of the structural properties occurs. The thickness induced effects such as orientational changes towards [200] become significant for thicker films. The growth rate induced effects leading to re-orientational changes in the growth pattern allow one to grow these films within a wider growth rate range ~50-150 nm/min so as to enhance the preferred growth along [200] and to achieve better electrical properties.

REFERENCES

[1] H.L. Hartnagel, A.L. Das, A.K. Jain, C. Jagadish, Semiconducting Transparent Thin Films, Institute of Physics Publishing, Bristol, 1995.
[2] C.G. Granqvist, *Sol. Energy Mater. Sol. Cells* 91 (2007) 1529.
[3] S. Calnan, A. N. Tiwari, *Thin Solid Films* 518 (2010) 1839.
[4] H. Kawazoe, H. Yanagi, K. Ueda, H. Hosono, *Mater. Res. Soc. Bull.* 25 (2000) 28.
[5] D. Ko, K. Poeppelmeier, D. Kammler, G. Gonzalez, T. Mason, D. Williamson, D. Young, T. Coutts, *J. Solid State Chem.* 163 (2002) 259.
[6] K. L. Chopra, S. Major, D. K. Pandya, *Thin Solid Films* 102 (1983) 1.
[7] C. Agashe, M. Takwale, V. Bhide, S. Mahamuni, S. Kulkarni, *J. Appl. Phys*. 70 (1991) 7382.
[8] C. Agashe, S. Major, *J. Mater. Sci*. 31 (1996) 2965.
[9] D. Goyal, C. Agashe, M.G. Takwale, V.G. Bhide, *J. Cryst. Growth* 130 (1993) 567.
[10] E. Shanthi, A. Banerjee, V. Dutta, K.L. Chopra, *J. Appl. Phys*. 53 (1982) 1615.
[11] C. Agashe, M. G. Takwale, B. R. Marathe, V. G. Bhide, *Sol. Energy Mater.* 17 (1988) 99.
[12] Joint Committee on Powder Diffraction Standards, *Powder Diffraction File* 5-0467.
[13] W.R. Runyan, Semiconductor Measurements and Instrumentation, McGraw-Hill, Kogakusha Ltd., Tokyo, 1975.
[14] G. Korotcenkov, M. DiBattista, J. Schwank, V. Brinzari, *Mater. Sci. Engg*. B77 (2000) 33.
[15] V. Vasu, A. Subrahmanyam, *Thin Solid Films* 189 (1990) 217.
[16] V. Vasu, A. Subrahmanyam, *Thin Solid Films* 193/194 (1990) 973.
[17] Powder Diffraction File, Joint Committee on Powder Diffraction Standards -, International Center for Diffraction Data, Swarthmore, PA, 1988, Card 21-1250.
[18] M. Fantini, I. Torriani, C. Constantino, *J. Cryst. Growth* 74 (1986) 439.
[19] K. L. Chopra, Thin Film Phenomena, McGraw-Hill, New-York NY, 1969.
[20] J. M. Poate, K. N. Tu, J. W. Mayer, Thin Films - Interdiffusion and Reactions, Wiley, New-York NY, 1978.
[21] D. Belanger, J. P. Dodelet, B. A. Lombos, J. I. Dickson, *J. Electrochem. Soc.* 132 (1985) 1398.
[22] G. Gordillo, L. C. Moreno, W. de la Cruz, P. Teheran, *Thin Solid Films* 252 (1994) 61.
[23] H. Demiryont, K. E. Nietering, *Sol. Energy Mater.* 19 (1989) 79.
[24] O. Tabata, in: J. Blocher Jr., H. Hintermann, L. Hall (Eds.), 5th International Conference on Chemical Vapor Deposition Proceedings, New York, U.S.A., 1975, p. 681.
[25] J.C. Viguié, J. Spitz, *J. Electrochem. Soc.* 122 (1975) 585.
[26] R. N. Ghoshtagore, *J. Electrochem. Soc.* 125 (1978) 110.
[27] S. Kaneko, I. Yagi, K. Murakami, M. Okuya, *Solid State Ionics* 141–142 (2001) 463.
[28] M. Fantini, I. Toriani, *Thin Solid Films* 138 (1986) 255.

In: Sprays: Types, Technology and Modeling
Editor: Maria C. Vella, pp. 325-335
ISBN 978-1-61324-345-9

Chapter 10

Flamelet Equations for Spray Combustion

Kun Luo*, ***Jianren Fan and Kefa Cen***
State Key Laboratory of Clean Energy Utilization, Zhejiang University,
Hangzhou 310027, P. R. China

Abstract

Various sets of new flamelet equations for spray combustion are derived in mixture fraction space. The spray droplets are assumed as points and described in an Lagrangian framework. The influence of a droplet on gas phase is represented by the mass, momentum, and energy source terms. The derived equations are different from the standard flamelet equations for gaseous combustion because of the droplet evaporation source term that was neglected in most previous studies. In addition, a general flamelet transformation for spray combustion is formulated, and the potential advantages of combined premixed and diffusion flamelet models for spay combustion is discussed.

PACS 47.70.Pq, 47.55. D-, 44.35.+c.

Keywords: Flamelet equations; Spray combustion; Mixture fraction; Droplet evaporation

1. Introduction

Numerical simulation of turbulent combustion has been the subject of numerous studies in the past decades. Significant advances have been achieved in flamelet models [18], PDF methods [24], conditional moment closure [7], and linear eddy modeling [11]. A comprehensive review on turbulent combustion modeling has been given by Pitsch [21] for applications in large-eddy simulation context.

Flamelet modeling assumes that combustion and heat release in a turbulent flame can be represented by the effects of one or more moving laminar flames which are embeded in the turbulent flow field. Based on this assumption that is usually true for realistic applications, the description of laminar flames, involving detailed chemical kinetic and molecular diffusion mechanisms, can be decoupled from the turbulent flow field and therefore conducted

*E-mail address: zjulk@zju.edu.cn

separately. By appropriate mapping, the solutions to fully three dimensional turbulent combustion can be obtained. One of the particular advantages of this approach is that it does not artificially alter physics or require user tuning. Because of this, flamelet modeling has been extensively studied and applied to various gaseous combustion [22, 23, 20, 19].

The standard flamelet model [18] has been extended to study turbulent spray flames in realistic combustors [13, 3]. In these studies, the flamelet libraries were created based on laminar, gaseous counterflow diffusion flames. Although consistent results with experimental data have been achieved, using models of gaseous combustion to describe spray flames leads to some modeling issues, as discussed by [1]. For example, using gaseous flamelet library would suggest that combustion can be accurately modeled independent of spray evaporation. Actually, spray evaporation and combustion are strongly coupled. The flamelet modeling should correctly reflect the inter-phase mass, momentum and energy couplings.

To account for the influence of spray evaporation on gaseous flamelets, a flamelet model for turbulent spray diffusion flames in the counterflow configuration using the laminar spray flamelet library [5] was developed [6]. However, to create the laminar spray flamelet library, additional dependent parameters related to spray droplets, such as the initial droplet radius, the initial spray velocity and the equivalence ratio at the spray inlet are involved. This brings challenges to the storage and revisit of the flamelet library, especially for high-fidelity simulation of realistic spray combustion. In addition, recent DNS studies [15, 2, 10] show that the spray combustion is typically composed of both premixed and diffusion flames, which means both premixed and diffusion flamelet libraries are needed for spray combustion modeling. Although conditional moment closure equations for spray combustion in two-fluid context have been recently derived [14], there are not formal flamelet equations for spray combustion to the author's knowledge. To clarify this, the flamelet equations for spray combustion are first derived based on a one-step chemistry model. Then a general flamelet transformation for spray combustion is formulated, and the potential applications as well as its advantages are also highlighted.

2. Primitive Equations for Spray Combustion

2.1. N-S Equations for Gas Phase

Let us assume the gas-phase hydrodynamics can be described using the variable-density, low-Mach number Navier-Stokes equations. The spray droplets are assumed as points and described in an Lagrangian framework. The influence of a droplet on gas phase is represented by a mass source term $\dot{S}_m$ in the continuity equation and a momentum source term $\dot{S}_i$ in the momentum equations. These equations read

$$\frac{\partial \rho}{\partial t} + \frac{\partial \rho u_i}{\partial x_i} = \dot{S}_m, \tag{1}$$

and

$$\frac{\partial \rho u_i}{\partial t} + \frac{\partial \rho u_i u_j}{\partial x_j} = -\frac{\partial p}{\partial x_i} + \frac{\partial \sigma_{ij}}{\partial x_j} + \dot{S}_i, \tag{2}$$

where

$$\sigma_{ij} = \mu \left(\frac{\partial u_i}{\partial x_j} + \frac{\partial u_j}{\partial x_i} \right) - \frac{2}{3} \mu \frac{\partial u_k}{\partial x_k} \delta_{ij}. \tag{3}$$

2.2. Governing Equations for Liquid Phase

In the Lagrangian framework, dispersed particles are known to follow the Basset-Boussinesq-Oseen equations. Neglecting inter-droplet collision and coalescence, the equations for the displacement $(x_{d,i})$ and the velocity $(u_{d,i})$ of a droplet can be written as

$$\frac{dx_{d,i}}{dt} = u_{d,i}, \tag{4}$$

$$\frac{du_{d,i}}{dt} = F_i, \tag{5}$$

where F_i is the force acting on the droplet. The governing equations for droplet mass m_d and temperature T_d depend on the specific evaporation model [12].

2.3. Transport Equations for Scalars

Depend on complexity of employed chemistry, different transport equations for species are required to be solved. For simplicity, here we will derive the flamelet equations for spray combustion based on a one-step chemistry model. However, this procedure is straightforward for detailed or reduced chemistry. For the one-step model [4], three scalar transport equations corresponding to the mass fractions of fuel (F), oxidizer (O) and product (P) are solved. Besides the chemical source terms $\dot{\omega}_i$, $i \in \{F, O, P\}$ that can be obtained from specific chemical schemes, there is a spray evaporation source term for fuel. If Fick's law for the diffusion flux and equal diffusivities for all species and temperature are assumed, the transport equations can be written as

$$\frac{\partial \rho Y_F}{\partial t} + \frac{\partial \rho Y_F u_j}{\partial x_j} = \frac{\partial}{\partial x_j} \left(\rho D \frac{\partial Y_F}{\partial x_j} \right) + \dot{\omega}_F + \dot{S}_m, \tag{6}$$

$$\frac{\partial \rho Y_O}{\partial t} + \frac{\partial \rho Y_O u_j}{\partial x_j} = \frac{\partial}{\partial x_j} \left(\rho D \frac{\partial Y_O}{\partial x_j} \right) + \dot{\omega}_O, \tag{7}$$

and

$$\frac{\partial \rho Y_P}{\partial t} + \frac{\partial \rho Y_P u_j}{\partial x_j} = \frac{\partial}{\partial x_j} \left(\rho D \frac{\partial Y_P}{\partial x_j} \right) + \dot{\omega}_P. \tag{8}$$

Furthermore, a transport equation for gas temperature can be derived from energy conservation equation as

$$\frac{\partial \rho T}{\partial t} + \frac{\partial \rho T u_j}{\partial x_j} = \frac{\partial}{\partial x_j} \left(\rho D \frac{\partial T}{\partial x_j} \right) + \dot{\omega}_T + \dot{S}_T, \tag{9}$$

where $\dot{S}_T$ is the energy exchange rate with the droplets, and $\dot{\omega}_T$ is the rate of heat release from chemical reaction and combustion. Note that although equal diffusivities of all species and unity Lewis number are assumed here, it is not necessary. Non-constant Lewis numbers can be considered by applying the method of [22].

3. Derived Flamelet Equations for Spray Combustion

3.1. Flamelet Equations for Premixed Spray Flames

In flamelet modeling, the so-called flamelet library is typically created based on solutions to a steady one-dimensional laminar premixed flame for premixed combustion and a two-dimensional laminar counterflow flame for diffusion combustion. In premixed spray combustion, the air and droplet are injected together. Following the classic steady one-dimensional laminar premixed flame configuration, the governing equations for premixed spray combustion in physical space can be simplified as

$$\frac{\partial \rho}{\partial t} + \frac{\partial \rho u}{\partial x} = \dot{S}_m, \tag{10}$$

$$\frac{\partial \rho Y_F}{\partial t} + \frac{\partial \rho Y_F u}{\partial x} = \frac{\partial}{\partial x}\left(\rho D \frac{\partial Y_F}{\partial x}\right) + \dot{\omega}_F + \dot{S}_m, \tag{11}$$

$$\frac{\partial \rho Y_O}{\partial t} + \frac{\partial \rho Y_O u}{\partial x} = \frac{\partial}{\partial x}\left(\rho D \frac{\partial Y_O}{\partial x}\right) + \dot{\omega}_O, \tag{12}$$

$$\frac{\partial \rho Y_P}{\partial t} + \frac{\partial \rho Y_P u}{\partial x} = \frac{\partial}{\partial x}\left(\rho D \frac{\partial Y_P}{\partial x}\right) + \dot{\omega}_P, \tag{13}$$

and

$$\frac{\partial \rho T}{\partial t} + \frac{\partial \rho T u}{\partial x} = \frac{\partial}{\partial x}\left(\rho D \frac{\partial T}{\partial x}\right) + \dot{\omega}_T + \dot{S}_T. \tag{14}$$

By imposing proper boundary conditions, this set of equations can be solved to get the structures of premixed spray flamelets and create corresponding flamelet library. Actually, this set of equations could be further simplified as

$$\rho_u s_{L,u} \frac{\partial Y_F}{\partial x} = \frac{\partial}{\partial x}\left(\rho D \frac{\partial Y_F}{\partial x}\right) + \dot{\omega}_F + (1 - Y_F)\,\dot{S}_m, \tag{15}$$

$$\rho_u s_{L,u} \frac{\partial Y_O}{\partial x} = \frac{\partial}{\partial x}\left(\rho D \frac{\partial Y_O}{\partial x}\right) + \dot{\omega}_O - Y_O \dot{S}_m, \tag{16}$$

$$\rho_u s_{L,u} \frac{\partial Y_P}{\partial x} = \frac{\partial}{\partial x}\left(\rho D \frac{\partial Y_P}{\partial x}\right) + \dot{\omega}_P - Y_P \dot{S}_m, \tag{17}$$

$$\rho_u s_{L,u} \frac{\partial T}{\partial x} = \frac{\partial}{\partial x}\left(\rho D \frac{\partial T}{\partial x}\right) + \dot{\omega}_T + \dot{S}_T - T \dot{S}_m. \tag{18}$$

when the spray flame reaches steady status. Here ρ_u and $s_{L,u}$ represent the unburned density and flame propagation speed, respectively.

3.2. Flamelet Equations for Diffusion Spray Flames

Although the premixed flamelet equations are formulated in physical space, the flamelet equations for diffusion combustion are usually expressed in mixture fraction space. Different from gaseous combustion, the traditionally defined mixture fraction is no longer a conservative variable in spray combustion because of local evaporation source,

$$\frac{\partial \rho Z}{\partial t} + \frac{\partial \rho Z u_j}{\partial x_j} = \frac{\partial}{\partial x_j}\left(\rho D \frac{\partial Z}{\partial x_j}\right) + \dot{S}_m. \tag{19}$$

Following [18] and [20], let us locally introduce a coordinate system attached to the surface of stoichiometric mixture. The coordinate system is introduced such that one coordinate x_1 is normal to the surface of stoichiometric mixture at a given instant. Then we perform a Crocco type coordinate transformation, such that x_1 is replaced by the mixture fraction Z, and the other coordinates remain. By definition, the new coordinate Z is locally normal to the surface of stoichiometric mixture. Using $Z_2 = x_2$, $Z_3 = x_3$, $\tau = t$ as other independent variables, the transformation rules can be obtained. For this transformation to be valid, x_1 has to be uniquely representable by the new coordinate Z, which is the case in many simplified laminar configurations but is not generally true for turbulent flows. For turbulent flows, we have to assume that the reaction zone is smaller than the small scales of the turbulence and limit the analysis to a small region around the reaction zone. This is the so-called flamelet assumption. Also the possible rotation of the new coordinate Z with respect to the original coordinate x_1 during the temporal development of the flow field will be neglected. Based on these assumptions, changes of scalars along Z_2 and Z_3 directions are expected to be of lower order compared to the Z direction, and thus could be neglected. This can be formally demonstrated by introducing a stretched coordinate [17]. Following the flamelet transformation rules, the transport equations for scalars can be transformed into mixture fraction space as

$$\rho\frac{\partial Y_F}{\partial \tau} - \rho\frac{\chi}{2}\frac{\partial^2 Y_F}{\partial Z^2} = \dot{\omega}_F - \left[Y_F + \frac{\partial Y_F}{\partial Z}(1-Z) - 1\right]\dot{S}_m, \tag{20}$$

$$\rho\frac{\partial Y_O}{\partial \tau} - \rho\frac{\chi}{2}\frac{\partial^2 Y_O}{\partial Z^2} = \dot{\omega}_O - \left[Y_O + \frac{\partial Y_O}{\partial Z}(1-Z)\right]\dot{S}_m, \tag{21}$$

$$\rho\frac{\partial Y_P}{\partial \tau} - \rho\frac{\chi}{2}\frac{\partial^2 Y_P}{\partial Z^2} = \dot{\omega}_P - \left[Y_P + \frac{\partial Y_P}{\partial Z}(1-Z)\right]\dot{S}_m, \tag{22}$$

and

$$\rho\frac{\partial T}{\partial \tau} - \rho\frac{\chi}{2}\frac{\partial^2 T}{\partial Z^2} = \dot{\omega}_T + \dot{S}_T - \left[T + \frac{\partial T}{\partial Z}(1-Z)\right]\dot{S}_m, \tag{23}$$

where τ is the Lagrangian-like time coordinate, and χ is the scalar dissipation rate.

This set of flamelet equations for diffusion spray combustion differs from the classic counterpart for gaseous combustion, at least in two ways. First, the scalar dissipation rate may be modified by spray evaporation that makes the presumed profile of scalar dissipation rate in gaseous combustion invalid. Actually, multiplying Eq. (19) by the gradient of the mixture fraction and using the standard flamelet transformation and assumptions, one can derive an equation for the scalar dissipation rate in mixture fraction space as

$$\frac{\partial \chi}{\partial \tau} + \frac{1}{4}\left(\frac{\partial \chi}{\partial Z}\right)^2 - 2a\chi = \frac{\chi}{2}\frac{\partial^2 \chi}{\partial Z^2} + \frac{2\chi}{\rho}\frac{\partial \dot{S}_m}{\partial Z} - \frac{\partial \chi}{\partial Z}\frac{\dot{S}_m}{\rho}, \tag{24}$$

where a denotes the rate of strain. This equation describes the evolution of the scalar dissipation rate as a function of mixture fraction and a flamelet Lagrangian-like time τ. It is clear that the spray evaporation modifies the scalar dissipation rate through the two source terms in the right hand side of the equation. This equation also provides a way for modeling the scalar dissipation rate in spray combustion. Second, in each species equation, there is a source term combining the evaporation, the mass fraction of the species and the derivative of the mass fraction of the species in mixture fraction space. In the temperature equation, even another heat exchange term $\dot{S}_T$ appears. This indicates that the spray evaporation is able to directly influence the local flamelets in mixture fraction space, and this effect should be reflected in any application of flamelet modeling of spray combustion.

Obviously, to correctly describe the local flamelets in spray combustion, the flamelet library should be created in terms of the solutions to the above flamelet equations with the influence of evaporation. However, most previous studies neglected the influence of evaporation on flamelets when created flamelet library. Since τ is a Lagrangian-like time coordinate, the above flamelet equations for spray combustion can be called as the Lagrangian Flamelet Model(LFM) [23, 20]. If the local Lagrangian-like time term is negligible and neglected, the steady flamelet equations for spray combustion are obtained.

As the Lagrangian-like time coordinate τ is defined in the new coordinate system (Z, Z_1, Z_2), the time derivative $\frac{\partial}{\partial \tau}$ has to be evaluated at constant mixture fraction Z. Following [23], the flamelet equations for spray combustion in the Eulerian system can be obtained as

$$\rho\frac{\partial Y_F}{\partial t} + \rho\mathbf{v}\cdot\nabla Y_F - \rho\frac{\chi}{2}\frac{\partial^2 Y_F}{\partial Z^2} = \dot{\omega}_F - \left[Y_F + \frac{\partial Y_F}{\partial Z}(1-Z) - 1\right]\dot{S}_m, \tag{25}$$

$$\rho\frac{\partial Y_O}{\partial t} + \rho\mathbf{v}\cdot\nabla Y_O - \rho\frac{\chi}{2}\frac{\partial^2 Y_O}{\partial Z^2} = \dot{\omega}_O - \left[Y_O + \frac{\partial Y_O}{\partial Z}(1-Z)\right]\dot{S}_m, \tag{26}$$

$$\rho\frac{\partial Y_P}{\partial t} + \rho\mathbf{v}\cdot\nabla Y_P - \rho\frac{\chi}{2}\frac{\partial^2 Y_P}{\partial Z^2} = \dot{\omega}_P - \left[Y_P + \frac{\partial Y_P}{\partial Z}(1-Z)\right]\dot{S}_m, \tag{27}$$

and

$$\rho\frac{\partial T}{\partial t} + \rho\mathbf{v}\cdot\nabla T - \rho\frac{\chi}{2}\frac{\partial^2 T}{\partial Z^2} = \dot{\omega}_T + \dot{S}_T - \left[T + \frac{\partial T}{\partial Z}(1-Z)\right]\dot{S}_m, \tag{28}$$

where t is the time in Eulerian coordinate, and $\mathbf{v}$ is a convection velocity. These equations can also be derived using a two-scale asymptotic analysis as suggested by [17]. This set of equations could be called as Extended Flamelet Model (EFM) for spray combustion, in contrast to the Lagrangian Flamelet Model. Pitsch [20] has demonstrated that the EFM could provide significant improvement over the LFM, because the locally fluctuations of the modeled scalar dissipation rate and unsteady response of the interaction of molecular mixing and chemistry are considered. If the local Eulerian time term is negligible and neglected, the steady flamelet equations for spray combustion are obtained.

If spray autoignition is important and of interest, the influence of curvature could be included in the unsteady Lagrangian flamelet equations by following [9]

$$\rho\frac{\partial Y_F}{\partial \tau} - \rho\frac{\chi}{2}\frac{\partial^2 Y_F}{\partial Z^2} + \rho\left(\frac{D\chi}{2}\right)^{0.5}\kappa\frac{\partial Y_F}{\partial Z} = \dot{\omega}_F - \left[Y_F + \frac{\partial Y_F}{\partial Z}(1-Z) - 1\right]\dot{S}_m, \quad (29)$$

$$\rho\frac{\partial Y_O}{\partial \tau} - \rho\frac{\chi}{2}\frac{\partial^2 Y_O}{\partial Z^2} + \rho\left(\frac{D\chi}{2}\right)^{0.5}\kappa\frac{\partial Y_O}{\partial Z} = \dot{\omega}_O - \left[Y_O + \frac{\partial Y_O}{\partial Z}(1-Z)\right]\dot{S}_m, \quad (30)$$

$$\rho\frac{\partial Y_P}{\partial \tau} - \rho\frac{\chi}{2}\frac{\partial^2 Y_P}{\partial Z^2} + \rho\left(\frac{D\chi}{2}\right)^{0.5}\kappa\frac{\partial Y_P}{\partial Z} = \dot{\omega}_P - \left[Y_P + \frac{\partial Y_P}{\partial Z}(1-Z)\right]\dot{S}_m, \quad (31)$$

and

$$\rho\frac{\partial T}{\partial \tau} - \rho\frac{\chi}{2}\frac{\partial^2 T}{\partial Z^2} + \rho\left(\frac{D\chi}{2}\right)^{0.5}\kappa\frac{\partial T}{\partial Z} = \dot{\omega}_T + \dot{S}_T - \left[T + \frac{\partial T}{\partial Z}(1-Z)\right]\dot{S}_m, \quad (32)$$

where κ is the curvature of mixture fraction isosurfaces

3.3. General Flamelet Transformation for Spray Combustion

One disadvantage of flamelet modeling is that it is regime dependent, and is therefore only applicable under certain conditions. To overcome this issue, Knudsen and Pitsch [8] have recently developed a general flamelet transformation that holds in the limits of both premixed and non-premixed combustion regimes. This method significantly enhances the predictive nature of flamelet modeling in that it removes the need for users to select their application. It is especially of significance for spray combustion, because both premixed and diffusion flames co-exist there. Essentially, this method relies on the so-called combustion regime index. But this combustion regime index may be changed by evaporation in spray conditions. Therefore, it is necessary to perform a general flamelet transformation for spray combustion.

Let us start from the FPV approach [19]. When perform a flamelet transformation to (Z, C) space, some issues appear because Z and C, as typically defined, are not statistically independent. This issue can be overcome by using the reaction progress parameter (RPP) approach. Although a variety of choices for the flamelet index Λ are possible, a simple definition can be obtained using the progress variable C as

$$\Lambda = C(Z_{\mathrm{st}}, T^*_{Z_{\mathrm{st}}}), \quad (33)$$

which is the value of the progress variable that occurs on a given flamelet ($T_{Z_{\mathrm{st}}} = T^*_{Z_{\mathrm{st}}}$) and at a given mixture fraction ($Z = Z_{\mathrm{st}}$). Because there is a unique value of Λ associated with each flamelet, dependencies on Λ disappear in the limit of diffusion combustion.

To transform a scalar equation from $(\mathbf{x}, t)$ space into $(Z(\mathbf{x}, t), \Lambda(\mathbf{x}, t), e(\mathbf{x}, t), \tau)$ space where e represent a direction that is normal to the gradients of Z and Λ along which changes in scalar fields are expected to be small, and τ is a Lagrangian-like time coordinate, the above flamelet transformation technique can be used here too. Applying the rules of [8] to

Eqs. (6)-(9) results in the following flamelet equations for spray combustion

$$\rho\frac{\partial Y_F}{\partial \tau} + \frac{\partial Y_F}{\partial \Lambda}\left[\rho\frac{\partial \Lambda}{\partial t} + \rho u_j\frac{\partial \Lambda}{\partial x_j} - \frac{\partial}{\partial x_j}\left(\rho D\frac{\partial \Lambda}{\partial x_j}\right)\right] - \rho\frac{\chi_Z}{2}\frac{\partial^2 Y_F}{\partial Z^2} - \rho\frac{\chi_\Lambda}{2}\frac{\partial^2 Y_F}{\partial \Lambda^2} - \rho\chi_{\Lambda,Z}\frac{\partial^2 Y_F}{\partial \Lambda \partial Z} = \dot{\omega}_F - \left[Y_F + \frac{\partial Y_F}{\partial Z}(1-Z) - 1\right]\dot{S}_m, \quad (34)$$

$$\rho\frac{\partial Y_O}{\partial \tau} + \frac{\partial Y_O}{\partial \Lambda}\left[\rho\frac{\partial \Lambda}{\partial t} + \rho u_j\frac{\partial \Lambda}{\partial x_j} - \frac{\partial}{\partial x_j}\left(\rho D\frac{\partial \Lambda}{\partial x_j}\right)\right] - \rho\frac{\chi_Z}{2}\frac{\partial^2 Y_O}{\partial Z^2} - \rho\frac{\chi_\Lambda}{2}\frac{\partial^2 Y_O}{\partial \Lambda^2} - \rho\chi_{\Lambda,Z}\frac{\partial^2 Y_O}{\partial \Lambda \partial Z} = \dot{\omega}_O - \left[Y_O + \frac{\partial Y_O}{\partial Z}(1-Z)\right]\dot{S}_m, \quad (35)$$

$$\rho\frac{\partial Y_P}{\partial \tau} + \frac{\partial Y_P}{\partial \Lambda}\left[\rho\frac{\partial \Lambda}{\partial t} + \rho u_j\frac{\partial \Lambda}{\partial x_j} - \frac{\partial}{\partial x_j}\left(\rho D\frac{\partial \Lambda}{\partial x_j}\right)\right] - \rho\frac{\chi_Z}{2}\frac{\partial^2 Y_P}{\partial Z^2} - \rho\frac{\chi_\Lambda}{2}\frac{\partial^2 Y_P}{\partial \Lambda^2} - \rho\chi_{\Lambda,Z}\frac{\partial^2 Y_P}{\partial \Lambda \partial Z} = \dot{\omega}_P - \left[Y_P + \frac{\partial Y_P}{\partial Z}(1-Z)\right]\dot{S}_m, \quad (36)$$

and

$$\rho\frac{\partial T}{\partial \tau} + \frac{\partial T}{\partial \Lambda}\left[\rho\frac{\partial \Lambda}{\partial t} + \rho u_j\frac{\partial \Lambda}{\partial x_j} - \frac{\partial}{\partial x_j}\left(\rho D\frac{\partial \Lambda}{\partial x_j}\right)\right] - \rho\frac{\chi_Z}{2}\frac{\partial^2 T}{\partial Z^2} - \rho\frac{\chi_\Lambda}{2}\frac{\partial^2 T}{\partial \Lambda^2} - \rho\chi_{\Lambda,Z}\frac{\partial^2 T}{\partial \Lambda \partial Z} = \dot{\omega}_T + \dot{S}_T - \left[T + \frac{\partial T}{\partial Z}(1-Z)\right]\dot{S}_m. \quad (37)$$

This set of equations represent a general flamelet transformation for spray combustion. The equations described the limits of premixed and diffusion combustion could be recovered from these equations. In the limit of diffusion combustion, along a local flamelet, Λ is constant and $\nabla\Lambda = 0$. As a result, the related terms disappear and the steady flamelet equations for spray combustion are recovered if unsteadiness is negligible. Although it is less straightforward to recover the one-dimensional premixed spray flame equations in general, the procedure is transparent when progress variable C is chosen as the transport scalar

$$\rho\frac{\partial C}{\partial \tau} + \frac{\partial C}{\partial \Lambda}\left[\rho\frac{\partial \Lambda}{\partial t} + \rho u_j\frac{\partial \Lambda}{\partial x_j} - \frac{\partial}{\partial x_j}\left(\rho D\frac{\partial \Lambda}{\partial x_j}\right)\right] - \rho\frac{\chi_Z}{2}\frac{\partial^2 C}{\partial Z^2} - \rho\frac{\chi_\Lambda}{2}\frac{\partial^2 C}{\partial \Lambda^2} - \rho\chi_{\Lambda,Z}\frac{\partial^2 C}{\partial \Lambda \partial Z} = \dot{\omega}_C - \left[C + \frac{\partial C}{\partial Z}(1-Z)\right]\dot{S}_m, \quad (38)$$

and $Z = Z_{st}$, as demonstrated by [8]. In these conditions, the mixture fraction field is homogeneous, $\nabla Z = 0$, $\chi_Z = 0$, $\chi_{\Lambda,Z} = 0$, and $\Lambda = C$. Then the equation of C becomes

$$\rho\frac{\partial C}{\partial t} + \frac{\partial}{\partial x_j}(\rho u_j C) - \frac{\partial}{\partial x_j}\left(\rho D\frac{\partial C}{\partial x_j}\right) = \dot{\omega}_C - C\dot{S}_m. \quad (39)$$

In a steady one-dimensional situation, using the burning velocity replaces u_j leads to the equation for premixed spray combustion.

Based on these exercise and neglect the cross-dissipation term as well as the $\frac{\partial C}{\partial \tau}$ term, Eq.(38) can be regrouped as

$$\begin{aligned}
&\frac{\partial C}{\partial \Lambda}\left[\rho\frac{\partial \Lambda}{\partial t} + (\rho\mathbf{u} - \rho_u S_{L,u}\mathbf{n})\cdot\nabla\Lambda - C\dot{S}_m\right]_1 \\
&\quad + \left(\frac{\partial C}{\partial \Lambda}\left[\rho_u S_{L,u}\left|\nabla\Lambda\right| - \nabla\cdot(\rho D\nabla\Lambda) + C\dot{S}_m\right] - \rho\frac{\chi_\Lambda}{2}\frac{\partial^2 C}{\partial \Lambda^2}\right)_2 \\
&\quad + \left(-\rho\frac{\chi_Z}{2}\frac{\partial^2 C}{\partial Z^2} + \left[C + \frac{\partial C}{\partial Z}(1-Z)\right]\dot{S}_m\right)_3 = \dot{\omega}_C \qquad (40)
\end{aligned}$$

where $\mathbf{n} = \nabla\Lambda/\left|\nabla|\Lambda\right|$ is the normal vector pointing in the direction of the gradient of Λ. In this equation, group 1 terms describe the unsteady physics, group 2 terms denote the steady premixed physics, and group 3 terms represent the steady diffusion physics. The $s_{L,u}$ term has been explicitly included in the equation to separate steady premixed behavior from unsteady premixed behavior [8].

Based on the above understanding, a time scale ratio $\Theta_{1,23}$ could be defined to locally determine the nature of spray combustion from the groups

$$\Theta_{1,23} = \frac{\frac{\partial C}{\partial \Lambda}\left[\rho\frac{\partial \Lambda}{\partial t} + (\rho\mathbf{u} - \rho_u s_{L,u}\mathbf{n})\cdot\nabla\Lambda - C\dot{S}_m\right]}{\max\left(\frac{\partial C}{\partial \Lambda}\left[\rho_u s_{L,u}\left|\nabla\Lambda\right| - \nabla\cdot(\rho D\nabla\Lambda) + C\dot{S}_m\right] - \rho\frac{\chi_\Lambda}{2}\frac{\partial^2 C}{\partial \Lambda^2}, -\rho\frac{\chi_Z}{2}\frac{\partial^2 C}{\partial Z^2} + \left[C + \frac{\partial C}{\partial Z}(1-Z)\right]\dot{S}_m\right)}. \qquad (41)$$

If $\Theta_{1,23} \ll 1$, the steady flamelet approximation is safely valid, since the unsteady terms in the flamelet equations are relatively small. Otherwise, unsteady phenomena, such as ignition, extinction, or re-ignition are important and should be treated carefully. Therefore, the parameter $\Theta_{1,23}$ could be evaluated locally in a simulation as part of an unsteady modeling approach to determine the calling of proper flamelet library.

If the steady flamelet approximation holds, another time scale ratio $\Theta_{2,3}$ describing the relative importance of the premixed and the diffusion combustion regimes can be defined as

$$\Theta_{2,3} = \frac{\frac{\partial C}{\partial \Lambda}\left[\rho_u s_{L,u}\left|\nabla\Lambda\right| - \nabla\cdot(\rho D\nabla\Lambda) + C\dot{S}_m\right] - \rho\frac{\chi_\Lambda}{2}\frac{\partial^2 C}{\partial \Lambda^2}}{-\rho\frac{\chi_Z}{2}\frac{\partial^2 C}{\partial Z^2} + \left[C + \frac{\partial C}{\partial Z}(1-Z)\right]\dot{S}_m}. \qquad (42)$$

If $\Theta_{2,3} > 1$, premixed terms are dominant and balance the majority of the chemical reaction budget, and the local combustion event is best described by the asymptotic premixed flamelet model. If $\Theta_{2,3} < 1$, diffusion terms are significant and the local combustion event is best described by the asymptotic diffusion flamelet model. If $\Theta_{2,3} = 1$, premixed terms and diffusion terms are of equal importance and either can be used to describe the local chemistry. Following [8], the above parameter could be referred to as the spray combustion regime index. To demonstrate that the prediction can be improved by using the combustion regime index, LES of a low-swirl burner in [8] is revisited. The difference here is that the same configuration is also simulated using only the steady diffusion flamelet library for comparison. Although the predictions of the fluid dynamics, such as the mean velocities and velocity fluctuations, are not distinguishable between the steady diffusion flamelet

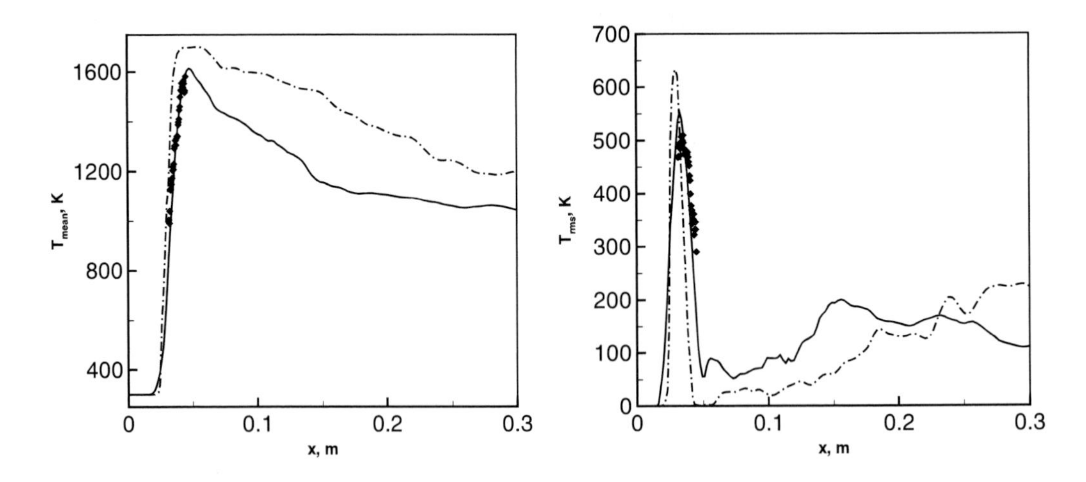

Figure 1. Distribution of mean and fluctuation temperature along the central line (Symbol: experimental data [16]; dash-dotted line: diffusion flamelet model; solid line: combined premixed and diffusion flamelet model).

model and the combined premixed and diffusion flamelet model, the prediction of the temperature and chemical heat release is obviously improved by the combined premixed and diffusion flamelet model, as demonstrated in Fig. 1. These comparisons clearly manifest the advantage of the combined premixed and diffusion flamelet model based on combustion regime index for partially premixed combustion. The potential application and validation of this model to spray combustion remain to be performed in the next step.

4. Summary

Different unsteady and steady flamelet equations for spray combustion are derived based on a one-step chemistry and flamelet assumption. A general flamelet transformation for spray combustion is formulated and discussed. A comparative study also demonstrates the importance of combustion regime index [8] in modeling partially premixed flame that is a typical form of spray combustion. However, spray evaporation can modify the combustion regime index, and application and validation of the model still need to be investigate in the future study.

Acknowledgments

Funding From a Foundation for the Author of National Excellent Doctoral Dissertation of PR China (2007B4) is appreciated.

References

[1] Y. Baba and R. Kurose, *J. Fluid Mech.* **612** (2008), 45–79.

[2] P. Domingo, L. Vervisch, and J. Réveillon, *Combust. Flame* **140** (2005), 172–195.

[3] H. El-Asrag and H. Pitsch, *Annual Research Briefs*, Center for Turbulence Research, NASA Ames/Stanford University, 2008, pp. 267–285.

[4] E. Fernandez-Tarrazo, A. L. Sanchez, A. Linan, and F. A. Williams, *Combust. Flame* **147** (2006), 32–38.

[5] W. A. Gutheil, E.and Sirignano, *Combust. Flame* **113** (1998), 92–97.

[6] E. Hollmann, C.and Gutheil, *Combust. Sci. Technol.* **135** (1998), 175–192.

[7] A. Y. Klimenko and R. W. Bilger, *Prog. Energ. Combust. Sci.* **25** (1999), 595–687.

[8] E. Knudsen and H. Pitsch, *Combust. Flame* **156** (2009), 678–696.

[9] C. Kortschik, S. Honnet, and N. Peters, *Combust. Flame* **142** (2005), 140–152.

[10] K. Luo, O. Desjardins, and H. Pitsch, *Annual Research Briefs*, Center for Turbulence Research, NASA Ames/Stanford University, 2008, pp. 1–13.

[11] P. A. McMurtry and S. Menon, *Energy & Fuels* **7** (1993), 817–826.

[12] R. S. Miller, K. Harstad, and J. Bellan, *Int. J. Multiphase Flow* **24** (1998), 1025–1055.

[13] P. Moin and S. Apte, *AIAA J.* **44** (2006), 698–708.

[14] M. Mortensen and R. W. Bilger, *Combust. Flame* **156** (2009), 62–72.

[15] M. Nakamura, F. Akamatsu, R. Kurose, and M. Katsuki, *Phys. Fluids* **17** (2005), 172–195.

[16] K.-J. Nogenmyr, P. Petersson, X. S. Bai, A. Nauert, J. Olofsson, C. Brackman, H. Seyfried, J. Zetterberg, Z. S. Li, M. Richter, A. Dreizler, M. Linne, and M. Aldén, *Proc. Comb. Inst.* **31** (2006), 1467–1475.

[17] N.Peters, Turbulent combustion, Cambridge Univ. Press, Cambridge, UK, 2000.

[18] N. Peters, *Prog. Energ. Combust. Sci.* **10** (1984), 319–339.

[19] C. D. Pierce and P. Moin, *J. Fluid Mech.* **504** (2004), 73–97.

[20] H. Pitsch, *Annual Research Briefs*, Center for Turbulence Research, NASA Ames/Stanford University, 2000, pp. 149–158.

[21] ———, *Annu. Rev. Fluid Mech.* **38** (2006), 453–482.

[22] H. Pitsch, M. Chen, and N. Peters, *Proc. Combust. Inst.* **27** (1998), 1057–1064.

[23] H. Pitsch, H.and Steiner, *Phys. Fluids* **12** (2000), 2541–2554.

[24] S. B. Pope, *Prog. Energ. Combust. Sci.* **11** (1985), 119–192.

INDEX

A

B

C

D

E

F

G

H

I

K

O

P

Q

R

S

T

U

V

W

X

Y

Z